Advances in Biochemistry Research

Volume I

Advances in Biochemistry Research Volume I

Edited by **Artie Weissberg**

R CALLISTO REFERENCE

New York

Published by Callisto Reference,
106 Park Avenue, Suite 200,
New York, NY 10016, USA
www.callistoreference.com

Advances in Biochemistry Research: Volume I
Edited by Artie Weissberg

International Standard Book Number: 978-1-63239-038-7 (Hardback)

Printed in the United States of America.

Contents

Preface

Metabolism is what distinguishes living organisms from non-living ones. There cannot be the possibility of life without it. Metabolism is the chemical process that takes place inside every living organism. The study of these chemical processes is known as biochemistry. Biochemistry also strives to understand how the biological molecules interact and give rise to the processes within living cells which are in turn related to the study and understanding of living organisms. It also deals with biomolecules like carbohydrates, nucleic acids, proteins and lipids which perform the dual action of providing the structure of the cell and performing many functions associated with life. Biochemistry gives a better understanding of organisms at molecular level.

Significant researches are being conducted in combination therapies for the treatment of advanced Melanoma. Some of the other important researches in biochemistry are on acetylcholinesterase biosensors for electrochemical detection of organophosphorus compounds, a quantitative structure-activity relationship and molecular modeling study on a series of heteroaryl and heterocyclyl-substituted imidazo, pyridine derivatives acting as acid pump antagonists, genetic basis for variation of metalloproteinase-associated biochemical activity in venom of the mojave rattlesnake (crotalus scutulatus), jatropha oil derived sophorolipids: production and characterization as laundry detergent additive and the role of ubiquitin-proteasome system in Huntington Disease whether it leads to proteasomes impaired initiators of disease, or comes to the rescue.

This book is the result of the efforts of veterans of the discipline. I wish to thank all the people involved in assisting in its completion at any step.

Editor

TCTP in Development and Cancer

Magdalena J. Koziol[1] and John B. Gurdon[2]

[1] Department of Genetics, Yale University School of Medicine, 333 Cedar Street, New Haven, CT 06510, USA
[2] Wellcome Trust/Cancer Research UK Gurdon Institute, University of Cambridge, Tennis Court Road, Cambridge CB2 1QN, UK

Correspondence should be addressed to Magdalena J. Koziol, magdalena.koziol@yale.edu

Academic Editor: Malgorzata Kloc

The translationally controlled tumor protein (TCTP) is highly conserved among animal species. It is widely expressed in many different tissues. It is involved in regulating many fundamental processes, such as cell proliferation and growth, apoptosis, pluripotency, and the cell cycle. Hence, it is not surprising that it is essential for normal development and, if misregulated, can lead to cancer. Provided herein is an overview of the diverse functions of TCTP, with a focus on development. Furthermore, we discuss possible ways by which TCTP misregulation or mutation could result in cancer.

1. Introduction

TCTP was first identified in tumor cells. Since its mRNA has all sequence and structural characteristics of translationally controlled mRNAs, it was named "Translationally Controlled Tumor Protein" [1, 2]. It is also known under many different names, such as histamine releasing factor (HRF), tumor protein translationally controlled (Tpt1), p23, and fortilin. The protein is highly conserved across different species [3], is ubiquitously expressed, but the level of the mRNA varies depending on the cell type [4, 5] and developmental stage [6]. A wide range of extracellular stimuli can rapidly regulate its mRNA level. Examples range from cytokines to calcium levels [7, 8]. Translational regulation of the mRNA adds another layer of TCTP level diversity [9].

TCTP expression seems to be highly regulated at many levels by many distinct mechanisms. It is not surprising that it is associated with an array of different biological activities, such as the cell cycle [3, 10], apoptosis [11–15], cytoskeleton [10, 16, 17], protein synthesis [18], immune response [19], development [6, 20–22], and cancer [11, 23, 24]. In recent years the protein has attracted most attention on account of its role in tumor reversion and its crucial role in development [21, 23]. In this paper we outline what is known so far about TCTP in development with the underlying molecular events and discuss how its misregulation might result in cancer.

2. TCTP Promotes Cell Proliferation and Growth

TCTP knockdown studies in *Drosophila* cause lethality in late first-instar larvae and result in reduced cell number, cell size, and organ size [21]. This indicates an effect on cell proliferation and growth, which is regulated mainly by the TOR pathway.

The TOR pathway is regulated by nutrient and energy availability, as well as hypoxia. It integrates signals from many pathways, such as insulin signaling, growth factors, and amino acids. It not only regulates cell growth and proliferation, but also cell motility, cell survival, protein synthesis, and transcription. The pathway is named after the Target of Rapamycin (TOR), a serine-threonine kinase, encoded by the FRAP1 gene. In the presence of growth-promoting signals, receptors on the cell membrane are activated that lead to the activation of the serine-threonine kinase Akt and ultimately TOR. In mammals, the protein TOR is either bound to the protein Raptor (complex TOR1) or to the protein Rictor (complex TOR2). The TOR1 complex is sensitive to the bacterial product rapamycin and is involved in mRNA translation and ribosome biogenesis. The other rapamycin-insensitive complex TOR2 regulates cell survival and the cytoskeleton (reviewed in [25]). It regulates the cytoskeleton by stimulating various proteins, for example, actin fibers [26]. It also phosphorylates the serine-threonine kinase Akt, which initially leads to TOR activation [27].

TOR1 is activated by an increase in nutrient levels, growth factors, and stress [28]. These extracellular signals activate a cascade of proteins within the cell, leading to the activation of the GTPase Rheb, that ultimately activates TOR1. TOR1 then targets various downstream factors, such as the serine-threonine kinase S6K and the protein 4EBP1 (reviewed in [25]). S6K is known to phosphorylate many proteins. One major target is the S6 ribosomal protein. When nutrients are sparse, the S6 ribosomal protein is bound to the eIF3 complex, which is involved in the initiation of translation by recognizing the 5′ cap structure of mRNAs. mRNAs that contain a 5′ polypyrimidine tract, referred to as 5′ TOP, are important targets of eIF3 translational activation [29, 30]. These transcripts generally encode further ribosomal proteins and translation elongation factors. When the availability of nutrients increases, TOR is activated, causing an increase of S6K. Ultimately, S6 becomes phosphorylated, which causes the eIF3 complex to be released resulting in the activation of translation [31]. Various mRNAs become translated, in particular the mRNAs with a 5′ TOP region that encode proteins involved in translation. This subsequently leads to the production proteins required for translation, overall leading to the amplification of translation.

The other TOR1 target, 4EBP1, is a translation repressor. 4EBP1 binds to the eukaryotic initiation factor 4E (eIF4E), which recruits 40S ribosomal subunits to the 5′ end of mRNAs to initiate translation. Interaction of 4EBP1 and eIF4E results in the inhibition of translation. Upon TOR1 activation, 4EBP1 is phosphorylated, resulting in the dissociation from eIF4E, allowing eIF4E to initiate translation [32].

The entire TOR1 cascade and the increased protein synthesis required the activation of Rheb. Studies in *Drosophila* showed that mutant Rheb resulted in smaller cell sizes and numbers, as observed in the absence of TCTP. It was then determined that TCTP associates with Rhed. This is likely to be conserved between species, as human TCTP was able to rescue *Drosophila* TCTP mutants [21]. This observation directly links TCTP with the TOR pathway, explaining its effect on cell proliferation and growth. In the absence of TCTP, Rhed is no longer active, leading to a decrease in TOR1 activity and ultimately a decrease in protein synthesis in response to external growth-stimulating stimuli. It is known that TCTP responds to many external stimuli. This suggests that the interaction of TCTP with the TOR1 complex might be the reason for the TCTP responsiveness to many external signals [7, 8]. It would be interesting to investigate the connection of TCTP with the TOR2 pathway. Since cells are smaller in TCTP mutants, it is likely that the TOR2 pathway that also regulates the cytoskeleton is involved. To test if TCTP has an effect on the TOR2 pathway, one could analyze the effect of TCTP in the presence of rapamycin. Since rapamycin inhibits the TOR1 activity, it is possible to investigate if the absence of TCTP still has a function on the cell size and growth. If this is the case, it would be interesting to analyze the level of the major TOR2 components in the absence of TCTP. Further studies, for example, with mutants

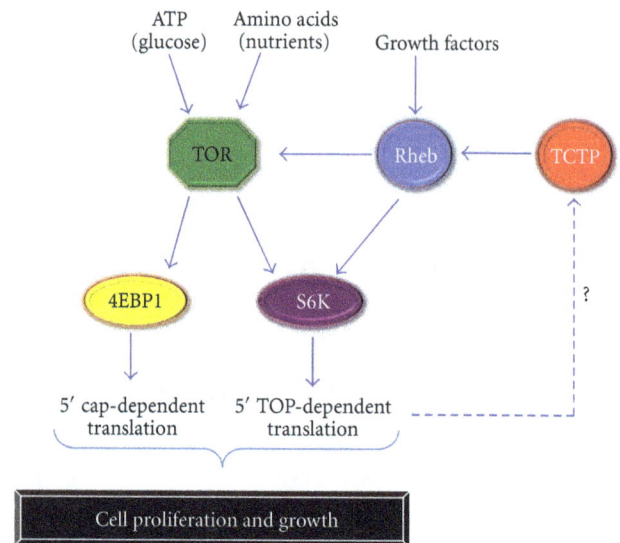

FIGURE 1: TCTP can activate the TOR pathway and promote cell proliferation and growth.

in a TCTP depleted background could help to elucidate if and where in the TOR2 pathway TCTP could act (Figure 1).

As described above, TOR1 activates the S6K kinase, which activates S6 and leads to translation of mRNAs, in particular of mRNAs that contain the 5′ TOP tract [29, 30]. TCTP mRNA itself contains the 5′-TOP domain [9]. This suggests that TOR1 might activate TCTP translation, via S6K and S6. Since TCTP activates TOR1 by binding to Rheb, the activator of TOR1, it is possible that TCTP not only activates TOR1, but also provides a positive feedback mechanism. Mutating the 5′ TOP domain of TCTP might help to determine if TCTP is actually activated by this route. If this is the case, overexpression of S6K should increase TCTP protein levels, which will in turn promote even more TOR1 activity. S6K could act as a major regulator of TOR1, since it also inactivates the repressor TOR1 by phosphorylation, suggesting a positive feedback mechanism [33]. TCTP could also act in this way, providing a positive feedback mechanism to upregulate TOR. When TCTP is high, it activates TOR1, which in turn leads to the phospporylation of S6 by S6K and increased translation. This might again lead to an increased TCTP protein level that increases TOR1 activation.

Even though abnormal cell proliferation and growth can be explained by TCTP interaction with the TOR pathway, this does not fully explain why development is ultimately arrested. This suggests that TCTP has also a major function that lies outside of the TOR pathway.

3. TCTP Inhibits Apoptosis

In the early development of mice, TCTP mRNA and protein levels are significantly increased from embryonic day E3 to E5, when they reach a maximum level. Selective depletion of TCTP in the uterus at E3 resulted in reduced numbers of implanted embryos compared to wild-type embryos [34]. In knockout mice, heterozygous mutants of TCTP had no obvious developmental effects, but homozygous mutants were lethal between E9.5 and E10.5 [22]. Severe abnormalities became most prominent at E5.5, which is when TCTP level is normally at its highest. Not only did the mice embryos appear smaller, but the epiblast that eventually develops into the fetus also contained a significantly lower cell number. The reason for this was determined to be a misregulation of apoptosis [22].

Apoptosis is a crucial part of the life of a multicellular organism. It is a highly regulated process, resulting in programmed cell death. Insufficient apoptosis may result in accumulation of mutations and uncontrolled cell proliferation, such as in cancer. Apoptosis can be induced by extracellular or intracellular signals and involves the activation of various regulatory proteins that activate the apoptotic pathway. This process is highly regulated, so that apoptosis is not induced unnecessarily, and can even be stopped if the need for apoptosis is no longer required. The intracellular apoptotic pathway is mainly regulated with the help of mitochondria, which supplies the cell with energy. A change in the permeability of the mitochondrial membrane can cause apoptotic proteins to leak into the cell. Pores called mitochondrial outer membrane permeabilization pores (MACs) regulate the permeability of the mitochondrial membrane to apoptotic proteins. Proteins belonging to the Bcl-2 protein family can regulate these MACs [35]. The protein Bax, when activated, dimerizes within the mitochondrial membrane. This dimerization promotes MAC pore formation, causing apoptotic proteins to enter the cell. In contrast, the proteins Bcl-2 and Mcl-1 inhibit MAC formation, preventing the influx of apoptotic proteins into the cell (reviewed in [36]). Apoptotic proteins that can be released via MACs into the cell are generally called small mitochondria-derived activator of caspases (SMACs). These can bind to inhibitors of apoptosis proteins (IAPs) within the cell. IAPs are usually bound to cysteine proteases that are referred to as caspases [37]. These caspases are enzymes that can degrade intracellular proteins, which ultimately cause the degradation of the entire cell. Often, these caspases need to be proteolytically cleaved in order to become active. In addition to SMACs, MAC pores also release the protein cytochrome c. Cytochrome c can then form a complex called apoptosome, by binding to ATP, the apoptotic protease activating factor1 (Apaf1) and procaspase-9. This results in the proteolytic cleavage of pro-caspase 9 into the enzymatically active form caspase 9, overall activating cellular degradation [38].

In a normal cell, the mitochondrial membrane is not permeable to SMACs. As a result, no cytochrome c is in the cell to activate caspase 9. Another class of inhibitor, the IAP proteins, are bound to caspases. Upon an apoptosis-inducing signal, the mitochondrial membrane is permeabilized, releasing SMACs and cytochrome c into the cell. SMACs bind IAPs, which release caspases, and cytochrome c converts caspase 9 to its active form (reviewed in [36]). This results in intracellular digestion and cell death. The necessity for different factors to be exported by the mitochondria shows the high level of regulation. This is not surprising, as a malfunction of the system would be detrimental to the cell.

TCTP also seems to play an important role in controlling the potentially suicidal pathway. It was found to inhibit the proapoptotic protein Bax that promotes MAC pore formation by dimerizing in the mitochondrial membrane. TCTP inserts itself into the mitochondrial membrane, preventing Bax from dimerizing [22]. This prevents MAC pore formation and inhibits any flux of apoptosis promoting factors into the cell [15]. Another study showed that TCTP also binds to Mcl-1. As discussed above, Mcl-1 inhibits MAC formation. Since binding of TCTP was found to stabilize Mcl-1, TCTP increases the block on MAC formation and ultimately prevents apoptosis [13, 14].

It remains to be investigated what happens to TCTP when apoptosis is initiated. It is possible that the TCTP protein is actively degraded or isolated from the system, or that the TCTP mRNA level decreases. In both cases, it is likely that a factor is required for TCTP inactivation. Pull-down studies and promoter analysis when apoptosis is induced could help to find important regulators of TCTP.

4. TCTP in Pluripotency and Nuclear Reprogramming

During development cells become committed and differentiate from one cell into many distinct cell types. Embryonic stem (ES) cells are pluripotent cells derived from the inner cell mass of the blastocysts of an early embryo. In contrast to committed or differentiated cells, pluripotent cells can differentiate into any fetal or adult cell type and are capable of self-renewal and unlimited proliferation [39]. These have tremendous potential in medicine, as ES cells could be differentiated into any cell type or even tissue of the body and be used for potential cell replacement therapies.

ES cells are characterized by a particular pattern of gene expression. For example, various genes are upregulated in ES cells and are frequently used as pluripotency markers. Oct4 seems to be an important regulator of pluripotency and differentiation [40]. It represses or activates expression of different genes, which occurs either directly by binding to promoter regions or indirectly by neutralising transcription activators [41]. Oct4, also known as Oct3, is a member of the POU transcription family [42]. These are transcription factors that bind via an octameric sequence to an AGTCAAAT consensus sequence [41]. The gene is expressed in early mammalian embryos, during gametogenesis, in ES cells [43], and occasionally in tumours [44]. After gastrulation, Oct4 becomes silent in mouse and human mammalian somatic cells [45]. In mouse oocytes, Oct4 mRNA is present as a maternal transcript [46] and it is downregulated when development proceeds [47]. It is essential, but not sufficient to maintain cells in an undifferentiated state [48]. During embryonic development, Oct4 is expressed in early blastomeres. Then, it becomes restricted to the inner cell mass, and is down regulated in the trophectoderm and primitive endoderm [47]. Oct4 is widely conserved. Homologues even exist in early amphibian development, where they also act as suppressors of cell fate commitment. Even though so far ES cells have not been derived from amphibians, the *Xenopus laevis* version of Oct4, Pou91, was able to fully support mouse ES cell self-renewal [49]. This suggests a similar function for Pou91 in pluripotency.

Pluripotency also requires other factors, for example, the leukaemia inhibitory factor (LIF). LIF is a key molecule required for self-renewal and pluripotency in mouse ES cells [50, 51], but not for monkey or human ES cells [52]. It is known to bind to the heterodimer LIF receptor—gp130 and to activate the transcription factor STAT3 by phosphorylation [53]. Interestingly, overexpression of the gene Nanog can bypass the requirement for LIF in mouse ES cells [54]. Nanog is also required for maintaining the undifferentiated state of early postimplantation embryos and ES cells [54, 55], making Nanog an important regulator of pluripotency. There are also other components required, such as bone morphogenic proteins (BMP) that activate the inhibitor of differentiation (Id), which represses differentiation [56]. Another important regulator is Sox2, which cooperatively binds the Oct4 protein and activates genes promoting pluripotency [57], but represses its inhibitors [58].

Despite obtaining the ES cells from blastocysts, ES or ES-cell-like cells can be obtained by nuclear reprogramming, a term introduced to describe the restoration of the embryonic pattern of gene expression [59]. Nuclear reprogramming was first demonstrated in nuclear transfer experiments. *Xenopus laevis* nuclei of differentiated cells were transplanted into enucleated frog eggs. This gave rise to normal fertile adult frogs, illustrating that differentiated cells can become reprogrammed and give rise to an entire new organism [60, 61]. Another way to reprogram nuclei was achieved when cells were fused to each other [62, 63]. Cell fusions with ES cells rejuvenated somatic cells that could differentiate into many different cell types. In these hybrids the silent gene Oct4 was reactivated [64]. Fusion experiments with an increased expression of the pluripotency gene Nanog increased nuclear reprogramming efficiency by 200-fold [65]. Nowadays, the most common way somatic cells are reprogrammed to an embryonic-like pattern of gene expression is by overexpressing different factors, such as Oct4, Sox2, c-Myc, and Klf4 under ES cell culture conditions [66]. Surprisingly, Nanog was not required, even though it seemed to promote nuclear reprogramming in cell fusion experiments [65]. These ES-like cells had normal ES cell morphology, a gene expression pattern typical for normal ES cells and could differentiate into all three germ layers. They were named iPS cells, induced pluripotent stem cells [66]. Even though the generation of iPS cells is a very convenient way to generate ES cells, this approach does not reveal the mechanism underlying nuclear reprogramming. Also, it does not identify novel factors that are involved in this process.

To better understand the process of nuclear reprogramming, nuclear transfer experiments of somatic cells into Xenopus oocytes were carried out. It was found that even human or mouse nuclei could be reprogrammed by frog oocytes and induce an ES cell or ES cell-like pattern of gene expression [67]. For example, genes such as Oct4, Nanog, and Sox2 became transcriptionally active upon nuclear transfer [67]. Using this system, novel molecules were isolated that interact with the promoter region of Oct4. One of these molecules was TCTP. Further functional assays revealed that it in fact TCTP changed the transcriptional level of Oct4 and even Nanog in human nuclei, genes essential for successful nuclear reprogramming [68]. A similar effect of TCTP was found in bovine oocytes, suggesting a conserved function of TCTP in activating pluripotency [69]. TCTP knockout mice have an abnormal number of cells in the epiblast [22]. The epiblast is formed from the inner cell mass of the blastocyst, from which ES cells can be obtained. Since TCTP activates the pluripotency genes Oct4 and Nanog, it is possible that, in the TCTP knockout mice, the epiblast does not develop normally due to misregulation of pluripotency genes such as Oct4 and Nanog.

It would be interesting to determine if TCTP activates also other pluripotency genes such as Sox2 and Klf4. TCTP might promote pluripotency in two different ways, namely, by (1) activating pluripotency genes and (2) inhibiting somatic gene expression. Genomewide studies in the absence of TCTP could help to determine what other genes TCTP regulates. Another important question is whether TCTP is sufficient for nuclear reprogramming and if its overexpression in somatic cells could replace the four reprogramming factors used to make iPS cells. Even if it does not replace these four factors, it could increase the generation of iPS cells, a currently very inefficient process.

Nuclear actin polymerization has been reported to be required for Oct4 activation in *Xenopus laevis* oocytes [70]. Since TCTP has been found to contain an actin-binding site [17], it is possible that it might interfere with pluripotency gene regulation by interfering with actin. Testing actin polymerization in the absence and presence of TCTP, as well as the effect on Oct4, would help to understand any possible interactions required to induce pluripotency. These experiments could also be analyzed Genomewide, which will greatly help to elucidate the underlying network required to establish pluripotency. Using TCTP as bait to pull down interaction partners together with Genomewide Chromatin Immunoprecipitation analysis of TCTP and its interaction partners will also contribute towards understanding how pluripotency is established.

Another protein that has been found to interact with TCTP in Xenopus oocytes is nucleoplasmin Npm1 [71]. Similar to TCTP knockout mice, mice deficient in Npm1 are embryonic lethal and have smaller embryo sizes [72]. Npm1 is a very abundant protein. In fertilized *Xenopus* eggs, it is involved in the decondensation and hence transcriptional activation of the paternal genome provided after normal fertilization by the sperm (reviewed in [73]). It is possible that TCTP not only activates pluripotency genes, but also that it has a role in paternal gene activation by interacting with Npm1. Disturbing the interaction of TCTP and Npm1 could show if TCTP is also involved in this process. But it is possible that pluripotency and paternal and maternal genome activation is actually not that different. After all, when the genome becomes transcriptionally active, it is set as such, so that it can proliferate and differentiate into an entire organism. Hence, zygotic genome activation could be regarded as nuclear reprogramming that occurs naturally in nature, without the need of nuclear transplantation, cell fusion experiments, or overexpression of a few transcription factors.

5. Cell Cycle Regulation of TCTP

The cell cycle describes the stages a cell has to go through to divide and duplicate its genome. In eukaryotes, the cell cycle is divided into four phases: (1) the G_1 phase, in which the cell grows and makes sure it is prepared for DNA replication, (2) the S or synthesis phase, where the DNA is duplicated, the (3) G_2 phase, in which the cell ensures it is ready for mitosis, and (4) the M phase, in which cell growth stops and the cell divides its DNA and other cellular components giving rise to two cells. There is also an additional phase, which is not part of the cell cycle, G_0, in which the cell has exited the cell cycle and has stopped dividing [74]. Since the cell cycle is crucial for the survival of the cell and generation of a multicellular organism, the process is highly controlled. There are many proteins that control each phase and that detect and repair genetic damage, as well as avoiding the propagation of mutations [74]. Any misregulation might result in uncontrolled cell proliferation and ultimately cancer. The key enzymes regulating the progression from one phase into the next are called cyclins and cyclin-dependent kinases. There are also many other proteins, such as the serine-threonine protein kinase polo-like kinase 1 (PLK1) and the protein checkpoint with forkhead and ring finger domains (CHFR).

The protein CHFR is an E3 ubiquitin ligase that can detect microtubule abnormalities. It delays the G_2 to mitosis transition when it is exposed to altered microtubules. Microtubules are part of the cytoskeleton and act in mitosis and move the duplicated genomes into the forming daughter cells. CHFR is usually present in an inactive form, unable to carry out ubiquitination. When microtubules are damaged, CHFR becomes activated [75]. CHFR then ubiquitinates PLK1 that results in PLK1 degradation [76]. The kinase PLK1 is required in the late G_2 and early mitotic phases. It regulates spindle assembly and centrosome maturation, which is a microtubule-organizing center. PLK1 phosphorylates and activates Cdc25C, which dephosphates and activates the cyclins required for mitosis, the cyclinB/cdc2 complex [77, 78]. Any loss of PLK1 can induce a block in cell cycle progression and lead to apoptosis. PLK1 overexpression is frequently observed in connection with centrosome abnormalities, improper segregation of chromosomes and tumor cells.

Although the TOR pathway might be indirectly involved in cell cycle regulation by responding to growth factors and energy levels and driving cell proliferation, TCTP seems to be involved more directly in the cell cycle. For example, TCTP expression is upregulated upon entry into the cell cycle, but when overexpressed, cell cycle progression is delayed [10]. TCTP also has a tubulin-binding site that allows it to bind to microtubules in a cell-cycle-dependent way. As a result, it is recruited to the mitotic spindle during metaphase, but is released at the M/G_1 transition [10]. Furthermore,

TCTP interacts with CHFR that interacts with microtubules [79]. Upon depolymerization of the microtubules, CHFR and TCTP interaction is diminished. It has been suggested that this might provide a mechanism by which CHFR senses microtubule abnormalities that results in CHFR activation, PLK1 degradation, and ultimately cell cycle arrest [79]. It would be interesting to determine if CHFR can bind with the same affinity to microtubules in the absence of TCTP, or if it is no longer sensitive to microtubule abnormalities in the absence of TCTP, confirming the proposed model. In addition to binding to CHFR, TCTP can be phosphorylated by the substrate of CHFR, PLK1 [80]. This presumably leads to a decrease in the affinity of TCTP for microtubules or CHFR. When PLK1 phosphorylation sites on TCTP are blocked, a dramatic increase in multinucleated cells is observed suggesting that the completion of mitosis is inhibited [81]. This suggests that TCTP is crucial in cell cycle regulation and that its phosphorylation by PLK1 is required for accurate exit from mitosis. In the TCTP mutants that cannot become phosphorylated, an increase in apoptosis is also observed [81]. Bearing in mind that TCTP is involved in apoptosis, it is possible that PLK1 acts *via* TCTP to inhibit apoptosis. TCTP phosphorylation by PLK1 causes cell cycle progression. It is possible that this modified TCTP might have inhibitory effects on the apoptotic pathway. In this way, TCTP could make sure that when cell cycle progresses no apoptosis is induced. In contrast, if it is not modified by PLK1 during mitosis, it might induce apoptosis via the different routes described above. It would be revealing to investigate the role of the modified TCTP protein in apoptosis.

6. TCTP in Cancer

TCTP has been associated with tumorigenesis and cancer since its discovery in tumor cells [1, 2]. It was not until tumor reversion screens that TCTP got attention as a key player in cancer (Figure 2) [11, 23]. Tumor reversion is a process by which some cancer cells lose their malignant phenotype. Studying this process might help to understand how cancer can be inhibited and ultimately lead to a cure. To understand this process on a molecular level, tumor cells were grown in the presence of the H1 parvovirus [23]. This virus preferentially kills tumor cells, which in turn allows for selection of cells that revert back to a normal, nonmalignant phenotype [82, 83]. To identify which genes are most likely to be involved in this process, the level of gene expression was compared between malignant and reverted state. The TCTP gene expression level showed the largest difference between malignant and reverted state. A high level associates with tumorigenesis and a low level with normal cell growth (124 times higher TCTP level in tumor cells versus revertants). This was confirmed in several different tumor cell lines, suggesting that it is a universal gene that is implicated in tumor reversal [23]. Furthermore, knockdown experiments of TCTP in various malignant cell lines increased tumor reversal by approximately 30% [11].

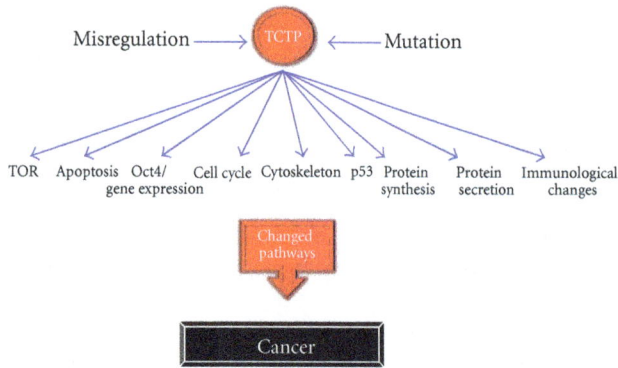

FIGURE 2: Pathways in which TCTP misregulation or mutations could cause cancer.

The p53 protein is one of the most famous tumor suppressors and is often referred to as the "guardian" of cancer. It is a transcription factor and regulates the transcription of various genes. It can activate the transcription of DNA repair genes when the DNA is damaged, through genes involved in cell cycle and initiate apoptosis by regulating genes such as Bax and Bcl-2 [84]. In response to stress such as DNA damage, it either induces repair genes to repair the damage, cell cycle arrest to prevent the replication of damaged DNA, or induces apoptosis to eliminate potentially malignant cells. Various signals are responsible for whether p53 induces repair, cell cycle arrest, or apoptosis (reviewed in [85]).

To better understand how TCTP levels control cancer, the interaction between TCTP and p53 has been studied in more detail. It was found that TCTP overexpression can lead to p53 degradation. This was accompanied by the observation that p53 was no longer able to induce apoptosis [24]. This suggested that TCTP is an important regulator in the p53 pathway and also links p53 with apoptosis.

MDM2 is a transcriptional target of p53. When overexpressed, MDM2 ubiquitinates and degrades p53, providing a negative feedback mechanism. TCTP was found to inhibit MDM2 autoubiquitination and to promote MDM2-mediated ubiquitination of p53, which ultimately leads to p53 degradation [86]. In addition, p53 was found to downregulate TCTP levels [23] and to promote TCTP exosome secretion [87, 88]. This shows that p53 and TCTP antagonize each other. Similar evidence comes from a different observation. The dsRNA-dependent protein kinase (PKR) increases p53 transcriptional function [89]. Mice depleted of PKR had altered TCTP protein levels. Further analysis showed that PKR directly interacts with TCTP mRNA. This interaction is required for PKR activation [9]. Hence, the presence of PKR might sequester TCTP mRNA and remove free TCTP mRNA from the RNA pool that would otherwise be available for translation. Hence, a higher level of PKR might be associated with lower TCTP protein levels. As PKR activates p53 and counteracts TCTP, it adds another layer of antagonistic control between TCTP and p53. The level of both p53 and TCTP might determine which pathway to choose, cell cycle arrest or apoptosis.

As outlined above, TCTP misregulation has an impact on the TOR pathway, apoptosis, reprogramming and cell cycle. All of these pathways can be implicated in cancer when they do not function correctly. In the TOR pathway, overexpression or mutations enhancing TCTP activity might result in increased TOR activation, leading to enhanced cell growth and ultimately tumor formation. Similarly, alterations in TCTP level might alter the ability of TCTP to inhibit apoptosis. Any TCTP misregulation might prevent damaged cells from being eliminated by apoptosis and in this way promote the survival of cells that might result in cancerous cells. Nuclear transfer experiments have shown that TCTP induces the transcription of pluripotency genes such as Oct4 and Nanog. An increased level of TCTP in normal cells may promote the formation of pluripotent-like gene expression. This might partly reprogram quiescent differentiated cells into pluripotent like proliferating cells. If in addition mutations accumulate in these cells, elevated levels of TCTP might enhance the propagation of these mutated cells. The higher the level of pluripotency transcripts is, the greater is the cell's malignant potential [90]. This suggests that this is also true for a higher TCTP level. Ultimately, this can result in cancer. Misregulation of TCTP might also impact cell cycle progression by interfering with PLK1. PLK1 is overexpressed in a range of human tumors, and PLK1 overexpression is associated with a bad cancer prognosis [91]. Since PLK1 phosphorylates TCTP that is required for cell cycle progression from mitosis, it is possible that an overexpression of PLK1 causes faster TCTP phosphorylation and cell cycle progression. This faster cell cycle progression might result in cell cycle progression even when mitosis is not complete. The resulting daughter cells could in this way inherit not fully replicated genomes. This might result in a vast amount of mutations that might result in cancer. Alternatively, TCTP could be mutated, being irresponsive to PLK1, and have the same effect.

Finally, TCTP is also known to be involved in protein synthesis by acting as a guanine nucleotide dissociation inhibitor for the elongation factor EF1A [18]. Any changes in TCTP level could influence many genes at once and change the status of a cell substantially. Similarly, changes in TCTP will also affect the immune response and ultimately might promote cancer development [19].

7. Conclusion

In summary, TCTP is highly conserved and abundant. It is involved in many key biological pathways, such as the TOR pathway, apoptosis, nuclear reprogramming, and cell cycle. It is highly regulated on a transcriptional, translational and protein level. As TCTP is involved in a wide array of biological functions, it is not surprising that any changes to TCTP might result in an array of abnormal phenotypes. Furthermore, abnormal cell proliferation, growth, and survival are probably the most important characteristics of cancer, all of which are regulated by TCTP. Due to this involvement, and its presence in many other pathways, TCTP might be a crucial target for cancer therapies. Some success has already been reported in this regard [11]. Bearing in mind that TCTP is also a histamine-releasing factor, inhibitors of this pathway were tested for their ability to decrease the number of tumor cells by inhibiting TCTP. In fact, many of such inhibitors were found to kill tumors [11]. However, further studies are required to better understand the function of TCTP in the pathways described before and maybe to reveal further functions. Overall, this will greatly contribute to the understanding of basic molecular pathways and provide further target sites for cancer therapies.

References

[1] G. Thomas, G. Thomas, and H. Luther, "Transcriptional and translational control of cytoplasmic proteins after serum stimulation of quiescent Swiss 3T3 cells," *Proceedings of the National Academy of Sciences of the United States of America*, vol. 78, no. 9, pp. 5712–5716, 1981.

[2] R. Yenofski, I. Bergmann, and G. Brawerman, "Messenger RNA species partially in a repressed state in mouse sarcoma ascites cell," *Proceedings of the National Academy of Sciences of the United States of America*, vol. 79, no. 19, pp. 5876–5880, 1982.

[3] F. Brioudes, A. M. Thierry, P. Chambrier, B. Mollereau, and M. Bendahmane, "Translationally controlled tumor protein is a conserved mitotic growth integrator in animals and plants," *Proceedings of the National Academy of Sciences of the United States of America*, vol. 107, no. 37, pp. 16384–16389, 2010.

[4] H. Thiele, M. Berger, A. Skalweit, and B. J. Thiele, "Expression of the gene and processed pseudogenes encoding the human and rabbit translationally controlled turnout protein (TCTP)," *European Journal of Biochemistry*, vol. 267, no. 17, pp. 5473–5481, 2000.

[5] E. Guillaume, C. Pineau, B. Evrard et al., "Cellular distribution of translationally controlled tumor protein in rat and human testes," *Proteomics*, vol. 1, no. 7, pp. 880–889, 2001.

[6] Z. Chen, H. Zhang, H. Yang, X. Huang, X. Zhang, and P. Zhang, "The expression of AmphiTCTP, a TCTP orthologous gene in amphioxus related to the development of notochord and somites," *Comparative Biochemistry and Physiology*, vol. 147, no. 3, pp. 460–465, 2007.

[7] S. Teshima, K. Rokutan, T. Nikawa, and K. Kishi, "Macrophage colony-stimulating factor stimulates synthesis and secretion of a mouse homolog of a human IgE-dependent histamine-releasing factor by macrophages in vitro and in vivo," *Journal of Immunology*, vol. 161, no. 11, pp. 6356–6366, 1998.

[8] A. Xu, A. R. Bellamy, and J. A. Taylor, "Expression of translationally controlled tumour protein is regulated by calcium at both the transcriptional and post-transcriptional level," *Biochemical Journal*, vol. 342, no. 3, pp. 683–689, 1999.

[9] U. A. Bommer, A. V. Borovjagin, M. A. Greagg et al., "The mRNA of the translationally controlled tumor protein P23/TCTP is a highly structured RNA, which activates the dsRNA-dependent protein kinase PKR," *RNA*, vol. 8, no. 4, pp. 478–496, 2002.

[10] Y. Gachet, S. Tournier, M. Lee, A. Lazaris-Karatzas, T. Poulton, and U. A. Bommer, "The growth-related, translationally controlled protein P23 has properties of a tubulin binding protein and associates transiently with microtubules during the cell cycle," *Journal of Cell Science*, vol. 112, no. 8, pp. 1257–1271, 1999.

[11] M. Tuynder, G. Fiucci, S. Prieur et al., "Translationally controlled tumor protein is a target of tumor reversion," *Proceedings of the National Academy of Sciences of the United States of America*, vol. 101, no. 43, pp. 15364–15369, 2004.

[12] F. Li, D. Zhang, and K. Fujise, "Characterization of fortilin, a novel antiapoptotic protein," *Journal of Biological Chemistry*, vol. 276, no. 50, pp. 47542–47549, 2001.

[13] H. Liu, H. W. Peng, Y. S. Cheng, H. S. Yuan, and H. F. Yang-Yen, "Stabilization and enhancement of the antiapoptotic activity of Mcl-1 by TCTP," *Molecular and Cellular Biology*, vol. 25, no. 8, pp. 3117–3126, 2005.

[14] Y. Yang, F. Yang, Z. Xiong et al., "An N-terminal region of translationally controlled tumor protein is required for its antiapoptotic activity," *Oncogene*, vol. 24, no. 30, pp. 4778–4788, 2005.

[15] L. Susini, S. Besse, D. Duflaut et al., "TCTP protects from apoptotic cell death by antagonizing bax function," *Cell Death and Differentiation*, vol. 15, no. 8, pp. 1211–1220, 2008.

[16] A. Burgess, J. C. Labbé, S. Vigneron et al., "Chfr interacts and colocalizes with TCTP to the mitotic spindle," *Oncogene*, vol. 27, no. 42, pp. 5554–5566, 2008.

[17] K. Tsarova, E. G. Yarmola, and M. R. Bubb, "Identification of a cofilin-like actin-binding site on translationally controlled tumor protein (TCTP)," *FEBS Letters*, vol. 584, no. 23, pp. 4756–4760, 2010.

[18] C. Cans, B. J. Passer, V. Shalak et al., "Translationally controlled tumor protein acts as a guanine nucleotide dissociation inhibitor on the translation elongation factor eEF1A," *Proceedings of the National Academy of Sciences of the United States of America*, vol. 100, no. 2, pp. 13892–13897, 2003.

[19] S. M. MacDonald, T. Rafnar, J. Langdon, and L. M. Lichtenstein, "Molecular identification of an IgE-Dependent histamine-releasing factor," *Science*, vol. 269, no. 5224, pp. 688–690, 1995.

[20] J. Z. Kubiak, F. Bazile, A. Pascal et al., "Temporal regulation of embryonic M-phases," *Folia Histochemica et Cytobiologica*, vol. 46, no. 1, pp. 5–9, 2008.

[21] Y. C. Hsu, J. J. Chern, Y. Cai, M. Liu, and K. W. Choi, "Drosophila TCTP is essential for growth and proliferation through regulation of dRheb GTPase," *Nature*, vol. 445, no. 7129, pp. 785–788, 2007.

[22] S. H. Chen, P. S. Wu, C. H. Chou et al., "A knockout mouse approach reveals that TCTP functions as an essential factor for cell proliferation and survival in a tissue- or cell type-specific manner," *Molecular Biology of the Cell*, vol. 18, no. 7, pp. 2525–2532, 2007.

[23] M. Tuynder, L. Susini, S. Prieur et al., "Biological models and genes of tumor reversion: cellular reprogramming through tpt1/TCTP and SIAH-1," *Proceedings of the National Academy of Sciences of the United States of America*, vol. 99, no. 23, pp. 14976–14981, 2002.

[24] S. B. Rho, J. H. Lee, M. S. Park et al., "Anti-apoptotic protein TCTP controls the stability of the tumor suppressor p53," *FEBS Letters*, vol. 585, no. 1, pp. 29–35, 2011.

[25] N. Hay and N. Sonenberg, "Upstream and downstream of mTOR," *Genes and Development*, vol. 18, no. 16, pp. 1926–1945, 2004.

[26] D. S. Dos, S. M. Ali, D. H. Kim et al., "Rictor, a novel binding partner of mTOR, defines a rapamycin-insensitive and raptor-independent pathway that regulates the cytoskeleton," *Current Biology*, vol. 14, no. 14, pp. 1296–1302, 2004.

[27] D. D. Sarbassov, D. A. Guertin, S. M. Ali, and D. M. Sabatini, "Phosphorylation and regulation of Akt/PKB by the rictor-mTOR complex," *Science*, vol. 307, no. 5712, pp. 1098–1101, 2005.

[28] D. H. Kim, D. D. Sarbassov, S. M. Ali et al., "mTOR interacts with raptor to form a nutrient-sensitive complex that signals to the cell growth machinery," *Cell*, vol. 110, no. 2, pp. 163–175, 2002.

[29] A. Dufner and G. Thomas, "Ribosomal S6 kinase signaling and the control of translation," *Experimental Cell Research*, vol. 253, no. 1, pp. 100–109, 1999.

[30] H. B. J. Jefferies, S. Fumagalli, P. B. Dennis, C. Reinhard, R. B. Pearson, and G. Thomas, "Rapamycin suppresses 5'TOP mRNA translation through inhibition of p70^{s6k}," *EMBO Journal*, vol. 16, no. 12, pp. 3693–3704, 1997.

[31] R. T. Peterson and S. L. Schreiber, "Translation control: connecting mitogens and the ribosome," *Current Biology*, vol. 8, no. 7, pp. R248–R250, 1998.

[32] A. Pause, G. J. Belsham, A. C. Gingras et al., "Insulin-dependent stimulation of protein synthesis by phosphorylation of a regulator of 5'-cap function," *Nature*, vol. 371, no. 6500, pp. 762–767, 1994.

[33] A. C. Gingras, S. P. Gygi, B. Raught et al., "Regulation of 4E-BP1 phosphorylation: a novel two-step mechanism," *Genes & Development*, vol. 13, pp. 1422–1437, 1999.

[34] S. Li, X. Chen, X. Liu, Y. Wang, and J. He, "Expression of translationally controlled tumor protein (TCTP) in the uterus of mice of early pregnancy and its possible significance during embryo implantation," *Human Reproduction*, vol. 26, no. 11, pp. 2972–2980, 2011.

[35] L. M. Dejean, S. Martinez-Caballero, S. Manon, and K. W. Kinnally, "Regulation of the mitochondrial apoptosis-induced channel, MAC, by BCL-2 family proteins," *Biochimica et Biophysica Acta*, vol. 1762, no. 2, pp. 191–201, 2006.

[36] S. W. G. Tait and D. R. Green, "Mitochondria and cell death: outer membrane permeabilization and beyond," *Nature Reviews Molecular Cell Biology*, vol. 11, no. 9, pp. 621–632, 2010.

[37] S. W. Fesik and Y. Shi, "Controlling the caspases," *Science*, vol. 294, no. 5546, pp. 1477–1478, 2001.

[38] S. B. Bratton and G. S. Salvesen, "Regulation of the Apaf-1-caspase-9 apoptosome," *Journal of Cell Science*, vol. 123, no. 19, pp. 3209–3214, 2010.

[39] P. J. Donovan and J. Gearhart, "The end of the beginning for pluripotent stem cells," *Nature*, vol. 414, no. 6859, pp. 92–97, 2001.

[40] M. Pesce and H. R. Schöler, "Oct-4: gatekeeper in the beginnings of mammalian development," *Stem Cells*, vol. 19, no. 4, pp. 271–278, 2001.

[41] G. J. Pan, Z. Y. I. Chang, H. R. Schöler, and D. Pei, "Stem cell pluripotency and transcription factor Oct4," *Cell Research*, vol. 12, no. 5-6, pp. 321–329, 2002.

[42] H. R. Scholer, S. Ruppert, N. Suzuki, K. Chowdhury, and P. Gruss, "New type of POU domain in germ line-specific protein Oct-4," *Nature*, vol. 344, no. 6265, pp. 435–439, 1990.

[43] H. R. Scholer, A. K. Hatzopoulos, R. Balling, N. Suzuki, and P. Gruss, "A family of octamer-specific proteins present during mouse embryogenesis: evidence for germline-specific expression of an Oct factor," *EMBO Journal*, vol. 8, no. 9, pp. 2543–2550, 1989.

[44] M. Monk and C. Holding, "Human embryonic genes re-expressed in cancer cells," *Oncogene*, vol. 20, no. 56, pp. 8085–8091, 2001.

[45] N. Kirchhof, J. W. Carnwath, E. Lemme, K. Anastassiadis, H. Scholer, and H. Niemann, "Expression pattern of Oct-4 in preimplantation embryos of different species," *Biology of Reproduction*, vol. 63, no. 6, pp. 1698–1705, 2000.

[46] H. R. Scholer, G. R. Dressler, R. Balling, H. Rohdewohld, and P. Gruss, "Oct-4: a germline-specific transcription factor mapping to the mouse t-complex," *EMBO Journal*, vol. 9, no. 7, pp. 2185–2195, 1990.

[47] S. L. Palmieri, W. Peter, H. Hess, and H. R. Scholer, "Oct-4 transcription factor is differentially expressed in the mouse embryo during establishment of the first two extraembryonic cell lineages involved in implantation," *Developmental Biology*, vol. 166, no. 1, pp. 259–267, 1994.

[48] J. Nichols, B. Zevnik, K. Anastassiadis et al., "Formation of pluripotent stem cells in the mammalian embryo depends on the POU transcription factor Oct4," *Cell*, vol. 95, no. 3, pp. 379–391, 1998.

[49] G. M. Morrison and J. M. Brickman, "Conserved roles for Oct4 homologues in maintaining multipotency during early vertebrate development," *Development*, vol. 133, no. 10, pp. 2011–2022, 2006.

[50] A. G. Smith, J. K. Heath, D. D. Donaldson et al., "Inhibition of pluripotential embryonic stem cell differentiation by purified polypeptides," *Nature*, vol. 336, no. 6200, pp. 688–690, 1988.

[51] R. L. Williams, D. J. Hilton, S. Pease et al., "Myeloid leukaemia inhibitory factor maintains the developmental potential of embryonic stem cells," *Nature*, vol. 336, no. 6200, pp. 684–687, 1988.

[52] J. A. Thomson, J. Itskovitz-Eldor, S. S. Shapiro et al., "Embryonic stem cell lines derived from human blastocysts," *Science*, vol. 282, pp. 1145–1147, 1998.

[53] S. Davis, T. H. Aldrich, N. Stahl et al., "LIFRβ and gp130 as heterodimerizing signal transducers of the tripartite CNTF receptor," *Science*, vol. 260, no. 5115, pp. 1805–1808, 1993.

[54] I. Chambers, D. Colby, M. Robertson et al., "Functional expression cloning of Nanog, a pluripotency sustaining factor in embryonic stem cells," *Cell*, vol. 113, no. 5, pp. 643–655, 2003.

[55] K. Mitsui, Y. Tokuzawa, H. Itoh et al., "The homeoprotein nanog is required for maintenance of pluripotency in mouse epiblast and ES cells," *Cell*, vol. 113, no. 5, pp. 631–642, 2003.

[56] Q. L. Ying, J. Nichols, I. Chambers, and A. Smith, "BMP induction of Id proteins suppresses differentiation and sustains embryonic stem cell self-renewal in collaboration with STAT3," *Cell*, vol. 115, no. 3, pp. 281–292, 2003.

[57] M. Nishimoto, A. Fukushima, A. Okuda, and M. Muramatsu, "The gene for the embryonic stem cell coactivator UTF1 carries a regulatory element which selectively interacts with a complex composed of Oct-3/4 and Sox-2," *Molecular and Cellular Biology*, vol. 19, no. 8, pp. 5453–5465, 1999.

[58] H. Niwa, J. I. Miyazaki, and A. G. Smith, "Quantitative expression of Oct-3/4 defines differentiation, dedifferentiation or self-renewal of ES cells," *Nature Genetics*, vol. 24, no. 4, pp. 372–376, 2000.

[59] E. M. De Robertis and J. B. Gurdon, "Gene activation in somatic nuclei after injection into amphibian oocytes," *Proceedings of the National Academy of Sciences of the United States of America*, vol. 74, no. 6, pp. 2470–2474, 1977.

[60] J. B. Gurdon, T. R. Elsdale, and M. Fischberg, "Sexually mature individuals of Xenopus laevis from the transplantation of

single somatic nuclei," *Nature*, vol. 182, no. 4627, pp. 64–65, 1958.

[61] J. B. Gurdon, "Adult frogs derived from the nuclei of single somatic cells," *Developmental Biology*, vol. 4, no. 2, pp. 256–273, 1962.

[62] H. Harris, O. J. Miller, G. Klein, P. Worst, and T. Tachibana, "Suppression of malignancy by cell fusion," *Nature*, vol. 223, no. 5204, pp. 363–368, 1969.

[63] N. Ringertz and R. E. Savage, *Cell Hybrids*, Academic Press, New York, NY, USA, 1976.

[64] M. Tada, Y. Takahama, K. Abe, N. Nakatsuji, and T. Tada, "Nuclear reprogramming of somatic cells by in vitro hybridization with ES cells," *Current Biology*, vol. 11, no. 19, pp. 1553–1558, 2001.

[65] J. Silva, I. Chambers, S. Pollard, and A. Smith, "Nanog promotes transfer of pluripotency after cell fusion," *Nature*, vol. 441, no. 7096, pp. 997–1001, 2006.

[66] K. Takahashi and S. Yamanaka, "Induction of pluripotent stem cells from mouse embryonic and adult fibroblast cultures by defined factors," *Cell*, vol. 126, no. 4, pp. 663–676, 2006.

[67] J. A. Byrne, S. Simonsson, P. S. Western, and J. B. Gurdon, "Nuclei of adult mammalian somatic cells are directly reprogrammed to oct-4 stem cell gene expression by amphibian oocytes," *Current Biology*, vol. 13, no. 14, pp. 1206–1213, 2003.

[68] M. J. Koziol, N. Garrett, and J. B. Gurdon, "Tpt1 activates transcription of oct4 and nanog in transplanted somatic nuclei," *Current Biology*, vol. 17, no. 9, pp. 801–807, 2007.

[69] T. Tani, H. Shimada, Y. Kato, and Y. Tsunoda, "Bovine oocytes with the potential to reprogram somatic cell nuclei have a unique 23-kDa protein, phosphorylated Transcriptionally Controlled Tumor Protein (TCTP)," *Cloning and Stem Cells*, vol. 9, no. 2, pp. 267–280, 2007.

[70] K. Miyamoto, V. Pasque, J. Jullien, and J. B. Gurdon, "Nuclear actin polymerization is required for transcriptional reprogramming of Oct4 by oocytes," *Genes and Development*, vol. 25, no. 9, pp. 946–958, 2011.

[71] H. Johansson, D. Vizlin-Hodzic, T. Simonsson, and S. Simonsson, "Translationally controlled tumor protein interacts with nucleophosmin during mitosis in ES cells," *Cell Cycle*, vol. 9, no. 11, pp. 2160–2169, 2010.

[72] S. Grisendi, R. Bernardi, M. Rossi et al., "Role of nucleophosmin in embryonic development and tumorigenesis," *Nature*, vol. 437, no. 7055, pp. 147–153, 2005.

[73] M. S. Lindström, "NPM1/B23: a multifunctional chaperone in ribosome biogenesis and chromatin remodeling," *Biochemistry Research International*, vol. 2011, Article ID 195209, 2011.

[74] G. M. Cooper, "Chapter 14: the eukaryotic cell cycle," in *The Cell: A Molecular Approach*, ASM Press, Washington, DC, USA, 2nd edition, 2000.

[75] D. M. Scolnick and T. D. Halazonetis, "Chfr defines a mitotic stress checkpoint that delays entry into metaphase," *Nature*, vol. 406, no. 6794, pp. 430–435, 2000.

[76] D. Kang, J. Chen, J. Wong, and G. Fang, "The checkpoint protein Chfr is a ligase that ubiquitinates Plk1 and inhibits Cdc2 at the G_2 to M transition," *Journal of Cell Biology*, vol. 156, no. 2, pp. 249–259, 2002.

[77] B. C. M. Van De Weerdt and R. H. Medema, "Polo-like kinases: a team in control of the division," *Cell Cycle*, vol. 5, no. 8, pp. 853–864, 2006.

[78] N. K. Soung, J. E. Park, L. R. Yu et al., "Plk1-dependent and -independent roles of an ODF2 splice variant, hCenexin1, at the centrosome of somatic cells," *Developmental Cell*, vol. 16, no. 4, pp. 539–550, 2009.

[79] A. Burgess, J. C. Labbé, S. Vigneron et al., "Chfr interacts and colocalizes with TCTP to the mitotic spindle," *Oncogene*, vol. 27, no. 42, pp. 5554–5566, 2008.

[80] U. Cucchi, L. M. Gianellini, A. De Ponti et al., "Phosphorylation of TCTP as a marker for polo-like kinase-1 activity in vivo," *Anticancer Research*, vol. 30, no. 12, pp. 4973–4985, 2010.

[81] F. R. Yarm, "Plk phosphorylation regulates the microtubule-stabilizing protein TCTP," *Molecular and Cellular Biology*, vol. 22, no. 17, pp. 6209–6221, 2002.

[82] H. W. Toolan, "Lack of oncogenic effect of the H-viruses for hamsters," *Nature*, vol. 214, no. 92, p. 1036, 1967.

[83] S. Mousset and J. Rommelaere, "Minute virus of mice inhibits cell transformation by simian virus 40," *Nature*, vol. 300, no. 5892, pp. 537–539, 1982.

[84] T. Riley, E. Sontag, P. Chen, and A. Levine, "Transcriptional control of human p53-regulated genes," *Nature Reviews Molecular Cell Biology*, vol. 9, no. 5, pp. 402–412, 2008.

[85] R. Vogt Sionov and Y. Haupt, "The cellular response to p53: the decision between life and death," *Oncogene*, vol. 18, no. 45, pp. 6145–6157, 1999.

[86] R. Amson, S. Pece, A. Lespagnol et al., "Reciprocal repression between p53 and TCTP," *Nature Medicine*, vol. 18, pp. 91–99, 2011.

[87] N. Amzallag, B. J. Passer, D. Allanic et al., "TSAP6 facilitates the secretion of translationally controlled tumor protein/histamine-releasing factor via a nonclassical pathway," *Journal of Biological Chemistry*, vol. 279, no. 44, pp. 46104–46112, 2004.

[88] A. Lespagnol, D. Duflaut, C. Beekman et al., "Exosome secretion, including the DNA damage-induced p53-dependent secretory pathway, is severely compromised in TSAP6/Steap3-null mice," *Cell Death and Differentiation*, vol. 15, no. 11, pp. 1723–1733, 2008.

[89] A. R. Cuddihy, S. Li, N. W. N. Tam et al., "Double-stranded-RNA-activated protein kinase PKR enhances transcriptional activation by tumor suppressor p53," *Molecular and Cellular Biology*, vol. 19, no. 4, pp. 2475–2484, 1999.

[90] S. Gidekel, G. Pizov, Y. Bergman, and E. Pikarsky, "Oct-3/4 is a dose-dependent oncogenic fate determinant," *Cancer Cell*, vol. 4, no. 5, pp. 361–370, 2003.

[91] M. A. T. M. Van Vugt and R. H. Medema, "Getting in and out of mitosis with Polo-like kinase-1," *Oncogene*, vol. 24, no. 17, pp. 2844–2859, 2005.

2

Ubiquitin C-Terminal Hydrolase L1 in Tumorigenesis

Jennifer Hurst-Kennedy, Lih-Shen Chin, and Lian Li

Department of Pharmacology and Center for Neurodegenerative Disease, Emory University School of Medicine, Atlanta, GA 30322, USA

Correspondence should be addressed to Lih-Shen Chin, chinl@pharm.emory.edu and Lian Li, lianli@pharm.emory.edu

Academic Editor: Dmitry Karpov

Ubiquitin carboxyl-terminal hydrolase L1 (UCH-L1, aka PGP9.5) is an abundant, neuronal deubiquitinating enzyme that has also been suggested to possess E3 ubiquitin-protein ligase activity and/or stabilize ubiquitin monomers *in vivo*. Recent evidence implicates dysregulation of UCH-L1 in the pathogenesis and progression of human cancers. Although typically only expressed in neurons, high levels of UCH-L1 have been found in many nonneuronal tumors, including breast, colorectal, and pancreatic carcinomas. UCH-L1 has also been implicated in the regulation of metastasis and cell growth during the progression of nonsmall cell lung carcinoma, colorectal cancer, and lymphoma. Together these studies suggest UCH-L1 has a potent oncogenic role and drives tumor development. Conversely, others have observed promoter methylation-mediated silencing of UCH-L1 in certain tumor subtypes, suggesting a potential tumor suppressor role for UCH-L1. In this paper, we provide an overview of the evidence supporting the involvement of UCH-L1 in tumor development and discuss the potential mechanisms of action of UCH-L1 in oncogenesis.

1. Introduction

Ubiquitin carboxyl-terminal hydrolase L1 (UCH-L1, aka PGP9.5) is an abundant neuronal protein consisting of 223 amino acids [1]. The best understood function of UCH-L1 is its deubiquitinating enzyme (DUB) activity that catalyzes hydrolysis of C-terminal esters and amides of ubiquitin (Ub) to generate monomeric Ub [2, 3]. In addition to its DUB activity, UCH-L1 has also been suggested to possess a putative, dimerization-dependent E3 ubiquitin-protein ligase activity and/or have a role in stabilizing Ub monomers *in vivo* [4, 5]. As a DUB, UCH-L1 facilitates Ub recycling and, therefore, can regulate the cellular pool of available Ub [6], giving UCH-L1 the capacity to modulate many ubiquitin-dependent cellular processes. Although its exact physiological function remains unclear, a growing body of evidence implicates UCH-L1 in the progression of human malignancies. Currently, the specific role of UCH-L1 in cancer pathogenesis is not known. UCH-L1 has been reported to be upregulated in several tumor tissues and cancer cell lines [7–13] and has been suggested to function as an oncogene in the progression of many cancers including

lymphoma [11], colorectal cancer [14], and nonsmall cell lung carcinoma [8]. Conversely, studies have been put forth designating UCH-L1 as a tumor suppressor in the pathogenesis of nasopharyngeal [15] and breast [16] cancer. Despite the controversy regarding the exact function of UCH-L1 in oncogenesis, these studies suggest that UCH-L1 is an important regulator of tumor formation and maturation. Here, we review the current knowledge of the function and mechanisms of actions of UCH-L1 in tumorigenesis.

2. Functions of UCH-L1 in the UPS

The ubiquitin-proteasome system (UPS) is a major intracellular proteolytic pathway that facilitates the degradation of normal cellular proteins as well as the clearance of misfolded and damaged proteins [17]. In the UPS, protein substrates are tagged with polymers of a 76-amino-acid polypeptide, ubiquitin (Ub), followed by recognition and degradation by the 26S proteasome. This process is facilitated by the sequential actions of at least three classes of enzymes: ubiquitin-activating enzymes (E1), ubiquitin-conjugating enzymes (E2), and ubiquitin-protein ligases (E3). First, an E1 activates Ub at the expense of ATP. Next, activated Ub is transferred to an E2 enzyme. Finally, an E3 specifically recognizes its protein substrate, which can be in its normal conformation or misfolded, and catalyzes the transfers of activated Ub from an E2 to the substrate. Successive addition of Ub to a lysine residue of a previously conjugated Ub results in the formation of a polyubiquitin chain. K48-linked polyubiquitin chains serve as a recognition signaling for proteasomal degradation. Once ubiquitinated substrates are transferred to the proteasome, DUBs remove the Ub chain, allowing for free Ub monomers to be recycled. Monoubiquitination and noncanonical polyubiquitination (e.g., K63 ubiquitin linkages) of proteins have been implicated in nonproteasomal cellular processes, including endocytosis, trafficking, cell signaling, DNA damage repair, and modifications of histones [17, 18].

UCH-L1 was first identified as a member of the ubiquitin carboxyl-terminal hydrolase (UCH) family of DUBs with cysteine protease activity in the late 1980s [1]. UCH-L1 is an abundant neuronal protein, comprising approximately 2% of total brain protein [1, 19]. Although low levels of UCH-L1 protein have been reported to be present in kidneys, breast epithelium, and reproductive tissues [20, 21], UCH-L1 is absent in most other tissues [1, 19, 22, 23]. UCH-L1 appears to play an important role in neurons, as mice lacking functional UCH-L1 have been reported to exhibit neuronal dysfunction and neurodegeneration [24, 25]. At the subcellular level, UCH-L1 is primarily found in the cytoplasm [26], but recent reports indicate that a subpopulation of UCH-L1 can be transiently localized to the nucleus [13, 27]. Biochemical studies revealed that UCH-L1 hydrolyzes Ub at its C-terminal glycine residue to generate monomeric Ub in vitro [1] and that this activity is dependent upon the catalytic residues C90 and H161 [2]. Analysis of UCH-L1 crystal structure indicates that these catalytic residues are not accessible to large polymers of Ub and suggest that UCH-L1 preferentially binds monomeric Ub and small adducts of Ub [28]. It is possible that substrate binding and/or the presence of cofactors may induce a conformational change, allowing UCH-L1 to process larger Ub chains. However, this has not yet been demonstrated in vitro or in vivo. Thus, UCH-L1 is best understood to function as a cysteine protease capable of hydrolyzing small Ub moieties.

Although the exact function of UCH-L1 is not fully understood, several studies suggest that UCH-L1 regulates the cellular pool of free Ub (Figure 1). First, UCH-L1 has been reported to cleave the ubiquitin gene products UbB and UbC and the ribosomal ubiquitin fusion protein UbA80 to generate monomeric Ub [29], leading to an increase in the level of free Ub (Figure 1). UCH-L1 may also elevate free Ub levels by facilitating recycling of Ub (Figure 1). Next, it has also been suggested that UCH-L1 plays a role in stabilizing Ub monomers by binding to monomeric Ub and preventing its lysosomal degradation (Figure 1) [4]. Association of UCH-L1 with monomeric Ub occurs independently of the catalytic C90 residue, indicating that mono-Ub binding is not dependent upon UCH-L1 hydrolase

activity [4]. The role of UCH-L1 in the regulation of the free Ub pool is also supported by the observation that levels of monomeric Ub are decreased in gracile axonal dystrophy (gad) mice, which lack functional UCH-L1 [4] In contrast to other DUBs, in vitro studies indicate that UCH-L1 does not directly catalyze the deubiquitination of ubiquitinated protein substrates [29]. Moreover, no in vivo UCH-L1 substrates have been identified thus far. Collectively, current evidence suggests that UCH-L1 functions to increase the cellular pool of free Ub by hydrolyzing small Ub chains and stabilizing monomeric Ub rather than by directly acting on polyubiquitinated substrates.

UCH-L1 has been reported to possess putative, dimerization-dependent E3 ligase activity in addition to its hydrolase function (Figure 1) [5]. In vitro studies show that dimeric UCH-L1 promotes K63-linked polyubiquitination of α-synuclein [5]. Unlike other E3 ligases, UCH-L1 E3 ligase activity was observed in the absence of ATP [5], which differs from the mechanism of conventional ubiquitination [17, 18]. It is currently not known whether UCH-L1 exhibits E3 ligase activity in vivo. Further investigation into UCH-L1 enzymatic function is needed to understand its role in health and disease.

3. UCH-L1 as a Positive Regulator of Tumorigenesis

Although UCH-L1 is almost exclusively expressed in neurons [1, 19], proteomic screens have revealed that UCH-L1 is present in many nonneuronal human tumors (Table 1) including adenocarcinoma [35], pancreatic ductal carcinoma [36], and squamous cell carcinoma [31]. Similarly, microarray profiling analyses show UCH-L1 mRNA is upregulated in several breast cancer tumor types [37] and medullary thyroid carcinoma tumors [38]. UCH-L1 mRNA has also been shown to be elevated in gall bladder and colorectal tumor tissues as a result of hypomethylation of the UCH-L1 promoter [39, 40]. High levels of UCH-L1 protein have also been observed in many human tumor-derived cell lines (Table 1) such as those cultured from lung [8], prostate [41, 42], and bladder tumors [43] as well as B-cell lymphomas [44] and osteosarcomas [45]. The presence of UCH-L1 in nonneuronal tumor tissues and cancer cell lines suggests that increased levels of UCH-L1 may promote oncogenic transformation and, therefore, point to a possible role for UCH-L1 as an oncogene in cancer pathogenesis.

The potential oncogenic function of UCH-L1 is supported by a number of clinical studies demonstrating that UCH-L1 expression level in tumors is inversely correlated with patient survivability [14, 36, 37]. High levels of UCH-L1 mRNA in breast tumors have been reported to be associated with poor prognosis in patients [37]. Likewise, elevated UCH-L1 mRNA in colorectal tumors is associated with higher incidence of tumor recurrence and shorter survival time [14]. Moreover, UCH-L1 expression in pancreatic ductal tumors is correlated with decreased patient survival [36]. Together, these data suggest that UCH-L1 is involved in tumor maturation.

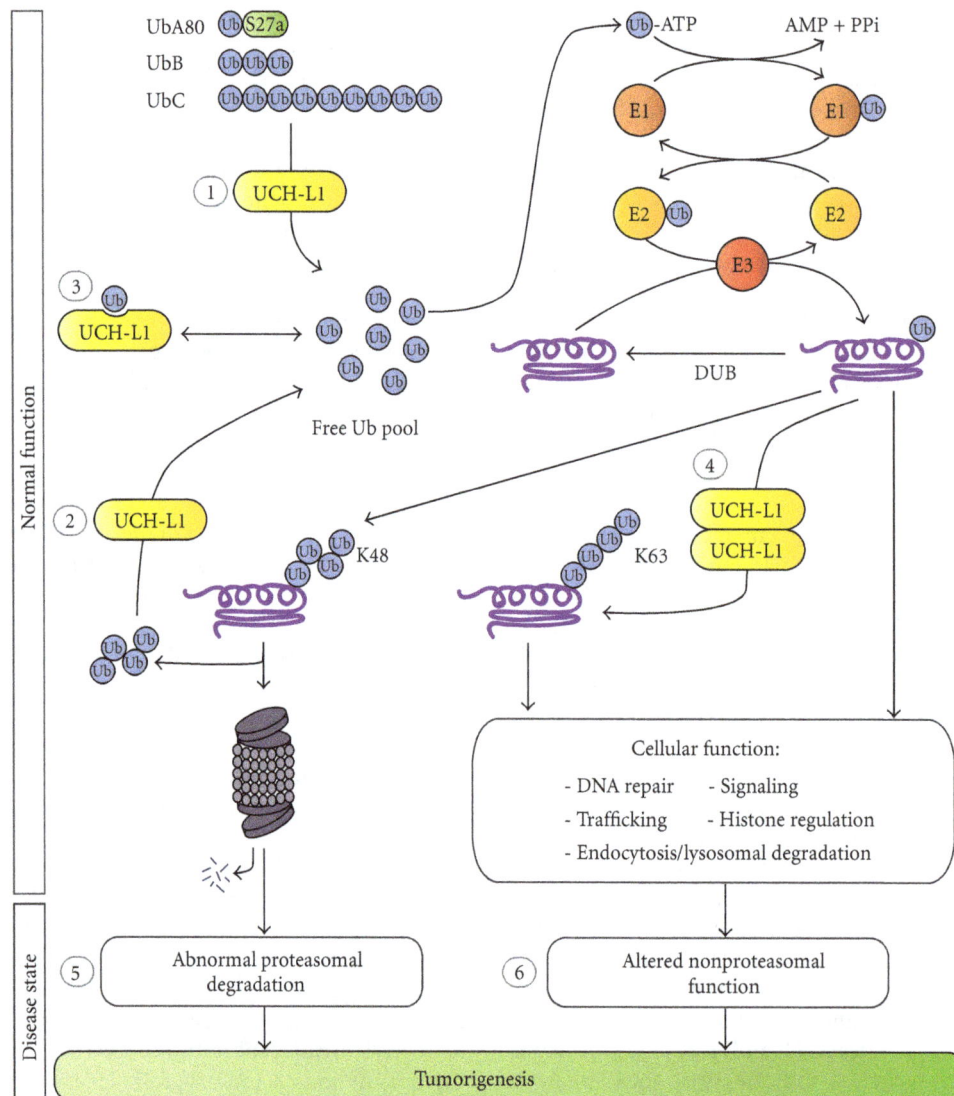

FIGURE 1: Molecular functions of ubiquitin c-terminal hydrolase L1. (1) UCH-L1 can hydrolyze ubiquitin pro-proteins to generate monomeric ubiquitin (Ub) [29]. (2) UCH-L1 may also facilitate Ub recycling by processing Ub chains. (3) UCH-L1 has been reported to stabilize monomeric Ub by binding to Ub and preventing its degradation by the lysosome [4]. Collectively, these functions (1, 2, and 3) give UCH-L1 control over the availability of free Ub and, therefore, the potential to influence many ubiquitination-dependent cellular processes, including proteasomal degradation, DNA damage repair, trafficking, cell signaling, endocytosis, and lysosomal degradation. (4) Dimerized UCH-L1 may possess ATP-independent E3 ligase activity that facilitates K63-linked polyubiquitination [5], although it is currently unclear whether this putative E3 ligase activity directly regulates ubiquitination of protein substrates *in vivo*. (5) Altered expression of UCH-L1 may cause changes to the free Ub pool, resulting in abnormal K48-linked polyubiquitination and proteasomal degradation. (6) Changes in the free Ub pool may also affect mono- and K63-linked polyubiquitination, leading to altered nonproteasomal functions and tumorigenesis.

To determine whether upregulation of UCH-L1 is a result of oncogenic transformation or itself a driving force of tumorigenesis, the direct involvement of UCH-L1 in cancer pathogenesis has been investigated. *In vitro* tumorigenesis studies show that UCH-L1 stimulates oncogenic transformation and invasion in nonsmall cell lung carcinoma [8] and colorectal cancer [14] cells, suggesting that UCH-L1 may function as an oncogene in these cancers. Furthermore, Hussain et al. have demonstrated that transgenic mice constitutively expressing UCH-L1 under the control of a CAGGS promoter form sporadic tumors in all tissues [11]. Of these tumors, lymphomas are the most prevalent [11].

Further investigation revealed that shRNA-mediated knock down of UCH-L1 in immortalized B cells decreased cell growth and viability, suggesting UCH-L1 promotes the development of lymphomas by inhibiting cell death and by stimulating proliferation [11]. Collectively, these data suggest UCH-L1 is a potent oncogene with the capacity to promote tumorigenesis in many different cell types.

Recently, it has been suggested that UCH-L1 promotes cancer cell motility and invasion, which may contribute to its oncogenic role. Overexpression of UCH-L1 in HCT8 colorectal cancer cells has been reported to enhance cell migration [9]. Additionally, Kim et al. have shown that

TABLE 1: Aberrant expression of UCH-L1 in tumor tissues and cancer cells.

Elevated UCH-L1	Down-Regulated UCH-L1
Malignant Tumors	
Squamous cell carcinoma [31]	Prostate tumors [46]
Medullary thyroid carcinoma tumors [38]	Primary breast cancer tumors [16]
Osteosarcoma [45]	Primary nasopharyngeal carcinoma [10]
Adenocarcinoma [35]	Colorectal carcinoma [47]
Metastatic colorectal cancer tumors [9]	Melanoma [48]
Breast cancer tumors [37]	Diffuse-type gastric cancer [34]
Pancreatic ductal carcinoma tumors [36]	
Parathyroid carcinoma [49]	
Transformed Cells	
SaOS-2 osteosarcoma cells [45]	LNCaP prostate cancer cells [50]
BLZ-211 and BLS-211 bladder cancer cells [43]	
BL30, X-50/7, KR4, Raji, KR4 B-cell lymphoma cells [44]	
HCT8 colorectal cancer cells [9]	
DU154 prostate cancer cells [41, 42]	
H157, W138, H358 lung carcinoma cells [8]	

siRNA-mediated knock down of UCH-L1 reduces H157 lung carcinoma cancer cells migration in vitro [8]. They further demonstrated that depletion of UCH-L1 attenuates lung metastasis in vivo in a murine xenograft model [8]. UCH-L1 stimulates prostate cancer cell migration and invasion as well by promoting epithelial-to-mesenchymal transition (EMT) [41]. UCH-L1 level also appears to be correlated with cancer cell metastatic capacity. While UCH-L1 is found in many lung carcinoma cell lines, it is further upregulated in high metastatic lines [8]. Likewise, low metastatic LNCaP and RWPE1 prostate cancer cells do not express UCH-L1, while high metastatic DU145 prostate cancer cells abundantly express UCH-L1 [41]. These studies suggest that UCH-L1 promotes cancer cell metastasis. Further studies are needed to determine how UCH-L1 regulates cell motility and invasion.

Despite growing evidence implicating UCH-L1 as a positive regulator of tumor growth and development, the mechanism by which UCH-L1 conveys oncogenesis is not fully understood. Many of the investigations into the role of UCH-L1 in cancer have focused on upregulation of UCH-L1 in tumor tissues and cancer cells. However, little is known about changes in UCH-L1 enzymatic activity during tumorigenesis. Although one group has observed a decrease UCH-L1 hydrolase activity in cervical carcinoma tissues and an increase in hydrolase activity in transformed keratinocytes [12], the role of UCH-L1 enzymatic function(s) in cancer is largely unknown. Furthermore, no evidence of genetic amplification of UCH-L1 or oncogenic mutations in UCH-L1 have been reported to date, although a Parkinson's disease-linked mutation has been identified [51]. Elucidation of UCH-L1 enzymatic activity in tumorigenesis and investigation into oncogenic genetic alterations of UCH-L1 may provide insights into the role of this enzyme in cancer pathogenesis.

4. UCH-L1 as a Potential Tumor Suppressor

In contrast to the body of literature identifying UCH-L1 as an oncogene, several reports have been put forth suggesting UCH-L1 acts as a tumor suppressor during the pathogenesis of certain cancers [10, 16, 46, 50]. Contrary to previous reports proposing UCH-L1 enhances the progression of prostate cancer [41, 42], two recent studies from Ummanni et al. suggest that UCH-L1 attenuates prostate tumor growth and maturation [46, 50]. UCH-L1 may possibly act as a tumor suppressor in breast cancer pathogenesis as well [16, 52]. In contrast to previous studies demonstrating that UCH-L1 is upregulated in breast tumors [21, 37], UCH-L1 mRNA expression was reported to be decreased in several breast carcinoma cell lines [16]. Moreover, ectopic expression of UCH-L1 in breast cancer cells caused a decrease in anchorage-independent cell growth and an increase apoptosis, suggesting UCH-L1 may act as a negative regulator of breast tumorigenesis [16, 52]. UCH-L1 has also been implicated in the suppression of nasopharyngeal carcinoma as UCH-L1 mRNA expression is decreased in many nasopharyngeal tumors [10]. Lastly, UCH-L1 promoter methylation is elevated in malignant prostate tumors [46], primary breast tumors [16], and nasopharyngeal carcinomas [10]. Similarly, several breast cancer [16] and gastric cancer cell lines [34] exhibit enhanced methylation of UCH-L1 promoter sequences, resulting in decreased UCH-L1 transcription (Table 1). Taken together, these observations suggest that UCH-L1 may function as a tumor suppressor, particularly in prostate [46, 50], breast [16, 52], and nasopharyngeal [10] carcinomas.

There are a number of possible reasons for the discrepancies in the observed oncogenic and tumor suppressor functions of UCH-L1 in tumorigenesis. First, studies suggesting UCH-L1 attenuates prostate cancer progression

[46, 50] focused on the behavior of low metastatic prostate cancer cells, while those implicating UCH-L1 as a positive regulator of prostate tumorigenesis [41, 42] investigated more mature prostate tumors and cell lines. Whether UCH-L1 elicits different effects as prostate tumors become more malignant remains to be investigated. Next, many studies have suggested UCH-L1 functions as a tumor suppressor based on observed decreases in UCH-L1 mRNA in tumor tissues and transformed cells [10, 16, 34, 47, 48]. However, it is not known whether differences in UCH-L1 transcription in these tumors and cells result in changes in UCH-L1 protein level and/or UCH-L1 enzymatic activity. Finally, UCH-L1 is absent or expressed at very low levels in all nonneuronal tissues [1, 19, 22, 23], raising an important question regarding the reported tumor suppressor role for UCH-L1: how can a reduction in UCH-L1 mRNA in tissues that normally express little to no UCH-L1 protein convey oncogenic transformation? To address this question, the normal expression pattern and/or physiological role of UCH-L1 in nonneuronal tissues need to be clarified.

5. Potential Mechanisms of Actions of UCH-L1 in Tumorigenesis

Currently, the precise mechanism(s) of UCH-L1-mediated tumorigenesis are not fully understood. Previous studies have identified several cancer-related signaling processes that are regulated by UCH-L1, which may contribute to its role in oncogenesis. In particular, UCH-L1 has been implicated in the regulation of cell cycle progression, cell survival, cell motility, and invasion (Figure 2).

5.1. UCH-L1 Enzymatic Activity and Oncogenesis. Disruption of UPS function has been implicated in cancer pathogenesis and progression [53] and many cancer-related cellular processes are controlled by ubiquitination, including cell division, growth factor signaling, DNA damage repair, and apoptosis [54–57]. As previously stated, UCH-L1 hydrolyzes small Ub molecules to generate free Ub and also stabilizes monomeric Ub [4, 29] (Figure 1). Through these functions, UCH-L1 can increase the free pool of Ub and, therefore, indirectly affect many ubiquitination-dependent cellular activities. In a pathogenic state, UCH-L1 dysfunction has the potential to alter the cellular levels of monomeric Ub, possibly causing global changes in protein ubiquitination. Therefore, aberrant UCH-L1 signaling may indirectly alter both the poly- and monoubiquitination of oncogenes and tumor suppressors, possibility leading to abnormal protein degradation and/or altered protein function and subsequent tumorigenesis (Figure 1).

UCH-L1 has also been reported to promote K63-linked polyubiquitination of α-synuclein through its putative, dimerization-dependent E3 ubiquitin ligase activity [5]. K63-linked polyubiquitination has been implicated in cancer-related cellular processes such as DNA damage repair and cell survival signaling [58, 59]. Although UCH-L1 E3 activity has not been observed *in vivo* and substrates other than α-synuclein have not been identified, alterations in

UCH-L1 function may disrupt K63-linked polyubiquitination, possibly altering nonproteasomal cellular processes to promote tumorigenesis (Figure 1). Further investigation is needed to clarify the potential E3 function of UCH-L1 as well as the normal physiological and oncogenic role of this enzymatic function.

5.2. A Possible Function for UCH-L1 in Cell Cycle Regulation. UCH-L1 has been shown to stimulate proliferation in transformed lymphocytes and cervical carcinoma cells [11, 27], while it promotes G1/S arrest in breast cancer cells [16]. Together, these studies imply that UCH-L1 contributes to cancer pathogenesis by regulating cell division, although the exact control that UCH-L1 confers on the cell cycle remains unclear. UCH-L1 has been shown to modulate the levels of several cell cycle regulators in cancer cells including cyclin D [31] and p53 [10]. Coimmunoprecipitation experiments conducted by Caballero et al. have shown that UCH-L1 also interacts with JAB1 (Jun-activation domain-binding protein 1). Binding of UCH-L1 to JAB1 promotes the nuclear export and subsequent proteasomal degradation of the cyclin dependent kinase inhibitor p27 [13], resulting in increased cell proliferation (Figure 2(a)). These observations suggest UCH-L1 controls cell cycle progression by modulating the availability of cell cycle regulatory proteins, possibly by altering their ubiquitination status. Recent evidence implicates UCH-L1 in the regulation of cell cycle progression via direct interactions with microtubules. Bheda et al. have demonstrated that UCH-L1 is tightly associated with microtubules during cell division in several transformed cell lines, and that siRNA-mediated knockdown of UCH-L1 reduces microtubule assembly and disassembly [27]. Interestingly, both 25 kDa and 50 kDa UCH-L1 species were associated with purified microtubules [27], suggesting UCH-L1 may act as a dimer to regulate microtubule dynamics. Taken together, these observations suggest that UCH-L1 regulates cell cycle progression by altering levels of cell cycle regulatory proteins and by controlling microtubule dynamics. However, further studies are needed to determine the specific manner by which UCH-L1 controls cell division. In particular, whether or not UCH-L1 controls ubiquitination of cell cycle regulators should be examined.

5.3. UCH-L1 and Cell Survival Signaling. Overactivation of the serine-threonine kinase Akt is a common hallmark of cancer pathogenesis [60]. Phosphorylation of Akt leads to activation of several signaling cascades that together promote cell survival by stimulating proliferation and inhibiting apoptosis. Pharmacological inhibitors of Akt kinase activity attenuate UCH-L1-mediated ECM invasion in nonsmall cell lung carcinoma cells [8]. Additionally, overexpression of UCH-L1 in these cells increases phosphorylation of the downstream Akt targets p38 and ERK1/2, suggesting that UCH-L1 promotes cell survival through Akt-dependent activation of MAPK signaling [8]. Lastly, overexpression of UCH-L1 in immortalized B cells also has been shown to increase Akt phosphorylation during lymphoma progression [11]. Together, these data suggest UCH-L1 elicits at least

(a)

(b)

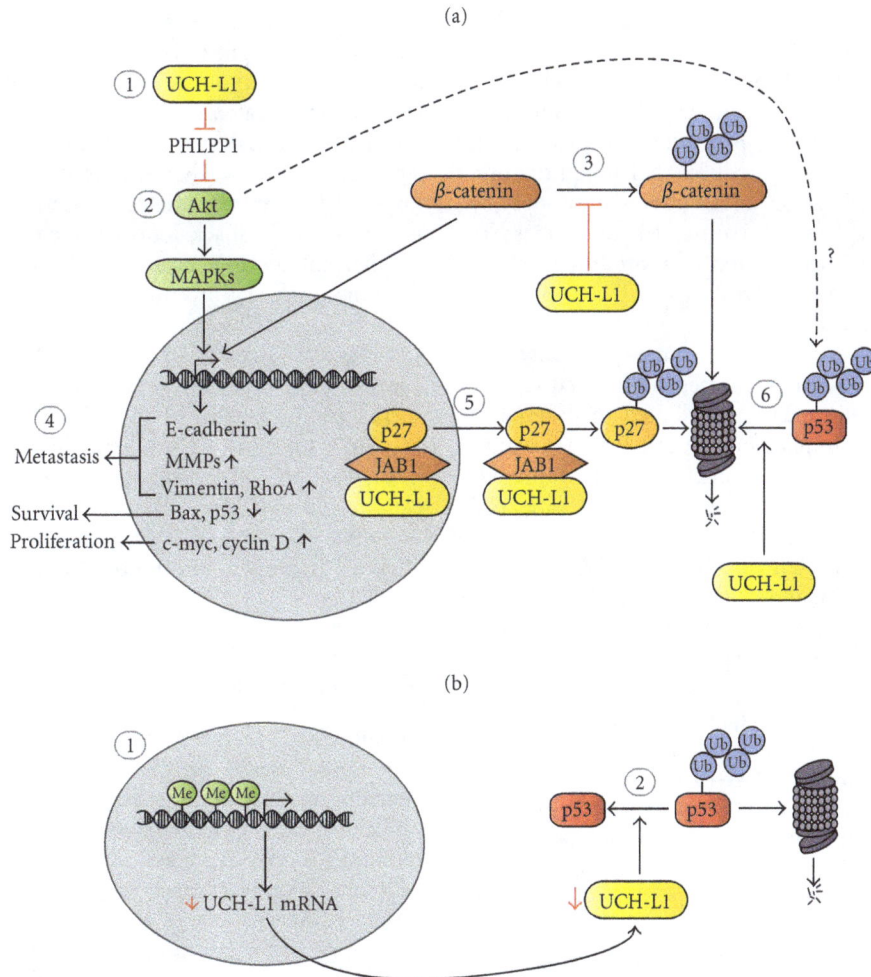

FIGURE 2: The potential roles of UCH-L1 in tumorigenesis. (a) UCH-L1 as a possible oncogene that promotes metastasis and cell growth. (1) UCH-L1 is up-regulated in several tumor tissues and cancer cell lines [7–13]. (2) Elevated UCH-L1 may stimulate Akt through inhibition of the phosphatase PHLLP1 [11], leading to increased MAPK signaling [8]. (3) UCH-L1 has been reported to decrease polyubiquitination and proteasomal degradation of β-catenin, resulting in enhanced β-catenin-mediated transcription [30]. (4) Increased β-catenin and Akt signaling could potentially cause changes in gene transcription that promote metastasis and proliferation and inhibit apoptosis, resulting in enhanced oncogenicity [31–33]. (5) UCH-L1 binds to JAB1 and promotes the nuclear export and subsequent proteasomal degradation of the cell cycle inhibitor p27 [13]. (6) Upregulation of UCH-L1 has been reported to promote proteasomal degradation of p53 [11], which may be a consequence of activation of Akt signaling. Reduction of p27 and p53 levels by UCH-L1 may attenuate cell cycle arrest, allowing for uncontrolled cell growth. (b) UCH-L1 as a putative tumor suppressor in certain cancer subtypes. (1) Reduction of UCH-L1 transcription via promoter methylation-silencing has been observed in certain cancer cells and tumor tissues (e.g., nasopharyngeal carcinomas [10] and gastric cancer cells [34]). (2) In these cancer types, it has been proposed that UCH-L1 promotes deubiquitination of p53 and inhibits its proteasomal degradation [10, 16]. Reduced UCH-L1 transcription due to promoter methylation may thus lead to increased degradation of p53, resulting in a reduction of p53-mediated transcription of tumor suppressing genes and enhanced tumorigenesis (see text for more details).

some of its cellular effects through Akt-dependent signaling and that stimulation of Akt by UCH-L1 is a potential mechanism of UCH-L1-mediated oncogenesis (Figure 2(a)). UCH-L1 promotes Akt signaling, in part, by reducing the level of the tumor suppressor PHLPP1 [11], a phosphatase that reverses Akt phosphorylation rendering Akt inactive. However, the mechanism by which UCH-L1 suppresses PHLPP1 merits further investigation as UCH-L1 does not alter PHLPP1 transcription or promote proteasomal degradation of PHLPP1. Furthermore, whether or not UCH-L1 modulates upstream activators of Akt remains to be

determined. Nevertheless, stimulation of Akt by UCH-L1 and the subsequent promotion of prosurvival signaling may contribute to the function of UCH-L1 in oncogenesis.

A number of studies suggest that UCH-L1 exerts its actions through regulation of the tumor suppressor p53. However, the specific manner in which UCH-L1 modulates p53 level and function remains controversial. UCH-L1 has been shown to promote the proteasomal degradation of p53 in HeLa cells [11], and microarray analyses conducted by Bheda et al. show that depletion of UCH-L1 in 293T cells increases the levels of many p53 target genes [32], suggesting UCH-L1 suppresses p53 signaling. On the other hand, overexpression of UCH-L1 was reported to increase p53 levels in MDA-MB-231 breast carcinoma cells [16] and HONE1 nasopharyngeal carcinoma cells [10]. Similarly, Li et al. have shown that over-expression of UCH-L1 in LNCaP prostate cancer cells reduces polyubiquitination of p53, leading to inhibition of degradation of p53 by the proteasome [50]. They also observed an increase in polyubiquitination and degradation of mdm2 in response to UCH-L1 over-expression, suggesting UCH-L1 suppresses mdm2 to stabilize p53 levels [50].

Further studies are needed to determine specifically how UCH-L1 modulates p53. Discrepancies in observed regulation of p53 by UCH-L1 may be attributed to differences in p53 status. Studies implicating UCH-L1 as a negative regulator of p53 [11, 32] were conducted in cells with wild-type p53 [61, 62]. On the contrary, UCH-L1 elevates p53 levels in MDA-MB-231 and HONE1 cells, which express DNA binding domain mutant p53 with little to no transcriptional activity [63–65] and LNCaP cells, which have also been reported to express DNA binding domain mutant p53 [66], although this is controversial [67].This suggests that UCH-L1 may regulate wild-type and DNA binding domain mutant p53 differently. P53 is frequently mutated in human cancers [63] and variation in p53 status may offer another possible explanation for why UCH-L1 has been reported to function as an oncogene and a tumor suppressor in different cancer cell lines and tumor types. It is possible that in cells with wild type p53, UCH-L1 promotes degradation p53, resulting in reduced p53 signaling and inhibition of cell death (Figure 2(a)). On the other hand, in other cell types with weakened p53 transcriptional activity, UCH-L1 may regulate nontranscriptional functions of p53 [68, 69] to promote apoptosis and attenuation of tumor growth (Figure 2(b)). It is also possible that UCH-L1 indirectly elicits control over p53 by modulating negative regulators of p53, such as mdm2, as suggested by Li et al. [10]. As p53 level and function are regulated, in part, by ubiquitination [58, 70, 71], investigation into modulation of p53 ubiquitination by UCH-L1 may offer additional insights into the role of UCH-L1 in tumorigenesis. However, while exploring the relationship between UCH-L1 and p53 ubiquitination, it is important to keep in mind that, despite hypotheses to the contrary [10, 16], it is unlikely that UCH-L1 directly deubiquitinates or ubiquitinates p53 based on what is known about UCH-L1 structure and function [28, 29]. Additionally, it might be possible that activation of Akt signaling by UCH-L1 [8, 11] might also contribute to its control over p53, as Akt is an established negative regulator of p53 activity [60, 72].

5.4. The Potential Role of UCH-L1 in Metastasis. UCH-L1 has been suggested to promote metastasis in colorectal, lung, and prostate cancer cells [8, 9, 41]. Cancer cell metastasis is often attributed to hyperactivation of β-catenin, a transcription factor that when over-activated promotes cell migration and invasion [73]. UCH-L1 overexpression has been shown decrease polyubiquitination and proteasomal degradation of β-catenin in HEK 293 cells, leading to stabilization of

TCF: β-catenin complexes and increased β-catenin-mediated transcription of prosurvival genes such as c-myc, c-jun, and survivin [30]. Although, it is unlikely UCH-L1 directly deubiquitinates β-catenin [28, 29], these observations suggest UCH-L1 may convey its oncogenic function through Wnt signaling pathways. UCH-L1 itself has been identified as a target of β-catenin-mediated transcription, suggesting there is a positive feedback loop between β-catenin and UCH-L1 that enhances metastasis [30]. One consequence of β-catenin signaling is promotion of epithelial-to-mesenchymal transition (EMT) [33]. Recently, it was shown that UCH-L1 enhances prostate cancer cell metastasis by increasing the expression of pro-EMT genes such as vimentin and matrix metalloproteinases (MMPs) and reducing transcription of the EMT suppressor E-cadherin [41]. Together, these data imply that UCH-L1 promotes cancer cell metastasis via β-catenin-induced EMT (Figure 2(a)). Therefore, therapeutic targeting of Wnt and EMT signaling may prove to be an effective treatment for tumors that express high levels of UCH-L1.

6. Conclusions

In summary, emerging evidence suggests that UCH-L1 is a potent oncogene that promotes tumor growth and development during the progression of many forms of cancer. However, the exact role of UCH-L1 in oncogenesis remains controversial, as UCH-L1 has been suggested to function as a tumor suppressor in certain tumor types. The observed involvement of UCH-L1 in the regulation of cell cycle progression, cell survival, and metastasis may explain its oncogenic role. However, further studies are needed to clarify the exact mechanisms of action of UCH-L1 in tumorigenesis. Continued investigation into the function of UCH-L1 in cancer may tell us whether or not UCH-L1 can be used as a diagnostic marker. UCH-L1 is upregulated in many cancer tissues and, therefore, high levels of UCH-L1, particularly in nonneuronal tissues, may serve as an early detection biomarker for tumors. Furthermore, UCH-L1 itself could be a potential therapeutic target, which may have benefits for the treatment of cancer. Elucidation of the role of UCH-L1 in cancer may lead to a better understanding of the molecular pathogenesis of tumors as well as potentially facilitate the development of novel cancer therapeutics and diagnostics tools.

Abbreviations

UCH-L1:	Ubiquitin C-terminal hydrolase L1
Ub:	Ubiquitin
UPS:	Ubiquitin-proteasome system
DUB:	Deubiquitinating enzyme
E1:	E1 Ubiquitin-activating enzyme
E2:	E2 Ubiquitin-conjugating enzyme
E3:	E3 Ubiquitin-protein ligase
Akt:	Protein Kinase B
MAPKs:	Mitogen-activated protein kinases
ERK1/2:	Extracellular signal-related kinases 1 and 2

PHLPP1: PH domain leucine-rich repeat protein
phosphatase 1
JAB1: Jun-activation domain-binding protein 1
EMT: Epithelial-to-mesenchymal transition
MMPs: Matrix metalloproteinases.

Acknowledgments

This work was supported by National Institutes of Health grants NS050650 (L. S. Chin), AG034126 (L. S. Chin), ES015813 (L. Li), and GM082828 (L. Li) and by Fellowships in Research and Science Teaching (Institutional Research and Academic Career Development Award, 5K12GM000680).

References

[1] K. D. Wilkinson, K. Lee, S. Deshpande, P. Duerksen-Hughes, J. M. Boss, and J. Pohl, "The neuron-specific protein PGP 9.5 is a ubiquitin carboxyl-terminal hydrolase," *Science*, vol. 246, no. 4930, pp. 670–673, 1989.

[2] C. N. Larsen, J. S. Price, and K. D. Wilkinson, "Substrate binding and catalysis by ubiquitin C-terminal hydrolases: identification of two active site residues," *Biochemistry*, vol. 35, no. 21, pp. 6735–6744, 1996.

[3] A. Case and R. L. Stein, "Mechanistic studies of ubiquitin C-terminal hydrolase L1," *Biochemistry*, vol. 45, no. 7, pp. 2443–2452, 2006.

[4] H. Osaka, Y. L. Wang, K. Takada et al., "Ubiquitin carboxy-terminal hydrolase L1 binds to and stabilizes monoubiquitin in neuron," *Human Molecular Genetics*, vol. 12, no. 16, pp. 1945–1958, 2003.

[5] Y. Liu, L. Fallon, H. A. Lashuel, Z. Liu, and P. T. Lansbury Jr, "The UCH-L1 gene encodes two opposing enzymatic activities that affect α-synuclein degradation and Parkinson's disease susceptibility," *Cell*, vol. 111, no. 2, pp. 209–218, 2002.

[6] B. J. Walters, S. L. Campbell, P. C. Chen et al., "Differential effects of Usp14 and Uch-L1 on the ubiquitin proteasome system and synaptic activity," *Molecular and Cellular Neuroscience*, vol. 39, no. 4, pp. 539–548, 2008.

[7] Y. Liu, H. A. Lashuel, S. Choi et al., "Discovery of inhibitors that elucidate the role of UCH-L1 activity in the H1299 lung cancer cell line," *Chemistry and Biology*, vol. 10, no. 9, pp. 837–846, 2003.

[8] H. J. Kim, Y. M. Kim, S. Lim et al., "Ubiquitin C-terminal hydrolase-L1 is a key regulator of tumor cell invasion and metastasis," *Oncogene*, vol. 28, no. 1, pp. 117–127, 2009.

[9] Y. Ma, M. Zhao, J. Zhong et al., "Proteomic profiling of proteins associated with lymph node metastasis in colorectal cancer," *Journal of Cellular Biochemistry*, vol. 110, no. 6, pp. 1512–1519, 2010.

[10] L. Li, Q. Tao, H. Jin et al., "The tumor suppressor UCHL1 forms a complex with p53/MDM2/ARF to promote p53 signaling and is frequently silenced in nasopharyngeal carcinoma," *Clinical Cancer Research*, vol. 16, no. 11, pp. 2949–2958, 2010.

[11] S. Hussain, O. Foreman, S. L. Perkins et al., "The de-ubiquitinase UCH-L1 is an oncogene that drives the development of lymphoma *in vivo* by deregulating PHLPP1 and Akt signaling," *Leukemia*, vol. 24, no. 9, pp. 1641–1655, 2010.

[12] U. Rolén, V. Kobzeva, N. Gasparjan et al., "Activity profiling of deubiquitinating enzymes in cervical carcinoma biopsies and cell lines," *Molecular Carcinogenesis*, vol. 45, no. 4, pp. 260–269, 2006.

[13] O. L. Caballero, V. Resto, M. Patturajan et al., "Interaction and colocalization of PGP9.5 with JAB1 and p27Kip1," *Oncogene*, vol. 21, no. 19, pp. 3003–3010, 2002.

[14] Y. Akishima-Fukasawa, Y. Ino, Y. Nakanishi et al., "Significance of PGP9.5 expression in cancer-associated fibroblasts for prognosis of colorectal carcinoma," *American Journal of Clinical Pathology*, vol. 134, no. 1, pp. 71–79, 2010.

[15] L. Li, Q. Tao, H. Jin et al., "The tumor suppressor UCHL1 forms a complex with p53/MDM2/ARF to promote p53 signaling and is frequently silenced in nasopharyngeal carcinoma," *Clinical Cancer Research*, vol. 16, no. 11, pp. 2949–2958, 2010.

[16] T. Xiang, L. Li, X. Yin et al., "The ubiquitin peptidase UCHL1 induces G0/G1 cell cycle arrest and apoptosis through stabilizing p53 and is frequently silenced in breast cancer," *PLoS One*, vol. 7, no. 1, Article ID e29783, 2012.

[17] L. Li and L.-S. Chin, "Impairment of the ubiquitin-proteasome system: a commnon pathogenic mechansim in neurodegenerative disorders," in *The Ubiquitin Proteasome System*, pp. 553–577, 2007.

[18] C. M. Pickart and D. Fushman, "Polyubiquitin chains: polymeric protein signals," *Current Opinion in Chemical Biology*, vol. 8, no. 6, pp. 610–616, 2004.

[19] P. O. G. Wilson, P. C. Barber, Q. A. Hamid et al., "The immunolocalization of protein gene product 9.5 using rabbit polyclonal and mouse monoclonal antibodies," *British Journal of Experimental Pathology*, vol. 69, no. 1, pp. 91–104, 1988.

[20] J. M. Bradbury and R. J. Thompson, "Immunoassay of the neuronal and neuroendocrine marker PGP 9.5 in human tissues," *Journal of Neurochemistry*, vol. 44, no. 2, pp. 651–653, 1985.

[21] U. Schumacher, B. S. Mitchell, and E. Kaiserling, "The neuronal marker protein gene product 9.5 (PGP 9.5) is phenotypically expressed in human breast epithelium, in milk, and in benign and malignant breast tumors," *DNA and Cell Biology*, vol. 13, no. 8, pp. 839–843, 1994.

[22] Y. Kajimoto, T. Hashimoto, Y. Shirai, N. Nishino, T. Kuno, and C. Tanaka, "cDNA cloning and tissue distribution of a rat ubiquitin carboxyl-terminal hydrolase PGP9.5," *Journal of Biochemistry*, vol. 112, no. 1, pp. 28–32, 1992.

[23] I. N. M. Day and R. J. Thompson, "UCHL1 (PGP 9.5): neuronal biomarker and ubiquitin system protein," *Progress in Neurobiology*, vol. 90, no. 3, pp. 327–362, 2010.

[24] M. Sakurai, M. Sekiguchi, K. Zushida et al., "Reduction in memory in passive avoidance learning, exploratory behaviour and synaptic plasticity in mice with a spontaneous deletion in the ubiquitin C-terminal hydrolase L1 gene," *European Journal of Neuroscience*, vol. 27, no. 3, pp. 691–701, 2008.

[25] K. Saigoh, Y. L. Wang, J. G. Suh et al., "Intragenic deletion in the gene encoding ubiquitin carboxy-terminal hydrolase in gad mice," *Nature Genetics*, vol. 23, no. 1, pp. 47–51, 1999.

[26] J. Lowe, H. McDermott, M. Landon, R. J. Mayer, and K. D. Wilkinson, "Ubiquitin carboxyl-terminal hydrolase (PGP 9.5) is selectively present in ubiquitinated inclusion bodies characteristic of human neurodegenerative diseases," *Journal of Pathology*, vol. 161, no. 2, pp. 153–160, 1990.

[27] A. Bheda, A. Gullapalli, M. Caplow, J. S. Pagano, and J. Shackelford, "Ubiquitin editing enzyme UCH L1 and microtubule dynamics: implication in mitosis," *Cell Cycle*, vol. 9, no. 5, pp. 980–994, 2010.

[28] T. E. Messick, N. S. Russell, A. J. Iwata et al., "Structural basis for ubiquitin recognition by the Otu1 ovarian tumor domain

protein," *Journal of Biological Chemistry*, vol. 283, no. 16, pp. 11038–11049, 2008.

[29] C. N. Larsen, B. A. Krantz, and K. D. Wilkinson, "Substrate specificity of deubiquitinating enzymes: ubiquitin C-terminal hydrolases," *Biochemistry*, vol. 37, no. 10, pp. 3358–3368, 1998.

[30] A. Bheda, W. Yue, A. Gullapalli et al., "Positive reciprocal regulation of ubiquitin C-terminal hydrolase L1 and β-catenin/TCF signaling," *PLoS ONE*, vol. 4, no. 6, Article ID e5955, 2009.

[31] A. Mastoraki, E. Ioannidis, A. Apostolaki, E. Patsouris, and K. Aroni, "PGP 9.5 and cyclin D1 coexpression in cutaneous squamous cell carcinomas," *International Journal of Surgical Pathology*, vol. 17, no. 6, pp. 413–420, 2009.

[32] A. Bheda, J. Shackelford, and J. S. Pagano, "Expression and functional studies of ubiquitin C-terminal hydrolase L1 regulated genes," *PLoS ONE*, vol. 4, no. 8, Article ID e6764, 2009.

[33] G. I. Rozenberg, K. B. Monahan, C. Torrice, J. E. Bear, and N. E. Sharpless, "Metastasis in an orthotopic murine model of melanoma is independent of RAS/RAF mutation," *Melanoma Research*, vol. 20, no. 5, pp. 361–371, 2010.

[34] K. Yamashita, H. L. Park, M. S. Kim et al., "PGP9.5 methylation in diffuse-type gastric cancer," *Cancer Research*, vol. 66, no. 7, pp. 3921–3927, 2006.

[35] G. Chen, T. G. Gharib, C. C. Huang et al., "Proteomic analysis of lung adenocarcinoma: identification of a highly expressed set of proteins in tumors," *Clinical Cancer Research*, vol. 8, no. 7, pp. 2298–2305, 2002.

[36] E. Tezel, K. Hibi, T. Nagasaka, and A. Nakao, "PGP9.5 as a prognostic factor in pancreatic cancer," *Clinical Cancer Research*, vol. 6, no. 12, pp. 4764–4767, 2000.

[37] Y. Miyoshi, S. Nakayama, Y. Torikoshi et al., "High expression of ubiquitin caboxy-terminal hydrolase-L1 and -L3 mRNA predicts early recurrence in patients with invasive breast cancer," *Cancer Science*, vol. 97, no. 6, pp. 523–529, 2006.

[38] T. Takano, A. Miyauchi, F. Matsuzuka et al., "PGP9.5 mRNA could contribute to the molecular-based diagnosis of medullary thyroid carcinoma," *European Journal of Cancer*, vol. 40, no. 4, pp. 614–618, 2004.

[39] Y. M. Lee, J. Y. Lee, M. J. Kim et al., "Hypomethylation of the protein gene product 9.5 promoter region in gallbladder cancer and its relationship with clinicopathological features," *Cancer Science*, vol. 97, no. 11, pp. 1205–1210, 2006.

[40] H. Mizukami, A. Shirahata, T. Goto et al., "PGP9.5 methylation as a marker for metastatic colorectal cancer," *Anticancer Research A*, vol. 28, no. 5, pp. 2697–2700, 2008.

[41] M. J. Jang, S. H. Baek, and J. H. Kim, "UCH-L1 promotes cancer metastasis in prostate cancer cells through EMT induction," *Cancer Letters*, vol. 302, no. 2, pp. 128–135, 2011.

[42] A. Leiblich, S. S. Cross, J. W. F. Catto, G. Pesce, F. C. Hamdy, and I. Rehman, "Human prostate cancer cells express neuroendocrine cell markers PGP 9.5 and chromogranin A," *Prostate*, vol. 67, no. 16, pp. 1761–1769, 2007.

[43] Y.-C. Yang, X. Li, and W. Chen, "Characterization of genes associated with different phenotypes of human bladder cancer cells," *Acta Biochimica et Biophysica Sinica*, vol. 38, no. 9, pp. 602–610, 2006.

[44] A. Bheda, W. Yue, A. Gullapalli, J. Shackelford, and J. S. Pagano, "PU.1-dependent regulation of UCH L1 expression in B-lymphoma cells," *Leukemia and Lymphoma*, vol. 52, no. 7, pp. 1336–1347, 2011.

[45] X. Liu, B. Zeng, J. Ma, and C. Wan, "Comparative proteomic analysis of osteosarcoma cell and human primary cultured osteoblastic cell," *Cancer Investigation*, vol. 27, no. 3, pp. 345–352, 2009.

[46] R. Ummanni, F. Mundt, H. Pospisil et al., "Identification of clinically relevant protein targets in prostate cancer with 2D-DIGE coupled mass spectrometry and systems biology network platform," *PLoS ONE*, vol. 6, no. 2, Article ID e16833, 2011.

[47] S. Fukutomi, N. Seki, K. Koda, and M. Miyazaki, "Identification of methylation-silenced genes in colorectal cancer cell lines: genomic screening using oligonucleotide arrays," *Scandinavian Journal of Gastroenterology*, vol. 42, no. 12, pp. 1486–1494, 2007.

[48] V. F. Bonazzi, D. J. Nancarrow, M. S. Stark et al., "Cross-platform array screening identifies COL1A2, THBS1, TNFRSF10D and UCHL1 as genes frequently silenced by methylation in melanoma," *PLoS ONE*, vol. 6, no. 10, Article ID e26121, 2011.

[49] M. A. Adam, B. R. Untch, and J. A. Olson Jr, "Parathyroid carcinoma: current understanding and new insights into gene expression and intraoperative parathyroid hormone kinetics," *Oncologist*, vol. 15, no. 1, pp. 61–72, 2010.

[50] R. Ummanni, E. Jost, M. Braig et al., "Ubiquitin carboxyl-terminal hydrolase 1 (UCHL1) is a potential tumour suppressor in prostate cancer and is frequently silenced by promoter methylation," *Molecular Cancer*, vol. 10, p. 129, 2011.

[51] K. Nishikawa, H. Li, R. Kawamura et al., "Alterations of structure and hydrolase activity of parkinsonism-associated human ubiquitin carboxyl-terminal hydrolase L1 variants," *Biochemical and Biophysical Research Communications*, vol. 304, no. 1, pp. 176–183, 2003.

[52] W. J. Wang, Q. Q. Li, J. D. Xu et al., "Over-expression of Ubiquitin carboxy terminal hydrolase-L1 induces apoptosis in breast cancer cells," *International Journal of Oncology*, vol. 33, no. 5, pp. 1037–1045, 2008.

[53] K. Newton and D. Vucic, "Ubiquitin ligases in cancer: ushers for degradation," *Cancer Investigation*, vol. 25, no. 6, pp. 502–513, 2007.

[54] M. Li, C. L. Brooks, F. Wu-Baer, D. Chen, R. Baer, and W. Gu, "Mono- versus polyubiquitination: differential control of p53 Fate by Mdm2," *Science*, vol. 302, no. 5652, pp. 1972–1975, 2003.

[55] C. K. Ea, L. Deng, Z. P. Xia, G. Pineda, and Z. J. Chen, "Activation of IKK by TNFα requires site-specific ubiquitination of RIP1 and polyubiquitin binding by NEMO," *Molecular Cell*, vol. 22, no. 2, pp. 245–257, 2006.

[56] M. E. Moynahan, J. W. Chiu, B. H. Koller, and M. Jasint, "Brca1 controls homology-directed DNA repair," *Molecular Cell*, vol. 4, no. 4, pp. 511–518, 1999.

[57] R. Mamillapalli, N. Gavrilova, V. T. Mihaylova et al., "PTEN regulates the ubiquitin-dependent degradation of the CDK inhibitor p27KIP1 through the ubiquitin E3 ligase SCFSKP2," *Current Biology*, vol. 11, no. 4, pp. 263–267, 2001.

[58] R. Wen, J. Li, X. Xu, Z. Cui, and W. Xiao, "Zebrafish Mms2 promotes K63-linked polyubiquitination and is involved in p53-mediated DNA-damage response," *DNA Repair*, vol. 11, no. 2, pp. 157–166, 2012.

[59] E. W. Harhaj and V. M. Dixit, "Regulation of NF-kappaB by deubiquitinases," *Immunological Reviews*, vol. 246, no. 1, pp. 107–124, 2012.

[60] A. Carnero, "The PKB/AKT pathway in cancer," *Current Pharmaceutical Design*, vol. 16, no. 1, pp. 34–44, 2010.

[61] S. Bamford, E. Dawson, S. Forbes et al., "The COSMIC (Catalogue of Somatic Mutations in Cancer) database and website," *British Journal of Cancer*, vol. 91, no. 2, pp. 355–358, 2004.

[62] T. A. Lehman, R. Modali, P. Boukamp et al., "p53 Mutations in human immortalized epithelial cell lines," *Carcinogenesis*, vol. 14, no. 5, pp. 833–839, 1993.

[63] J. Bartek, R. Iggo, J. Gannon, and D. P. Lane, "Genetic and immunochemical analysis of mutant p53 in human breast cancer cell lines," *Oncogene*, vol. 5, no. 6, pp. 893–899, 1990.

[64] S. L. Hoe and C. K. Sam, "Mutational analysis of p53 and RB2/p130 genes in Malaysian nasopharyngeal carcinoma samples: a preliminary report," *The Malaysian Journal of Pathology*, vol. 28, no. 1, pp. 35–39, 2006.

[65] S. Kato, S. Y. Han, W. Liu et al., "Understanding the function-structure and function-mutation relationships of p53 tumor suppressor protein by high-resolution missense mutation analysis," *Proceedings of the National Academy of Sciences of the United States of America*, vol. 100, no. 14, pp. 8424–8429, 2003.

[66] S. G. Chi, R. W. DeVere, F. J. Meyers, D. B. Siders, F. Lee, and P. H. Gumerlock, "p53 in prostate cancer: frequent expressed transition mutations," *Journal of the National Cancer Institute*, vol. 86, no. 12, pp. 926–933, 1994.

[67] A. Van Bokhoven, M. Varella-Garcia, C. Korch et al., "Molecular characterization of human prostate carcinoma cell lines," *Prostate*, vol. 57, no. 3, pp. 205–225, 2003.

[68] B. S. Sayan, A. E. Sayan, R. A. Knight, G. Melino, and G. M. Cohen, "p53 is cleaved by caspases generating fragments localizing to mitochondria," *Journal of Biological Chemistry*, vol. 281, no. 19, pp. 13566–13573, 2006.

[69] D. Speidel, H. Helmbold, and W. Deppert, "Dissection of transcriptional and non-transcriptional p53 activities in the response to genotoxic stress," *Oncogene*, vol. 25, no. 6, pp. 940–953, 2006.

[70] L. Collavin, A. Lunardi, and G. Del Sal, "P53-family proteins and their regulators: hubs and spokes in tumor suppression," *Cell Death and Differentiation*, vol. 17, no. 6, pp. 901–911, 2010.

[71] A. M. Bode and Z. Dong, "Post-translational modification of p53 in tumorigenesis," *Nature Reviews Cancer*, vol. 4, no. 10, pp. 793–805, 2004.

[72] J. Downward, "PI 3-kinase, Akt and cell survival," *Seminars in Cell and Developmental Biology*, vol. 15, no. 2, pp. 177–182, 2004.

[73] Y. Fu, S. Zheng, N. An et al., "β-Catenin as a potential key target for tumor suppression," *International Journal of Cancer*, vol. 129, no. 7, pp. 1541–1551, 2011.

3

The Increased Activity of Liver Lysosomal Lipase in Nonalcoholic Fatty Liver Disease Contributes to the Development of Hepatic Insulin Resistance

Monika Cahova,[1] **Helena Dankova,**[1] **Eliska Palenickova,**[1] **Zuzana Papackova,**[1]
Radko Komers,[2, 3] **Jana Zdychova,**[2, 4] **Eva Sticova,**[5] **and Ludmila Kazdova**[1]

[1] *Department of Metabolism and Diabetes, Institute for Clinical and Experimental Medicine, Videnska 1958/9,*
 14021 Prague 4, Czech Republic
[2] *Diabetes Center, Institute for Clinical and Experimental Medicine, 14021 Prague 4, Czech Republic*
[3] *Division of Nephrology and Hypertension, Oregon Health and Science University, Portland, OR 97239-3098, USA*
[4] *Department of Medicinal and Clinical Chemistry, University of Heidelberg, 69117 Heidelberg, Germany*
[5] *Laboratory of Experimental Hepatology, Institute for Clinical and Experimental Medicine, 14021 Prague 4, Czech Republic*

Correspondence should be addressed to Monika Cahova, monika.cahova@ikem.cz

Academic Editor: Todd B. Reynolds

We tested the hypothesis that TAG accumulation in the liver induced by short-term high-fat diet (HFD) in rats leads to the dysregulation of endogenous TAG degradation by lysosomal lipase (LIPA) via lysosomal pathway and is causally linked with the onset of hepatic insulin resistance. We found that LIPA could be translocated between qualitatively different depots (light and dense lysosomes). In contrast to dense lysosomal fraction, LIPA associated with light lysosomes exhibits high activity on both intracellular TAG and exogenous substrate and prandial- or diet-dependent regulation. On standard diet, LIPA activity was upregulated in fasted and downregulated in fed animals. In the HFD group, we demonstrated an increased TAG content, elevated LIPA activity, enhanced production of diacylglycerol, and the abolishment of prandial-dependent LIPA regulation in light lysosomal fraction. The impairment of insulin signalling and increased activation of PKCε was found in liver of HFD-fed animals. Lipolysis of intracellular TAG, mediated by LIPA, is increased in steatosis probably due to the enhanced formation of phagolysosomes. Consequent overproduction of diacylglycerol may represent the causal link between HFD-induced hepatic TAG accumulation and hepatic insulin resistance via PKCε activation.

1. Introduction

NAFLD (nonalcoholic fatty liver disease) is often associated with insulin resistance (IR) and type 2 diabetes [1]. High-fat diet-induced liver triacylglycerol (TAG) accumulation results in the hepatic IR even after three days of administration and without significant impairment of insulin-mediated peripheral glucose disposal [2]. However, the mechanism by which hepatic fat accumulation might lead to the hepatic insulin resistance is far from being clearly understood [3]. The TAG metabolism in the liver is subject to a highly sensitive regulation in order to fulfil the actual needs of the organism. It has been shown that the liver is a site of continuous lipolysis of endogenous TAG and partial reesterification of released free fatty acids (FFA) back to the intracellular lipid storage pool [4]. The rate of intracellular lipolysis is 2-3 times greater than required to maintain the observed rate of TAG secretion [5].

Nevertheless, in spite of intensive research in this field, there are many uncertainties concerning the enzyme(s) responsible for intracellular TAG degradation. One possible candidate is lysosomal lipase (LIPA) [6]. It belongs to a group of more than 50 acid hydrolases that are characterised by low pH optimum (4.5–5). Because these enzymes require a pH range that is incompatible with the neutral cytoplasma milieu, they are sequestered in specific cytoplasmic particles

The Increased Activity of Liver Lysosomal Lipase in Nonalcoholic Fatty Liver Disease Contributes to the Development of
Hepatic Insulin Resistance

21

termed lysosomes [7]. Due to a large variety of lysosomal enzymes (including proteases, lipases, glycosidases, and nucleases), lysosomes mediate complete breakdown of many types of molecules and confer upon this organelle its high degradative capacity [8]. Lysosomal enzymes are synthesized in endoplasmic reticulum, sequestered into specialised regions of Golgi apparatus, and bud out and detach as small vesicles called primary lysosomes [9]. Substrates can reach lysosomes via heterophagy (including exocytosis and phagocytosis), in which cargo originates at the plasma membrane or extracellularly, or via autophagy, for cargo located in the cytosol. Material designed for degradation is temporally stored in digestively inactive organelles termed phagosomes. Only after fusion of phagosome with primary lysosome and acidification of intralysosomal space could the internalized material be degraded and the degradation products released back into cytoplasm [10]. The lysosomal pathway was originally associated with removal of organelles and degradation of proteins [11]. Only recently the critical role of this pathway in metabolism and storage of intracellular lipids has been discovered [12]. Hayase and Tappel [13] showed that lysosomal lipase is capable of hydrolyzing triacylglycerols and that the dominant products of lysosomal lipase action on TAG molecule are diacylglycerol (DAG) and one molecule of fatty acid. DAG is a known activator of classic and novel isoforms of protein kinase C (PKC), and DAG concentrations have closely paralleled insulin resistance in other models [14, 15]. While PKCs, in general, have been implicated in the pathogenesis of insulin resistance in many tissues, Samuel et al. [16, 17] delineated the specific role of one particular isoform, PKCε, in the development of fat-induced insulin resistance in the liver.

We hypothesized that steatosis-associated hepatic IR is causally linked with alteration of endogenous TAG degradation in NAFLD. To address this issue, the activity of LIPA and the production of TAG breakdown intermediates were determined in animals with normal insulin sensitivity and with hepatic IR induced by a two-week administration of a high-fat diet. We identified the steatosis-associated changes in the regulation of LIPA activity based on the alteration in its intracellular distribution, and we proposed the mechanism by which it can contribute to the establishment of hepatic IR.

2. Materials and Methods

2.1. Animals and Experimental Protocol. Male rats were kept in temperature-controlled room at 12 : 12 h light-dark cycle. Animals had free access to drinking water and diet if not stated otherwise. All experiments were performed in agreement with the Animal Protection Law of the Czech Republic 311/1997 which is in compliance with Principles of Laboratory Animal Care (NIH publication no. 85-23, revised 1985) and were approved by the ethical committee of the Institute for Clinical and Experimental Medicine. Starting at age 3 months (b.wt. 300 ± 20 g), all animals were fed either HFD (70 cal% as saturated fat, 20 cal% as protein, and 10 cal% as carbohydrate) or standard laboratory chow diet (SD) for 2 weeks. The groups labelled SD fed or HFD fed

had free access to the diet until decapitation (10–11 am), and the groups designated as SD fasted or HFD fasted were deprived of food for the last 24 hours. Glucose tolerance was determined as the rate of disappearance of glucose from circulation after a single dose of glucose (3 g/kg b.wt.) administered intragastrically to overnight-fasted animals.

2.2. Preparation of Lysosomal and Phagolysosomal Fractions. The lysosomes and phagolysosomes represent a heterogeneous population of organelles. 20% (wt/vol) homogenate was prepared by homogenization of liver tissue in 0.25 M sucrose; 0.001 M EDTA pH = 7.4; heparin 7 IU/m, 1 mM PMSF, leupeptin 10 μg/mL, and aprotinin 10 μg/mL by Teflon pestle homogenizer. The crude impurities were removed by brief centrifugation at 850 g. The fat cake was removed carefully in order to prevent contamination of liquid fraction. An aliquot of the homogenate was kept at 4°C until lipase assay (maximum 2 hour), the rest was centrifuged for 10 000 g 20 min 4°C, and the resulting pellet and supernatant were separated. The supernatant contains preferentially the less dense lysosomes with higher TAG content (light lysosomes), and the pellet is formed by more dense particles (dense lysosomes).

2.3. Assay of Triacylglycerol Lipase Activity on Exogenous Substrate. The optimal conditions for the lipase assay (substrate concentration, reaction temperature, and linear range of the assay) were determined in the pilot experiments. The data are provided in supplements (a–c). 4% liver homogenate or lysosomal subfractions prepared from the fresh tissue under iso-osmotic conditions were used for the assay. The reaction medium (92.5 kBq ^{3}H triolein, 100 μM triolein, 110 μM lecithin, 0.15 M NaCl, and 0.1 M acetate buffer pH 4.5) was emulsified by sonication (Hielsler sonicator UP200S). The assay itself was performed under hypoosmotic conditions (50 mM sucrose) in order to ensure the release of the enzyme sequestered within the lysosomes. The liver homogenate or isolated fractions were incubated for 60 minutes at 30°C. The released fatty acids were extracted according to [18] and counted for radioactivity.

2.4. Assay of Triglyceride Lipase Activity on Endogenous Substrate. This approach takes advantage of the coordinated changes in the intracellular localisation of LIPA and its intracellular substrate. The optimal conditions for the lipase assay were determined in the pilot experiments. The data are provided in supplements (e, f). The liver homogenate and subcellular fractions were prepared as described above under iso-osmotic conditions that prevent the disruption of lysosomes. The lysis of lysosomes was induced only after separation of fractions during the assay. 20% homogenate was mixed 1 : 1 with 0.2 M acetate buffer pH = 4.5 and incubated for 60 min in 30°C in shaking water bath. The reaction mixture was extracted in chloroform-methanol, and phases were separated by 1 M NaCl.

Aliquots of lower chlorophorm phase were separated for further determination of FFA and DAG content. An aliquot of chlorophorm phase was evaporated, and 100 μL of Krebs-Ringer phosphate buffer (pH = 7.6) containing 6% FFA-free BSA was added. The tubes were incubated in shaking incubator at 37°C for 2 hours. FFA concentration in final KRF/BSA solution was measured using commercially available kit. In order to check the efficiency of FFA solubilisation, the emptied tubes were washed with fresh KRB + 6% BSA, and then 100 μL of chlorophorm was added. An aliquot was separated by TLC, but no substantial traces of FFA were detected.

2.5. Determination of DAG Content. This method is based on the phosphorylation of DAG in the sample to DAG-3-phosphate using γ^{35}-ATP followed by quantification of radioactivity in a chlorophorm extract. Lipids from liver tissue or incubation mixture were extracted in chlorophorm-methanol and an aliquot of chlorophorm phase was evaporated under the stream of nitrogen. The sample was than solubilised by sonication in detergent buffer (7.5% n-octyl-β-D-glucopyranoside, 5 mM cardiolipin, and 1 mM DETAPAC). Reaction buffer (50 mM imidazole/HCl, pH = 6.6, 50 mM NaCl, 12.5 mM $MgCl_2$, and 1 mM EGTA), diacylglycerol kinase, and γ^{35}-ATP were added and incubated 30 min in 25°C. Lipids were extracted into chlorophorm-methanol, phases were separated with 1% $HClO_4$, and the exact volume of lower chlorophorm phase was determined. An aliquot was evaporated, resolved in 5% chlorophorm-methanol, and separated by TLC. Individual populations of lipids were visualised by iodine vapours, the bands corresponding to DAG were scraped off, and the radioactivity was determined by scintillation counting.

2.6. Incubation of Liver Slices In Vitro. The production of β-hydroxybutyrate from liver slices *in vitro* was measured in the absence of exogenous FFA. Liver slices (width approximately 1 mm) were quickly dissected and incubated for 2 hours in Krebs Ringer bicarbonate buffer with 5 mmol/L glucose, 2% bovine serum albumin, gaseous phase 95% O_2, and 5% CO_2. All incubations were carried out at 37°C in sealed vials in a shaking water bath. The aliquots of the incubation medium were stored frozen until the further analysis.

2.7. Electrophoretic Separation and Immunodetection. The homogenate, light lysosomal fraction, and dense lysosomal fraction prepared as described above were used for the assessment of LIPA protein content. A separate group of rats were used to assess the impact of hepatic fat accumulation on the insulin signalling pathway. The animals were either deprived of food for 24 hours (fasted) or had free access to food, and insulin (6 U/kg i.p.) was administered 30 min prior decapitation (fed + insulin). Liver samples (200 mg) were harvested *in situ* and stored in liquid nitrogen until further utilization. The homogenate was prepared by Ultra-Turax homogenizer (IKA Worke, Staufen, Germany) in homogenization buffer (150 mM NaCl, 2 mM EDTA, 50 mM TRIS, 20 mM glycerolphosphate, 1 mM Na_3VO_4, 2 mM sodium pyrophosphate, 1 mM PMSF, leupeptin 10 μg/mL, and aprotinin 10 μg/mL). The homogenate was used for determination of mTOR and Akt phosphorylation. The proteins were

separated by electrophoretic separation under denaturing conditions and electroblotted onto PVDF membranes. The level of phosphorylation of Akt and mTOR kinases was assessed by immunodetection using specific phospho-Akt (Ser473) antibody and phospho-mTOR (Ser2448) antibody, respectively. The total expression of Akt and mTOR protein was determined on the same membrane after striping and reblotting using specific antibodies. All these antibodies were purchased from Cell Signalling Technology, (Boston, MA). The immunodetection of LIPA protein was performed using mouse monoclonal (9G7F12) antibody to lysosomal acid lipase (Abcam, Cambridge, UK). The loading control was performed using rabbit polyclonal antibody to beta actin (Abcam, Cambridge, UK). The bands were visualized using ECL and quantified using FUJI LAS-3000 imager (FUJI FILM, Japan) and Quantity One software (Biorad, Hercules, CA).

2.8. PKC Membrane Translocation. The liver homogenate was prepared as described above. The total membrane and cytosolic fractions were prepared by centrifugation of the homogenate at 100 000 g. Solubilisation of membrane fraction was carried out in 1% Triton X-100, 0.1% SDS, and 0.5% deoxycholate. After electrophoretic separation and blotting, the PKCε was detected using anti-PKCε antibody (Sigma, St. Louis, USA). PKC translocation was expressed as the ratio of arbitrary units of membrane bands over the cytosol bands.

2.9. Biochemical Analysis. TAG content in liver homogenate or phagolysosomal fraction was determined after the extraction according to Folch et al. [19]. The glycogen content was determined in fat-free dry mass after hydrolysis in 30% KOH and expressed as a glucose equivalent (μmoles per g dry weight).

FFA, insulin, TAG and glucose serum content, and β-hydroxybutyrate production were determined using commercially available kits (FFA: FFA half microtest, Roche Diagnostics GmbH Mannheim, Germany; triglycerides and glucose: Pliva-Lachema, Brno CR; insulin: Mercodia, Uppsala, Sweden; β-hydroxybutyrate: RanBut, RANDOX Crumlin, UK).

2.10. Statistical Analysis. Data are presented as mean ± SEM. Statistical analysis was performed using Kruskal-Wallis test with multiple comparisons ($n = 5$–7). Differences were considered statistically significant at the level of $P < 0.05$.

3. Results

3.1. The Effect of HFD on Physical and Metabolic Parameters. The two-week period of HFD resulted in higher body weight and increased fat accumulation determined as epididymal fat pad : body weight ratio (Table 1). The impairment of glucose metabolism was indicated by increased fasting glycemia, increased fasting insulinemia, and impaired glucose tolerance measured by oral glucose tolerance test and expressed as $AUC_{1-180 \, min}$. The alterations in glucose metabolism were

The Increased Activity of Liver Lysosomal Lipase in Nonalcoholic Fatty Liver Disease Contributes to the Development of Hepatic Insulin Resistance

23

TABLE 1: The effect of HFD on physical and metabolic parameters.

	Standard diet		High-fat diet	
	Fasted	Fed	Fasted	Fed
Body weight (g)		331 ± 6.9		379 ± 10.7^x
Epididymal fat pad/b.w. (g/100 g)	0.9 ± 0.07	1 ± 0.06	$1.4 \pm 0.07^*$	$1.4 \pm 0.04^\#$
Glycemia (mmol/L)	5.1 ± 0.1	$7.9 \pm 0.5^+$	$5.9 \pm 0.2^*$	$8.1 \pm 0.3^+$
Insulinemia (pmol/L)	56 ± 15	$135 \pm 21^+$	$125 \pm 10^*$	127 ± 18
AUC$_{0-180}$ (mmol glucose/L)		1168 ± 26.6		1325 ± 32.9^x
Serum Tg (mmol/L)	0.7 ± 0.1	$1.4 \pm 0.08^+$	0.7 ± 0.02	$1.4 \pm 0.1^+$
Serum FFA (mmol/L)	0.7 ± 0.05	$0.4 \pm 0.02^+$	0.6 ± 0.08	$0.45 \pm 0.07^+$
ALT (μkat/L)		1.2 ± 0.1		1.2 ± 0.1
AST (μkat/L)		4.3 ± 0.6		3.9 ± 0.3
β-hydroxybutyrate (μmol/L)	1.67 ± 0.05	$0.05 \pm 0.01^+$	$3.2 \pm 0.25^*$	$0.28 \pm 0.05^{\#,+}$

Data are given as means $\pm$ SEM, $n = 7$. $^xP < 0.05$ HFD versus SD group; $^+P < 0.05$ fasted versus fed animals; $^*P < 0.05$ SD- versus HFD-fasted animals; $^\#P < 0.05$ SD- versus HFD-fed animals.

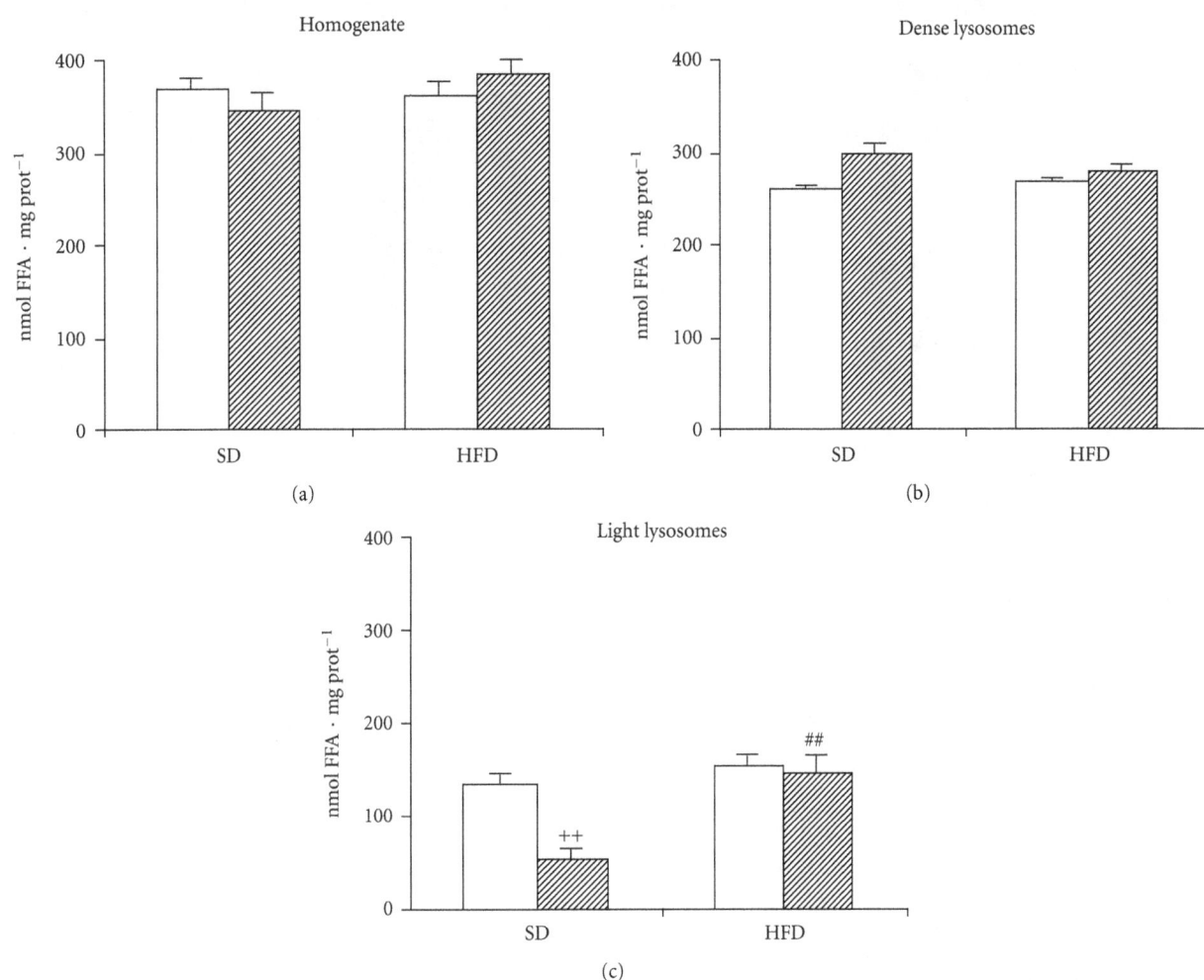

FIGURE 1: The effect of HFD on the LIPA activity measured as FFA release from artificial substrate (^{3}H-triolein). (a) Homogenate; (b) dense lysosomes; (c) light lysosomes. The lipase activity was measured as the release of fatty acids at pH $= 4.5$ from ^{3}H-triolein. Open bars = fasted animals; hatched bars = fed animals. $^{++}P < 0.01$ fed versus fasted; $^{\#\#}P < 0.01$ HFD fed versus SD fed.

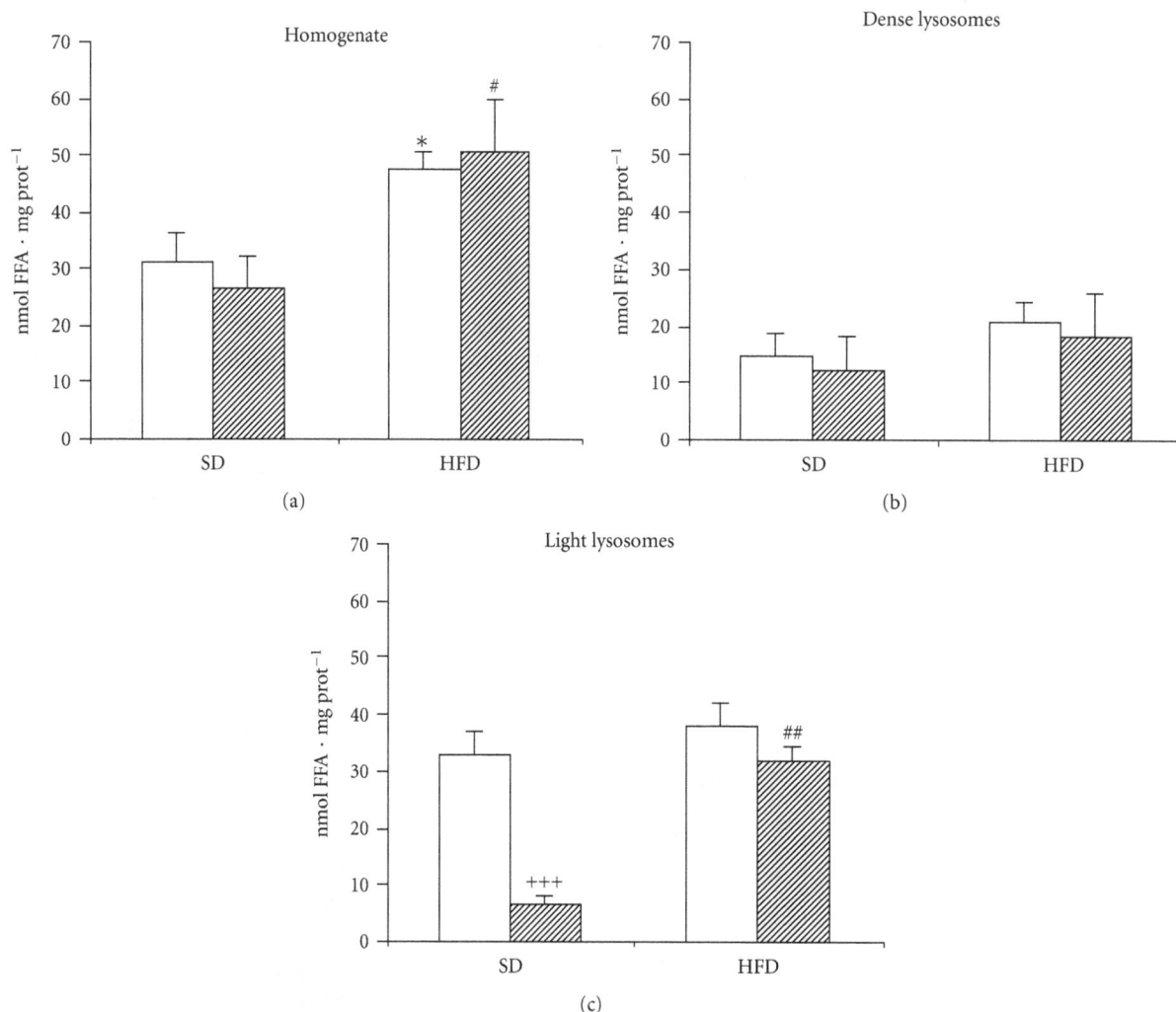

FIGURE 2: The effect of HFD on the LIPA activity measured as FFA release from endogenous TAG. (a) Homogenate; (b) dense lysosomes; (c) light lysosomes. 10% liver homogenate, light or dense lysosomal fraction were incubated 60 min at pH = 4.5. At the end of incubation, the released FFAs were quantified as described in Section 2. The graph shows the difference between FFA concentration in the sample at the beginning and at the end of the incubation. Open bars = fasted animal; hatched bars = fed animals. $^{+++}P < 0.001$ fed versus fasted; $^{*}P < 0.05$ HFD fasted versus SD fasted; $^{#}P < 0.05$, $^{##}P < 0.01$ HFD fed versus SD fed.

not accompanied by dyslipidemia. Serum β-hydroxybutyrate concentration was significantly elevated in both HFD-fed as well as HFD-fasted group compared to corresponding SD groups what indicates increased utilisation of fatty acids for ketogenesis in the liver. Short-term HFD administration did not alter ALT and AST serum concentrations.

As expected, compared to the SD group, the HFD-administered animals accumulated increased amount of TAG (fasted: 14.6 ± 1.4 versus 3.2 ± 0.2; $P < 0.001$; fed: 16.2 ± 2.5 versus $2.9 \pm 0.2 \, \mu mol/g$; $P < 0.001$) and DAG (fasted: 138 ± 17 versus 83 ± 12; $P < 0.01$; fed: 145 ± 19 versus $53 \pm 9 \, nmol/g$; $P < 0.001$) in the liver. The insulin-stimulated increase of glycogen content in liver was lower in HFD compared to SD animals (fasted: 27 ± 7 versus 41 ± 9 n.s.; fed: 123 ± 10 versus $261 \pm 15 \, \mu mol/g$; $P < 0.001$).

3.2. The Effect of HFD on Lysosomal Lipase Activity.
In order to determine maximal LIPA activity in liver homogenate and in particular lysosomal subpopulations, we employed

emulsified ^{3}H-labeled triolein as a substrate. In this experimental setting, the substrate is present in excess, and the only limiting factor is the amount of enzyme. The total LIPA activity measured in the whole homogenate was not affected either by prandial status (fasted or fed state) or diet intervention (SD or HFD) (Figure 1(a)). Similar results were observed in the fraction of dense lysosomes that represent primary lysosomes (Figure 1(b)). On ^{3}H-triolein as a substrate, we found most of total LIPA activity in this fraction. The LIPA activity determined in light lysosomes represents only minor portion of total activity, but unlike homogenate or dense lysosomes, it responds to different metabolic states (Figure 1(c)). In the SD group, it is elevated in fasting and depressed in fed state. HFD feeding abolished the prandial regulation of LIPA activity especially due to its upregulation in fed state.

A separate set of experiment was designed in order to evaluate the contribution of LIPA associated with dense and

The Increased Activity of Liver Lysosomal Lipase in Nonalcoholic Fatty Liver Disease Contributes to the Development of Hepatic Insulin Resistance

25

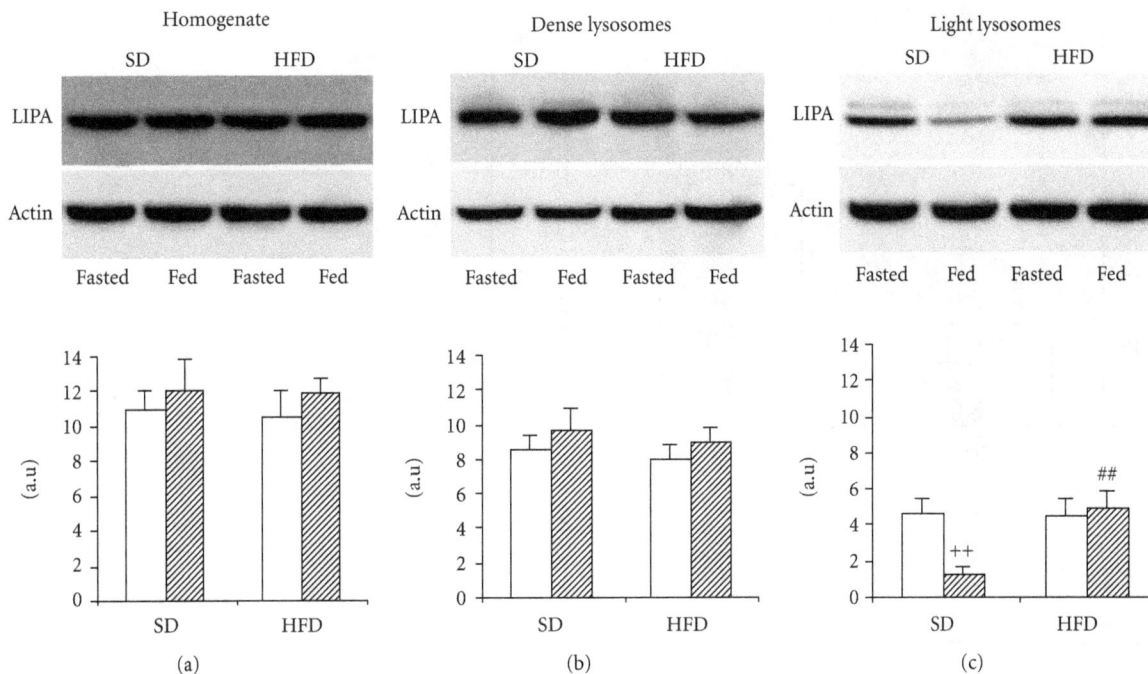

FIGURE 3: The effect of HFD on the LIPA protein expression. (a) Homogenate; (b) dense lysosomes; (c) light lysosomes. Representative Western blots are shown above each graph (f = fasted, F = fed). The results are expressed as arbitrary units after normalisation to the actin expression (loading control). Open bars = fasted animals; hatched bars = fed animals. $^{++}P < 0.01$ fed versus fasted; $^{##}P < 0.01$ HFD fed versus SD fed.

light lysosomes to the degradation of intracellular TAG. In this experimental design, the intracellular TAGs contained in particular fraction are the only source of substrate, and the intensity of lipolysis depends not only on the amount of enzyme but also on the amount of substrate available in the sample (Figures 2(a), 2(b), and 2(c)). Compared with the same experiments carried on ^{3}H-triolein, we found two differences. First, HFD administration led to a significant increase of total LIPA activity measured in homogenate. Second, after separation of lysosomal subpopulations, LIPA activity associated with light lysosomes was higher than those associated with dense lysosomes. This observation could be explained by the previous "*in vivo*" translocation of both the substrate (TAG droplets) and the enzyme (LIPA) into light lysosomal fraction (phagolysosomes). In accordance with this presumption, we found higher TAG content in phagolysosomal fraction in HFD compared with SD group (fasted: 2.3 ± 0.2 versus 4.1 ± 0.7; fed: 1.02 ± 0.3 versus $5.6 \pm 0.48\,\mu\text{mol}\cdot\text{mg prot}^{-1}$). Similarly with the results obtained on ^{3}H-triolein, the effect of fasting was manifested only in SD group and only in light lysosomal fraction. HFD feeding resulted into the elevation of LIPA activity in light lysosomes and into the abolishment of prandial regulation. Taken together, our results indicate that in the liver most of the enzyme is present in inactive form in dense (primary) lysosomes, and the physiologically active portion of the enzyme could be determined in light lysosomal fraction.

3.3. The Effect of HFD on Lysosomal Lipase Protein Distribution. In order to distinguish whether the higher LIPA

activity found in the light lysosomal fraction in HFD group is consequent to the increased amount of enzyme in this fraction or only to the increased availability of the substrate, we determined the amount of LIPA protein in liver homogenate and in particular fractions. We found that the LIPA protein content in homogenate (Figure 3(a)) and dense lysosomal fraction (Figure 3(b)) is similar in both fasted and fed animals and that it is not affected by short-term HFD administration. In contrast to these findings, the abundance of LIPA protein in light lysosomal fraction is lower than in homogenate or dense lysosomes, but it varies according to several factors (Figure 3(c)). In SD group, it strongly depends on prandial status. In SD-fasted rats, LIPA protein abundance in this fraction is significantly higher compared with their fed counterparts. Short-term HFD diet has no effect on the content of LIPA protein in fasted animals, but it significantly increases its amount in the fed ones. Consequently, the prandial-dependent regulation is completely abolished in HFD group.

3.4. Diacylglycerol Production in Incubated Liver Homogenate. DAG is one of the major products of LIPA action on TAG molecule as this enzyme has lower affinity to DAG or monoacylglycerol compared with its affinity to TAG [13]. In our experimental conditions (incubation of liver homogenate or isolated fraction in pH $= 4.5$), DAG could not be further utilised for TAG biosynthesis, and the difference in DAG concentrations at the end and at the beginning of incubation represents the net DAG production from TAG degradation. Nevertheless, we cannot exclude some

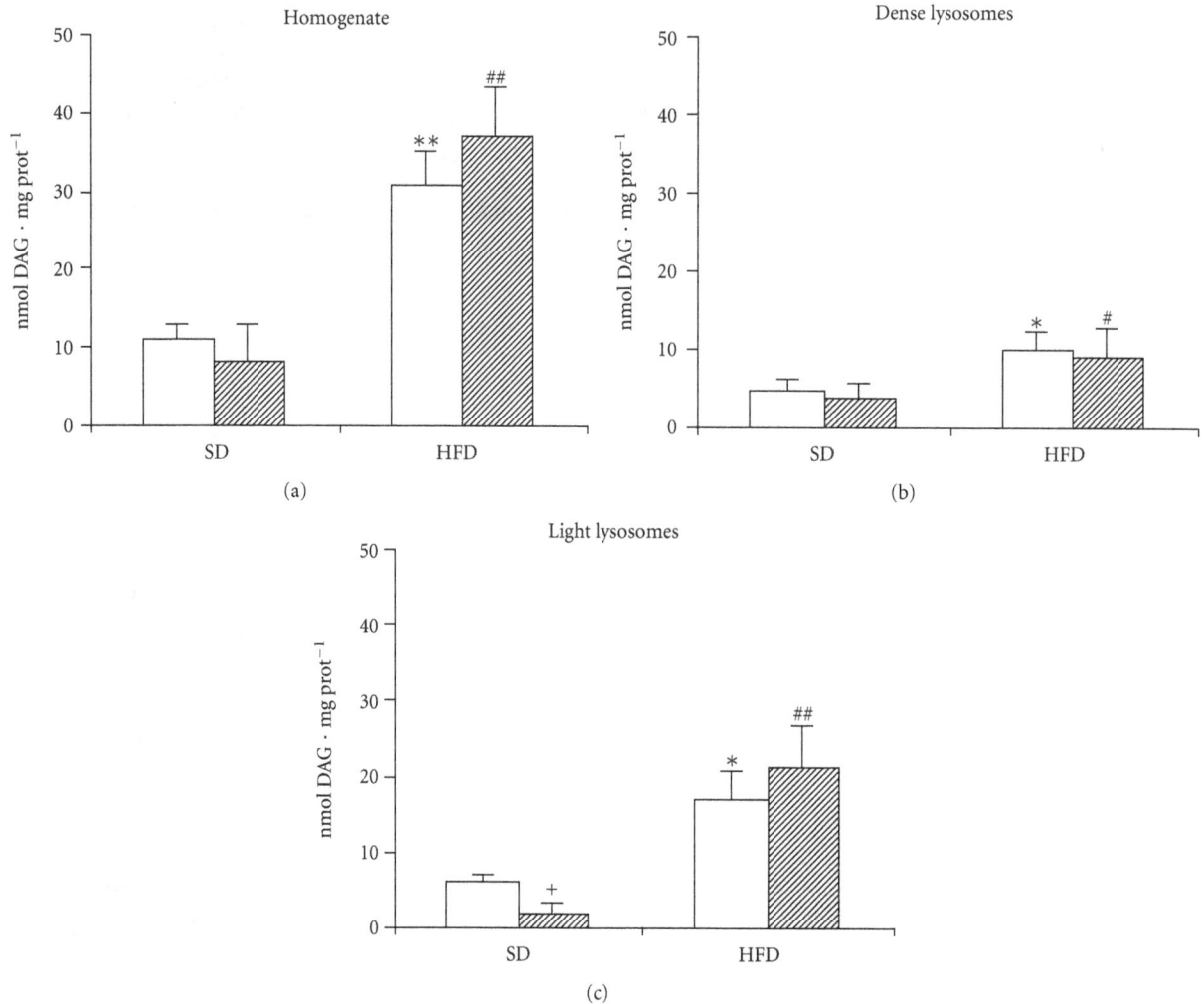

FIGURE 4: The effect of HFD on the DAG production from intracellular TAG *in vitro*. (a) Homogenates; (b) dense lysosomes; (c) light lysosomes. 10% liver homogenate, dense lysosomal or light lysosomal fractions were incubated 60 min at pH = 4.5. At the end of incubation, DAG was extracted into chlorophorm-methanol and quantified as described in Section 2. The graph shows the difference between DAG concentration in the sample at the beginning and at the end of the incubation. Open bars = fasted animals; hatched bars = fed animals. $^{+}P < 0.05$ fed versus fasted; $^{*}P < 0.05$, $^{**}P < 0.01$ HFD fasted versus SD fasted; $^{#}P < 0.05$, $^{##}P < 0.01$ HFD fed versus SD fed.

degradation of DAG by lysosomal carboxylesterases. In homogenate, DAG production in SD group was significantly lower compared with those in HFD, and it was prandial dependent, that is, elevated in fasting and downregulated in fed state. In HFD group, a significant DAG production was detected in both fasted and fed animals (Figure 4(a)). The stimulatory effect of HFD on DAG production was found in both dense (Figure 4(b)) and light (Figure 4(c)) lysosomes. In HFD group, approximately 60% of DAG formation occurred in light lysosomal fraction, and in contrast to the SD group, it was independent of prandial status.

3.5. The Effect of HFD on Ketogenesis In Vitro.
LIPA is expressed not only in hepatocytes but also in many other cell types including Kupffer cells present in the liver. In order to address the issue whether the above-mentioned changes in LIPA activity could be ascribed to hepatocytes, we measured ketone bodies production from liver slices *in vitro*

(Figure 5). Ketogenesis is the metabolic pathway occurring exclusively in hepatocytes and tightly reflects the intracellular TAG metabolism. We found an elevated ketogenesis due to the HFD administration what under these experimental set up implicates the accentuation of TAG hydrolysis. When liver slices were incubated in the absence of exogenous FFA, HFD-fasted group exhibited significantly higher β-hydroxybutyrate production compared with SD fasted. A similar trend was found also in fed animals.

3.6. The Effect of HFD on Key Components of Insulin Signalling Pathway.
To determine the effect of HFD on hepatic insulin sensitivity, the activation of key components of insulin signalling pathway was measured by immunodetection of their phosphorylation status. As shown in Figure 6(a), the insulin-stimulated phosphorylation of Akt kinase was significantly impaired in HFD-compared to SD group. Similar

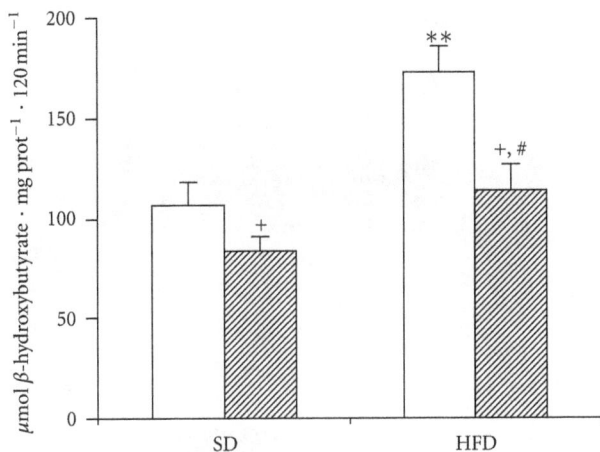

FIGURE 5: The effect of HFD on β-hydroxybutyrate production from liver slices *in vitro*. Liver slices were incubated in oxygenated KRB without exogenous fatty acids. Open bars = fasted animals; hatched bars = fed animals. $^{+}P < 0.05$ fed versus fasted; $^{**}P < 0.01$ HFD versus SD fasted; $^{\#}P < 0.05$; $^{\#}P < 0.05$ HFD fed versus SD fed.

results were obtained for mTOR (Figure 6(b)) suggesting an impairment of insulin signal transduction.

3.7. The Effect of HFD on PKCε Activity. Determination of the relative abundance of the particular PKC isoform in the membrane and cytosol fractions reflects PKCε activation. An increase in the membrane to cytosol fraction ratio was used as an indicator of PKCε activation. As shown in Figure 7, PKCε was significantly activated in the liver of the HFD-administered animals.

4. Discussion

In the present study, we provide evidence that in steatosis the increased degradation of TAG mediated by LIPA and associated with the increased production of DAG may be one of the mechanisms determining the rapid onset of hepatic IR. Our hypothesis is based on following findings. First, alterations in LIPA activity associated with different metabolic states are based on prandial-dependent transloca-tion of the enzyme from the inactive pool of dense lysosomes into light lysosomal fraction, and it is upregulated in fasting and downregulated in the fed state. After short-term HFD administration, this prandial-dependent regulation of LIPA activity is abolished. The fed state-associated downregula-tion of LIPA activity is impaired, and the portion of the active enzyme is permanently increased. These changes were demonstrated on both endogenous TAG and exogenous substrate (emulsified ^{3}H-triolein). Second, in steatosis, the production of TAG degradation intermediates, FFA and DAG, by lysosomal lipase was significantly elevated. Finally, we proved an increased PKCε activation together with the defects in the insulin-signalling cascade in the fatty liver. Taken together, these data indicate that the enhanced activity of LIPA in HFD-fed animals and following overproduction of

PKCε activator DAG contribute to the establishment of HFD-induced IR. We have previously shown that LIPA is involved in the degradation of intracellular TAG in the liver [20]. The essential role of LIPA for hydrolysis of TAG is supported by findings of Du et al. [21] who reported that LIPA knock-out mice (*Lipa* $^{-/-}$) exhibited progressive hepatosplenomegaly and massive TAG accumulation in the liver.

Our data indicate that the principal factor regulating the LIPA activity is not the total amount of the enzyme itself but rather its intracellular localisation. We did not find any significant differences in LIPA mRNA expression in response to either fasting or diet intervention (not shown), and in accordance with this, we did not find any difference in total LIPA protein content determined in the whole homogenate. According to Seglen and Solheim [22], active phagolysosomes have a lower density than the small, inactive lysosomes, allowing their separation by differential centrifugation. Based on this observation, we separated the total lysosomes into two subpopulations according to their density. We expected that the active lysosomes containing the TAG substrate would remain in the less dense fraction (light lysosomes), while the inactive lysosomes would sediment (dense lysosomes). In our experimental setting, the effect of fasting or HFD was manifested predominantly in light lysosomal fraction what supports the physiological relevance of this methodology. In SD group, we observed a signif-icant prandial-dependent regulation, LIPA activity being upregulated in fasted and downregulated in fed animals. HFD feeding was associated with a significant elevation of LIPA protein content and LIPA activity in light lysosomal fraction particularly in fed animals and consequently with the abolishment of prandial-dependent regulation of LIPA activity. Similar trends were observed on both exo- and endogenous substrates. The changes in LIPA activity were reflected by the corresponding changes in LIPA protein con-tent in light lysosomal fraction. The interesting conclusions come from the comparison of LIPA activity in the light and dense lysosomal fractions determined on either ^{3}H-triolein or intracellular TAG. The activity measured on ^{3}H-triolein depends only on the amount of the enzyme present in the particular fraction as the substrate is available in excess. In contrast, when intracellular TAGs are the only source of substrate, the activity in particular fractions depends on the coordinated translocation of the enzyme and the substrate. The main difference in LIPA activity determined by these two approaches was found in the distribution of LIPA activity among dense and light lysosomal fractions. In dense lysosomes, we found high LIPA activity on ^{3}H-triolein but only low LIPA activity on intracellular TAG. This difference indicates that dense lysosomal fraction contains an enzyme that is not active in physiological situation but that could be activated after addition of the arteficial substrate. The LIPA activity determined in light lysosomes represented the bulk of total LIPA activity on intracellular TAG substrate but only minor portion of total activity determined on ^{3}H-triolein. It is possible to speculate that the LIPA activity on endogenous substrate quantitatively reflects the formation of activated lysosomes, that is particles containing both the substrate and the enzyme. Taken together, these data indicate that LIPA

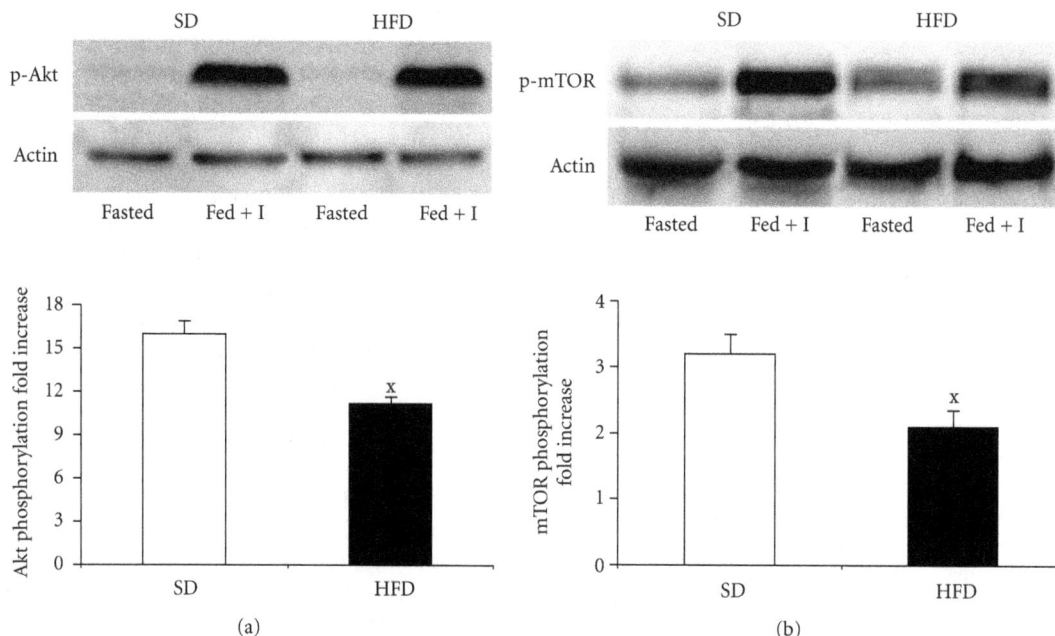

FIGURE 6: Alterations of insulin signalling cascade associated with hepatic fat accumulation. All results are expressed as a fold increase in the insulin-stimulated state relative to the basal state. Representative Western blots are shown above each graph. (a) Fold increase in Akt (Ser473) phosphorylation, (b) fold increase in mTOR (Ser2448) phosphorylation. The basal level of protein phosphorylation was determined in the homogenate prepared from the liver of 24 hours fasted animals. The effect of insulin was determined in identically processed samples from animals which had free access to food and 40 min prior to decapitation were administered insulin 6 U/kg. The total protein (Akt or mTOR) expression was determined after striping the membrane and reblotting with anti-Akt or anti-mTOR antibody. Values represent means $\pm$ S.E.M. of 7 animals. $^{x}P < 0.05$.

associated with light lysosomes represents the physiologically active enzyme.

We suppose that in NAFLD, characterised by high TAG intracellular content, one of the factors determining the phagolysosomal formation may be the substrate availability itself. The increased amount of intracellular lipid droplets in steatosis could promote the phagolysosome formation and stimulate the lysosomal lipolysis. Only recently, Singh et al. [23] described direct involvement of autophagy and lysosomal pathway in the degradation of intracellular lipid droplets in the liver. They found that lipid droplets can enter the autophagic degradation pathway in the same manner as proteins and damaged organelles via formation of autophago(lipo)somes that further fuses with primary lysosomes. As the only known lysosomal enzyme with lipolytic activity is LIPA, we believe that our results are in accordance with findings of Singh et al..

The ketone body formation tightly reflects the liver lipid metabolism. Debeer et al. [24] demonstrated that both ketogenesis and FFA oxidation are a particularly good markers of lysosomal TAG degradation. We observed higher β-hydroxybutyrate concentration in serum and higher ketone body production from isolated liver slices in the absence of exogenous fatty acids in HFD group. We conclude that these data provide indirect evidence that confirms the stimulatory effect of short-term HFD on lysosomal lipolysis.

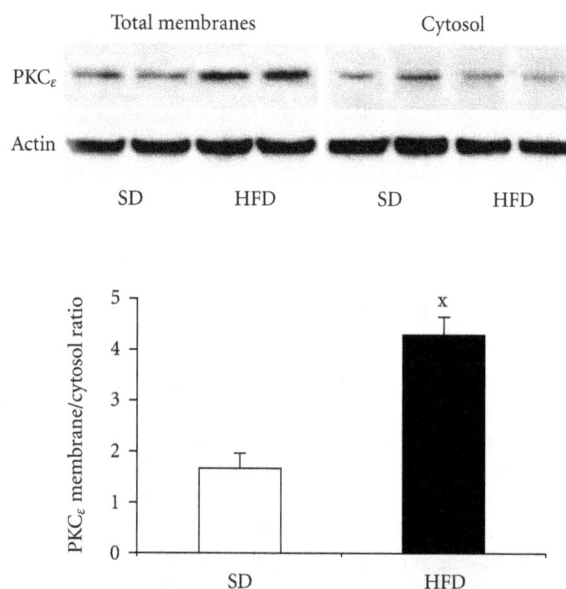

FIGURE 7: The effect of hepatic fat accumulation on PKCε activation. Representative Western blot is shown in the upper part of the figure; TM, total membrane fraction, C, cytosol fraction. The PKCε membrane to cytosol ratio is shown in the graph. The relative densities of the bands in the membrane fraction were compared with corresponding ones in cytosol fraction in order to obtain the measurable parameter of activation. Values represent means $\pm$ S.E.M. of 7 animals. $^{x}P < 0.05$.

Concomitantly occurring stimulation of lipolysis and the accumulation of endogenous TAG after HFD administration seem to be contradictory. However, in hepatocytes, a significant portion of FFA released from intracellular TAG (approximately 70%) is reesterified back [25]. HFD impairs VLDL secretion [26], and most of FFA reenter the intracellular storage pool. We have previously reported that DGAT-1 expression is increased in fatty liver what indicates enhanced esterification of fatty acids and may result in the intensification of lipolytic/reesterification cycle in hepatocytes [20]. The increased lipolysis thus does not result in decreased TAG content but only in higher TAG turnover.

In the liver, PKCε, member of novel PKCs subfamily, is involved in the development of HFD-induced IR [16, 17]. Samuel et al. showed that fat-induced hepatic IR may result from activation of PKCε and its downstream targets. Nevertheless, the nature of the signal that activates PKCε has not been fully explained. Systemic increase in FFA serum levels, as one possible underlying factor, has not been described after HFD administration. Another candidate, 1,2-sn-DAG, is an important intracellular signalling molecule, and it is the known activator of novel PKCs isoform family [27]. The increased DAG content due to the increased flux through TAG synthetic pathway and the following PKCε activation was described in skeletal muscle in HFD-administered animals [28]. However, DAG is also an intermediate in TAG degradation pathway that, in contrast to muscle, is quite active in the liver. Our findings suggest that DAG originating from the increased lipolytic activity of LIPA and accentuated TAG breakdown could act as PKCε activator in fatty liver. This hypothesis is supported by the fact that in fatty liver LIPA is activated specifically in the fed state, and possible PKCε activator is available during the period of insulin action.

In conclusion, we found that short-term HFD-induced TAG accumulation in the liver is associated with the increased degradation of intracellular TAG by lysosomal lipase and with higher production of lipolytic products—DAG and FFA. Our findings suggest that the elevated DAG production by LIPA activated by increased supply of dietary lipids may represent the causal link between dietary fat-induced hepatic TAG accumulation and hepatic IR via the PKCε activation. In the light of these findings, lysosomal lipolysis may represent a new promising therapeutic target.

Acknowledgments

The auther thank Dr. Milan Jirsa PhD for critical reading of the paper. This study was supported by Grant no. NS 9696-3 from the Ministry of Health of the Czech Republic.

References

[1] R. P. Mensink, J. Plat, and P. Schrauwen, "Diet and nonalcoholic fatty liver disease," Current Opinion in Lipidology, vol. 19, no. 1, pp. 25–29, 2008.

[2] E. W. Kraegen, P. W. Clark, A. B. Jenkins, E. A. Daley, D. J. Chisholm, and L. H. Storlien, "Development of muscle insulin resistance after liver insulin resistance in high-fat-fed rats," Diabetes, vol. 40, no. 11, pp. 1397–1403, 1991.

[3] G. Marchesini, M. Brizi, A. M. Morselli-Labate et al., "Association of nonalcoholic fatty liver disease with insulin resistance," American Journal of Medicine, vol. 107, no. 5, pp. 450–455, 1999.

[4] D. L. Lankester, A. M. Brown, and V. A. Zammit, "Use of cytosolic triacylglycerol hydrolysis products and of exogenous fatty acid for the synthesis of triacylglycerol secreted by cultured rat hepatocytes," Journal of Lipid Research, vol. 39, no. 9, pp. 1889–1895, 1998.

[5] D. Wiggins and G. F. Gibbons, "The lipolysis/esterification cycle of hepatic triacylglycerol. Its role in the secretion of very-low-density lipoprotein and its response to hormones and sulphonylureas," Biochemical Journal, vol. 284, part 2, pp. 457–462, 1992.

[6] H. Vavřínková and B. Mosinger, "Effect of glucagon, catecholamines and insulin on liver acid lipase and acid phosphatase," Biochimica et Biophysica Acta, vol. 231, no. 2, pp. 320–326, 1971.

[7] C. de Duve, "The lysosome turns fifty," Nature Cell Biology, vol. 7, no. 9, pp. 847–849, 2005.

[8] J. Dice, Lysosomal Pathways of Protein Degradation, Moleculat Biology Intelligence Unit, Landes Bioscience, Austin, Tex, USA, 2000.

[9] Y. Tanaka, R. Harada, M. Himeno, and K. Kato, "Biosynthesis, processing, and intracellular transport of lysosomal acid phosphatase in rat hepatocytes," Journal of Biochemistry, vol. 108, no. 2, pp. 278–286, 1990.

[10] G. B. Gordon, L. R. Miller, and K. G. Bensch, "Studies on the intracellular digestive process in mammalian tissue culture cells," The Journal of Cell Biology, vol. 25, pp. 41–55, 1965.

[11] E. Knecht, C. Aguado, J. Cárcel et al., "Intracellular protein degradation in mammalian cells: recent developments," Cellular and Molecular Life Sciences, vol. 66, no. 15, pp. 2427–2443, 2009.

[12] M. J. Czaja, "Autophagy in health and disease. 2. Regulation of lipid metabolism and storage by autophagy: pathophysiological implications," American Journal of Physiology, vol. 298, no. 5, pp. C973–C978, 2010.

[13] K. Hayase and A. L. Tappel, "Specificity and other properties of lysosomal lipase of rat liver," The Journal of Biological Chemistry, vol. 245, no. 1, pp. 169–175, 1970.

[14] S. Neschen, K. Morino, L. E. Hammond et al., "Prevention of hepatic steatosis and hepatic insulin resistance in mitochondrial acyl-CoA:glycerol-sn-3-phosphate acyltransferase 1 knockout mice," Cell Metabolism, vol. 2, no. 1, pp. 55–65, 2005.

[15] D. B. Savage, S. C. Cheol, V. T. Samuel et al., "Reversal of diet-induced hepatic steatosis and hepatic insulin resistance by antisense oligonucleotide inhibitors of acetyl-CoA carboxylases 1 and 2," The Journal of Clinical Investigation, vol. 116, no. 3, pp. 817–824, 2006.

[16] V. T. Samuel, Z. X. Liu, X. Qu et al., "Mechanism of hepatic insulin resistance in non-alcoholic fatty liver disease," The Journal of Biological Chemistry, vol. 279, no. 31, pp. 32345–32353, 2004.

[17] V. T. Samuel, Z. X. Liu, A. Wang et al., "Inhibition of protein kinase Cε prevents hepatic insulin resistance in nonalcoholic fatty liver disease," The Journal of Clinical Investigation, vol. 117, no. 3, pp. 739–745, 2007.

[18] P. Belfrage and M. Vaughan, "Simple liquid-liquid partition system for isolation of labeled oleic acid from mixtures with glycerides," Journal of Lipid Research, vol. 10, no. 3, pp. 341–344, 1969.

[19] J. Folch, M. Lees, and G. H. Sloane Stanley, "A simple method for the isolation and purification of total lipides from animal tissues," *The Journal of Biological Chemistry*, vol. 226, no. 1, pp. 497–509, 1957.

[20] M. Cahová, H. Daňková, E. Páleníčková, Z. Papáčková, and L. Kazdová, "The autophagy-lysosomal pathway is involved in TAG degradation in the liver: the effect of high-sucrose and high-fat diet," *Folia Biologica*, vol. 56, no. 4, pp. 173–182, 2010.

[21] H. Du, M. Heur, M. Duanmu et al., "Lysosomal acid lipase-deficient mice: depletion of white and brown fat, severe hepatosplenomegaly, and shortened life span," *Journal of Lipid Research*, vol. 42, no. 4, pp. 489–500, 2001.

[22] P. O. Seglen and A. E. Solheim, "Conversion of dense lysosomes into light lysosomes during hepatocytic autophagy," *Experimental Cell Research*, vol. 157, no. 2, pp. 550–555, 1985.

[23] R. Singh, S. Kaushik, Y. Wang et al., "Autophagy regulates lipid metabolism," *Nature*, vol. 458, no. 7242, pp. 1131–1135, 2009.

[24] L. J. Debeer, J. Thomas, P. J. De Schepper, and G. P. Mannaerts, "Lysosomal triacylglycerol lipase and lipolysis in isolated rat hepatocytes," *The Journal of Biological Chemistry*, vol. 254, no. 18, pp. 8841–8846, 1979.

[25] G. F. Gibbons and D. Wiggins, "Intracellular triacylglycerol lipase: its role in the assembly of hepatic very-low-density lipoprotein (VLDL)," *Advances in Enzyme Regulation*, vol. 35, pp. 179–198, 1995.

[26] L. Oussadou, G. Griffaton, and A. D. Kalopissis, "Hepatic VLDL secretion of genetically obese Zucker rats is inhibited by a high-fat diet," *American Journal of Physiology*, vol. 271, no. 6, pp. E952–E964, 1996.

[27] Y. Nishizuka, "Protein kinase C and lipid signaling for sustained cellular responses," *The FASEB Journal*, vol. 9, no. 7, pp. 484–496, 1995.

[28] C. Schmitz-Peiffer, C. L. Browne, N. D. Oakes et al., "Alterations in the expression and cellular localization of protein kinase C isozymes ε and θ are associated with insulin resistance in skeletal muscle of the high-fat-fed rat," *Diabetes*, vol. 46, no. 2, pp. 169–178, 1997.

4

Mitotic Kinases and p53 Signaling

Geun-Hyoung Ha[1] and Eun-Kyoung Yim Breuer[1, 2]

[1] *Department of Radiation Oncology, Stritch School of Medicine, Loyola University Chicago, Maywood, IL 60153, USA*
[2] *Department of Molecular Pharmacology and Therapeutics, Stritch School of Medicine, Loyola University Chicago, Maywood, IL 60153, USA*

Correspondence should be addressed to Eun-Kyoung Yim Breuer, eubreuer@lumc.edu

Academic Editor: Mandi M. Murph

Mitosis is tightly regulated and any errors in this process often lead to aneuploidy, genomic instability, and tumorigenesis. Deregulation of mitotic kinases is significantly associated with improper cell division and aneuploidy. Because of their importance during mitosis and the relevance to cancer, mitotic kinase signaling has been extensively studied over the past few decades and, as a result, several mitotic kinase inhibitors have been developed. Despite promising preclinical results, targeting mitotic kinases for cancer therapy faces numerous challenges, including safety and patient selection issues. Therefore, there is an urgent need to better understand the molecular mechanisms underlying mitotic kinase signaling and its interactive network. Increasing evidence suggests that tumor suppressor p53 functions at the center of the mitotic kinase signaling network. In response to mitotic spindle damage, multiple mitotic kinases phosphorylate p53 to either activate or deactivate p53-mediated signaling. p53 can also regulate the expression and function of mitotic kinases, suggesting the existence of a network of mutual regulation, which can be positive or negative, between mitotic kinases and p53 signaling. Therefore, deciphering this regulatory network will provide knowledge to overcome current limitations of targeting mitotic kinases and further improve the results of targeted therapy.

1. Introduction

Mitosis involves a highly orchestrated and fine-tuned sequence of events to properly transfer genetic information to the next generation by cell division [1, 2]. It is usually divided into five phases (prophase, prometaphase, metaphase, anaphase, and telophase) based on structure and behavior of the spindle and chromosomes, and cytokinesis begins at the end of mitosis [1, 3]. This whole process must be tightly regulated to prevent improper segregation of chromosomes [4, 5]. For this reason, cells employ a surveillance mechanism, known as the "spindle checkpoint" to ensure high fidelity of chromosome segregation in mitosis by sending a "wait signal" and thus delaying anaphase until all the chromosomes are properly aligned on the spindle apparatus (reviewed in [6]). When cells fail to delay anaphase in response to activation of spindle checkpoint, it will lead to an earlier anaphase onset, possibly causing chromosome instability, aneuploidy, and tumorigenesis [7–11].

Aneuploidy, an abnormal number of chromosomes, is a characteristic feature of cancer cells and a common cause of many genetic diseases [12, 13]. Aneuploid cells occur by an improper segregation of the chromosomes during cell division [12, 13]. The most common cause of aneuploidy is mitotic errors due to defects in "proper" mitotic kinase signaling in multiple cell cycle checkpoints, resulting in unfaithful chromosome segregation [12, 14, 15].

Multiple phosphorylation and proteolysis events play important roles in the regulation of mitotic progression and cytokinesis [1, 2]. Numerous proteins involved in these posttranslational events have been identified, including kinases and cysteine proteases [16–18]. One of the best understood kinases in the regulation of mitosis is cyclin-dependent kinase 1 (Cdk1) [2]. Cdks are highly conserved serine/threonine protein kinases that regulate cell cycle progression and subsequent cell division in eukaryotic cells and ubiquitously expressed throughout the cell cycle (reviewed in [19]). Among all Cdk family members, only five of them,

Cdk1, Cdk2, Cdk3, Cdk4, and Cdk6, have been implicated in controlling cell cycle [20, 21]. While other Cdks are mainly involved in the early phase of cell division, Cdk1 plays a key role in several mitotic processes [2, 21, 22]. The regulation of Cdk1 has been extensively reviewed elsewhere [23–25]. Briefly, during the G2/M transition, the activation of the mitotic kinase Cdk1/Cyclin B phosphorylates a variety of substrates, such as a kinesin-related motor protein Eg5 [26], lamin [27], and condensin [28], to initiate mitotic entrance and control its progression and mitotic exit [2, 26, 27, 29]. The kinase activity appears in late G2 and peaks at metaphase [30]. At the end of the metaphase, the anaphase promoting complex (APC) (also known as cyclosome, APC/C), which is an E3 ubiquitin ligase [31], recruits cyclin B for ubiquitination and degradation to allow mitosis to proceed [32, 33]. Therefore, it is undoubtful that the perfect regulation of Cdk1/cyclin B activity is critical for normal mitotic progression. Since the discovery of Cdks, much attention has been given to the other mitotic kinases, such as Aurora kinases, Polo-like kinases (Plks), monopolar spindle 1 (Mps1), benzimidazoles 1 homolog (Bub1), and Bub1-related kinase 1 (BubR1), due to their pivotal roles in mitosis [16] as well as the relevance to cancer. Studies indicate that Aurora kinases and Plks are mainly involved in regulating the centrosome cycle and mitotic spindle formation, while Mps1, Bub1, and BubR1 regulate the spindle assembly checkpoint [34, 35]. Therefore, the tight regulation of their kinase activities is required for proper mitotic progression, which is essential for maintaining genomic integrity [5].

Many studies have reported that deregulation of these mitotic kinases causes mitotic failure and aneuploidy and is closely associated with genomic instability and tumorigenesis [2, 36–38]. To defend against tumorigenesis caused by mitotic failure and guard genome stability, cells have utilized tumor suppressors, such as p53 [39] and BRCA1 [40] in a mitotic regulatory network. Because of its importance, tremendous efforts have been made to better understand the role of the functional crosstalk between mitotic kinases and tumor suppressors during mitosis. The p53 is one of the most frequently mutated or deleted genes in human cancers and plays a role in many cellular processes, including cell growth, differentiation, senescence, and DNA repair (reviewed in [41]). In addition, p53 is a key decision maker between cell cycle arrest and apoptosis in response to DNA damage [42, 43]. The loss-of-function of p53 can trigger an increase in genome instability and cancer predisposition, suggesting that p53 is essential for the maintenance of genome stability (reviewed in [44]). The human p53 is located on chromosome 17 (17p13) and consists of an N-terminal transactivation domain, a central specific DNA-binding domain and a C-terminal domain, containing a tetramerization domain and regulatory region [45]. At least 20 phosphorylation sites exist in human p53 [46] and importantly, several N-terminal phosphorylation sites, such as Ser-15 [47], Thr-18 [48], and Ser-20 [49] are critical for preventing oncogenic E3 ligase MDM2-mediated p53 ubiquitination and degradation [50]. On the other hand, phosphorylation at C-terminal and a few N-terminal sites, such as Ser-362/366 [51] and Thr-55 [52] often suppresses

its tumor suppressive function by destabilizing p53. These findings suggest that phosphorylation events may play significant roles in regulating p53 protein stability and function.

Under normal circumstances, cells induce the p53-dependent transcriptional activation, cell cycle arrest, and apoptosis in response to mitotic defects or DNA damage [53, 54]. However, cells lacking functional p53 due to deregulation of mitotic kinases, such as Aurora A [55], Plk1 [56], and Bub1 [57], do not undergo these cellular events and thus lead to genome instability, resulting in aneuploidy [15]. Phosphorylation of p53 by Mps1 [58] and BubR1 [59] stabilizes p53 and appears to antagonize the function of Aurora A, Plk1, and Bub1 in p53 signaling. Studies have shown that p53 can also regulate the expression and function of these kinases [60–64], suggesting that there may be mutual regulatory interactions between mitotic kinases and p53 in a mitotic signaling network (Figure 1).

In this paper, we will specifically focus on the classic mitotic kinases, including Aurora kinases, Plks, Bub1, Mps1, and BubR1, and their roles in regulating p53 protein stability and activity.

2. Negative Regulation of p53

2.1. Aurora Kinases. Aurora kinases belong to a highly conserved family of serine/threonine kinases crucial for chromosome segregation, condensation, and spindle assembly [1]. The first Aurora kinase was discovered in *Drosophila melanogaster* mutants having defects in mitotic spindle-pole formation [65]. Subsequently, homologues of Aurora kinases have been identified in various species. In budding yeast, there is a single Aurora kinase, known as increase-in-ploidy 1 (Ipl1) [66]. The *Ipl1* gene is essential for maintaining genome stability through its roles in chromosome segregation, spindle checkpoint, mitotic spindle disassembly, and cytokinesis [67, 68]. *Caenorhabditis elegans* has two Aurora kinases, Aurora/Ipl1-related-1 and -2 (AIR-1 and AIR-2), and they are thought to be key regulators of mitotic spindle assembly and dynamics [69, 70]. Three members of Aurora kinase family, Aurora A, B, and C, have been identified in mammalian cells [1]. The Aurora kinase family share a highly conserved C-terminal catalytic domain and a short N-terminal domain [71], and function in the regulation of mitosis and cytokinesis [72]. Deregulation of Aurora kinases causes a defect in spindle assembly, checkpoint function, and cell division, leading to chromosome missegregation or polyploidization [73]. Not surprisingly, overexpression of Aurora kinases is often found in a variety of human cancers [74–76]. Since the discovery of Aurora kinases, many efforts have been made to improve our understanding of their biological and physiological function in mitosis and the regulatory mechanisms relevant to cancer.

Aurora A is ubiquitously expressed in proliferating cells and its activity is tightly regulated through the cell cycle [77]. Both the expression level and kinase activity of Aurora A are significantly increased from the late G2 through the M phase [74, 78] and become low during interphase [79]. Aurora A plays a key role in mitotic spindle formation, centrosome

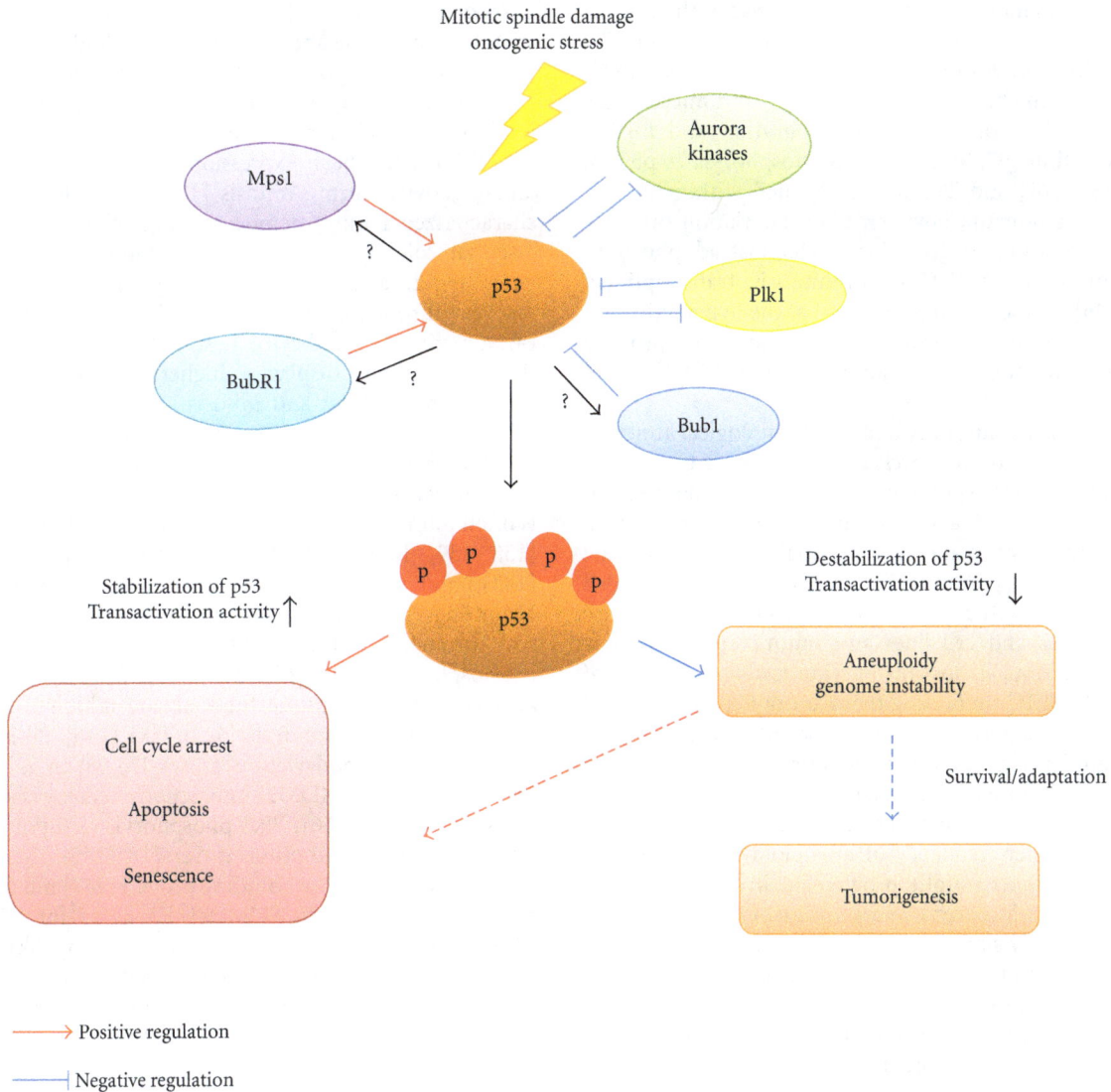

FIGURE 1: A model for regulatory networks of mitotic kinases controlling p53 signaling.

maturation [80], and activation of cell cycle regulators, such as Plk1 [81, 82] and Cdk1 [83]. Deregulated expression and activity of Aurora A can generate aneuploidy phenotype due to centrosome amplification and spindle multipolarity [84]. Numerous substrates of Aurora A have been identified, including p53 [85], human enhancer of filamentation 1 (HEF1) [86], TPX2 [87], Ajuba [88], Plk1 [81], BRCA1 [89], and transforming acidic coiled-coil 3 (TACC3) [90]. Human p53 is directly phosphorylated by Aurora A at two sites, Ser-215 [85] and Ser-315 [55], *in vitro* and *in vivo*. Phosphorylation of Ser-215 but not Ser-315 inhibits p53 DNA binding and its transactivational activity [85], whereas phosphorylation of Ser-315 induces MDM2-mediated p53 ubiquitination and subsequent degradation [55]. These findings suggest that Aurora A-mediated phosphorylation of p53 plays a negative regulatory role in p53 protein stability and its downstream signaling pathways. In response to DNA damage, p53 interacts with the heterogeneous nuclear ribonucleoprotein K (hnRNPK), a transcriptional coactivator of p53, and induces the p53 signaling pathway [91]. hnRNPK is phosphorylated on Ser-379 by Aurora A and this phosphorylation disrupts its interaction with p53 [92], suggesting that Aurora A can indirectly/negatively regulate p53 function via hnRNPK phosphorylation. Interestingly, a recent study shows that Aurora A can positively regulate p53 protein expression levels and *vice versa* [60]. In addition, *Xenopus* p53 can block *Xenopus* Aurora A's ability to transform cells [61], further supporting the existence of crosstalk between Aurora A and p53.

Aurora B is a member of the chromosome passenger complex (CPC), a key regulator of chromosome segregation, histone modification, and cytokinesis during mitosis [93, 94]. The CPC is composed of Aurora B and its nonenzymatic regulatory subunits inner centromere protein (INCENP), Borealin and Survivin [94], required for the activity, localization, and stability of Aurora B [93]. Aurora B governs the spindle assembly checkpoint and manages the correct chromosome segregation and cytokinesis during mitosis [72, 95]. Inhibition of Aurora B results in a failure of mitosis due to defects in chromosome segregation and microtubule dynamics [96], leading to endoreduplication and further polyploidization [97, 98]. Aurora B phosphorylates p53 on Ser-183, Ser-269, and Thr-284, all located within the p53 DNA binding domain; however, phosphorylation on these sites does not lead to degradation of p53, instead, phosphorylation on Ser-269 and Thr-284 inhibits its transcriptional activity [46]. These findings suggest that the hyperactivation or overexpression of Aurora A and B may compromise p53's tumor suppressive function via its destabilization and inactivation.

In contrast to Aurora A and B, the biological function of Aurora C has not been well-defined. Aurora C was first discovered in mouse sperm and eggs using a kinase screen [99]. While Aurora A and B are ubiquitously expressed in many different tissues and cells, especially actively dividing cells [98, 100, 101], Aurora C is predominantly expressed in the testis [99, 102], but not in other normal mouse somatic tissues and cell lines and mitotic spermatogonia [103]. In addition, its loss-of-function leads to a failure of meiosis [103, 104], indicating that Aurora C plays a critical role in meiosis. Recent studies show that Aurora B and C have similar structural and functional properties [105]. Inhibition of Aurora C causes aneuploidy, just like Aurora B, and furthermore, simultaneous inhibition of Aurora B and C causes a higher frequency of aneuploidy [105]. Aurora C can also support mitotic progression in the absence of Aurora B [105]. Moreover, overexpression of Aurora C causes abnormal cell division due to amplified centrosomes and micronucleation [101, 106], suggesting that Aurora C may be involved in mitosis as well. Unlike Aurora A and B, the role of Aurora C in the regulation of p53 protein stability and function has not been reported yet.

2.2. Polo-Like Kinase 1 (Plk 1).

Plks are a family of highly conserved serine/threonine protein kinases [107] named after the *polo* gene of *Drosophila melanogaster*, whose mutation causes a high frequency of abnormal mitosis and meiosis [108]. Subsequently, its homologues have been found in other species, including Cdc5 in *Saccharomyces cerevisiae*, [109], Plo1p in *Schizosaccharomyces pombe* [110], Plc1, Plc2, and Plc3 in *Caenorhabditis elegans* [111, 112], and Plx1, Plx2, and Plx3 in *Xenopus laevis* [113–115]. In mammals, five Plks have been identified: Plk1 (also known as serine/threonine-protein kinase 13, STPK13), Plk2 (also known as serum-inducible kinase, SNK), Plk3 (also known as fibroblast-growth-factor-inducible kinase, FNK; proliferation-related kinase, PRK; or cytokine-inducible kinase, CNK), Plk4 (also known as SNK akin kinase, SAK or serine/threonine-protein kinase 18, STK18), and Plk5 [116–125]. All Plks are abundantly expressed in tissues exhibiting high levels of mitotic activity [120] and share two conserved domains, an N-terminal Ser/Thr kinase domain and a C-terminal polo-box domain (PBD) [107, 126].

It is now widely recognized that Plks are key regulators of mitosis, meiosis, and cytokinesis [107, 127, 128] as well as

DNA damage response [107, 123, 126]. Deregulation of Plks leads to centrosome abnormalities, aneuploidy, and genomic instability [129], possibly leading to cancer development [130]. This may explain why deregulated expression of Plks is often detected in many types of cancer (reviewed in [37]).

Plk1 reaches peak expression during G2/M phase and kinase activity during mitosis [128, 129]. Plk1 is the best characterized family member among others and plays an essential role in centrosome maturation and separation [131], spindle assembly and formation [110], G2 checkpoint recovery through activating cyclin-dependent kinase [132], mitotic exit [113], and cytokinesis [133]. Studies have shown that cancer cells display a higher dependency on Plk1 for cell proliferation and mitosis [134, 135] than primary cells [136]. Deregulated expression and activity of Plk1 generate abnormal centrosomes [129] and initiate malignant transformation [137]. Not surprisingly, deregulation of Plk1 is often found in many types of cancer, including melanoma [138, 139], lung [140], head and neck [141, 142], breast [143], and ovarian cancer [144] with poor prognosis. Mounting evidence suggests that Plk1 negatively regulates p53 through direct and indirect mechanisms [145]. p53 is phosphorylated by Plk1 *in vitro* and its transcriptional activity and proapoptotic function are inhibited by direct interaction and phosphorylation of Plk1 [146]. Plk1 can also inhibit p53 phosphorylation at Ser-15, which is required for blocking p53-MDM2 interaction, thereby facilitating p53's degradation [56]. Plk1 phosphorylates topoisomerase I-binding protein (Topors) at Ser-718 [145]. Topors is a p53 and topoisomerase I binding protein [147] and functions as both ubiquitin and SUMO-1 E3 ligase for p53 [148, 149]. Phosphorylation of Topors on Ser-718 by Plk1 inhibits sumoylation of p53, whereas ubiquitination and subsequent degradation of p53 is enhanced, thereby suppressing p53 function [145]. G2 and S-phase-expressed 1 (GTSE1) is critical for G2 checkpoint recovery [150, 151] and negatively regulates transactivational and apoptotic activity of p53 [150, 152]. Phosphorylation of GTSE1 on Ser-435 by Plk1 promotes its nuclear localization and subsequently, shuttles p53 out from the nucleus to the cytoplasm [151, 152], leading to p53 degradation and inactivation during G2 checkpoint recovery [151]. Plk1, p53, and Cdc25C have shown to form a complex [56, 153]. Plk1 phosphorylates Cdc25C on Ser-198 [132, 154] and presumably, this phosphorylation may contribute to p53 destabilization [56, 153]. Interestingly, there is evidence that p53 can serve as a negative regulator of Plk1 by binding to the promoter of Plk1 and thus inhibiting its activity [62, 63].

The *Plk2* and *Plk3* are serum-inducible immediate early response genes [155] and activated near the G1/S phase transition [118, 156]. Evidence suggested that both *Plk2* and *Plk3* function as tumor suppressors in the p53-mediated signaling pathways to protect cell from DNA damage or oxidative stress (reviewed in [157]). Activation of Plk2 is required for centrosome duplication [156] and may have an important role in replication stress checkpoint signaling through the interaction with Chk1, Chk2, and p53 [158]. Plk2 appears to be a transcriptional target of p53 and its expression is induced after DNA damage in a p53-dependent manner

[159]. Promoter analysis has shown the possible existence of p53 binding homology element (p53RE) in the basal promoter of *Plk2* and furthermore, *Plk2* is transcriptionally regulated by p53RE in human thyroid cells [160].

Plk3 plays an important role in the regulation of mitosis and DNA damage checkpoint [161, 162]. Its kinase activity peaks during late S and G2 phase [116]. The gene expression signature of *Plk3* has shown deregulated expression of *Plk3* in various types of cancers [122, 163], such as head and neck squamous cell carcinomas [164] and colon cancer [165]. Overexpression of Plk3 suppresses cell proliferation [166] and induces chromosome condensation [167]. In response to DNA damage, Plk3 is activated in an ATM-dependent manner [162] and subsequently, mediates ATM-dependent Chk2 phosphorylation and activation [161, 162]. Plk3 also inhibits entry into mitosis by phosphorylating Cdc25C [168, 169] and induces p53-dependent apoptosis [169]. In addition, Plk3 interacts with and phosphorylates p53 at Ser-20 [169], thereby preventing the interaction between p53 and MDM2, with the effect of stabilizing p53.

Plk4 shares relatively little sequence homology with other members of Plks [170]. Plk4 is essential for centrosome duplication [171, 172] and mouse embryonic development [173]. Its protein expression peaks during mitosis [174]. The loss-of-function of Plk4 causes a failure of cell division, possibly leading to aneuploidy and polyploidy, which may in turn contribute to tumorigenesis [171]. Plk4 interacts with proteins involved in the cellular response to DNA damage, such as p53 [175], Cdc25C [176], and Chk2 [177], suggesting that Plk4 may play an important role in the DNA damage response signaling [178]. Plk4 also binds to and phosphorylates p53 [173, 175, 178], possibly affecting protein stability and activity of p53 [178], although phosphorylation site(s) are currently unknown. Overexpression of Plk4 promotes centriole overduplication [172] and is found in human colon cancer [179].

A fifth member of the Plk family, Plk5, is mainly expressed in differentiated tissues, such as the brain, eye, and ovary [180], whereas it is undetectable in proliferating tissues [181]. Plk5 is involved in the process of neurite formation [181] and DNA damage response [123], rather than mitotic process. Nucleotide sequence analysis of *Plk5* shows that the promoter region of *Plk5* contains several p53 binding motifs; however, no such regulatory mechanisms have yet been found [123]. Interestingly, recent studies demonstrated that Plk5 is significantly downregulated by promoter hypermethylation in human brain tumors and its overexpression suppresses cell proliferation and malignant transformation by *Ras* oncogene, suggesting that *Plk5* may function as a tumor suppressor gene in brain cancer [123, 181].

2.3. Budding Uninhibited by Benzimidazoles 1 Homolog (Bub1). Bub1 belongs to a small group of serine/threonine kinases that play multiple roles in chromosome segregation and spindle checkpoint during mitosis [182]. Bub1 was originally identified in genetic screens of *Saccharomyces cerevisiae* along with mitotic arrest-deficient 1, 2, and 3 (Mad1, Mad2, and Mad3 (BubR1) in mammals), Bub3, and Mps1 [183, 184]. All of these proteins play critical roles in the mitotic checkpoint signaling [183, 184]. Deregulated Bub1 expression and its kinase activity have been associated with chromosomal instability, aneuploidy, and several forms of human cancer [185–187]. APC/C is involved in controlling sister chromatid separation and mitotic exit [188]. Bub1 ensures that activation of APC/C is delayed until all the chromosomes have achieved proper bipolar connections to the mitotic spindle, by phosphorylating Cdc20, a key regulator of APC/C activity [189]. Phosphorylation of H2A on Ser-121 by Bub1 in fission yeast prevents chromosome instability via maintenance and localization of Sgo1 (Shugoshin), a protector of centromeric cohesion [190–192]. Bub1 interacts with p53 at kinetochores in response to mitotic spindle damage and negatively regulates p53-mediated cell death [57]. It has shown that SV40 large T antigen (LT) phosphorylates p53 on Ser-37 in a Bub1-binding manner [193]. In addition, purified Bub1 directly phosphorylates p53 on Ser-37 *in vitro*, possibly inducing cellular senescence [193]. An interesting observation has been reported that the loss of both Bub1 and p53 causes a failure in p53-mediated cell death signaling, thereby leading to the accumulation of cells with aneuploidy and polyploidy [194].

3. Positive Regulation of p53 Activation

3.1. Monopolar Spindle 1 (Mps1). Mps1 has an essential role in centrosome duplication, checkpoint signaling, cytokinesis, and development in organisms from yeast to mammalian [195–197]. Kinases structurally related to human Mps1 were identified in various organisms, including Mph1p in *Schizosaccharomyces pombe* [198], PPK1 in *Arabidopsis thaliana* [199], xMps1in *Xenopus laevis* [200] and mMps1 in mouse [201]. Mps1 acts as a dual-specificity protein kinase that can phosphorylate serine/threonine as well as tyrosine residues [198, 202] and is highly expressed during mitosis [203]. Deregulation of Mps1 causes a high frequency of chromosome missegregation and aneuploidy [203, 204] and fails to induce apoptosis in response to spindle damage [196]. The kinase activity of Mps1 is critical for maintaining chromosome stability by phosphorylating other protein substrates [205, 206]. For instance, Mps1 is crucial for Aurora B activity and chromosome alignment by phosphorylating Borealin/Dasra B, a member of CPC that regulates Aurora B [205]. In addition, Mps1 phosphorylates Blm, which is a bloom syndrome product and a member of the RecQ helicases [207], at Ser-144 [206]. Blm phosphorylation by Mps1 is important for the faithful chromosome segregation [206]. Mps1 phosphorylates p53 at Thr-18, and this phosphorylation is critical for the stabilization of p53 by interfering with MDM2 binding [58]. Mps1-mediated p53 phosphorylation is also required for the activation of p53-dependent postmitotic checkpoint [58]; thus, inhibition of Mps1 kinase activity causes a defective postmitotic checkpoint and chromosome instability [58, 208]. These findings suggest that Mps1-mediated phosphorylation and subsequent stabilization of p53 may play an important role in the activation of p53 after spindle damage as well as the prevention of aneuploidy/polyploidy [58, 208]. Interestingly,

TABLE 1: Mitotic kinases-mediated p53 phosphorylation and the possible consequences.

Mitotic kinases	Phosphorylation sites	Outcome	References
Aurora kinases			
Aurora A	Ser-215	Inhibition of DNA binding and transcriptional activity	[85]
	Ser-315	Protein destabilization	[55]
Aurora B	Ser-183	Unknown	[46]
	Ser-269/Thr284	Inhibition of transcriptional activity	
Aurora C		Unknown	
Polo-like kinases			
Plk1	Unknown	Inhibition of transcriptional and proapoptotic activity	[146]
Plk2		Unknown	
Plk3	Ser-20	Protein stabilization	[169]
Plk4	Unknown	Possibly affecting protein stabilization and transcriptional activation	[178]
Plk5		Unknown	
SAC kinases			
Bub1	Ser-37	Possibly inducing cellular senescence	[193]
Mps1	Thr-18	p53 stabilization	[58]
		p53-dependent postmitotic checkpoint activation	
BubR1	Unknown	p53 stabilization	[59]

a recent study shows that increased expression of Mps1 is associated with an increased *p53* mutation, a basal-like phenotype of breast cancer and a poor prognosis outcome [209]. These findings suggest that both the expression and function of Mps1 and p53 are highly correlated and critical for effective and faithful mitosis to maintain genome stability.

3.2. Bub1-Related Kinase 1 (BubR1). BubR1 is the mammalian homolog of yeast Mad3 and Bub1 [185, 210]. It has shown to play an essential role in mitotic checkpoint activation and subsequent apoptotic events to prevent the adaptation of abnormal and unstable mitotic cells with chromosome instability [59, 211]. During mitotic checkpoint activation, BubR1 directly binds to APC/C and Cdc20 and subsequently, inhibits the E3 ligase activity of APC/C by blocking the binding of Cdc20 to APC [212], suggesting that BubR1 plays an essential role in stabilization of kinetochores-microtubule attachment [213]. Several studies have shown that BubR1 deficiency causes a loss of checkpoint control, abnormal mitosis, genomic instability, and tumorigenesis as well as a compromised response to DNA damage [214]. For instance, mice with *BubR1* haploinsufficiency display a genetic instability phenotype due to underlying defects in DNA repair and chromosomal segregation [215]. Moreover, the complete loss of BubR1 leads to early embryonic lethality [216]. The reduced protein level of BubR1 promotes cellular senescence in mouse embryonic fibroblasts [217]. Increasing evidence suggests that a positive regulatory loop between p53 and BubR1 exits [218]. BubR1 interacts with and phosphorylates p53, thereby stabilizing p53 in response to spindle damage [59]. The expression level of p53 protein is reduced in BubR1-deficient cells, possibly leading to malignant transformation [214]. In *p53*-null cells, inhibition of BubR1 expression enhances chromosomal instability and polyploidy; conversely, overexpression of BubR1 restores the

checkpoint function, suppresses centrosome amplification, and selectively eliminates cells with amplified centrosomes [64]. Interestingly, BubR1 transcription and expression are largely controlled by p53 [64].Despite of its important function, mutations of *BubR1* in cancers are very rare [1, 219].

4. Conclusions

Thanks to advances in proteomics technology, many of the substrates for mitotic kinases have been identified, such as those listed above; however, the functional significance of these phosphorylation events has not been explored thoroughly. Therefore, dissecting the functional consequences of mitotic kinase-mediated phosphorylation should be given high priority to better understand their roles in mitosis.

It appears that there is a very well-organized interactive feedback loop between p53 and mitotic kinases in cell cycle progression. p53 tightly and negatively regulates the expression and activity of mitotic kinases, such as Aurora A, Plk1, and Bub1, thereby inhibiting cell proliferation and survival signaling in normal mitosis [61–64]. Protein stability and transcriptional and apoptotic activity of p53 can be also negatively regulated by mitotic kinases-mediated phosphorylation of p53 (summarized in Table 1) [55, 56, 85, 146]. On the other hand, Mps1 and BubR1 are thought to be positive regulators of p53 and may have an important role in antagonizing the function of Aurora kinases, Plk1, and Bub1 in the regulation of p53 signaling during mitosis [209, 220]. When this critical feedback loop is disrupted (e.g., by mutation of *p53* or deregulation of mitotic kinases), p53 cannot be activated when damage occurs to the mitotic spindle, thereby inducing mitotic slippage and preventing apoptosis (Figure 1) [221, 222]. Based on these studies, we speculate that the status of both mitotic kinases and p53 may be critical for cell fate decisions in mitotic cells.

Despite promising preclinical data of targeting mitotic kinases for cancer therapy, many challenges still remain to be overcome, such as safety issues and selection of patient population. Studies have demonstrated that current mitotic inhibitors that target mitotic kinases have major side effects because mitotic kinases are mainly expressed in actively proliferating cells (both normal dividing cells and cancer cells) [2]. Therefore, selecting the right drugs and doses for right patients may be the key to successful cancer therapy.

Studies have shown that depletion/inhibition of Aurora A, Aurora B, Plk1, or Bub1 induces cellular senescence or cell death in a p53-dependent or -independent but p73-dependent manner in many different cell types [217, 223–231]. Importantly, p53-deficient/mutated cells are more sensitive to inhibitors targeting Aurora kinases or Plk1 than cells with wild-type p53, due to a significant increase in cellular senescence and cell death [227, 231, 232], suggesting that patients with p53 deficiency and mutations may benefit from inhibitors targeting Aurora kinases, Plk1, or Bub1. Mps1 and BubR1-mediated p53 phosphorylation are required for p53 activation to properly induce cell death in a p53-dependent manner in response to mitotic spindle damage [58, 59, 209]. Inhibition of Mps1 or BubR1 appears to be disabling a p53-mediated cell death signaling pathway, possibly leading to accumulation of aneuploid/polyploid cells in response to mitotic spindle damage or oncogene-induced DNA damage [59, 217]. Moreover, a recent study shows that depletion/inhibition of Mps1 fails to kill p53-deficient/mutated cells more efficiently than cells expressing wild-type p53 [233], suggesting that Mps1 or BubR1 inhibition may offer a better therapeutic benefit for cancer patients expressing wild-type p53. These finding suggest that the status of p53 is a very attractive maker capable of selecting patients who will benefit from these mitotic kinase inhibitors.

Acknowledgment

The authors thank Dr. M. Denning for helpful discussion and critical reading of the paper.

References

[1] E. A. Nigg, "Mitotic kinases as regulators of cell division and its checkpoints," *Nature Reviews Molecular Cell Biology*, vol. 2, no. 1, pp. 21–32, 2001.

[2] T. L. Schmit and N. Ahmad, "Regulation of mitosis via mitotic kinases: new opportunities for cancer management," *Molecular Cancer Therapeutics*, vol. 6, no. 7, pp. 1920–1931, 2007.

[3] J. A. P. Fidalgo, D. Roda, S. Roselló, E. Rodríguez-Braun, and A. Cervantes, "Aurora kinase inhibitors: a new class of drugs targeting the regulatory mitotic system," *Clinical and Translational Oncology*, vol. 11, no. 12, pp. 787–798, 2009.

[4] A. Janssen, G. J. P. L. Kops, and R. H. Medema, "Elevating the frequency of chromosome mis-segregation as a strategy to kill tumor cells," *Proceedings of the National Academy of Sciences of the United States of America*, vol. 106, no. 45, pp. 19108–19113, 2009.

[5] L. Y. Lu and X. Yu, "The balance of Polo-like kinase 1 in tumorigenesis," *Cell Division*, vol. 4, article no. 4, 2009.

[6] R. D. Gardner and D. J. Burke, "The spindle checkpoint: two transitions, two pathways," *Trends in Cell Biology*, vol. 10, no. 4, pp. 154–158, 2000.

[7] L. H. Hartwell and T. A. Weinert, "Checkpoints: controls that ensure the order of cell cycle events," *Science*, vol. 246, no. 4930, pp. 629–634, 1989.

[8] L. H. Hartwell and M. B. Kastan, "Cell cycle control and cancer," *Science*, vol. 266, no. 5192, pp. 1821–1828, 1994.

[9] J. Basu, H. Bousbaa, E. Logarinho et al., "Mutations in the essential spindle checkpoint gene bub1 cause chromosome missegregation and fail to block apoptosis in *Drosophila*," *Journal of Cell Biology*, vol. 146, no. 1, pp. 13–28, 1999.

[10] M. Malumbres and M. Barbacid, "To cycle or not to cycle: a critical decision in cancer," *Nature Reviews Cancer*, vol. 1, no. 3, pp. 222–231, 2001.

[11] R. Rajaraman, D. L. Guernsey, M. M. Rajaraman, and S. R. Rajaraman, "Stem cells, senescence, neosis and self-renewal in cancer," *Cancer Cell International*, vol. 6, article no. 25, 2006.

[12] X. Fang and P. Zhang, "Aneuploidy and tumorigenesis," *Seminars in Cell and Developmental Biology*, vol. 22, no. 6, pp. 595–601, 2011.

[13] S. Sen, "Aneuploidy and cancer," *Current Opinion in Oncology*, vol. 12, no. 1, pp. 82–88, 2000.

[14] Y. H. Chi and K. T. Jeang, "Aneuploidy and cancer," *Journal of Cellular Biochemistry*, vol. 102, no. 3, pp. 531–538, 2007.

[15] M. Ciciarello, R. Mangiacasale, M. Casenghi et al., "p53 displacement from centrosomes and p53-mediated G1 arrest following transient inhibition of the mitotic spindle," *Journal of Biological Chemistry*, vol. 276, no. 22, pp. 19205–19213, 2001.

[16] S. L. Warner, P. J. Gray, and D. D. Von Hoff, "Tubulin-associated drug targets: Aurora kinases, polo-like kinases, and others," *Seminars in Oncology*, vol. 33, no. 4, pp. 436–448, 2006.

[17] S. Hauf, I. C. Waizenegger, and J. M. Peters, "Cohesin cleavage by separase required for anaphase and cytokinesis in human cells," *Science*, vol. 293, no. 5533, pp. 1320–1323, 2001.

[18] S. Honda, T. Marumoto, T. Hirota et al., "Activation of m-calpain is required for chromosome alignment on the metaphase plate during mitosis," *Journal of Biological Chemistry*, vol. 279, no. 11, pp. 10615–10623, 2004.

[19] H. Hochegger, S. Takeda, and T. Hunt, "Cyclin-dependent kinases and cell-cycle transitions: does one fit all?" *Nature Reviews Molecular Cell Biology*, vol. 9, no. 11, pp. 910–916, 2008.

[20] M. Malumbres and M. Barbacid, "Mammalian cyclin-dependent kinases," *Trends in Biochemical Sciences*, vol. 30, no. 11, pp. 630–641, 2005.

[21] A. Viera, J. S. Rufas, I. Martínez, J. L. Barbero, S. Ortega, and J. A. Suja, "CDK2 is required for proper homologous pairing, recombination and sex-body formation during male mouse meiosis," *Journal of Cell Science*, vol. 122, no. 12, pp. 2149–2159, 2009.

[22] S. L. Harvey, G. Enciso, N. Dephoure, S. P. Gygi, J. Gunawardena, and D. R. Kellogg, "A phosphatase threshold sets the level of Cdk1 activity in early mitosis in budding yeast," *Molecular Biology of the Cell*, vol. 22, no. 19, pp. 3595–3608, 2011.

[23] J. Bloom and F. R. Cross, "Multiple levels of cyclin specificity in cell-cycle control," *Nature Reviews Molecular Cell Biology*, vol. 8, no. 2, pp. 149–160, 2007.

[24] D. O. Morgan, "Principles of CDK regulation," *Nature*, vol. 374, no. 6518, pp. 131–134, 1995.

[25] A. Lindqvist, V. Rodríguez-Bravo, and R. H. Medema, "The decision to enter mitosis: feedback and redundancy in the mitotic entry network," *Journal of Cell Biology*, vol. 185, no. 2, pp. 193–202, 2009.

[26] A. Blangy, H. A. Lane, P. D'Hérin, M. Harper, M. Kress, and E. A. Nigg, "Phosphorylation by p34cdc2 regulates spindle association of human Eg5, a kinesin-related motor essential for bipolar spindle formation in vivo," *Cell*, vol. 83, no. 7, pp. 1159–1169, 1995.

[27] M. Peter, J. Nakagawa, M. Doree, J. C. Labbe, and E. A. Nigg, "In vitro disassembly of the nuclear lamina and M phase-specific phosphorylation of lamins by cdc2 kinase," *Cell*, vol. 61, no. 4, pp. 591–602, 1990.

[28] K. Kimura, M. Hirano, R. Kobayashi, and T. Hirano, "Phosphorylation and activation of 13S condensin by Cdc2 in vitro," *Science*, vol. 282, no. 5388, pp. 487–490, 1998.

[29] K. U. Hong, H. J. Kim, H. S. Kim et al., "Cdk1-cyclin B1-mediated phosphorylation of tumor-associated microtubule-associated protein/cytoskeleton-associated protein 2 in mitosis," *Journal of Biological Chemistry*, vol. 284, no. 24, pp. 16501–16512, 2009.

[30] P. Salaun, Y. Rannou, and P. Claude, "Cdk1, plks, auroras, and neks: the mitotic bodyguards," *Advances in Experimental Medicine and Biology*, vol. 617, pp. 41–56, 2008.

[31] C. Acquaviva and J. Pines, "The anaphase-promoting complex/cyclosome: APC/C," *Journal of Cell Science*, vol. 119, no. 12, pp. 2401–2404, 2006.

[32] A. D. Rudner and A. W. Murray, "Phosphorylation by Cdc28 activates the Cdc20-dependent activity of the anaphase-promoting complex," *Journal of Cell Biology*, vol. 149, no. 7, pp. 1377–1390, 2000.

[33] P. Clute and J. Pines, "Temporal and spatial control of cyclin B1 destruction in metaphase," *Nature Cell Biology*, vol. 1, no. 2, pp. 82–87, 1999.

[34] M. Malumbres and M. Barbacid, "Cell cycle kinases in cancer," *Current Opinion in Genetics and Development*, vol. 17, no. 1, pp. 60–65, 2007.

[35] S. M. A. Lens, E. E. Voest, and R. H. Medema, "Shared and separate functions of polo-like kinases and aurora kinases in cancer," *Nature Reviews Cancer*, vol. 10, no. 12, pp. 825–841, 2010.

[36] H. T. Ma and R. Y. C. Poon, "How protein kinases co-ordinate mitosis in animal cells," *Biochemical Journal*, vol. 435, no. 1, pp. 17–31, 2011.

[37] N. Takai, R. Hamanaka, J. Yoshimatsu, and I. Miyakawa, "Polo-like kinases (Plks) and cancer," *Oncogene*, vol. 24, no. 2, pp. 287–291, 2005.

[38] J. Fu, M. Bian, Q. Jiang, and C. Zhang, "Roles of aurora kinases in mitosis and tumorigenesis," *Molecular Cancer Research*, vol. 5, no. 1, pp. 1–10, 2007.

[39] A. Tritarelli, E. Oricchio, M. Ciciarello et al., "p53 localization at centrosomes during mitosis and postmitotic checkpoint are ATM-dependent and require Serine 15 phosphorylation," *Molecular Biology of the Cell*, vol. 15, no. 8, pp. 3751–3757, 2004.

[40] S. Jin, H. Gao, L. Mazzacurati et al., "BRCA1 interaction of centrosomal protein Nlp is required for successful mitotic progression," *Journal of Biological Chemistry*, vol. 284, no. 34, pp. 22970–22977, 2009.

[41] K. H. Vousden and X. Lu, "Live or let die: the cell's response to p53," *Nature Reviews Cancer*, vol. 2, no. 8, pp. 594–604, 2002.

[42] Y. Tang, J. Luo, W. Zhang, and W. Gu, "Tip60-dependent acetylation of p53 modulates the decision between cell-cycle arrest and apoptosis," *Molecular Cell*, vol. 24, no. 6, pp. 827–839, 2006.

[43] M. Oren, "Decision making by p53: life, death and cancer," *Cell Death and Differentiation*, vol. 10, no. 4, pp. 431–442, 2003.

[44] M. Oren and V. Rotter, "Mutant p53 gain-of-function in cancer," *Cold Spring Harbor perspectives in biology*, vol. 2, no. 2, article a001107, 2010.

[45] L. J. Ko and C. Prives, "p53: puzzle and paradigm," *Genes and Development*, vol. 10, no. 9, pp. 1054–1072, 1996.

[46] L. Wu, C. A. Ma, Y. Zhao, and A. Jain, "Aurora B interacts with NIR-p53, leading to p53 phosphorylation in its DNA-binding domain and subsequent functional suppression," *Journal of Biological Chemistry*, vol. 286, no. 3, pp. 2236–2244, 2011.

[47] S. Y. Shieh, M. Ikeda, Y. Taya, and C. Prives, "DNA damage-induced phosphorylation of p53 alleviates inhibition by MDM2," *Cell*, vol. 91, no. 3, pp. 325–334, 1997.

[48] V. Böttger, A. Böttger, C. Garcia-Echeverria et al., "Comparative study of the p53-mdm2 and p53-MDMX interfaces," *Oncogene*, vol. 18, no. 1, pp. 189–199, 1999.

[49] T. Unger, T. Juven-Gershon, E. Moallem et al., "Critical role for Ser20 of human p53 in the negative regulation of p53 by Mdm2," *EMBO Journal*, vol. 18, no. 7, pp. 1805–1814, 1999.

[50] F. Toledo and G. M. Wahl, "Regulating the p53 pathway: in vitro hypotheses, in vivo veritas," *Nature Reviews Cancer*, vol. 6, no. 12, pp. 909–923, 2006.

[51] Y. Xia, R. C. Padre, T. H. De Mendoza, V. Bottero, V. B. Tergaonkar, and I. M. Verma, "Phosphorylation of p53 by IκB kinase 2 promotes its degradation by β-TrCP," *Proceedings of the National Academy of Sciences of the United States of America*, vol. 106, no. 8, pp. 2629–2634, 2009.

[52] H. H. Li, A. G. Li, H. M. Sheppard, and X. Liu, "Phosphorylation on Thr-55 by TAF1 mediates degradation of p53: a role for TAF1 in cell G1 progression," *Molecular Cell*, vol. 13, no. 6, pp. 867–878, 2004.

[53] J. S. Lanni and T. Jacks, "Characterization of the p53-dependent postmitotic checkpoint following spindle disruption," *Molecular and Cellular Biology*, vol. 18, no. 2, pp. 1055–1064, 1998.

[54] A. J. Minn, L. H. Boise, and C. B. Thompson, "Expression of Bcl-x(L) and loss of p53 can cooperate to overcome a cell cycle checkpoint induced by mitotic spindle damage," *Genes and Development*, vol. 10, no. 20, pp. 2621–2631, 1996.

[55] H. Katayama, K. Sasai, H. Kawai et al., "Phosphorylation by aurora kinase A induces Mdm2-mediated destabilization and inhibition of p53," *Nature Genetics*, vol. 36, no. 1, pp. 55–62, 2004.

[56] J. Chen, G. Dai, Y. Q. Wang et al., "Polo-like kinase 1 regulates mitotic arrest after UV irradiation through dephosphorylation of p53 and inducing p53 degradation," *FEBS Letters*, vol. 580, no. 15, pp. 3624–3630, 2006.

[57] F. Gao, J. F. Ponte, M. Levy et al., "hBub1 negatively regulates p53 mediated early cell death upon mitotic checkpoint activation," *Cancer Biology and Therapy*, vol. 8, no. 7, 2009.

[58] Y. F. Huang, M. D. T. Chang, and S. Y. Shieh, "TTK/hMps1 mediates the p53-dependent postmitotic checkpoint by phosphorylating p53 at Thr18," *Molecular and Cellular Biology*, vol. 29, no. 11, pp. 2935–2944, 2009.

[59] G. H. Ha, K. H. Baek, H. S. Kim et al., "p53 activation in response to mitotic spindle damage requires signaling via

BubR1-mediated phosphorylation," *Cancer Research*, vol. 67, no. 15, pp. 7155–7164, 2007.

[60] L. J. Warnock, S. A. Raines, and J. Milner, "Aurora A mediates cross-talk between N- and C-terminal post-translational modifications of p53," *Cancer Biology and Therapy*, vol. 12, no. 12, pp. 1059–1068, 2011.

[61] P. A. Eyers and J. L. Maller, "Regulation of Xenopus Aurora A Activation by TPX2," *Journal of Biological Chemistry*, vol. 279, no. 10, pp. 9008–9015, 2004.

[62] P. S. Kho, Z. Wang, L. Zhuang et al., "p53-regulated transcriptional program associated with genotoxic stress-induced apoptosis," *Journal of Biological Chemistry*, vol. 279, no. 20, pp. 21183–21192, 2004.

[63] L. McKenzie, S. King, L. Marcar et al., "p53-dependent repression of polo-like kinase-1 (PLK1)," *Cell Cycle*, vol. 9, no. 20, pp. 4200–4212, 2010.

[64] T. Oikawa, M. Okuda, Z. Ma et al., "Transcriptional control of BubR1 by p53 and suppression of centrosome amplification by BubR1," *Molecular and Cellular Biology*, vol. 25, no. 10, pp. 4046–4061, 2005.

[65] D. M. Glover, M. H. Leibowitz, D. A. McLean, and H. Parry, "Mutations in aurora prevent centrosome separation leading to the formation of monopolar spindles," *Cell*, vol. 81, no. 1, pp. 95–105, 1995.

[66] C. S. M. Chan and D. Botstein, "Isolation and characterization of chromosome-gain and increase-in-ploidy mutants in yeast," *Genetics*, vol. 135, no. 3, pp. 677–691, 1993.

[67] S. Buvelot, S. Y. Tatsutani, D. Vermaak, and S. Biggins, "The budding yeast Ipl1/Aurora protein kinase regulates mitotic spindle disassembly," *Journal of Cell Biology*, vol. 160, no. 3, pp. 329–339, 2003.

[68] S. Biggins and A. W. Murray, "The budding yeast protein kinase Ipl1/Aurora allows the absence of tension to activate the spindle checkpoint," *Genes and Development*, vol. 15, no. 23, pp. 3118–3129, 2001.

[69] J. M. Schumacher, N. Ashcroft, P. J. Donovan, and A. Golden, "A highly conserved centrosomal kinase, AIR-1, is required for accurate cell cycle progression and segregation of developmental factors in *Caenorhabditis elegans* embryos," *Development*, vol. 125, no. 22, pp. 4391–4402, 1998.

[70] J. M. Schumacher, A. Golden, and P. J. Donovan, "AIR-2: an Aurora/Ipl1-related protein kinase associated with chromosomes and midbody microtubules is required for polar body extrusion and cytokinesis in *Caenorhabditis elegans* embryos," *Journal of Cell Biology*, vol. 143, no. 6, pp. 1635–1646, 1998.

[71] R. Glet and C. Prigent, "The non-catalytic domain of the *Xenopus laevis* aurora A kinase localises the protein to the centrosome," *Journal of Cell Science*, vol. 114, no. 11, pp. 2095–2104, 2001.

[72] M. Carmena, S. Ruchaud, and W. C. Earnshaw, "Making the Auroras glow: regulation of Aurora A and B kinase function by interacting proteins," *Current Opinion in Cell Biology*, vol. 21, no. 6, pp. 796–805, 2009.

[73] P. Meraldi, R. Honda, and E. A. Nigg, "Aurora kinases link chromosome segregation and cell division to cancer susceptibility," *Current Opinion in Genetics and Development*, vol. 14, no. 1, pp. 29–36, 2004.

[74] H. Zhou, J. Kuang, L. Zhong et al., "Tumour amplified kinase STK15/BTAK induces centrosome amplification, aneuploidy and transformation," *Nature Genetics*, vol. 20, no. 2, pp. 189–193, 1998.

[75] S. Sen, H. Zhou, and R. A. White, "A putative serine/threonine kinase encoding gene BTAK on chromosome 20q13 is amplified and overexpressed in human breast cancer cell lines," *Oncogene*, vol. 14, no. 18, pp. 2195–2200, 1997.

[76] J. R. Bischoff, L. Anderson, Y. Zhu et al., "A homologue of *Drosophila* aurora kinase is oncogenic and amplified in human colorectal cancers," *EMBO Journal*, vol. 17, no. 11, pp. 3052–3065, 1998.

[77] A. A. Dar, L. W. Goff, S. Majid, J. Berlin, and W. El-Rifai, "Aurora kinase inhibitors—rising stars in cancer therapeutics?" *Molecular Cancer Therapeutics*, vol. 9, no. 2, pp. 268–278, 2010.

[78] K. Sasai, J. M. Parant, M. E. Brandt et al., "Targeted disruption of Aurora A causes abnormal mitotic spindle assembly, chromosome misalignment and embryonic lethality," *Oncogene*, vol. 27, no. 29, pp. 4122–4127, 2008.

[79] C. A. Johnston, K. Hirono, K. E. Prehoda, and C. Q. Doe, "Identification of an Aurora-A/Pins^LINKER/ Dlg spindle orientation pathway using induced cell polarity in S2 cells," *Cell*, vol. 138, no. 6, pp. 1150–1163, 2009.

[80] E. Kress and M. Gotta, "Aurora A in cell division: kinase activity not required," *Nature Cell Biology*, vol. 13, no. 6, pp. 638–639, 2011.

[81] L. Macůrek, A. Lindqvist, D. Lim et al., "Polo-like kinase-1 is activated by aurora A to promote checkpoint recovery," *Nature*, vol. 455, no. 7209, pp. 119–123, 2008.

[82] A. Seki, J. A. Coppinger, C. Y. Jang, J. R. Yates, and G. Fang, "Bora and the kinase Aurora A cooperatively activate the kinase Plk1 and control mitotic entry," *Science*, vol. 320, no. 5883, pp. 1655–1658, 2008.

[83] R. D. Van Horn, S. Chu, L. Fan et al., "Cdk1 activity is required for mitotic activation of Aurora A during G 2/M transition of human cells," *Journal of Biological Chemistry*, vol. 285, no. 28, pp. 21849–21857, 2010.

[84] P. Meraldi, R. Honda, and E. A. Nigg, "Aurora-A overexpression reveals tetraploidization as a major route to centrosome amplification in p53-/- cells," *EMBO Journal*, vol. 21, no. 4, pp. 483–492, 2002.

[85] Q. Liu, S. Kaneko, L. Yang et al., "Aurora-A abrogation of p53 DNA binding and transactivation activity by phosphorylation of serine 215," *Journal of Biological Chemistry*, vol. 279, no. 50, pp. 52175–52182, 2004.

[86] E. N. Pugacheva and E. A. Golemis, "HEF1-Aurora A interactions: points of dialog between the cell cycle and cell attachment signaling networks," *Cell Cycle*, vol. 5, no. 4, pp. 384–391, 2006.

[87] R. Bayliss, T. Sardon, I. Vernos, and E. Conti, "Structural basis of Aurora-A activation by TPX2 at the mitotic spindle," *Molecular Cell*, vol. 12, no. 4, pp. 851–862, 2003.

[88] T. Hirota, N. Kunitoku, T. Sasayama et al., "Aurora-A and an interacting activator, the LIM protein Ajuba, are required for mitotic commitment in human cells," *Cell*, vol. 114, no. 5, pp. 585–598, 2003.

[89] M. Ouchi, N. Fujiuchi, K. Sasai et al., "BRCA1 phosphorylation by Aurora-A in the regulation of G2 to M transition," *Journal of Biological Chemistry*, vol. 279, no. 19, pp. 19643–19648, 2004.

[90] K. Kinoshita, T. L. Noetzel, L. Pelletier et al., "Aurora A phosphorylation of TACC3/maskin is required for centrosome-dependent microtubule assembly in mitosis," *Journal of Cell Biology*, vol. 170, no. 7, pp. 1047–1055, 2005.

[91] A. Moumen, P. Masterson, M. J. O'Connor, and S. P. Jackson, "hnRNP K: an HDM2 target and transcriptional coactivator

of p53 in response to DNA damage," *Cell*, vol. 123, no. 6, pp. 1065–1078, 2005.

[92] K.-W. Hsueh, S.-L. Fu, C.-Y. F. Huang, and C.-H. Lin, "Aurora-A phosphorylates hnRNPK and disrupts its interaction with p53," *FEBS Letters*, vol. 585, no. 17, pp. 2671–2675, 2011.

[93] G. Vader, R. H. Medema, and S. M. A. Lens, "The chromosomal passenger complex: guiding Aurora-B through mitosis," *Journal of Cell Biology*, vol. 173, no. 6, pp. 833–837, 2006.

[94] S. Ruchaud, M. Carmena, and W. C. Earnshaw, "The chromosomal passenger complex: one for all and all for one," *Cell*, vol. 131, no. 2, pp. 230–231, 2007.

[95] H. B. Mistry, D. E. MacCallum, R. C. Jackson, M. A. J. Chaplain, and F. A. Davidson, "Modeling the temporal evolution of the spindle assembly checkpoint and role of Aurora B kinase," *Proceedings of the National Academy of Sciences of the United States of America*, vol. 105, no. 51, pp. 20215–20220, 2008.

[96] M. J. Kallio, M. L. McCleland, P. T. Stukenberg, and G. J. Gorbsky, "Inhibition of Aurora B kinase blocks chromosome segregation, overrides the spindle checkpoint, and perturbs microtubule dynamics in mitosis," *Current Biology*, vol. 12, no. 11, pp. 900–905, 2002.

[97] C. Ditchfield, V. L. Johnson, A. Tighe et al., "Aurora B couples chromosome alignment with anaphase by targeting BubR1, Mad2, and Cenp-E to kinetochores," *Journal of Cell Biology*, vol. 161, no. 2, pp. 267–280, 2003.

[98] S. Hauf, R. W. Cole, S. La Terra et al., "The small molecule Hesperadin reveals a role for Aurora B in correcting kinetochore-microtubule attachment and in maintaining the spindle assembly checkpoint," *Journal of Cell Biology*, vol. 161, no. 2, pp. 281–294, 2003.

[99] T. C. Tseng, "Protein kinase profile of sperm and eggs: cloning and characterization of two novel testis-specific protein kinases (AIE1, AIE2) related to yeast and fly chromosome segregation regulators," *DNA and Cell Biology*, vol. 17, no. 10, pp. 823–833, 1998.

[100] H. Katayama, W. R. Brinkley, and S. Sen, "The Aurora kinases: role in cell transformation and tumorigenesis," *Cancer and Metastasis Reviews*, vol. 22, no. 4, pp. 451–464, 2003.

[101] J. Khan, F. Ezan, J.-Y. Crémet et al., "Overexpression of active aurora-C kinase results in cell transformation and tumour formation," *PLoS One*, vol. 6, no. 10, Article ID e26512, 2011.

[102] M. Bernard, P. Sanseau, C. Henry, A. Couturier, and C. Prigent, "Cloning of STK13, a third human protein kinase related to *Drosophila* Aurora and budding yeast Ipl1 that maps on chromosome 19q13.3-ter," *Genomics*, vol. 53, no. 3, pp. 406–409, 1998.

[103] K. T. Yang, S. K. Li, C. C. Chang et al., "Aurora-C kinase deficiency causes cytokinesis failure in meiosis I and production of large polyploid oocytes in mice," *Molecular Biology of the Cell*, vol. 21, no. 14, pp. 2371–2383, 2010.

[104] K. Dieterich, R. Zouari, R. Harbuz et al., "The Aurora Kinase C c.144delC mutation causes meiosis I arrest in men and is frequent in the North African population," *Human Molecular Genetics*, vol. 18, no. 7, pp. 1301–1309, 2009.

[105] S. D. Slattery, M. A. Mancini, B. R. Brinkley, and R. M. Hall, "Aurora-C kinase supports mitotic progression in the absence of Aurora-B," *Cell Cycle*, vol. 8, no. 18, pp. 2984–2994, 2009.

[106] H. L. Chen, C. J. C. Tang, C. Y. Chen, and T. K. Tang, "Overexpression of an Aurora-C kinase-deficient mutant disrupts

the Aurora-B/INCENP complex and induces polyploidy," *Journal of Biomedical Science*, vol. 12, no. 2, pp. 297–310, 2005.

[107] F. A. Barr, H. H. W. Silljé, and E. A. Nigg, "Polo-like kinases and the orchestration of cell division," *Nature Reviews Molecular Cell Biology*, vol. 5, no. 6, pp. 429–440, 2004.

[108] C. E. Sunkel and D. M. Glover, "polo, a mitotic mutant of *Drosophila* displaying abnormal spindle poles," *Journal of Cell Science*, vol. 89, p. 1, 1988.

[109] K. Kitada, A. L. Johnson, L. H. Johnston, and A. Sugino, "A multicopy suppressor gene of the Saccharomyces cerevisiae G1 cell cycle mutant gene dbf4 encodes a protein kinase and is identified as CDC5," *Molecular and Cellular Biology*, vol. 13, no. 7, pp. 4445–4457, 1993.

[110] H. Ohkura, I. M. Hagan, and D. M. Glover, "The conserved Schizosaccharomyces pombe kinase plo1, required to form a bipolar spindle, the actin ring, and septum, can drive septum formation in G1 and G2 cells," *Genes and Development*, vol. 9, no. 9, pp. 1059–1073, 1995.

[111] D. Chase, C. Serafinas, N. Ashcroft et al., "The Polo-like kinase PLK-1 is required for nuclear envelope breakdown and the completion of meiosis in *Caenorhabditis elegans*," *Genesis*, vol. 26, no. 1, pp. 26–41, 2000.

[112] D. Chase, A. Golden, G. Heidecker, and D. K. Ferris, "*Caenorhabditis elegans* contains a third polo-like kinase gene," *Mitochondrial DNA*, vol. 11, no. 3-4, pp. 327–334, 2000.

[113] P. Descombes and E. A. Nigg, "The polo-like kinase Plx1 is required for M phase exit and destruction of mitotic regulators in Xenopus egg extracts," *EMBO Journal*, vol. 17, no. 5, pp. 1328–1335, 1998.

[114] A. Kumagai and W. G. Dunphy, "Purification and molecular cloning of Plx1, a Cdc25-regulatory kinase from Xenopus egg extracts," *Science*, vol. 273, no. 5280, pp. 1377–1380, 1996.

[115] P. I. Duncan, N. Pollet, C. Niehrs, and E. A. Nigg, "Cloning and characterization of Plx2 and Plx3, two additional Polo-like kinases from *Xenopus laevis*," *Experimental Cell Research*, vol. 270, no. 1, pp. 78–87, 2001.

[116] Q. Wang, S. Xie, J. Chen et al., "Cell cycle arrest and apoptosis induced by human Polo-like kinase 3 is mediated through perturbation of microtubule integrity," *Molecular and Cellular Biology*, vol. 22, no. 10, pp. 3450–3459, 2002.

[117] G. Kauselmann, M. Weiler, P. Wulff et al., "The polo-like protein kinases Fnk and Snk associate with a Ca^{2+}- and integrin-binding protein and are regulated dynamically with synaptic plasticity," *EMBO Journal*, vol. 18, no. 20, pp. 5528–5539, 1999.

[118] S. Ma, M. A. Liu, Y. L. O. Yuan, and R. L. Erikson, "The serum-inducible protein kinase Snk is a G1 phase polo-like kinase that is inhibited by the calcium- and integrin-binding protein CIB," *Molecular Cancer Research*, vol. 1, no. 5, pp. 376–384, 2003.

[119] C. Fode, B. Motro, S. Yousefi, M. Heffernan, and J. W. Dennis, "Sak, a murine protein-serine/threonine kinase that is related to the *Drosophila* polo kinase and involved in cell proliferation," *Proceedings of the National Academy of Sciences of the United States of America*, vol. 91, no. 14, pp. 6388–6392, 1994.

[120] R. M. Golsteyn, S. J. Schultz, J. Bartek, A. Ziemiecki, T. Ried, and E. A. Nigg, "Cell cycle analysis and chromosomal localization of human Plk1, a putative homologue of the mitotic kinases *Drosophila* polo and Saccharomyces cerevisiae Cdc5," *Journal of Cell Science*, vol. 107, no. 6, pp. 1509–1517, 1994.

[121] R. Hamanaka, S. Maloid, M. R. Smith, C. D. O'Connell, D. L. Longo, and D. K. Ferris, "Cloning and characterization of human and murine homologues of the *Drosophila* polo serine-threonine kinase," *Cell Growth and Differentiation*, vol. 5, no. 3, pp. 249–257, 1994.

[122] B. Li, B. Ouyang, H. Pan et al., "prk, A cytokine-inducible human protein serine/threonine kinase whose expression appears to be down-regulated in lung carcinomas," *Journal of Biological Chemistry*, vol. 271, no. 32, pp. 19402–19408, 1996.

[123] Z. Andrysik, W. Z. Bernstein, L. Deng et al., "The novel mouse Polo-like kinase 5 responds to DNA damage and localizes in the nucleolus," *Nucleic Acids Research*, vol. 38, no. 9, Article ID gkq011, pp. 2931–2943, 2010.

[124] U. Holtrich, G. Wolf, A. Brauninger et al., "Induction and down-regulation of PLK, a human serine/threonine kinase expressed in proliferating cells and tumors," *Proceedings of the National Academy of Sciences of the United States of America*, vol. 91, no. 5, pp. 1736–1740, 1994.

[125] U. Holtrich, G. Wolf, J. Yuan et al., "Adhesion induced expression of the serine/threonine kinase Fnk in human macrophages," *Oncogene*, vol. 19, no. 42, pp. 4832–4839, 2000.

[126] D. M. Lowery, D. Lim, and M. B. Yaffe, "Structure and function of Polo-like kinases," *Oncogene*, vol. 24, no. 2, pp. 248–259, 2005.

[127] F. Eckerdt and K. Strebhardt, "Polo-like kinase 1: target and regulator of anaphase-promoting complex/cyclosome-dependent proteolysis," *Cancer Research*, vol. 66, no. 14, pp. 6895–6898, 2006.

[128] R. Hamanaka, M. R. Smith, P. M. O'Connor et al., "Polo-like kinase is a cell cycle-regulated kinase activated during mitosis," *Journal of Biological Chemistry*, vol. 270, no. 36, pp. 21086–21091, 1995.

[129] F. Eckerdt, J. Yuan, and K. Strebhardt, "Polo-like kinases and oncogenesis," *Oncogene*, vol. 24, no. 2, pp. 267–276, 2005.

[130] S. Reagan-Shaw and N. Ahmad, "Polo-like kinase (Plk) 1 as a target for prostate cancer management," *IUBMB Life*, vol. 57, no. 10, pp. 677–682, 2005.

[131] W. Zhang, L. Fletcher, and R. J. Muschel, "The role of polo-like kinase 1 in the inhibition of centrosome separation after ionizing radiation," *Journal of Biological Chemistry*, vol. 280, no. 52, pp. 42994–42999, 2005.

[132] A. K. Roshak, E. A. Capper, C. Imburgia, J. Fornwald, G. Scott, and L. A. Marshall, "The human polo-like kinase, PLK, regulates cdc2/cyclin B through phosphorylation and activation of the cdc25C phosphatase," *Cellular Signalling*, vol. 12, no. 6, pp. 405–411, 2000.

[133] M. Carmena, M. G. Riparbelli, G. Minestrini et al., "*Drosophila* polo kinase is required for cytokinesis," *Journal of Cell Biology*, vol. 143, no. 3, pp. 659–671, 1998.

[134] A. G. Renner, C. Dos Santos, C. Recher et al., "Polo-like kinase 1 is overexpressed in acute myeloid leukemia and its inhibition preferentially targets the proliferation of leukemic cells," *Blood*, vol. 114, no. 3, pp. 659–662, 2009.

[135] M. Schmidt, H. P. Hofmann, K. Sanders, G. Sczakiel, T. L. Beckers, and V. Gekeler, "Molecular alterations after Polo-like kinase 1 mRNA suppression versus pharmacologic inhibition in cancer cells," *Molecular Cancer Therapeutics*, vol. 5, no. 4, pp. 809–817, 2006.

[136] M. Raab, S. Kappel, A. Krämer et al., "Toxicity modelling of Plk1-targeted therapies in genetically engineered mice and cultured primary mammalian cells," *Nature Communications*, vol. 2, no. 1, article no. 395, 2011.

[137] M. R. Smith, M. L. Wilson, R. Hamanaka et al., "Malignant transformation of mammalian cells initiated by constitutive expression of the Polo-like kinase," *Biochemical and Biophysical Research Communications*, vol. 234, no. 2, pp. 397–405, 1997.

[138] K. Strebhardt, L. Kneisel, C. Linhart, A. Bernd, and R. Kaufmann, "Prognostic value of pololike kinase expression in melanomas," *Journal of the American Medical Association*, vol. 283, no. 4, pp. 479–480, 2000.

[139] L. Kneisel, K. Strebhardt, A. Bernd, M. Wolter, A. Binder, and R. Kaufmann, "Expression of polo-like kinase (PLK1) in thin melanomas: a novel marker of metastatic disease," *Journal of Cutaneous Pathology*, vol. 29, no. 6, pp. 354–358, 2002.

[140] G. Wolf, R. Elez, A. Doermer et al., "Prognostic significance of polo-like kinase (PLK) expression in non-small cell lung cancer," *Oncogene*, vol. 14, no. 5, pp. 543–549, 1997.

[141] R. Knecht, R. Elez, M. Oechler, C. Solbach, C. Von Ilberg, and K. Strebhardt, "Prognostic significance of polo-like kinase (PLK) expression in squamous cell carcinomas of the head and neck," *Cancer Research*, vol. 59, no. 12, pp. 2794–2797, 1999.

[142] R. Knecht, C. Oberhauser, and K. Strebhardt, "PLK (polo-like kinase), a new prognostic marker for oropharyngeal carcinomas," *International Journal of Cancer*, vol. 89, no. 6, pp. 535–536, 2000.

[143] G. Wolff, R. Hildenbrand, C. Schwar et al., "Polo-like kinase: a novel marker of proliferation: correlation with estrogen-receptor expression in human breast cancer," *Pathology Research and Practice*, vol. 196, no. 11, pp. 753–759, 2000.

[144] W. Weichert, C. Denkert, M. Schmidt et al., "Polo-like kinase isoform expression is a prognostic factor in ovarian carcinoma," *British Journal of Cancer*, vol. 90, no. 4, pp. 815–821, 2004.

[145] X. Yang, H. Li, Z. Zhou et al., "Plk1-mediated phosphorylation of topors regulates p53 stability," *Journal of Biological Chemistry*, vol. 284, no. 28, pp. 18588–18592, 2009.

[146] K. Ando, T. Ozaki, H. Yamamoto et al., "Polo-like kinase 1 (Plk1) inhibits p53 function by physical interaction and phosphorylation," *Journal of Biological Chemistry*, vol. 279, no. 24, pp. 25549–25561, 2004.

[147] S. Weger, E. Hammer, and R. Heilbronn, "Topors, a p53 and topoisomerase I binding protein, interacts with the adeno-associated virus (AAV-2) Rep78/68 proteins and enhances AAV-2 gene expression," *Journal of General Virology*, vol. 83, no. 3, pp. 511–516, 2002.

[148] R. Rajendra, D. Malegaonkar, P. Pungaliya et al., "Topors functions as an E3 ubiquitin ligase with specific E2 enzymes and ubiquitinates p53," *Journal of Biological Chemistry*, vol. 279, no. 35, pp. 36440–36444, 2004.

[149] E. Hammer, R. Heilbronn, and S. Weger, "The E3 ligase Topors induces the accumulation of polysumoylated forms of DNA topoisomerase I in vitro and in vivo," *FEBS Letters*, vol. 581, no. 28, pp. 5418–5424, 2007.

[150] M. Monte, R. Benetti, G. Buscemi, P. Sandy, G. Del Sal, and C. Schneider, "The cell cycle-regulated protein human GTSE-1 controls DNA damage-induced apoptosis by affecting p53 function," *Journal of Biological Chemistry*, vol. 278, no. 32, pp. 30356–30364, 2003.

[151] X. S. Liu, H. Li, B. Song, and X. Liu, "Polo-like kinase 1 phosphorylation of G2 and S-phase-expressed 1 protein is essential for p53 inactivation during G2 checkpoint recovery," *EMBO Reports*, vol. 11, no. 8, pp. 626–632, 2010.

[152] M. Monte, R. Benetti, L. Collavin, L. Marchionni, G. Del Sal, and C. Schneider, "hGTSE-1 expression stimulates cytoplasmic localization of p53," *Journal of Biological Chemistry*, vol. 279, no. 12, pp. 11744–11752, 2004.

[153] T. L. Schmit, W. Zhong, M. Nihal, and N. Ahmad, "Polo-like kinase 1 (Plk1) in non-melanoma skin cancers," *Cell Cycle*, vol. 8, no. 17, pp. 2697–2702, 2009.

[154] F. Toyoshima-Morimoto, E. Taniguchi, and E. Nishida, "Plk1 promotes nuclear translocation of human Cdc25C during prophase," *EMBO Reports*, vol. 3, no. 4, pp. 341–348, 2002.

[155] D. L. Simmons, B. G. Neel, R. Stevens, G. Evett, and R. L. Erikson, "Identification of an early-growth-response gene encoding a novel putative protein kinase," *Molecular and Cellular Biology*, vol. 12, no. 9, pp. 4164–4169, 1992.

[156] S. Warnke, S. Kemmler, R. S. Hames et al., "Polo-like kinase-2 is required for centriole duplication in mammalian cells," *Current Biology*, vol. 14, no. 13, pp. 1200–1207, 2004.

[157] K. Strebhardt, "Multifaceted polo-like kinases: drug targets and antitargets for cancer therapy," *Nature Reviews Drug Discovery*, vol. 9, no. 8, pp. 643–660, 2010.

[158] E. M. Matthew, T. J. Yen, D. T. Dicker et al., "Replication stress, defective S-phase checkpoint and increased death in Plk2-deficient human cancer cells," *Cell Cycle*, vol. 6, no. 20, pp. 2571–2578, 2007.

[159] T. F. Burns, P. Fei, K. A. Scata, D. T. Dicker, and W. S. El-Deiry, "Silencing of the novel p53 target gene Snk/Plk2 leads to mitotic catastrophe in paclitaxel (Taxol)-exposed cells," *Molecular and Cellular Biology*, vol. 23, no. 16, pp. 5556–5571, 2003.

[160] Y. Shimizu-Yoshida, K. Sugiyama, T. Rogounovitch et al., "Radiation-inducible hSNK gene is transcriptionally regulated by p53 binding homology element in human thyroid cells," *Biochemical and Biophysical Research Communications*, vol. 289, no. 2, pp. 491–498, 2001.

[161] E. M. Bahassi, C. W. Conn, D. L. Myer et al., "Mammalian Polo-like kinase 3 (Plk3) is a multifunctional protein involved in stress response pathways," *Oncogene*, vol. 21, no. 43, pp. 6633–6640, 2002.

[162] E. M. Bahassi, D. L. Myer, R. J. McKenney, R. F. Hennigan, and P. J. Stambrook, "Priming phosphorylation of Chk2 by polo-like kinase 3 (Plk3) mediates its full activation by ATM and a downstream checkpoint in response to DNA damage," *Mutation Research - Fundamental and Molecular Mechanisms of Mutagenesis*, vol. 596, no. 1-2, pp. 166–176, 2006.

[163] J. A. Winkles and G. F. Alberts, "Differential regulation of polo-like kinase 1, 2, 3, and 4 gene expression in mammalian cells and tissues," *Oncogene*, vol. 24, no. 2, pp. 260–266, 2005.

[164] W. Dai, Y. Li, B. Ouyang et al., "PRK, a cell cycle gene localized to 8p21, is downregulated in head and neck cancer," *Genes Chromosomes and Cancer*, vol. 27, no. 3, pp. 332–336, 2000.

[165] W. Dai, T. Liu, Q. Wang, C. V. Rao, and B. S. Reddy, "Downregulation of PLK3 gene expression by types and amount of dietary fat in rat colon tumors," *International journal of oncology*, vol. 20, no. 1, pp. 121–126, 2002.

[166] M. Iida, T. Sasaki, and H. Komatani, "Overexpression of Plk3 causes morphological change and cell growth suppression in ras pathway-activated cells," *Journal of Biochemistry*, vol. 146, no. 4, pp. 501–507, 2009.

[167] C. W. Conn, R. F. Hennigan, W. Dai, Y. Sanchez, and P. J. Stambrook, "Incomplete cytokinesis and induction of apoptosis by overexpression of the mammalian polo-like kinase, Plk31," *Cancer Research*, vol. 60, no. 24, pp. 6826–6831, 2000.

[168] B. Ouyang, W. Li, H. Pan, J. Meadows, I. Hoffmann, and W. Dai, "The physical association and phosphorylation of Cdc25C protein phosphatase by Prk," *Oncogene*, vol. 18, no. 44, pp. 6029–6036, 1999.

[169] S. Xie, H. Wu, Q. Wang et al., "Plk3 Functionally Links DNA Damage to Cell Cycle Arrest and Apoptosis at Least in Part via the p53 Pathway," *Journal of Biological Chemistry*, vol. 276, no. 46, pp. 43305–43312, 2001.

[170] J. E. Sillibourne and M. Bornens, "Polo-like kinase 4: the odd one out of the family," *Cell Division*, vol. 5, article no. 25, 2010.

[171] M. Bettencourt-Dias, A. Rodrigues-Martins, L. Carpenter et al., "SAK/PLK4 is required for centriole duplication and flagella development," *Current Biology*, vol. 15, no. 24, pp. 2199–2207, 2005.

[172] A. J. Holland, W. Lan, S. Niessen, H. Hoover, and D. W. Cleveland, "Polo-like kinase 4 kinase activity limits centrosome overduplication by autoregulating its own stability," *Journal of Cell Biology*, vol. 188, no. 2, pp. 191–198, 2010.

[173] M. A. Ko, C. O. Rosario, J. W. Hudson et al., "Plk4 haploinsufficiency causes mitotic infidelity and carcinogenesis," *Nature Genetics*, vol. 37, no. 8, pp. 883–888, 2005.

[174] C. Fode, C. Binkert, and J. W. Dennis, "Constitutive expression of murine Sak-a suppresses cell growth and induces multinucleation," *Molecular and Cellular Biology*, vol. 16, no. 9, pp. 4665–4672, 1996.

[175] C. J. Swallow, M. A. Ko, N. U. Siddiqui, J. W. Hudson, and J. W. Dennis, "Sak/Plk4 and mitotic fidelity," *Oncogene*, vol. 24, no. 2, pp. 306–312, 2005.

[176] S. Bonni, M. L. Ganuelas, S. Petrinac, and J. W. Hudson, "Human Plk4 phosphorylates Cdc25C," *Cell Cycle*, vol. 7, no. 4, pp. 545–547, 2008.

[177] S. Petrinac, M. L. Ganuelas, S. Bonni, J. Nantais, and J. W. Hudson, "Polo-like kinase 4 phosphorylates Chk2," *Cell Cycle*, vol. 8, no. 2, pp. 327–329, 2009.

[178] A. Morettin, A. Ward, J. Nantais, and J. W. Hudson, "Gene expression patterns in heterozygous Plk4 murine embryonic fibroblasts," *BMC Genomics*, vol. 10, article no. 319, 2009.

[179] J. C. Macmillan, J. W. Hudson, S. Bull, J. W. Dennis, and C. J. Swallow, "Comparative expression of the mitotic regulators SAK and PLK in colorectal cancer," *Annals of Surgical Oncology*, vol. 8, no. 9, pp. 729–740, 2001.

[180] G. De Cárcer, G. Manning, and M. Malumbres, "From Plk1 to Plk5: functional evolution of Polo-like kinases," *Cell Cycle*, vol. 10, no. 14, pp. 2255–2262, 2011.

[181] G. De Cárcer, B. Escobar, A. M. Higuero et al., "Plk5, a polo box domain-only protein with specific roles in neuron differentiation and glioblastoma suppression," *Molecular and Cellular Biology*, vol. 31, no. 6, pp. 1225–1239, 2011.

[182] P. Meraldi and P. K. Sorger, "A dual role for Bub1 in the spindle checkpoint and chromosome congression," *EMBO Journal*, vol. 24, no. 8, pp. 1621–1633, 2005.

[183] M. A. Hoyt, L. Totis, and B. T. Roberts, "S. cerevisiae genes required for cell cycle arrest in response to loss of microtubule function," *Cell*, vol. 66, no. 3, pp. 507–517, 1991.

[184] R. Li and A. W. Murray, "Feedback control of mitosis in budding yeast," *Cell*, vol. 66, no. 3, pp. 519–531, 1991.

[185] D. P. Cahill, C. Lengauer, J. Yu et al., "Mutations of mitotic checkpoint genes in human cancers," *Nature*, vol. 392, no. 6673, pp. 300–303, 1998.

[186] A. Gemma, M. Seike, Y. Seike et al., "Somatic mutation of the hBUB1 mitotic checkpoint gene in primary lung cancer," *Genes Chromosomes and Cancer*, vol. 29, no. 3, pp. 213–218, 2000.

[187] Y. R. Hon, L. C. Ron, C. L. We, and H. C. Ji, "hBUB1 defects in leukemia and lymphoma cells," *Oncogene*, vol. 21, no. 30, pp. 4673–4679, 2002.

[188] L. A. Díaz-Martínez and H. Yu, "Running on a treadmill: dynamic inhibition of APC/C by the spindle checkpoint," *Cell Division*, vol. 2, article no. 23, 2007.

[189] Z. Tang, H. Shu, D. Oncel, S. Chen, and H. Yu, "Phosphorylation of Cdc20 by Bub1 provides a catalytic mechanism for APC/C inhibition by the spindle checkpoint," *Molecular Cell*, vol. 16, no. 3, pp. 387–397, 2004.

[190] Z. Tang, Y. Sun, S. E. Harley, H. Zou, and H. Yu, "Human Bub1 protects centromeric sister-chromatid cohesion through Shugoshin during mitosis," *Proceedings of the National Academy of Sciences of the United States of America*, vol. 101, no. 52, pp. 18012–18017, 2004.

[191] S. A. Kawashima, Y. Yamagishi, T. Honda, K. I. Lshiguro, and Y. Watanabe, "Phosphorylation of H2A by Bub1 prevents chromosomal instability through localizing shugoshin," *Science*, vol. 327, no. 5962, pp. 172–177, 2010.

[192] T. S. Kitajima, S. A. Kawashima, and Y. Watanabe, "The conserved kinetochore protein shugoshin protects centromeric cohesion during meiosis," *Nature*, vol. 427, no. 6974, pp. 510–517, 2004.

[193] G. L. Williams, T. M. Roberts, and O. V. Gjoerup, "Bub1: escapades in a cellular world," *Cell Cycle*, vol. 6, no. 14, pp. 1699–1704, 2007.

[194] F. Gao, J. F. Ponte, P. Papageorgis et al., "hBub1 deficiency triggers a novel p53 mediated early apoptotic checkpoint pathway in mitotic spindle damaged cells," *Cancer Biology and Therapy*, vol. 8, no. 7, pp. 627–635, 2009.

[195] H. A. Fisk, C. P. Mattison, and M. Winey, "Human Mps1 protein kinase is required for centrosome duplication and normal mitotic progression," *Proceedings of the National Academy of Sciences of the United States of America*, vol. 100, no. 25, pp. 14875–14880, 2003.

[196] V. M. Stucke, H. H. W. Silljé, L. Arnaud, and E. A. Nigg, "Human Mps1 kinase is required for the spindle assembly checkpoint but not for centrosome duplication," *EMBO Journal*, vol. 21, no. 7, pp. 1723–1732, 2002.

[197] M. G. Fischer, S. Heeger, U. Häcker, and C. F. Lehner, "The mitotic arrest in response to hypoxia and of polar bodies during early embryogenesis requires *Drosophila* Mps1," *Current Biology*, vol. 14, no. 22, pp. 2019–2024, 2004.

[198] X. He, M. H. Jones, M. Winey, and S. Sazer, "mph1, a member of the Mps1-like family of dual specificity protein kinases, is required for the spindle checkpoint in *S. pombe*," *Journal of Cell Science*, vol. 111, no. 12, pp. 1635–1647, 1998.

[199] O. Poch, E. Schwob, F. De Fraipont, A. Camasses, R. Bordonne, and R. P. Martin, "RPK1, an essential yeast protein kinase involved in the regulation of the onset of mitosis, shows homology to mammalian dual-specificity kinases," *Molecular and General Genetics*, vol. 243, no. 6, pp. 641–653, 1994.

[200] A. Abrieu, L. Magnaghi-Jaulin, J. A. Kahana et al., "Mps1 is a kinetochore-associated kinase essential for the vertebrate mitotic checkpoint," *Cell*, vol. 106, no. 1, pp. 83–93, 2001.

[201] H. A. Fisk and M. Winey, "The mouse Mps1p-like kinase regulates centrosome duplication," *Cell*, vol. 106, no. 1, pp. 95–104, 2001.

[202] E. Lauze, B. Stoelcker, F. C. Luca, E. Weiss, A. R. Schutz, and M. Winey, "Yeast spindle pole body duplication gene MPS1 encodes an essential dual specificity protein kinase," *EMBO Journal*, vol. 14, no. 8, pp. 1655–1663, 1995.

[203] K. D. Tardif, A. Rogers, J. Cassiano et al., "Characterization of the cellular and antitumor effects of MPI-0479605, a small-molecule inhibitor of the mitotic kinase Mps1," *Molecular Cancer Therapeutics*, vol. 10, no. 12, pp. 2267–2275, 2011.

[204] P. D. Straight, T. H. Giddings, and M. Winey, "Mps1p regulates meiotic spindle pole body duplication in addition to having novel roles during sporulation," *Molecular Biology of the Cell*, vol. 11, no. 10, pp. 3525–3537, 2000.

[205] N. Jelluma, A. B. Brenkman, N. J. F. van den Broek et al., "Mps1 phosphorylates borealin to control Aurora B activity and chromosome alignment," *Cell*, vol. 132, no. 2, pp. 233–246, 2008.

[206] M. Leng, D. W. Chan, H. Luo, C. Zhu, J. Qin, and Y. Wang, "MPS1-dependent mitotic BLM phosphorylation is important for chromosome stability," *Proceedings of the National Academy of Sciences of the United States of America*, vol. 103, no. 31, pp. 11485–11490, 2006.

[207] N. Selak, C. Z. Bachrati, I. Shevelev et al., "The Bloom's syndrome helicase (BLM) interacts physically and functionally with p12, the smallest subunit of human DNA polymerase δ," *Nucleic Acids Research*, vol. 36, no. 16, pp. 5166–5179, 2008.

[208] R. K. Dorer, S. Zhong, J. A. Tallarico, W. H. Wong, T. J. Mitchison, and A. W. Murray, "A small-molecule inhibitor of Mps1 blocks the spindle-checkpoint response to a lack of tension on mitotic chromosomes," *Current Biology*, vol. 15, no. 11, pp. 1070–1076, 2005.

[209] J. Daniel, J. Coulter, J. H. Woo, K. Wilsbach, and E. Gabrielson, "High levels of the Mps1 checkpoint protein are protective of aneuploidy in breast cancer cells," *Proceedings of the National Academy of Sciences of the United States of America*, vol. 108, no. 13, pp. 5384–5389, 2011.

[210] B. Ouyang, Z. Lan, J. Meadows et al., "Human Bub1: a putative spindle checkpoint kinase closely linked to cell proliferation," *Cell Growth and Differentiation*, vol. 9, no. 10, pp. 877–885, 1998.

[211] H. J. Shin, K. H. Baek, A. H. Jeon et al., "Dual roles of human BubR1, a mitotic checkpoint kinase, in the monitoring of chromosomal instability," *Cancer Cell*, vol. 4, no. 6, pp. 483–497, 2003.

[212] M. Sczaniecka, A. Feoktistova, K. M. May et al., "The spindle checkpoint functions of Mad3 and Mad2 depend on a Mad3 KEN box-mediated interaction with Cdc20-anaphase-promoting complex (APC/C)," *Journal of Biological Chemistry*, vol. 283, no. 34, pp. 23039–23047, 2008.

[213] M. A. Lampson and T. M. Kapoor, "The human mitotic checkpoint protein BubR1 regulates chromosome-spindle attachments," *Nature Cell Biology*, vol. 7, no. 1, pp. 93–98, 2005.

[214] Y. Fang, T. Liu, X. Wang et al., "BubR1 is involved in regulation of DNA damage responses," *Oncogene*, vol. 25, no. 25, pp. 3598–3605, 2006.

[215] W. Dai, Q. Wang, T. Liu et al., "Slippage of mitotic arrest and enhanced tumor development in mice with bubr1 haploinsufficiency," *Cancer Research*, vol. 64, no. 2, pp. 440–445, 2004.

[216] Q. Wang, T. Liu, Y. Fang et al., "BUBR1 deficiency results in abnormal megakaryopoiesis," *Blood*, vol. 103, no. 4, pp. 1278–1285, 2004.

[217] J. Lee, C. G. Lee, K. W. Lee, and C. W. Lee, "Cross-talk between BubR1 expression and the commitment to differentiate in adipose-derived mesenchymal stem cells," *Experimental and Molecular Medicine*, vol. 41, no. 12, pp. 873–879, 2009.

[218] R. Tomasini, T. W. Mak, and G. Melino, "The impact of p53 and p73 on aneuploidy and cancer," *Trends in Cell Biology*, vol. 18, no. 5, pp. 244–252, 2008.

[219] A. Gupta, S. Inaba, O. K. Wong, G. Fang, and J. Liu, "Breast cancer-specific gene 1 interacts with the mitotic checkpoint kinase BubR1," *Oncogene*, vol. 22, no. 48, pp. 7593–7599, 2003.

[220] S. J. E. Suijkerbuijk, M. H. J. Van Osch, F. L. Bos, S. Hanks, N. Rahman, and G. J. P. L. Kops, "Molecular causes for BUBR1 dysfunction in the human cancer predisposition syndrome mosaic variegated aneuploidy," *Cancer Research*, vol. 70, no. 12, pp. 4891–4900, 2010.

[221] M. V. Blagosklonny, "Mitotic arrest and cell fate: why and how mitotic inhibition of transcription drives mutually exclusive events," *Cell Cycle*, vol. 6, no. 1, pp. 70–74, 2007.

[222] H. Tsuiki, M. Nitta, M. Tada, M. Inagaki, Y. Ushio, and H. Saya, "Mechanism of hyperploid cell formation induced by microtubule inhibiting drug in glioma cell lines," *Oncogene*, vol. 20, no. 4, pp. 420–429, 2001.

[223] Y. Niikura, H. Ogi, K. Kikuchi, and K. Kitagawa, "BUB3 that dissociates from BUB1 activates caspase-independent mitotic death (CIMD)," *Cell Death and Differentiation*, vol. 17, no. 6, pp. 1011–1024, 2010.

[224] K. Kitagawa and Y. Niikura, "Caspase-independent mitotic death (CIMD)," *Cell Cycle*, vol. 7, no. 8, pp. 1001–1005, 2008.

[225] N. N. Kreis, K. Sommer, M. Sanhaji et al., "Long-term downregulation of Polo-like kinase 1 increases the cyclin-dependent kinase inhibitor p21WAF1/CIP1," *Cell Cycle*, vol. 8, no. 3, pp. 460–472, 2009.

[226] T. Abbas, E. Shibata, J. Park, S. Jha, N. Karnani, and A. Dutta, "CRL4Cdt2 regulates cell proliferation and histone gene expression by targeting PR-Set7/Set8 for degradation," *Molecular Cell*, vol. 40, no. 1, pp. 9–21, 2010.

[227] Y. Degenhardt, J. Greshock, S. Laquerre et al., "Sensitivity of cancer cells to Plk1 inhibitor GSK461364A is associated with loss of p53 function and chromosome instability," *Molecular Cancer Therapeutics*, vol. 9, no. 7, pp. 2079–2089, 2010.

[228] X. Liu and R. L. Erikson, "Polo-like kinase (Plk)1 depletion induces apoptosis in cancer cells," *Proceedings of the National Academy of Sciences of the United States of America*, vol. 100, no. 10, pp. 5789–5794, 2003.

[229] H.-J. Kim, J. H. Cho, H. Quan, and J.-R. Kim, "Down-regulation of Aurora B kinase induces cellular senescence in human fibroblasts and endothelial cells through a p53-dependent pathway," *FEBS Letters*, vol. 585, no. 22, pp. 3569–3576, 2011.

[230] J. J. Huck, M. Zhang, A. McDonald et al., "MLN8054, an inhibitor of Aurora A kinase, induces senescence in human tumor cells both in vitro and in vivo," *Molecular Cancer Research*, vol. 8, no. 3, pp. 373–384, 2010.

[231] A. A. Dar, A. Belkhiri, J. Ecsedy, A. Zaika, and W. El-Rifai, "Aurora kinase A inhibition leads to p73-dependent apoptosis in p53-deficient cancer cells," *Cancer Research*, vol. 68, no. 21, pp. 8998–9004, 2008.

[232] S. Sur, R. Pagliarini, F. Bunz et al., "A panel of isogenic human cancer cells suggests a therapeutic approach for cancers with inactivated p53," *Proceedings of the National Academy of Sciences of the United States of America*, vol. 106, no. 10, pp. 3964–3969, 2009.

[233] M. Jemaa, I. Vitale, O. Kepp et al., "Selective killing of p53-deficient cancer cells by SP600125," *EMBO Molecular Medicine*. In press.

5

Biological Effects of Mammalian Translationally Controlled Tumor Protein (TCTP) on Cell Death, Proliferation, and Tumorigenesis

Michiyo Nagano-Ito and Shinichi Ichikawa

Laboratory for Animal Cell Engineering, Niigata University of Pharmacy and Applied Life Sciences (NUPALS), 265-1 Higashijima, Akiha-ku, Niigata-shi, Niigata 956-8603, Japan

Correspondence should be addressed to Shinichi Ichikawa, shin@nupals.ac.jp

Academic Editor: Malgorzata Kloc

Translationally controlled tumor protein (TCTP) is a highly conserved protein found in eukaryotes, across animal and plant kingdoms and even in yeast. Mammalian TCTP is ubiquitously expressed in various tissues and cell types. TCTP is a multifunctional protein which plays important roles in a number of cell physiological events, such as immune responses, cell proliferation, tumorigenicity, and cell death, including apoptosis. Recent identification of TCTP as an antiapoptotic protein has attracted interest of many researchers in the field. The mechanism of antiapoptotic activity, however, has not been solved completely, and TCTP might inhibit other types of cell death. Cell death (including apoptosis) is closely linked to proliferation and tumorigenesis. In this context, we review recent findings regarding the role of TCTP in cell death, proliferation, and tumorigenesis and discuss the mechanisms.

1. Introduction

Translationally controlled tumor protein (TCTP) was initially identified as a factor implicated in cell growth [1, 2]. TCTP has also been termed histamine releasing factor (HRF), fortilin, P21, P23, TPT-1, and Q23. This protein was named TCTP because its mRNA was controlled at the translational level [3–5]. Although TCTP is found ubiquitously in tissues and cell types, its expression is relatively low in lung and colon, and cell lines derived from normal cells such as a mouse fibroblast NIH-3T3 and human embryonic kidney HEK293T cells [6]. Because of its multifunctional properties, TCTP has attracted the attention of an increasing number of researchers in many fields (reviewed in [7]). TCTP plays important roles in a number of cell physiological events in cancer, cell proliferation, stress response, gene regulation, and heat shock response [8–13]. TCTP was also shown to possess an extracellular function, that is, histamine release [14].

Tumorigenicity, proliferation, and cell death, including apoptosis, are closely related functions. Uncontrolled or promoted proliferation and loss of cell death are general properties of tumor cells. In this paper, we will focus on mammalian TCTP and discuss its physiological functions, emphasizing cell death, proliferation, and tumorigenesis.

2. Properties of TCTP

Human [5] and murine [3] TCTP cDNAs were isolated and their sequences determined more than 20 years ago. Human TCTP cDNA encodes a protein with a calculated molecular mass of 19 kDa (172 amino acids). Sequence analyses revealed that TCTP is a highly conserved protein lacking homology to any other protein. TCTP has been found in a wide range of eukaryotes, including yeast, plants, and animals, suggesting it originated in the distant evolutionary past. Since immune systems are restricted to animals, its function in histamine release has been acquired only recently in evolution. Ubiquitous expression of TCTP in mammalian tissues suggests its importance in normal physiological functions. In fact, a gene-targeting approach revealed that TCTP is an essential protein in mice since knockouts deficient in this protein die at embryonic stage day E9.5-E10.5 [15]. However, studies with mouse embryonic fibroblast (MEF) cells showed that TCTP is not essential for cell survival *in vitro* [15]. The intracellular localization of TCTP is predominantly in the cytosol and nucleus [16] although it functions as an antiapoptotic protein in mitochondria. TCTP is a hydrophilic protein and does not contain any hydrophobic transmembrane domains or any localization signals to an organelle [6]. Translocation of TCTP to the nucleus under certain conditions such as oxidative stress was reported recently. However, TCTP does not contain a nuclear localization signal and the mechanism of translocation remains to be solved [17].

3. TCTP Interacts with Many Kinds of Proteins

To exert various physiological functions, TCTP interacts with many other proteins, including translation elongation factors eEF1A and eEF-B-β [18], tubulin [19], actin [20], myeloid cell leukemia protein-1 (MCL1) [6, 16], Bcl-xL [21], p53 [22], and Na, K-ATPase [12]. TCTP can also bind to itself, forming homodimers [11], and this binding is required for the cytokine-like activity of this protein during allergic responses [23]. However, it is not known whether the dimerization of TCTP is necessary for its other functions.

4. How Does TCTP Protect Cells from Death?

It is well known that TCTP protects cells from death. Although many mechanisms have been proposed, details remain to be identified.

4.1. TCTP Could Directly Reduce Cellular Stress. TCTP expression increases in response to a variety of cell stresses and stimuli, and in some cases, TCTP could directly reduce stress, protecting cells from death (Figure 1). The first case we describe deals with its protection of cells from heat shock-induced cell death. TCTP is markedly upregulated in a variety of cells following thermal shock. Recent studies demonstrated that TCTP is a heat shock protein and serves as a molecular chaperone. TCTP binds to denatured proteins, refolds them, and also interacts with native proteins and protects them from denaturation [24]. Although no strong homology with other proteins has been found, recent studies revealed relationships with guanine nucleotide-free chaperones, the Mss4/Dss4 family of proteins that binds to the GDP/GTP-free form of Rab [25]. This fact also supports TCTP's function as a chaperone.

The second case is cell death induced by an influx of Ca^{2+}. The level of TCTP is controlled by the intracellular Ca^{2+} concentration and elevation of Ca^{2+} also induces TCTP mRNA in cells [26]. Binding of TCTP to Ca^{2+} was demonstrated for the first time using *Trypanosoma brucei* protein [27] and later with the human protein [28]. Thapsigargin raises cytosolic Ca^{2+} by blocking the ability of the cells to pump calcium into the ER, which depletes its Ca^{2+} stores. This depletion can secondarily activate plasma membrane calcium channels,

allowing an influx of Ca^{2+} into the cytosol, thereby initiating apoptosis. The lack of TCTP resulted in exaggerated elevation of Ca^{2+} in thapsigargin-challenged cells [29]. Elevation of the intracellular Ca^{2+} level beyond the normal range could injure mitochondrial membranes and lead to release of cytochrome C and AIF, resulting in apoptosis. Graidist's group also demonstrated that Ca^{2+} binding of TCTP is required for protection of the cells against thapsigargin-induced apoptosis. They hypothesized that TCTP exerts its antiapoptotic function by serving as a Ca^{2+} scavenger. On the other hand, thapsigargin is also known to induce ER stress, in which unfolded proteins are accumulated in the organelle. Thapsigargin reduces Ca^{2+} concentration in the ER and suppresses small molecule Ca^{2+}-dependent chaperones in the organelle, allowing accumulation of aberrant proteins, which in turn eventually leads cells to undergo apoptosis. Thus, TCTP might also protect cells from ER stress-induced apoptosis by inhibiting the signal pathway.

The last case is oxidative stress. TCTP from the parasite *Brugia malayi* has antioxidant functions and when it was overexpressed in *Escherichia coli*, it protected the cells from hydrogen peroxide-induced cell death [30]. Although TCTP used in this experiment originated in the parasite and was expressed in bacteria, this result suggests that TCTP itself might serve as antioxidant and could neutralize ROS in mammalian cells.

4.2. TCTP Inhibits Apoptosis. Many types of cellular stresses induce apoptosis via the mitochondrial pathway and TCTP is able to inhibit this type of apoptosis by regulating the relevant signal pathways (Figure 1). TCTP protects cells from apoptosis triggered by serum deprivation [6], or treatment with etoposide, taxol, or 5-fluorouracil [21, 31]. Mitochondria contain proapoptotic proteins such as apoptosis inducing factor (AIF), Smac/DIABLO, and cytochrome C. In the course of apoptosis, these proteins are released from mitochondria following the formation of the permeability transition pore in the membrane by the action of proapoptotic Bcl-2 family proteins such as Bax and BH3. Other Bcl-2 family members such as Bcl-2, MCL1, and Bcl-xL are known to suppress apoptosis by binding and inactivating the proapoptotic proteins. Among Bcl-2 family proteins, MCL1 is a unique protein. Unlike other Bcl-2 family proteins, MCL1 is not constitutively expressed and is induced by various stimuli. It was demonstrated that TCTP specifically associates with MCL1 [8, 16], which has the ability to stabilize TCTP [16]. In contrast to this result, another research group showed that TCTP stabilized MCL1 by suppressing its degradation by blocking its ubiquitination [8]. In their experimental conditions, MCL1 did not stabilize TCTP. The discrepancies of the results obtained from the two research groups are presumably due to the use of different cell lines and experimental conditions. TCTP and MCL1 are also capable of functioning as antiapoptotic proteins independently of each other [31]. Bcl-xL is another antiapoptotic Bcl-2 family protein that interacts with TCTP. In this case, binding between the BH3 domain of Bcl-xL and the N-terminal region of TCTP is required for

Biological Effects of Mammalian Translationally Controlled Tumor Protein (TCTP) on
Cell Death, Proliferation, and Tumorigenesis

47

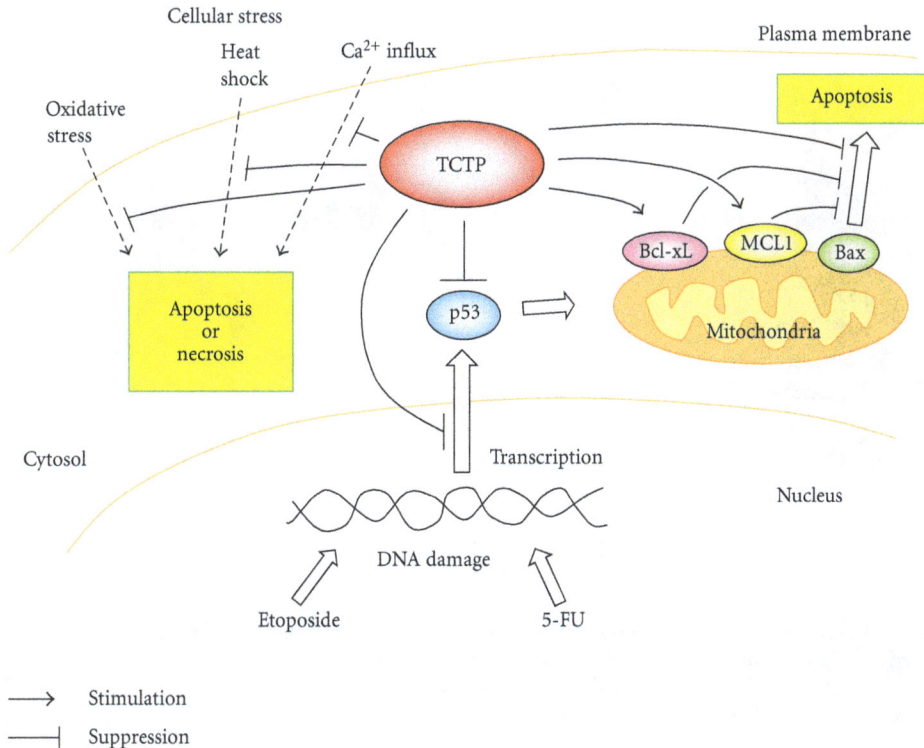

FIGURE 1: TCTP protects cells from cell death. TCTP inhibits cell death induced by oxidative stress, heat shock, or influx of Ca^{2+}. In addition, TCTP can protect cells from apoptosis triggered by treatment with genotoxic reagent such as etoposide and 5-fluorouracil. TCTP inhibits apoptosis by stabilizing antiapoptotic Bcl-2 family proteins, MCL1 and Bcl-xL and by inhibiting activation of proapoptotic Bcl-2 family protein, Bax. Moreover, TCTP inhibits p53-dependent apoptosis by downregulating the protein.

the antiapoptotic activity of TCTP [8]. The BH3 domain is responsible for hetero- and homodimerization between antiapoptotic and proapoptotic Bcl-2 family proteins. TCTP also interferes with dimerization of the proapoptotic Bcl-2 family protein Bax [32]. The crystal structure of TCTP was solved and a structural similarity with that of Bax [32] was found despite lack of amino acid sequence homology. This similarity suggests localization of TCTP to mitochondrial membranes. Dimerization of Bax is required for its apoptotic activity and TCTP blocks the formation of Bax homodimers by inserting into mitochondrial membranes (reviewed in [33]). Although TCTP inhibits apoptosis induced by Bax, unlike MCL1 and Bcl-xL, TCTP does not bind Bax directly.

TCTP affects the tumor suppressor p53 (Figures 1 and 2). The mutation in p53 is found in about half of all cancers and dysfunction of the protein is one of the main causes of cancer development. p53 is also a potent mediator of cellular responses against various cellular stresses including genotoxic insults. In addition, overexpression of p53 induces apoptosis in cancer cells. TCTP was shown to bind p53 and prevent apoptosis by destabilizing the protein in a human lung carcinoma cell line A549 [22]. TCTP also represses transcription of p53 [34]. These facts also indicate the ability of TCTP to promote transformation by reducing p53 function.

4.3. Oxidative Stress-Induced Cell Death and TCTP. Intrinsic reactive oxygen species (ROS) such as hydrogen peroxide, superoxide, and hydroxyl radicals are generated in cells in the course of normal metabolism, including electron transport and various oxidase reactions. Oxidative stress induced by ROS has been implicated in aging and in the pathophysiology of various diseases such as diabetes, cancer, and Parkinson's disease (reviewed in [35]). These diseases are, at least in part, caused by ROS-mediated cell death in tissues. Although the effect of TCTP on apoptosis first attracted attention, TCTP might regulate other types of cell death. The types of cell death induced by oxidative stress depend on the cell lines and experimental conditions. In most cases, however, cell death caused by oxidative stress leads to necrosis rather than apoptosis. Types of hydrogen peroxide-induced cell death differ depending on cell types, and conditions of hydrogen peroxide treatment and cell culture. High concentrations of hydrogen peroxide inhibited apoptosis in T-lymphoma Jurkat cells by lowering intracellular ATP levels (necessary for apoptosome formation), and this might also be the case in other cell lines [36]. In the course of isolating cDNAs which protect cells from hydrogen peroxide, we found for the first time that TCTP could inhibit cell death induced by oxidative stress [37]. Overexpression of TCTP protected hydrogen peroxide-induced cell death in a Chinese hamster ovary cell line, CHO-K1; however, cell death was not typical apoptosis. Although the cells showed apoptosis-like morphological changes after hydrogen peroxide treatment, their genomic DNA did not show DNA ladder pattern formation [37]. Presumably the cells stopped apoptotic signaling after cytochrome C release from mitochondria and

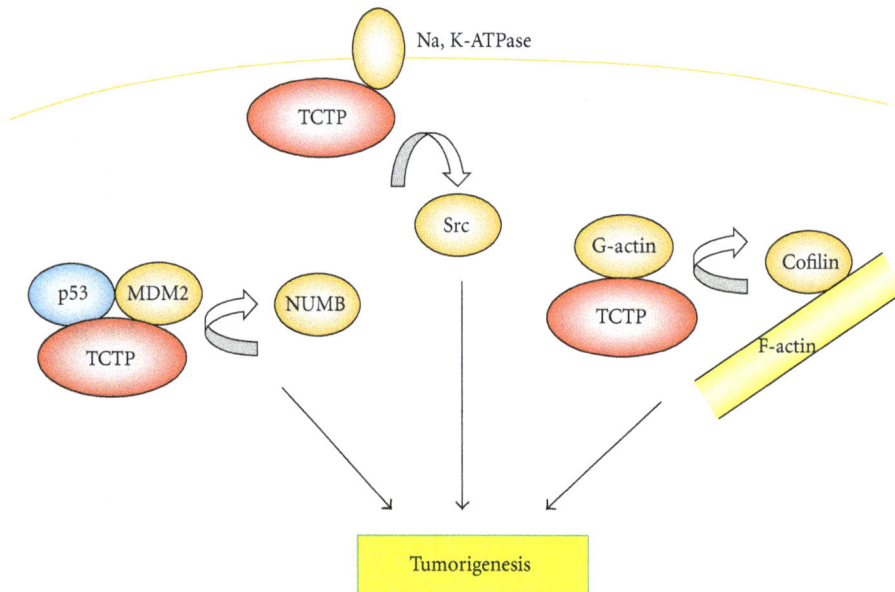

FIGURE 2: TCTP functions as an oncogene. NUMB forms a tricomplex with p53 and the E3 ubiquitin ligase MDM2, thereby preventing ubiquitination followed by degradation of p53. TCTP promotes p53 degradation by competing with NUMB for MDM2 binding. TCTP binds to Na, K-ATPase and, as a result, it releases Src binding to Na, K-ATPase and activate it. TCTP can release cofilin binding to G-actin by competing with and replacing cofilin. The increase of free cofilin then promotes the binding of the protein to F-actin and exerts its functions.

were subjected to secondary necrosis. Recent studies showed the existence of programmed necrosis (necroptosis) which is physiological cell death regulated by its signal pathway (reviewed in [38]). TCTP might inhibit the signal pathways of physiologically regulated necrosis. As mentioned in the previous section, it is also possible that TCTP itself acts as an antioxidant and reduces oxidative stress induced by hydrogen peroxide. This protective effect of TCTP against oxidative stress is presumably an intrinsic function in malignant breast cancer cells. Treatment with hydrogen peroxide upregulated TCTP level in T4-2 malignant breast cancer cells, but not in their parental S-1 cells that are nonmalignant [39]. TCTP upregulation was also observed in another breast cancer cell line, MDB-MB-231 after treatment with hydrogen peroxide or arsenic trioxide [40], which leads to ROS generation.

In conclusion, oxidative stress upregulates cellular TCTP levels leading to cellular protection against death. However, hydrogen peroxide treatment did not upregulate TCTP in another tumorigenic cell line (CHO-K1), although over-expression of TCTP protects cells from hydrogen peroxide [37]. The mechanisms by which oxidative stress upregulates TCTP are not known. Interestingly, TCTP translocates from cytosol to the nucleus in a keratinocyte cell line (HaCat) where it binds the vitamin D3 receptor [17]. Thus far, the physiological meaning of TCTP binding to the vitamin D3 receptor is not clear. However, this interesting phenomenon suggests that TCTP might regulate transcription of genes in response to oxidative stress. As the upregulation mediated by hydrogen peroxide is restricted to malignant cancer cells, protein factors controlling its expression could be suitable targets for cancer drug discovery. Primary culture cells of mouse embryonic fibroblasts (MEF) from TCTP knockout and control mice manifested similar proliferative activities

and apoptotic sensitivities to various stimuli including hydrogen peroxide treatment [15]. These results suggest that prevention of cell death by TCTP is restricted to certain cell types such as transformed cancer cells. This hypothesis was supported by the fact that the depletion of TCTP by siRNA induced apoptosis via caspases 8 and 3 in human prostate cancer cell line LNCaP [41]. Interestingly, Mmi1P, a yeast ortholog of mammalian TCTP that binds microtubules, translocates from the cytosol to mitochondria following mild oxidative stress stimuli. In contrast to its mammalian counterpart, Mmi P has an apoptotic function in yeast cells [42].

5. Tumorigenicity and TCTP

Several lines of evidence indicate that TCTP can induce oncogenic transformation. Transformation of normal cells into tumor cells requires a series of genetic changes. Since TCTP is overexpressed in many types of cancer cells and silencing of the gene decreases the viability of the cells [6], it was postulated that TCTP functions as an oncogene. Tuynder et al. developed unique systems to select cells with a reverted phenotype using H-1 parvovirus which preferentially kills tumor cells [19, 43]. TCTP was found to be downregulated in reverted cells with a normal phenotype. In addition, silencing of TCTP with antisense DNA or siRNA revealed a reverted tumor phenotype, supporting this idea [19, 43, 44]. These results suggest that TCTP is directly involved in malignant transformation. Although the mechanisms of TCTP-dependent transformation are not known, it could be the result of p53 destabilization as noted in the previous section. Another line of evidence also indicates regulation of p53 by TCTP and *vice versa*. NUMB is a protein known to be a regulator of p53. It forms a tricomplex with p53

Biological Effects of Mammalian Translationally Controlled Tumor Protein (TCTP) on
Cell Death, Proliferation, and Tumorigenesis

49

and the E3 ubiquitin ligase MDM2, thereby preventing ubiquitination followed by degradation of p53 [45]. TCTP promotes p53 degradation by competing with NUMB for MDM2 binding (Figure 2). On the other hand, p53 directly represses transcription of TCTP. Thus, TCTP and p53 form a reciprocal negative regulation loop [34]. This fact also suggests that TCTP might inhibit p53-dependent apoptosis by downregulating the protein.

The most important properties of tumor and cancer cells are unregulated cell proliferation and avoidance of cell death. Inhibition or gene silencing by TCTP siRNA reduces viability and induces apoptosis in cancer cells, including human prostate cancer cells [41]. TCTP might also be involved in the malignancy of tumors by interacting with actin at the cofilin binding site. Cofilin is an actin binding protein and has the ability to regulate the cell cycle (reviewed in [46]) and promote metastasis [47]. TCTP competes with cofilin at the cofilin-binding site of actin. Although cofilin can bind to both monomeric (G-actin) and filamentous actin (F-actin), it exerts its functions by binding to and changing the twist of F-actin. On the other hand, TCTP has a higher affinity with G-actin than F-actin. TCTP can release cofilin binding to G-actin by competing with and replacing cofilin. The increase of free cofilin then promotes the binding of the protein to F-actin and exerts its functions (Figure 2) [20]. Recent studies also revealed that TCTP induces transformation in human breast epithelial cells through activation of a protooncogene product Src [48]. TCTP binds to the $\alpha 1$ subunit of Na, K-ATPase and, as a result, it releases Src binding to Na, K-ATPase. This TCTP-mediated Src release activates Src and promotes various tumor progression signal pathways (Figure 2) [48].

6. TCTP Regulates Cell Proliferation

Since TCTP is highly expressed in actively dividing cells [28, 49], one might expect TCTP to modulate physiological functions during cell proliferation. TCTP has the ability to bind microtubules during G1-, S-, G2-, and M-phases of the cell cycle. It associates with the metaphase spindle, but is detached from the spindle after metaphase [50]. TCTP is phosphorylated by the polo-like kinase Plk, which is likely to cause detachment of TCTP from the mitotic spindle [9]. Since the TCTP level is upregulated during entry into the cell cycle, the protein is believed to be important for cell growth and division. TCTP overexpression in mammalian cells results in cell cycle retardation, microtubule stabilization, and alteration of cell morphology [49]. Furthermore, TCTP mutated in the phosphorylation sites for Plk disrupts the completion of mitosis, indicating the importance of TCTP phosphorylation in normal cell cycle regulation [9]. The fact that increased TCTP levels slow cell cycle progression is unexpected because high levels of TCTP expression are generally observed in actively dividing cells and the discrepancy has yet to be explained.

TCTP might regulate proliferation through the target of rapamycin (TOR) pathway. TOR is a Ser/Thr kinase that regulates proliferation and metabolism in response to nutrients, hormones, and growth factors. In case of *Drosophila, Drosophila* TCTP (dTCTP) binds to nucleotide free form of a small GTPase, *Drosophila* Ras homolog enriched in brain (dRheb), and stimulates GDP-GTP exchange of dRheb. As a result, dTCTP activates the TOR signaling pathways. In fact, tissue-specific reduction of dTCTP *in vivo* resulted in smaller organs with reduction of both cell size and cell number [51]. This might be also the case in mammals [52].

7. Concluding Remarks

We have reviewed recent findings on biological effects of mammalian TCTP, focusing on inhibition of cell death, regulation of proliferation, and tumorigenesis. Although many hypotheses have been proposed, mechanistic explanations of TCTP on phenomena are still elusive. Presumably, TCTP is able to modulate multiple protein targets simultaneously and as a result, it exerts effects. Further comprehensive studies are necessary to clarify the detailed mechanisms. Recent studies also suggest protective functions of TCTP against cell death other than apoptosis. The mechanisms of TCTP's action on the cell death is interesting and important issues in future studies.

References

[1] G. Thomas, G. Thomas, and H. Luther, "Transcriptional and translational control of cytoplasmic proteins after serum stimulation of quiescent Swiss 3T3 cells," *Proceedings of the National Academy of Sciences of the United States of America*, vol. 78, no. 9, pp. 5712–5716, 1981.

[2] R. Yenofski, I. Bergmann, and G. Brawerman, "Messenger RNA species partially in a repressed state in mouse sarcoma ascites cell," *Proceedings of the National Academy of Sciences of the United States of America*, vol. 79, no. 19, pp. 5876–5880, 1982.

[3] S. T. Chitpatima, S. Makrides, R. Bandyopadhyay, and G. Brawerman, "Nucleotide sequence of a major messenger RNA for a 21 kilodalton polypeptide that is under translational control in mouse tumor cells," *Nucleic Acids Research*, vol. 16, no. 5, p. 2350, 1988.

[4] H. Böhm, R. Benndorf, M. Gaestel et al., "The growth-related protein p23 of the Ehrlich ascites tumor: translational control, cloning and primary structure," *Biochemistry International*, vol. 19, no. 2, pp. 277–286, 1989.

[5] B. Gross, M. Gaestel, H. Bohm, and H. Bielka, "cDNA sequence coding for a translationally controlled human tumor protein," *Nucleic Acids Research*, vol. 17, no. 20, p. 8367, 1989.

[6] F. Li, D. Zhang, and K. Fujise, "Characterization of fortilin, a novel antiapoptotic protein," *The Journal of Biological Chemistry*, vol. 276, no. 50, pp. 47542–47549, 2001.

[7] U. A. Bommer and B. J. Thiele, "The translationally controlled tumour protein (TCTP)," *International Journal of Biochemistry and Cell Biology*, vol. 36, no. 3, pp. 379–385, 2004.

[8] H. Liu, H. W. Peng, Y. S. Cheng, H. S. Yuan, and H. F. Yang-Yen, "Stabilization and enhancement of the antiapoptotic activity of Mcl-1 by TCTP," *Molecular and Cellular Biology*, vol. 25, no. 8, pp. 3117–3126, 2005.

[9] F. R. Yarm, "Plk phosphorylation regulates the microtubule-stabilizing protein TCTP," *Molecular and Cellular Biology*, vol. 22, no. 17, pp. 6209–6221, 2002.

[10] S. M. MacDonald, J. Bhisutthibhan, T. A. Shapiro et al., "Immune mimicry in malaria: *Plasmodium falciparum* secretes a functional histamine-releasing factor homologue in vitro and in vivo," *Proceedings of the National Academy of Sciences of the United States of America*, vol. 98, no. 199, pp. 10829–10832, 2001.

[11] T. Yoon, J. Jung, M. Kim, Kang Man Lee, Eung Chil Choi, and K. Lee, "Identification of the self-interaction of rat TCTP/IgE-dependent histamine-releasing factor using yeast two-hybrid system," *Archives of Biochemistry and Biophysics*, vol. 384, no. 2, pp. 379–382, 2000.

[12] J. Jung, M. Kim, M. J. Kim et al., "Translationally controlled tumor protein interacts with the third cytoplasmic domain of Na,K-ATPase α subunit and inhibits the pump activity in HeLa cells," *The Journal of Biological Chemistry*, vol. 279, no. 48, pp. 49868–49875, 2004.

[13] C. Cans, B. J. Passer, V. Shalak et al., "Translationally controlled tumor protein acts as a guanine nucleotide dissociation inhibitor on the translation elongation factor eEF1A," *Proceedings of the National Academy of Sciences of the United States of America*, vol. 100, no. 2, pp. 13892–13897, 2003.

[14] S. M. MacDonald, T. Rafnar, J. Langdon, and L. M. Lichtenstein, "Molecular identification of an IgE-Dependent histamine-releasing factor," *Science*, vol. 269, no. 5224, pp. 688–690, 1995.

[15] S. H. Chen, P. S. Wu, C. H. Chou et al., "A knockout mouse approach reveals that TCTP functions as an essential factor for cell proliferation and survival in a tissue- or cell type-specific manner," *Molecular Biology of the Cell*, vol. 18, no. 7, pp. 2525–2532, 2007.

[16] D. Zhang, F. Li, D. Weidner, Z. H. Mnjoyan, and K. Fujise, "Physical and functional interaction between myeloid cell leukemia 1 protein (MCL1) and fortilin. The potential role of MCL1 as a fortilin chaperone," *The Journal of Biological Chemistry*, vol. 277, no. 40, pp. 37430–37438, 2002.

[17] R. Rid, K. Önder, A. Trost et al., "H_2O_2-dependent translocation of TCTP into the nucleus enables its interaction with VDR in human keratinocytes: TCTP as a further module in calcitriol signalling," *Journal of Steroid Biochemistry and Molecular Biology*, vol. 118, no. 1-2, pp. 29–40, 2010.

[18] J. M. Langdon, B. M. Vonakis, and S. M. MacDonald, "Identification of the interaction between the human recombinant histamine releasing factor/translationally controlled tumor protein and elongation factor-1 delta (also known as eElongation factor-1B beta)," *Biochimica et Biophysica Acta*, vol. 1688, no. 3, pp. 232–236, 2004.

[19] M. Tuynder, L. Susini, S. Prieur et al., "Biological models and genes of tumor reversion: cellular reprogramming through tpt1/TCTP and SIAH-1," *Proceedings of the National Academy of Sciences of the United States of America*, vol. 99, no. 23, pp. 14976–14981, 2002.

[20] K. Tsarova, E. G. Yarmola, and M. R. Bubb, "Identification of a cofilin-like actin-binding site on translationally controlled tumor protein (TCTP)," *FEBS Letters*, vol. 584, no. 23, pp. 4756–4760, 2010.

[21] Y. Yang, F. Yang, Z. Xiong et al., "An N-terminal region of translationally controlled tumor protein is required for its antiapoptotic activity," *Oncogene*, vol. 24, no. 30, pp. 4778–4788, 2005.

[22] S. B. Rho, J. H. Lee, M. S. Park et al., "Anti-apoptotic protein TCTP controls the stability of the tumor suppressor p53," *FEBS Letters*, vol. 585, no. 1, pp. 29–35, 2011.

[23] M. Kim, H. J. Min, H. Y. Won et al., "Dimerization of translationally controlled tumor protein is essential for its cytokine-like activity," *PLoS ONE*, vol. 4, no. 7, Article ID e6464, 2009.

[24] M. Gnanasekar, G. Dakshinamoorthy, and K. Ramaswamy, "Translationally controlled tumor protein is a novel heat shock protein with chaperone-like activity," *Biochemical and Biophysical Research Communications*, vol. 386, no. 2, pp. 333–337, 2009.

[25] P. Thaw, N. J. Baxter, A. M. Hounslow, C. Price, J. P. Waltho, and C. J. Craven, "Structure of TCTP reveals unexpected relationship with guanine nucleotide-free chaperones," *Nature Structural Biology*, vol. 8, no. 8, pp. 701–704, 2001.

[26] A. Xu, A. R. Bellamy, and J. A. Taylor, "Expression of translationally controlled tumour protein is regulated by calcium at both the transcriptional and post-transcriptional level," *Biochemical Journal*, vol. 342, no. 3, pp. 683–689, 1999.

[27] N. G. Haghighat and L. Ruben, "Purification of novel calcium binding proteins from *Trypanosoma brucei*: properties of 22-, 24- and 38-kilodalton proteins," *Molecular and Biochemical Parasitology*, vol. 51, no. 1, pp. 99–110, 1992.

[28] J. C. Sanchez, D. Schaller, F. Ravier et al., "Translationally controlled tumor protein: a protein identified in several nontumoral cells including erythrocytes," *Electrophoresis*, vol. 18, no. 1, pp. 150–155, 1997.

[29] P. Graidist, M. Yazawa, M. Tonganunt et al., "Fortilin binds Ca^{2+} and blocks Ca^{2+}-dependent apoptosis in vivo," *Biochemical Journal*, vol. 408, no. 2, pp. 181–191, 2007.

[30] M. Gnanasekar and K. Ramaswamy, "Translationally controlled tumor protein of Brugia malayi functions as an antioxidant protein," *Parasitology Research*, vol. 101, no. 6, pp. 1533–1540, 2007.

[31] P. Graidist, A. Phongdara, and K. Fujise, "Antiapoptotic protein partners fortilin and MCL1 independently protect cells from 5-fluorouracil-induced cytotoxicity," *The Journal of Biological Chemistry*, vol. 279, no. 39, pp. 40868–40875, 2004.

[32] L. Susini, S. Besse, D. Duflaut et al., "TCTP protects from apoptotic cell death by antagonizing bax function," *Cell Death and Differentiation*, vol. 15, no. 8, pp. 1211–1220, 2008.

[33] A. M. Petros, E. T. Olejniczak, and S. W. Fesik, "Structural biology of the Bcl-2 family of proteins," *Biochimica et Biophysica Acta*, vol. 1644, no. 2-3, pp. 83–94, 2004.

[34] R. Amson, S. Pece, A. Lespagnol et al., "Reciprocal repression between P53 and TCTP," *Nature Medicine*, vol. 18, no. 1, pp. 91–99, 2011.

[35] B. Halliwel and J. M. C. Gutteridge, *Free Radicals in Biology and Medicine*, Oxford University Press, New York, NY, USA, 1999.

[36] Y. Saito, K. Nishio, Y. Ogawa et al., "Turning point in apoptosis/necrosis induced by hydrogen peroxide," *Free Radical Research*, vol. 40, no. 6, pp. 619–630, 2006.

[37] M. Nagano-Ito, A. Banba, and S. Ichikawa, "Functional cloning of genes that suppress oxidative stress-induced cell death: TCTP prevents hydrogen peroxide-induced cell death," *FEBS Letters*, vol. 583, no. 8, pp. 1363–1367, 2009.

[38] P. Vandenabeele, L. Galluzzi, T. Vanden Berghe, and G. Kroemer, "Molecular mechanisms of necroptosis: an ordered cellular explosion," *Nature Reviews Molecular Cell Biology*, vol. 11, no. 10, pp. 700–714, 2010.

[39] Y. Yan, V. M. Weaver, and I. A. Blair, "Analysis of protein expression during oxidative stress in breast epithelial cells using a stable isotope labeled proteome internal standard," *Journal of Proteome Research*, vol. 4, no. 6, pp. 2007–2014, 2005.

[40] M. Lucibello, A. Gambacurta, M. Zonfrillo et al., "TCTP is a critical survival factor that protects cancer cells from oxidative

Biological Effects of Mammalian Translationally Controlled Tumor Protein (TCTP) on
Cell Death, Proliferation, and Tumorigenesis

51

stress-induced cell-death," *Experimental Cell Research*, vol. 317, no. 17, pp. 2479–2489, 2011.

[41] M. Gnanasekar, S. Thirugnanam, G. Zheng, A. Chen, and K. Ramaswamy, "Gene silencing of translationally controlled tumor protein (TCTP) by siRNA inhibits cell growth and induces apoptosis of human prostate cancer cells," *International Journal of Oncology*, vol. 34, no. 5, pp. 1241–1246, 2009.

[42] M. Rinnerthaler, S. Jarolim, G. Heeren et al., "MMI1 (YKL056c, TMA19), the yeast orthologue of the translationally controlled tumor protein (TCTP) has apoptotic functions and interacts with both microtubules and mitochondria," *Biochimica et Biophysica Acta*, vol. 1757, no. 5-6, pp. 631–638, 2006.

[43] M. Tuynder, G. Fiucci, S. Prieur et al., "Translationally controlled tumor protein is a target of tumor reversion," *Proceedings of the National Academy of Sciences of the United States of America*, vol. 101, no. 43, pp. 15364–15369, 2004.

[44] F. Arcuri, S. Papa, A. Carducci et al., "Translationally controlled tumor protein (TCTP) in the human prostate and prostate cancer cells: expression, distribution, and calcium binding activity," *Prostate*, vol. 60, no. 2, pp. 130–140, 2004.

[45] I. N. Colaluca, D. Tosoni, P. Nuciforo et al., "NUMB controls p53 tumour suppressor activity," *Nature*, vol. 451, no. 7174, pp. 76–80, 2008.

[46] J. van Rheenen, J. Condeelis, and M. Glogauer, "A common cofilin activity cycle in invasive tumor cells and inflammatory cells," *Journal of Cell Science*, vol. 122, no. 3, pp. 305–311, 2009.

[47] M. Sidani, D. Wessels, G. Mouneimne et al., "Cofilin determines the migration behavior and turning frequency of metastatic cancer cells," *Journal of Cell Biology*, vol. 179, no. 4, pp. 777–791, 2007.

[48] J. Jung, H. Y. Kim, M. Kim, K. Sohn, M. Kim, and K. Lee, "Translationally controlled tumor protein induces human breast epithelial cell transformation through the activation of Src," *Oncogene*, vol. 30, no. 19, pp. 2264–2274, 2011.

[49] H. Thiele, M. Berger, A. Skalweit, and B. J. Thiele, "Expression of the gene and processed pseudogenes encoding the human and rabbit translationally controlled turnout protein (TCTP)," *European Journal of Biochemistry*, vol. 267, no. 17, pp. 5473–5481, 2000.

[50] Y. Gachet, S. Tournier, M. Lee, A. Lazaris-Karatzas, T. Poulton, and U. A. Bommer, "The growth-related, translationally controlled protein P23 has properties of a tubulin binding protein and associates transiently with microtubules during the cell cycle," *Journal of Cell Science*, vol. 112, no. 8, pp. 1257–1271, 1999.

[51] Y. C. Hsu, J. J. Chern, Y. Cai, M. Liu, and K. W. Choi, "Drosophila TCTP is essential for growth and proliferation through regulation of dRheb GTPase," *Nature*, vol. 445, no. 7129, pp. 785–788, 2007.

[52] X. Dong, B. Yang, Y. Li, C. Zhong, and J. Ding, "Molecular basis of the acceleration of the GDP-GTP exchange of human Ras homolog enriched in brain by human translationally controlled tumor protein," *The Journal of Biological Chemistry*, vol. 284, no. 35, pp. 23754–23764, 2009.

Inhibition of the Mitochondrial Permeability Transition for Cytoprotection: Direct *versus* Indirect Mechanisms

Cécile Martel, Le Ha Huynh, Anne Garnier, Renée Ventura-Clapier, and Catherine Brenner

LabEx LERMIT, INSERM U769, Faculté de Pharmacie, Université Paris-Sud, 5 Rue J.-B. Clément,
92290 Châtenay-Malabry, France

Correspondence should be addressed to Catherine Brenner, catherine.brenner-jan@u-psud.fr

Academic Editor: Etienne Jacotot

Mitochondria are fascinating organelles, which fulfill multiple cellular functions, as diverse as energy production, fatty acid β oxidation, reactive oxygen species (ROS) production and detoxification, and cell death regulation. The coordination of these functions relies on autonomous mitochondrial processes as well as on sustained cross-talk with other organelles and/or the cytosol. Therefore, this implies a tight regulation of mitochondrial functions to ensure cell homeostasis. In many diseases (e.g., cancer, cardiopathies, nonalcoholic fatty liver diseases, and neurodegenerative diseases), mitochondria can receive harmful signals, dysfunction and then, participate to pathogenesis. They can undergo either a decrease of their bioenergetic function or a process called mitochondrial permeability transition (MPT) that can coordinate cell death execution. Many studies present evidence that protection of mitochondria limits disease progression and severity. Here, we will review recent strategies to preserve mitochondrial functions via direct or indirect mechanisms of MPT inhibition. Thus, several mitochondrial proteins may be considered for cytoprotective-targeted therapies.

1. Introduction

Mitochondria are intracellular organelles, whose first discovered function is energy production by oxidative phosphorylation [1]. Depending on the mammalian tissue, mitochondria may have additional functions, such as β oxidation, heat production, reactive oxygen species (ROS) metabolism, and cell death coordination. However, since the emergence of the concept of mitochondrial control of cell death in the 95's (for recent reviews: [2, 3]), it became evident that mitochondria participate to various types of cell death, that are, apoptosis, necrosis, oncosis and mitotic catastrophy via mitochondrial membrane permeabilization (MMP), release of proapoptotic factors contained in the intermembrane space to the cytosol and possibly fission, even if mitochondrial fragmentation is not sufficient *per se* to induce cell death [4, 5]. Mitochondrial dysfunction has been associated with a series of human diseases such as cancer, cardiopathies, nonalcoholic fatty liver diseases, neurodegenerative diseases, and aging. When due to genetic dysfunction, the diseases have been systematically

characterized in animal models [6, 7]. Thus, mitochondrial impairment can be linked either to the metabolic function of these organelles, their role in cell death, or both. In addition, in chronic pathologies, such as cardiac volume overload-induced hypertrophy [8], mitochondrial dysfunction precedes cell loss by apoptosis and necrosis, meaning that both dysfunctions can be separated chronologically during the progression of the disease. This is also observed in the pathogenesis of nonalcoholic steatohepatitis, whatever its initial cause, as extensively reviewed [9]. Hepatic mitochondrial dysfunction would lead to apoptosis or necrosis depending on the energy status of the cell [9].

Metabolic impairment manifests by decreased ATP synthesis capacity, enhanced ROS production due to electron leak from the respiratory chain, change in intracellular pH and frequently, by morphological alterations of mitochondrial network. For instance, heart failure which is defined as the inability of the heart to keep up with the demands and to provide adequate blood flow to other organs such as the brain, liver, and kidneys is accompanied by a decrease in energy production and energy transfer capacity [10]. This leads to a decrease in energy charge of the myocardium that has been described as a prognostic factor in dilated cardiomyopathies [11]. This metabolic impairment also affects the peripheric circulation and was shown to involve decreased mitochondrial biogenesis [10].

MMP corresponds to multiple events that irreversibly lead to cell death [2, 12]. These lethal events are nonexclusive, some of them can occur independently, whereas others are intimately linked. Thus, MMP refers to protein translocation from cytosol to outer membrane (OM), rupture of outer mitochondrial membrane, loss of inner membrane potential ($\Delta\Psi m$), cristae remodeling and release of intermembrane space proteins such as cytochrome c or apoptosis-inducing factor (AIF). For instance, upon various stress, Bax or tBid, which reside in the cytosol, can translocate to mitochondrial membranes, oligomerize with mitochondrial proteins to form large channels allowing cytochrome c release and activation of the intrinsic apoptosis signaling cascade (for review: [2]).

In many pathophysiological models, but not all, MMP also involves the so-called opening of the permeability transition pore complex (PTPC), which mediates a nonselective permeabilization of the IM and OM to molecules of molecular mass (MM) under 1.5 kDa (see below for more details) [2, 12]. Thus, in chemotherapy-treated tumor cell lines and ischemic neuronal cells, Bax can interact with the adenine nucleotide translocase (ANT) and/or the voltage-dependent anion channel (VDAC) to promote MMP and cell death.

Here, we will review direct and indirect mechanisms or means to protect mitochondrial functions via a closure of PTPC and a prevention of mitochondrial permeability transition (MPT). The discussion of MPT regulatory mechanisms will be based on selected articles focusing on heart diseases and cancer.

2. Mitochondrial Membrane Permeability and PTPC

Mitochondrial membrane permeability is strictly controlled in unicellular and multicellular organisms harboring these organelles. The OM is believed to be freely permeable to ions and metabolites via entry *through* protein channels (e.g., voltage-dependent anion channel, VDAC), whereas the inner membrane (IM) is considered as impermeable. Thus, the entry and exit of ions or metabolites trough the IM are mediated by integral proteins such as the members of the mitochondrial carrier family [13]. The prototypic protein of this family is ANT or ADP/ATP carrier, which mediates the stoichiometric exchange of ADP and ATP between the matrix and the intermembrane space [14]. Moreover, osmotic movements of water accompany solutes transport from cytosol to matrix, but the molecular basis of these transports is still largely unknown [15]. When excessive stimulation by endogenous signals (excessive ROS, calcium (Ca^{2+}) overload, protease activation, lipid accumulation etc) or activation of harmful signaling pathways (e.g., kinases/phosphatases,

proteases, Bax/-Bid-mediated pathways etc.) occur, mitochondria undergo the MPT, a phenomenon that consists in a sudden increase in IM permeability to small molecules. MPT is a phenomenon first studied in isolated beef heart mitochondria in response to Ca^{2+} overload [16]. Thus, the response of isolated mitochondria to doses of Ca^{2+} is a nonspecific increase of the permeability of the IM, resulting in entry of water and solutes, loss of $\Delta\Psi m$, matrix swelling, and simultaneous uncoupling of oxidative phosphorylation (Figure 1). Of note, the doses of Ca^{2+} depend largely on the tissue origin of mitochondria and the amount of Ca^{2+} present in the buffers. Ultimately, MPT is accompanied by matrix swelling and OM ruptures as shown by transmission electron microscopy [17–19]. This phenomenon can be blocked by Ca^{2+} chelation, ATP, Mg^{2+}, and cyclosporin A (CsA) *in vitro* as well as *in vivo* [20–22].

MPT can be followed experimentally by the loss of absorbance of a suspension of isolated mitochondria by spectrophotometry and by the loss of $\Delta\Psi m$ using suitable fluorescent probes (e.g. tetramethylrhodamine methyl ester (TMRM), rhodamine 123 (Rhod123), 5, 5′, 6, 6′-tetrachloro-1, 1′, 3, 3′-tetraethylbenzimidazol-carbocyanine iodide (JC-1)) [19, 23]. The main interests of the use of isolated organelles are to monitor mitochondrial responses that are directly induced by compounds independently of other cellular compartments and the possibility to automate the measure in the perspective of pharmacological studies [19, 23].

One major pitfall is that, whereas isolation procedures are believed to be nondestructive for liver and cell lines mitochondria [24, 25], mitochondrial responses of isolated mitochondria from skeletal muscle and heart that may rely on the cell architecture are (obviously) lost [26].

Another pitfall is the cross-contamination of the mitochondrial fraction with other cellular compartments and purity of preparations must be checked carefully. MPT can also be measured by imaging with the fluorescent probe calcein in the presence of cobalt in living cell as various as hepatocytes, astrocytes and cardiomyocytes [27–31]. The principle is that calcein (molecular weight, 620 Da) can diffuse into the whole cell, whereas cobalt, a fluorescence quencher, cannot enter into the mitochondrial matrix and diffuses into the rest of the cell. Thus, in physiological conditions mitochondria appear fluorescent and following MPT, the quenching of calcein by cobalt triggers a decrease in fluorescence. For instance, HeLa cells treated by thapsigargin, a SERCA pump inhibitor or A23187, a Ca^{2+} ionophore [32], undergo MPT as shown by a significant decrease in calcein fluorescence due to IM permeabilization and cobalt quenching [30]. MPT has also been monitored in whole heart by 2-deoxy[^{3}H]glucose entrapment technique [33].

Of note, the full demonstration that the process is mediated by PTPC opening requires its inhibition by pretreatment of cells or isolated mitochondria by CsA, the well-known cyclophilin D (CypD) ligand [20, 34]. Moreover, silencing of CypD by siRNA to prevent the induction of MPT is becoming mandatory *in cellulo*, since the genetic demonstration that CypD is critical for MPT and cell death [35].

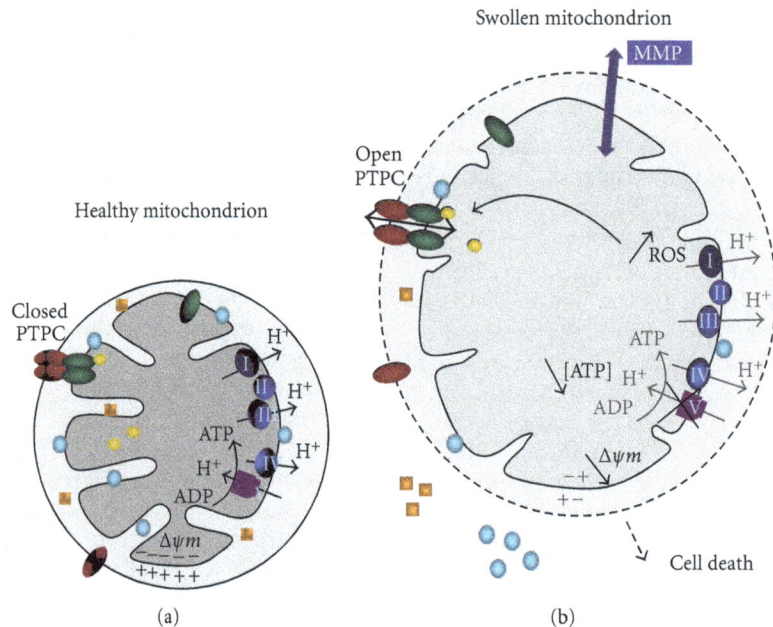

FIGURE 1: Scheme of mitochondrial alterations following mitochondrial membrane permeabilization (MMP). In this model, in response to the opening of the permeability transition pore (PTPC; green and red ellipses, corresponding to ANT and VDAC resp.), swollen mitochondria exhibit an increase in volume, a more translucide matrix with less cristae and a permeabilized outer membrane. Cytochrome c and apoptosis-inducing factor (AIF) (blue circles and yellow squares), normally confined into the intermembrane space, are released trough ruptures in the outer membrane. The transmembrane inner potential ($\Delta\Psi m$) is dissipated in response to the arrest of the function of the respiratory complexes (I to V), which contributes to an inhibition of ATP biosynthesis. Altogether, these alterations are lethal, irreversible and lead to cell death.

3. Direct Mechanisms of MPT Inhibition

PTPC is defined as a voltage-dependent polyprotein complex, which in certain conditions might form a nonselective channel at contact sites between both mitochondrial membranes [36, 37]. By definition, mitochondria contain all the proteins necessary for MPT induction and then MPT does not necessitate any neosynthesis. Since the initial PTPC identification by electrophysiology, the molecular identity of this pore and its regulators is still controversial [38–40]. ANT, VDAC and CypD, the three former PTPC candidates, have their own functions, irrespective of their association within PTPC or other putative polyprotein complexes such as the ATP synthasome [41] and Bcl-2 family member oligomers [3, 12]. Thus, some of the unknown members of PTPC may have their own role in metabolism (e.g., kinase, peptidyl-prolyl isomerase, deshydrogenase), transport (e.g., mitochondrial carrier, channel) or structure (e.g., dynamic machinery, cytoskeleton, AKAP proteins). This means that lethal MPT needs a stimulation to occur and in pathophysiological conditions, this is mainly achieved by Ca^{2+} and ROS. Whatever its composition, the PTPC is a widespread phenomenon occurring in many diseases. Although it has been the subject of intense research and therapeutic developments in cancer with the search for PTPC inducers [42], it has emerged only recently as a promising target for cytoprotection in various diseases such as neurodegenerative, cardiovascular and metabolic diseases. PTPC can be modulated directly by a large panel of pharmacological agents, by post-translational modifications and by cooperation with other proteins that may have a major impact on the cell life as discussed below.

3.1. Pharmacological Inhibition of PTPC. Using isolated mitochondria from various sources, an impressive body of literature reports that many molecules or compounds modulate the PTPC in response to Ca^{2+}, ROS, or a disease. Thus, some compounds can activate MPT, whereas a more limited number of them can prevent the opening of PTPC. Some compounds have known mitochondrial targets such as ANT, VDAC, CypD, and translocator protein 18 kDa (Table 1). To summarize, the most investigated inhibitor of PTPC is CsA, which modulates CypD, as discussed below.

One example with future therapeutic applications is cardiac ischaemia/reperfusion injury. During an acute myocardial infarction (AMI), tissue injury occurring after reperfusion represents a significant amount of the whole, irreversible damage. Ischaemia and reperfusion cause a wide array of functional and structural alterations of mitochondria. Some of these responses are directly under the control of the highly conserved transcriptional complex HIF-1 and result from a modulation in expression of genes involved in glycolysis, glucose metabolism, mitochondrial function, cell survival, apoptosis, and resistance to oxidative stress [43].

PTPC opening plays a crucial role in this specific component of myocardial infarction. Strong support for this concept has recently been provided by the reduced infarct size observed in mice lacking CypD [35]. Thus targeting

TABLE 1: List of mitochondrial permeability transition (MPT) inhibitors. CypD, Cyclophilin D; VDAC, voltage-dependent anion channel; ANT, adenine nucleotide translocase; UQ, ubiquinone.

MPT inhibitor	Target	Model	References
Cyclosporin A	CypD binding with ANT	*In vivo*, isolated mitochondria, cells	[44, 45]
Sanglifehrin A	CypD	*In vivo*, isolated mitochondria, cells	[46, 47]
Bongkrekic acid	ANT	Isolated mitochondria, cells	[48–50]
ADP/ATP	ANT	Isolated mitochondria, *in vitro*	[51, 52]
NADH/NAD+	VDAC	Isolated mitochondria, *in vitro*	[53–55]
DIDS	VDAC	Isolated mitochondria, *in vitro*, cells	[52–58]
glutamate	VDAC	Isolated mitochondria, *in vitro*	[59–61]
Ro 68–3400	ANT or PiC, not VDAC1	Isolated mitochondria, *in vitro*	[61–63]
UQ(0)	ANT or PiC	Isolated mitochondria, *in vitro*	[64, 65]
S15176	unknown, in IM	*In vivo*, isolated mitochondria	[66, 67]
Sildenafil	unknown	*In vivo*, isolated mitochondria	[68]
Debio-025	CypD	*In vivo*, isolated mitochondria	[69]
TRO19622	VDAC, translocator protein 18 kDa	*In vivo*, isolated mitochondria	[70]
Carbon monoxide	ANT, unknown	Isolated mitochondria, cells	[71]
Antamanide	CypD	Isolated mitochondria, cells	[72]

PTPC appears a relevant strategy to reduce ischaemic damages at reperfusion. A large body of evidence has shown that it is possible to reduce infarct size and to protect the heart after an infarct by a postconditioning or pharmacological strategy. Brief episodes of myocardial ischemia-reperfusion employed during reperfusion after a prolonged ischaemic insult may attenuate the total ischaemia-reperfusion injury. Recently, CsA has been shown to dramatically reduce infarct size in many animal species and in human. Recent proof-of-concept clinical trials support the idea that targeting MPT by either coronary intervention postconditioning or CsA can reduce infarct size and improve the recovery of contractile function after reperfusion [21, 73]. Such a strategy was also applied to ischaemia-reperfusion damages in other tissues and cells like vascular endothelial cells [74], hepatocytes [75, 76], and neurons [77–79]. Moreover, some attempts to target CsA to the mitochondrial compartment by conjugation to the lipophilic triphenylphosphonium cation proved to be promising in cytoprotection from glucose and oxygen deprivation in neurons [80], in cardiomyocytes [81], and in other various organs [82].

3.2. Role of Mitochondrial Kinases to Prevent PTPC Opening. Several protective signal pathways involving multiple kinases have been shown to converge on mitochondria and the PTPC [83] to promote cell survival (Table 2). For instance, in cardiomyocytes, pools of kinases such as Akt, protein kinase C-ε (PKCε), extracellular-regulated kinases (ERK), glycogen synthase kinase-3 beta (GSK-3β), and hexokinases (HK) I and II, are localized in or on mitochondria in addition to the cytosol. These mitochondria-associated protein kinases may integrate cytosolic stimuli and in turn, enhance tolerance of myocytes to injury.

Moreover, systematic proteomic approaches revealed that some of these kinases might form hetero-oligomers with putative components of the PTPC. Briefly, GSK-3β and HKs are directly responsible for inhibition of opening of the PTPC and, thus, for myocyte protection from necrosis [96]. As a result, postconditioning, which leads to GSK-3β inhibition, allows the myocardial salvage from reperfusion injury by modulating MPT [86]. In the context of anticancer chemotherapy, β-adrenergic receptors (β-ARs) modulate anthracycline response through crosstalk with multiple signaling pathways. β2-ARs are cardioprotective during exposure to oxidative stress induced by doxorubicin (DOX). DOX cardiotoxicity is mediated in part through a Ca^{2+}-dependent triggering of MPT as clearly shown by a 41% reduction of DOX-induced mortality by CsA [97]. β2-ARs activate prosurvival kinases and attenuate mitochondrial dysfunction caused by oxidative stress. Accordingly, the invalidation of β2-ARs enhances cardiotoxicity via negative regulation of survival kinases and enhancement of intracellular Ca^{2+}, thus predisposing the mitochondria to opening of the PTPC [97].

Moreover, in cancer cell lines, activation of mitochondrial ERK protects cancer cells from death through inhibition of the MPT [88]. ERK inhibition enhanced GSK-3β-dependent phosphorylation of the pore regulator CypD, whereas GSK-3β inhibition protected from PTPC opening.

By different molecular mechanisms, some kinases such as creatine kinase (CK) and HK have also cytoprotective effects and prevent PTPC opening. Depending on the tissue, which supports their expression, these kinases may be cytoprotective via a role in energy transfer to metabolites such as creatine and glucose (Table 2).

3.3. Stabilization of Mitochondrial Membrane Permeability by Bcl-2 Family Members. The Bcl-2 family is composed of more than 25 proteins implicated in the control of life-or-death decision [98]. This protein family has been particularly studied in cancer, which led to the classification of Bcl-2 and

TABLE 2: List of kinases contributing to a closure of PTP via phosphorylation mechanisms or protein-protein interaction. HK, hexokinase, CK, creatine kinase, PKG, protein kinase G, PKA, protein kinase A, PKC, protein kinase C, ERK, extracellular signal-regulated kinase, GSK3, glucose-regulated kinase 3, PI3K, phosphoinositol3 kinase, and Akt/PKB, protein kinase B.

Kinase	Effect	Target/pathway	Model	References
Akt/PKB, PI3K	Indirect	GSK3 via PI3K or eNos/PKG pathways	Cells	[84, 85]
GSK3	Direct	VDAC, ANT, CypD	Isolated mitochondria, cells, *in vivo*, *in vitro*	[8, 37, 86, 87]
ERK	Indiret	GSK3 via PI3K pathway	Cells	[85, 88]
PKA	Direct	VDAC	Isolated mitochondria	[89]
PKC epsilon	Direct	VDAC	Isolated mitochondria, *in vivo*	[90]
PKG	Direct	Unknown	Isolated mitochondria, *in vivo*	[91, 92]
CK	Local regulation of ATP/creatine pools	Energetic metabolism	CK-expressing tissues	[93, 94]
HK	Local regulation of glucose/ATP pools	Energetic metabolism	Isolated mitochondria, cells, *in vitro*	[57, 74, 87, 95]

Bax as oncogenes and tumor-suppressors, respectively. Some members (e.g., Bax/Bad proteins, BH3-only proteins) favor apoptosis, whereas other members, such as Bcl-2 and Bcl-XL, prevent apoptosis. Moreover, it has been shown that the effects of Bcl-2 family proteins on mitochondria in cancer cells are linked to clinical responses to chemotherapy [99].

The cytoprotective mechanisms of Bcl-2 family members are multiple and include direct mitochondrial effects [3, 12]. Thus, Bcl-2 contributes to the stabilization of the mitochondrial membrane permeability, inhibition of $\Delta\Psi$m loss and cytochrome c release and, at least in tumor cells, to the stimulation of oxidative phosphorylation [100–103]. Thus, direct protein-protein interactions between Bcl-2 family members, but also with several constitutive mitochondrial proteins such as ANT (IM), VDAC (OM), or FoF1-ATP synthase (IM) have been evidenced [104]. Bcl-2 cooperates directly with ANT, to prevent PTPC opening and to inhibit cell death [101, 105, 106]. In addition, Bcl-2 blocks the Ca^{2+}-induced channel function of ANT and favors ADP/ATP translocase function, which positively impacts the intracellular levels of ATP [107]. Similarly, Bcl-2 and Bcl-XL directly interact with VDAC modulating its channel function [108]. Bcl-XL would bind also to the mitochondrial F0F1-ATP synthase and regulate metabolic efficiency in neurons [104]. Accordingly, recombinant proteins of some members of the Bcl-2 family directly modulate PTPC in isolated mitochondria from various sources, that is, liver and cancer cells [25, 109]. Some specific regions of Bcl-2 (e.g., BH3, BH4 domains) are responsible for these effects and as expected, peptides corresponding to these regions proved to modulate apoptosis *in cellulo* or in isolated mitochondria, again indicating a direct targeting of mitochondria [25]. Finally, *in vivo*, the BH4 domain of Bcl-XL exerts antiapoptotic effects and attenuates ischaemia/reperfusion injury through anti-apoptotic mechanism in rat hearts [110, 111].

4. Indirect Mechanisms, Which Lead to an Increased Resistance of PTPC Opening

By definition, several indirect mechanisms may lead to blockade of PTPC opening via modulation of $\Delta\Psi$m, mitochondrial mass regulation, redox state, fusion/fission processes and calcium retention capacity. This may be due to modification in protein expression, in posttranslational modifications and in their interactome, which consequently affect signaling pathways. Below, we will analyze three indirect mechanisms of PTPC protection that have recently been elucidated.

4.1. Anti-Oxidant Protection. Mitochondria are major sites of ROS production, which may contribute to the development of various diseases including cardiovascular diseases and aging. Several studies have thus described the effects of antioxidant administration in the context of cardiac and liver pathologies in mice [112]. It is widely admitted that natural antioxidants such as resveratrol and curcumin have beneficial effects against ischaemia/reperfusion damages to mitochondria and cells in rat liver or heart [113, 114]. These effects are complex since resveratrol has been proposed to have multiple intracellular targets such as AMPK, SIRT1 and Nrf2, which can influence the transcriptome to increase the anti-oxidant defense (e.g., catalase, GPx, and GCLC), and other genes such eNOS and PGC1α, which favor an increase in mitochondrial mass and bioenergetics and decrease in apoptosis and inflammation (for review: [115]).

Resveratrol treatment exerts beneficial protective effects on survival, endothelium-dependent relaxation, and cardiac contractility and mitochondrial function, suggesting that resveratrol or metabolic activators could be a relevant therapy in hypertension-induced heart failure [116]. Similarly, in the heart, curcumin another polyphenol with antioxidant properties showed cardioprotective effects in catecholamine induced cardiotoxicity through prevention of mitochondrial damage, PTPC opening [117], and ventricular dysfunction [118] and in protecting rat myocardium against ischaemic insult by decreasing oxidative stress [119].

Another promising example of compound is MitoQ10, an ubiquinone derivative, which is a mitochondria-targeted antioxidant [120]. It has proven to be useful for protecting endothelial function and attenuating cardiac hypertrophy in stroke-prone hypertensive rats [121]. Moreover, MitoQ10 potently inhibits cocain-induced cardiac damage via a restoration of oxygen consumption and a stabilization of ROS levels, specifically in interfibrillar mitochondria [122]. However, when used on isolated cardiac mitochondria, MitoQ10 can be enhanced in a dose-dependent-manner MPT in the presence of the prooxidant tert-butyl hydoperoxide (t-BHP) and suboptimal doses of Ca^{2+} (Figure 2), although it acts as an antioxidant on rat liver mitochondria [123]. This underscores the duality of anti- and prooxidant compounds, whose effects can depend either on the dose, the redox state of the cell, the tissue, and/or the mode of administration. This probably explains, at least in part, the failure of anti-oxidants to protect efficiently the heart function in clinical trials, as recently reviewed in [124].

4.2. Estrogens Protection. Sex and gender influence the onset and the progression of many human diseases, notably age-related diseases. Thus, estrogens, mainly 17β-estradiol, may have pleiotropic effects depending on the tissue [125]. For instance, certain cardiovascular diseases, such as myocardial hypertrophy and heart failure, differ clearly in their clinical manifestation and prognosis between women and men [126]. As a consequence, hormonal mechanisms underlying sex and gender differences are currently under intense investigation.

Animal and cellular models have been particularly instrumental to better understand estrogen protection at the level of mitochondria [127]. For instance, in cerebral circulation, estrogens mediate an enhancement of vasodilator capacity, suppression of vascular inflammation and increase of mitochondrial efficiency [125, 128, 129]. This effect is, at least in part, due to an increase in mitochondrial biogenesis via gene expression modulation [129] and a decrease in superoxide production [125]. Accordingly, chronic estrogen treatment increases mitochondrial capacity for oxidative phosphorylation while decreasing production of ROS. In breast and lung cancer cells, long-term estradiol treatment activates transcription of NRF-1 and increases mitochondrial biogenesis [130].

Moreover, mitochondrial effects on PTPC might be mediated indirectly by estrogen receptors, α and β, present in nucleus, plasma membrane, endoplasmic reticulum and even mitochondria [131]. In the context of ischaemia-reperfusion injury, it is widely admitted that estrogens protects from myocardial damage via an inhibition of PTPC function. Notably, estradiol may activate the signaling cascade which involves Akt, NO synthase, guanylyl cyclase and protein kinase G, which results in blockade of MPT-induced release

of cytochrome c from mitochondria, respiratory inhibition and caspase activation [131]. As a result, estrogens effects are multifactorial, mostly indirect. Even if some estrogen-like molecules can be effective on isolated mitochondria, a precise target of estrogen within PTPC is still unknown [132]. Thus, intriguingly, estrogens may prevent Ca^{2+}-induced cytochrome c release in isolated heart mitochondria, but not mitochondrial swelling [133].

Estrogens also protect from chemotherapy-induced cardiomyopathy in ovariectomized rats. Again, effects on the anti-oxidant cellular defenses have been proposed as one of the target mechanism of estrogen [134].

4.3. Exercise Protection. Exercise training has proven to be beneficial in chronic diseases including heart failure, obesity, diabetes or metabolic syndrome. Because endurance training improves symptoms and quality of life and decreases mortality rate and hospitalization, it is increasingly recognized as a beneficial practice for these patients. Adaptation to endurance training mainly involves energetic remodeling in skeletal and cardiac muscles [135]. The mechanisms involved in the beneficial effects of exercise training are far from being understood. Skeletal muscles adapt to repeated prolonged exercise by marked quantitative and qualitative changes in mitochondria. Endurance training promotes an increase in mitochondrial volume density and mitochondrial proteins by activating mitochondrial biogenesis [136]. Exercise training decreases apoptotic processes, and protects mitochondrial function from oxidative stress and other cardiac insults [137, 138]. Exercise training results in a reduced sensitivity to PTPC opening in heart mitochondria and confers mitochondrial protection. Moreover, even acute exercise protects against cardiac mitochondrial dysfunction, preserving mitochondrial phosphorylation capacity and attenuating DOX-induced decreased tolerance to PTPC opening [139]. Proposed mechanisms to explain the cardioprotective effects of exercise are mediated, at least partially, by redox changes and include the induction of myocardial heat shock proteins, improved cardiac antioxidant capacity and/or elevation of other cardioprotective molecules [137].

5. Conclusion and Open Questions

In the last decade, direct and indirect approaches to protect mitochondrial functions via PTPC modulation have been explored. However, it is still too early to decipher the most efficient strategies in term of cytoprotection. Nevertheless, recent studies and research advances have propelled mitochondria on the scene front of new therapeutic strategies. However, a contradiction emerges between the need to kill tumor cells in cancer therapy and to protect other cells from injuries. Even more worrying is the fact that many anticancer therapies have mitochondrial toxicity that becomes dramatic when highly oxidative nondividing cells like cardiomyocytes are concerned. Indeed, mitochondria are the main target when cardiotoxicity of anticancer drugs is concerned [44, 140]. Thus one challenging issue of cytoprotection directed

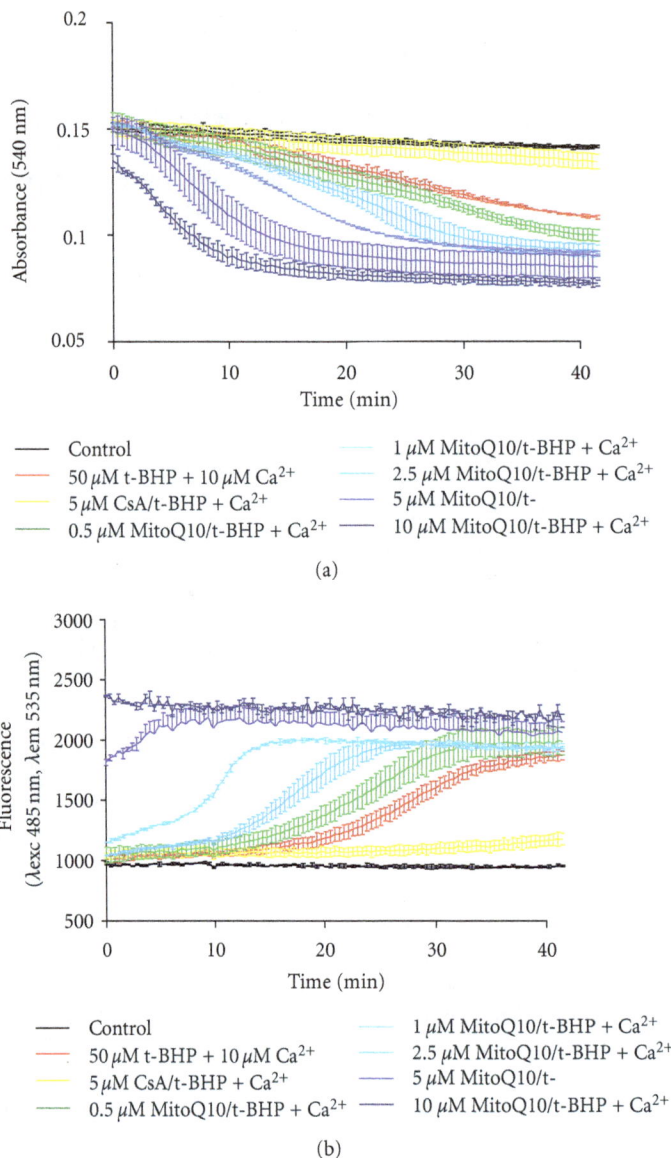

FIGURE 2: MitoQ10 stimulates MPT in isolated cardiac mitochondria. (a) MitoQ10 increases the mitochondrial swelling induced by an oxidant stress. Mitochondria (25 μg of proteins) have been pretreated by the indicated doses of MitoQ10 and treated by 50 μM t- BHP + 10 μM Ca^{2+}. Absorbance at 540 nm has been registered for 60 min at 37°C. (b) MitoQ10 increases the mitochondrial depolarization induced by an oxidant stress. Mitochondria (25 μg of proteins) have been loaded with 2 μM Rhodamine 123, pretreated by the indicated doses of MitoQ10, and treated by 50 μM t-BHP + 10 μM Ca^{2+}. Fluorescence has been registered for 60 min at 37°C.

to mitochondria would be to uncover new molecules or treatments that would selectively target cancer cells without affecting cardiac mitochondria. This should stimulate new studies devoted to increase our basic knowledge of the mechanisms and the tissue specificity of PTPC opening and mitochondrial function. At the same time, this will open the possibility to search for new drugs with tissue-specific effects on mitochondria. Finally, another challenge that basic and clinical research will face in the future is the notion of sex and gender influence that might be decisive for the treatment of many severe diseases.

Abbreviations

ANT: Adenine nucleotide translocase
β-Ars: β-adrenergic receptors
Ca^{2+}: Calcium
CK: Creatine kinase
CsA: Cyclosporin A
CypD: Cyclophilin D
ΔΨm: Mitochondrial inner membrane potential
DOX: Doxorubicin
ERK: Extracellular-signal-regulated kinase

GSK-3β: Glycogen synthase kinase-3 beta
HK: Hexokinase
IM: Inner membrane
JC-1:   5, 5′, 6, 6′ -tetrachloro-1, 1′, 3, 3′
    -tetraethylbenzimidazol-carbocyanine
    iodine
MMP: Mitochondrial membrane
    permeabilization
MPT: Mitochondrial permeability transition
OM: Outer membrane
PTPC: permeability transition pore complex
Rhod123:   Rhodamine 123
ROS: Reactive oxygen species
TMRM: Tetramethylrhodamine methyl ester
VDAC: Voltage-dependent anion channel.

Acknowledgments

C. Brenner and R. V. Clapier are senior scientists at the Centre National de la Recherche Scientifique. A. Garnier, C. Brenner and R. V. Clapier are supported by LabEx LERMIT. C. Brenner is supported by ANR (ANR-08PCVI 0008-01). C. Martel received a fellowship from Association pour la Recherche sur le Cancer. L. H. Huynh received a fellowship from INSERM.

References

[1] A. L. Lehninger, "The transfer of energy in oxidative phosphorylation," *Bulletin de la Société de Chimie Biologique*, vol. 46, pp. 1555–1575, 1964.

[2] G. Kroemer, L. Galluzzi, and C. Brenner, "Mitochondrial membrane permeabilization in cell death," *Physiological Reviews*, vol. 87, no. 1, pp. 99–163, 2007.

[3] J.-C. Martinou and R. Youle, "Mitochondria in apoptosis: Bcl-2 family members and mitochondrial dynamics," *Developmental Cell*, vol. 21, no. 1, pp. 92–101, 2011.

[4] G. Majno and I. Joris, "Apoptosis, oncosis, and necrosis: an overview of cell death," *American Journal of Pathology*, vol. 146, no. 1, pp. 3–15, 1995.

[5] K. N. Papanicolaou, G. A. Ngoh, E. R. Dabkowski et al., "Cardiomyocyte deletion of mitofusin-1 leads to mitochondrial fragmentation and improves tolerance to ROS-induced mitochondrial dysfunction and cell death," *American Journal of Physiology*, vol. 302, no. 1, pp. H167–H179, 2012.

[6] D. C. Wallace, "Mitochondrial diseases in man and mouse," *Science*, vol. 283, no. 5407, pp. 1482–1488, 1999.

[7] A. Ramachandran, S. Jha, and D. J. Lefer, "Review paper: pathophysiology of myocardial reperfusion injury: the role of genetically engineered mouse models," *Veterinary Pathology*, vol. 45, no. 5, pp. 698–706, 2008.

[8] J. Matas, N. Tien Sing Young, C. Bourcier-Lucas et al., "Increased expression and intramitochondrial translocation of cyclophilin-D associates with increased vulnerability of the permeability transition pore to stress-induced opening during compensated ventricular hypertrophy," *Journal of Molecular and Cellular Cardiology*, vol. 46, no. 3, pp. 420–430, 2009.

[9] K. Begriche, A. Igoudjil, D. Pessayre, and B. Fromenty, "Mitochondrial dysfunction in NASH: causes, consequences and possible means to prevent it," *Mitochondrion*, vol. 6, no.

1, pp. 1–38, 2006.

[10] R. Ventura-Clapier, A. Garnier, V. Veksler, and F. Joubert, "Bioenergetics of the failing heart," *Biochimica et Biophysica Acta*, vol. 1813, no. 7, pp. 1360–1372, 2011.

[11] S. Neubauer, "The failing heart-An engine out of fuel," *New England Journal of Medicine*, vol. 356, no. 11, pp. 1140–1151, 2007.

[12] I. R. Indran, G. Tufo, S. Pervaiz, and C. Brenner, "Recent advances in apoptosis, mitochondria and drug resistance in cancer cells," *Biochimica et Biophysica Acta*, vol. 1807, no. 6, pp. 735–745, 2011.

[13] F. Palmieri, "The mitochondrial transporter family (SLC25): physiological and pathological implications," *Pflugers Archiv European Journal of Physiology*, vol. 447, no. 5, pp. 689–709, 2004.

[14] M. Klingenberg, "The ADP and ATP transport in mitochondria and its carrier," *Biochimica et Biophysica Acta*, vol. 1778, no. 10, pp. 1978–2021, 2008.

[15] P. Gena, E. Fanelli, C. Brenner, M. Svelto, and G. Calamita, "News and views on mitochondrial water transport," *Frontiers in Bioscience*, vol. 14, pp. 4189–4198, 2009.

[16] D. R. Hunter, R. A. Haworth, and J. H. Southard, "Relationship between configuration, function, and permeability in calcium treated mitochondria," *Journal of Biological Chemistry*, vol. 251, no. 16, pp. 5069–5077, 1976.

[17] M. G. Vander Heiden, N. S. Chandel, E. K. Williamson, P. T. Schumacker, and C. B. Thompson, "Bcl-x(L) regulates the membrane potential and volume homeostasis of mitochondria," *Cell*, vol. 91, no. 5, pp. 627–637, 1997.

[18] E. Jacotot, L. Ravagnan, M. Loeffler et al., "The HIV-1 viral protein R induces apoptosis via a direct effect on the mitochondrial permeability transition pore," *Journal of Experimental Medicine*, vol. 191, no. 1, pp. 33–45, 2000.

[19] A. S. Belzacq-Casagrande, C. Martel, C. Pertuiset, A. Borgne-Sanchez, E. Jacotot, and C. Brenner, "Pharmacological screening and enzymatic assays for apoptosis," *Frontiers in Bioscience*, vol. 14, pp. 3550–3562, 2009.

[20] A. R. Halestrap, C. R. Connern, E. J. Griffiths, and P. M. Kerr, "Cyclosporin A binding to mitochondrial cyclophilin inhibits the permeability transition pore and protects hearts from ischaemia/reperfusion injury," *Molecular and Cellular Biochemistry*, vol. 174, no. 1-2, pp. 167–172, 1997.

[21] C. Piot, P. Croisille, P. Staat et al., "Effect of cyclosporine on reperfusion injury in acute myocardial infarction," *New England Journal of Medicine*, vol. 359, no. 5, pp. 473–481, 2008.

[22] A. L. Nieminen, T. G. Petrie, J. J. Lemasters, and W. R. Selman, "Cyclosporin A delays mitochondrial depolarization induced by N-methyl-D-aspartate in cortical neurons: evidence of the mitochondrial permeability transition," *Neuroscience*, vol. 75, no. 4, pp. 993–997, 1996.

[23] J. R. Blattner, L. He, and J. J. Lemasters, "Screening assays for the mitochondrial permeability transition using a fluorescence multiwell plate reader," *Analytical Biochemistry*, vol. 295, no. 2, pp. 220–226, 2001.

[24] C. Frezza, S. Cipolat, and L. Scorrano, "Organelle isolation: functional mitochondria from mouse liver, muscle and cultured filroblasts," *Nature Protocols*, vol. 2, no. 2, pp. 287–295, 2007.

[25] N. Buron, M. Porceddu, M. Brabant et al., "Use of human cancer cell lines mitochondria to explore the mechanisms of BH3 peptides and ABT-737-induced mitochondrial membrane permeabilization," *PloS One*, vol. 5, no. 3, article e9924, 2010.

[26] M. Picard, T. Taivassalo, G. Gouspillou, and R. T. Hepple, "Mitochondria: isolation, structure and function," *Journal of Physiology*, vol. 589, no. 18, pp. 4413–4421, 2011.

[27] V. Petronilli, G. Miotto, M. Canton et al., "Transient and long-lasting openings of the mitochondrial permeability transition pore can be monitored directly in intact cells by changes in mitochondrial calcein fluorescence," *Biophysical Journal*, vol. 76, no. 2, pp. 725–734, 1999.

[28] G. Zahrebelski, A. L. Nieminen, K. Al-Ghoul, T. Qian, B. Herman, and J. J. Lemasters, "Progression of subcellular changes during chemical hypoxia to cultured rat hepatocytes: a laser scanning confocal microscopic study," *Hepatology*, vol. 21, no. 5, pp. 1361–1372, 1995.

[29] A. L. Nieminen, A. K. Saylor, S. A. Tesfai, B. Herman, and J. J. Lemasters, "Contribution of the mitochondrial permeability transition to lethal injury after exposure of hepatocytes to t-butylhydroperoxide," *Biochemical Journal*, vol. 307, no. 1, pp. 99–106, 1995.

[30] E. N. Dedkova and L. A. Blatter, "Measuring mitochondrial function in intact cardiac myocytes," *Journal of Molecular and Cellular Cardiology*, vol. 52, no. 1, pp. 48–61, 2012.

[31] K. V. Rama Rao, A. R. Jayakumar, and M. D. Norenberg, "Ammonia neurotoxicity: role of the mitochondrial permeability transition," *Metabolic Brain Disease*, vol. 18, no. 2, pp. 113–127, 2003.

[32] L. Galluzzi, N. Zamzami, T. De La Motte Rouge, C. Lemaire, C. Brenner, and G. Kroemer, "Methods for the assessment of mitochondrial membrane permeabilization in apoptosis," *Apoptosis*, vol. 12, no. 5, pp. 803–813, 2007.

[33] E. J. Griffiths and A. P. Halestrap, "Mitochondrial non-specific pores remain closed during cardiac ischaemia, but open upon reperfusion," *Biochemical Journal*, vol. 307, no. 1, pp. 93–98, 1995.

[34] A. P. Halestrap and C. Brenner, "The adenine nucleotide translocase: a central component of the mitochondrial permeability transition pore and key player in cell death," *Current Medicinal Chemistry*, vol. 10, no. 16, pp. 1507–1525, 2003.

[35] C. P. Baines, R. A. Kaiser, N. H. Purcell et al., "Loss of cyclophilin D reveals a critical role for mitochondrial permeability transition in cell death," *Nature*, vol. 434, no. 7033, pp. 658–662, 2005.

[36] M. Crompton, "Mitochondrial intermembrane junctional complexes and their role in cell death," *Journal of Physiology*, vol. 529, no. 1, pp. 11–21, 2000.

[37] D. Brdiczka, G. Beutner, A. Rück, M. Dolder, and T. Wallimann, "The molecular structure of mitochondrial contact sites. Their role in regulation of energy metabolism and permeability transition," *BioFactors*, vol. 8, no. 3-4, pp. 235–242, 1998.

[38] A. P. Halestrap, "What is the mitochondrial permeability transition pore?" *Journal of Molecular and Cellular Cardiology*, vol. 46, no. 6, pp. 821–831, 2009.

[39] F. Ricchelli, J. Šileikyte, and P. Bernardi, "Shedding light on the mitochondrial permeability transition," *Biochimica et Biophysica Acta*, vol. 1807, no. 5, pp. 482–490, 2011.

[40] I. Szabo and M. Zoratti, "The mitochondrial megachannel is the permeability transition pore," *Journal of Bioenergetics and Biomembranes*, vol. 24, no. 1, pp. 111–117, 1992.

[41] Y. H. Ko, M. Delannoy, J. Hullihen, W. Chiu, and P. L. Pedersen, "Mitochondrial ATP synthasome: cristae-enriched membranes and a multiwell detergent screening assay yield dispersed single complexes containing the ATP synthase and carriers for Pi and ADP/ATP," *Journal of Biological Chemistry*,

[42] E. Jacotot, A. Deniaud, A. Borgne-Sanchez et al., "Therapeutic peptides: targeting the mitochondrion to modulate apoptosis," *Biochimica et Biophysica Acta*, vol. 1757, no. 9-10, pp. 1312–1323, 2006.

vol. 278, no. 14, pp. 12305–12309, 2003.

[43] G. Loor and P. T. Schumacker, "Role of hypoxia-inducible factor in cell survival during myocardial ischemia-reperfusion," *Cell Death and Differentiation*, vol. 15, no. 4, pp. 686–690, 2008.

[44] M. Tokarska-Schlattner, M. Zaugg, C. Zuppinger, T. Wallimann, and U. Schlattner, "New insights into doxorubicin-induced cardiotoxicity: the critical role of cellular energetics," *Journal of Molecular and Cellular Cardiology*, vol. 41, no. 3, pp. 389–405, 2006.

[45] K. M. Broekemeier, M. E. Dempsey, and D. R. Pfeiffer, "Cyclosporin A is a potent inhibitor of the inner membrane permeability transition in liver mitochondria," *Journal of Biological Chemistry*, vol. 264, no. 14, pp. 7826–7830, 1989.

[46] E. J. Griffiths and A. P. Halestrap, "Protection by cyclosporin A of ischemia/reperfusion-induced damage in isolated rat hearts," *Journal of Molecular and Cellular Cardiology*, vol. 25, no. 12, pp. 1461–1469, 1993.

[47] S. J. Clarke, G. P. McStay, and A. P. Halestrap, "Sanglifehrin A acts as a potent inhibitor of the mitochondrial permeability transition and reperfusion injury of the heart by binding to cyclophilin-D at a different site from cyclosporin A," *Journal of Biological Chemistry*, vol. 277, no. 38, pp. 34793–34799, 2002.

[48] D. J. Hausenloy, M. R. Duchen, and D. M. Yellon, "Inhibiting mitochondrial permeability transition pore opening at reperfusion protects against ischaemia-reperfusion injury," *Cardiovascular Research*, vol. 60, no. 3, pp. 617–625, 2003.

[49] P. J. Henderson and H. A. Lardy, "Bongkrekic acid. An inhibitor of the adenine nucleotide translocase of mitochondria," *Journal of Biological Chemistry*, vol. 245, no. 6, pp. 1319–1326, 1970.

[50] M. Narita, S. Shimizu, T. Ito et al., "Bax interacts with the permeability transition pore to induce permeability transition and cytochrome c release in isolated mitochondria," *Proceedings of the National Academy of Sciences of the United States of America*, vol. 95, no. 25, pp. 14681–14686, 1998.

[51] N. Zamzami, C. E. Hamel, C. Maisse et al., "Bid acts on the permeability transition pore complex to induce apoptosis," *Oncogene*, vol. 19, no. 54, pp. 6342–6350, 2000.

[52] S. A. Novgorodov, T. I. Gudz, Y. M. Milgrom, and G. P. Brierley, "The permeability transition in heart mitochondria is regulated synergistically by ADP and cyclosporin A," *Journal of Biological Chemistry*, vol. 267, no. 23, pp. 16274–16282, 1992.

[53] R. A. Haworth and D. R. Hunter, "Control of the mitochondrial permeability transition pore by high-affinity ADP binding at the ADP/ATP translocase in permeabilized mitochondria," *Journal of Bioenergetics and Biomembranes*, vol. 32, no. 1, pp. 91–96, 2000.

[54] A. C. Lee, M. Zizi, and M. Colombini, "β-NADH decreases the permeability of the mitochondrial outer membrane to ADP by a factor of 6," *Journal of Biological Chemistry*, vol. 269, no. 49, pp. 30974–30980, 1994.

[55] A. Deniaud, C. Rossi, A. Berquand et al., "Voltage-dependent anion channel transports calcium ions through biomimetic membranes," *Langmuir*, vol. 23, no. 7, pp. 3898–3905, 2007.

[56] A. Deniaud, O. Sharaf El Dein, E. Maillier et al., "Endoplasmic reticulum stress induces calcium-dependent permeability transition, mitochondrial outer membrane permeabilization and apoptosis," *Oncogene*, vol. 27, no. 3, pp. 285–299, 2008.

[57] F. P. Thinnes, H. Florke, H. Winkelbach et al., "Channel active mammalian porin, purified from crude membrane fractions of human B lymphocytes or bovine skeletal muscle, reversibly binds the stilbene-disulfonate group of the chloride channel blocker DIDS," *Biological Chemistry Hoppe-Seyler*, vol. 375, no. 5, pp. 315–322, 1994.

[58] V. Shoshan-Barmatz, N. Keinan, S. Abu-Hamad, D. Tyomkin, and L. Aram, "Apoptosis is regulated by the VDAC1 N-terminal region and by VDAC oligomerization: release of cytochrome c, AIF and Smac/Diablo," *Biochimica et Biophysica Acta*, vol. 1797, no. 6-7, pp. 1281–1291, 2010.

[59] M. P. Rigobello, F. Turcato, and A. Bindoli, "Inhibition of rat liver mitochondrial permeability transition by respiratory substrates," *Archives of Biochemistry and Biophysics*, vol. 319, no. 1, pp. 225–230, 1995.

[60] D. Gincel, S. Silberberg, and V. Shoshan-Barmatz, "Modulation of the voltagedependent anion channel (VDAC) by glutamate," *Journal of Bioenergetics and Biomembranes*, vol. 32, pp. 571–583, 2000.

[61] D. Gincel and V. Shoshan-Barmatz, "Glutamate interacts with VDAC and modulates opening of the mitochondrial permeability transition pore," *Journal of Bioenergetics and Biomembranes*, vol. 36, no. 2, pp. 179–186, 2004.

[62] A. M. Cesura, E. Pinard, R. Schubenel et al., "The voltage-dependent anion channel is the target for a new class of inhibitors of the mitochondrial permeability transition pore," *Journal of Biological Chemistry*, vol. 278, no. 50, pp. 49812–49818, 2003.

[63] A. Krauskopf, O. Eriksson, W. J. Craigen, M. A. Forte, and P. Bernardi, "Properties of the permeability transition in VDAC1$^{-/-}$ mitochondria," *Biochimica et Biophysica Acta*, vol. 1757, no. 5-6, pp. 590–595, 2006.

[64] A. W. C. Leung, P. Varanyuwatana, and A. P. Halestrap, "The mitochondrial phosphate carrier interacts with cyclophilin D and may play a key role in the permeability transition," *Journal of Biological Chemistry*, vol. 283, no. 39, pp. 26312–26323, 2008.

[65] E. Fontaine, F. Ichas, and P. Bernardi, "A ubiquinone-binding site regulates the mitochondrial permeability transition pore," *Journal of Biological Chemistry*, vol. 273, no. 40, pp. 25734–25740, 1998.

[66] A. Elimadi, R. Sapena, A. Settaf, H. Le Louet, J. P. Tillement, and D. Morin, "Attenuation of liver normothermic ischemia—reperfusion injury by preservation of mitochondrial functions with S-15176, a potent trimetazidine derivative," *Biochemical Pharmacology*, vol. 62, no. 4, pp. 509–516, 2001.

[67] A. Elimadi, V. Jullien, J. P. Tillement, and D. Morin, "S-15176 inhibits mitochondrial permeability transition via a mechanism independent of its antioxidant properties," *European Journal of Pharmacology*, vol. 468, no. 2, pp. 93–101, 2003.

[68] A. Ascah, M. Khairallah, F. Daussin et al., "Stress-induced opening of the permeability transition pore in the dystrophin-deficient heart is attenuated by acute treatment with sildenafil," *American Journal of Physiology*, vol. 300, no. 1, pp. H144–H153, 2011.

[69] D. P. Millay, M. A. Sargent, H. Osinska et al., "Genetic and pharmacologic inhibition of mitochondrial-dependent necrosis attenuates muscular dystrophy," *Nature Medicine*, vol. 14, no. 4, pp. 442–447, 2008.

[70] T. Bordet, B. Buisson, M. Michaud et al., "Identification and characterization of cholest-4-en-3-one, oxime (TRO19622), a novel drug candidate for amyotrophic lateral sclerosis," *Journal of Pharmacology and Experimental Therapeutics*, vol. 322, no. 2, pp. 709–720, 2007.

[71] C. S. F. Queiroga, A. S. Almeida, C. Martel, C. Brenner, P. M. Alves, and H. L. A. Vieira, "Glutathionylation of adenine nucleotide translocase induced by carbon monoxide prevents mitochondrial membrane permeabilization and apoptosis," *Journal of Biological Chemistry*, vol. 285, no. 22, pp. 17077–17088, 2010.

[72] L. Azzolin, N. Antolini, A. Calderan et al., "Antamanide, a derivative of amanita phalloides, is a novel inhibitor of the mitochondrial permeability transition pore," *PLoS One*, vol. 6, no. 1, article e16280, 2011.

[73] N. Mewton, P. Croisille, G. Gahide et al., "Effect of cyclosporine on left ventricular remodeling after reperfused myocardial infarction," *Journal of the American College of Cardiology*, vol. 55, no. 12, pp. 1200–1205, 2010.

[74] M. Okorie, D. Bhavsar, D. Ridout et al., "Postconditioning protects against human endothelial ischaemia-reperfusion injury via subtype-specific KATP channel activation and is mimicked by inhibition of the mitochondrial permeability transition pore," *European Heart Journal*, pp. 1266–1274, 2011.

[75] J. S. Kim, L. He, T. Qian, and J. J. Lemasters, "Role of the mitochondrial permeability transition in apoptotic and necrotic death after ischemia/reperfusion injury to hepatocytes," *Current Molecular Medicine*, vol. 3, no. 6, pp. 527–535, 2003.

[76] T. Qian, A. L. Nieminen, B. Herman, and J. J. Lemasters, "Mitochondrial permeability transition in pH-dependent reperfusion injury to rat hepatocytes," *American Journal of Physiology*, vol. 273, no. 6, pp. C1783–C1792, 1997.

[77] H. Friberg, M. Ferrand-Drake, F. Bengtsson, A. P. Halestrap, and T. Wieloch, "Cyclosporin A, but not FK 506, protects mitochondria and neurons against hypoglycemic damage and implicates the mitochondrial permeability transition in cell death," *Journal of Neuroscience*, vol. 18, no. 14, pp. 5151–5159, 1998.

[78] L. Khaspekov, H. Friberg, A. Halestrap, I. Viktorov, and T. Wieloch, "Cyclosporin A and its nonimmunosuppressive analogue N-Me-Val-4-cyclosporin A mitigate glucose/oxygen deprivation-induced damage to rat cultured hippocampal neurons," *European Journal of Neuroscience*, vol. 11, no. 9, pp. 3194–3198, 1999.

[79] S. Matsumoto, H. Friberg, M. Ferrand-Drake, and T. Wieloch, "Blockade of the mitochondrial permeability transition pore diminishes infarct size in the rat after transient middle cerebral artery occlusion," *Journal of Cerebral Blood Flow and Metabolism*, vol. 19, no. 7, pp. 736–741, 1999.

[80] S. Malouitre, H. Dube, D. Selwood, and M. Crompton, "Mitochondrial targeting of cyclosporin A enables selective inhibition of cyclophilin-D and enhanced cytoprotection after glucose and oxygen deprivation," *Biochemical Journal*, vol. 425, no. 1, pp. 137–148, 2010.

[81] H. Dube, D. Selwood, S. Malouitre, M. Capano, M. I. Simone, and M. Crompton, "A mitochondrial-targeted cyclosporin A with high binding affinity for cyclophilin D yields improved cytoprotection of cardiomyocytes," *Biochemical Journal*, vol. 441, no. 3, pp. 901–907, 2012.

[82] C. M. Porteous, A. Logan, C. Evans et al., "Rapid uptake of

lipophilic triphenylphosphonium cations by mitochondria in vivo following intravenous injection: implications for mitochondria-specific therapies and probes," *Biochimica et Biophysica Acta*, vol. 1800, no. 9, pp. 1009–1017, 2010.

[83] T. Miura, M. Tanno, and T. Sato, "Mitochondrial kinase signalling pathways in myocardial protection from ischaemia/reperfusion-induced necrosis," *Cardiovascular Research*, vol. 88, no. 1, pp. 7–15, 2010.

[84] J. Feng, E. Lucchinetti, P. Ahuja, T. Pasch, J. C. Perriard, and M. Zaugg, "Isoflurane postconditioning prevents opening of the mitochondrial permeability transition pore through inhibition of glycogen synthase kinase 3β," *Anesthesiology*, vol. 103, no. 5, pp. 987–995, 2005.

[85] S. Javadov, V. Rajapurohitam, A. Kilić, A. Zeidan, A. Choi, and M. Karmazyn, "Anti-hypertrophic effect of NHE-1 inhibition involves GSK-3β-dependent attenuation of mitochondrial dysfunction," *Journal of Molecular and Cellular Cardiology*, vol. 46, no. 6, pp. 998–1007, 2009.

[86] L. Gomez, M. Paillard, H. Thibault, G. Derumeaux, and M. Ovize, "Inhibition of GSK3β by postconditioning is required to prevent opening of the mitochondrial permeability transition pore during reperfusion," *Circulation*, vol. 117, no. 21, pp. 2761–2768, 2008.

[87] J. G. Pastorino, J. B. Hoek, and N. Shulga, "Activation of glycogen synthase kinase 3β disrupts the binding of hexokinase II to mitochondria by phosphorylating voltage-dependent anion channel and potentiates chemotherapy-induced cytotoxicity," *Cancer Research*, vol. 65, no. 22, pp. 10545–10554, 2005.

[88] A. Rasola, M. Sciacovelli, F. Chiara, B. Pantic, W. S. Brusilow, and P. Bernardi, "Activation of mitochondrial ERK protects cancer cells from death through inhibition of the permeability transition," *Proceedings of the National Academy of Sciences of the United States of America*, vol. 107, no. 2, pp. 726–731, 2010.

[89] P. Pediaditakis, J. S. Kim, L. He, X. Zhang, L. M. Graves, and J. J. Lemasters, "Inhibition of the mitochondrial permeability transition by protein kinase A in rat liver mitochondria and hepatocytes," *Biochemical Journal*, vol. 431, no. 3, pp. 411–421, 2010.

[90] C. P. Baines, C. X. Song, Y. T. Zheng et al., "Protein kinase Cε interacts with and inhibits the permeability transition pore in cardiac mitochondria," *Circulation Research*, vol. 92, no. 8, pp. 873–880, 2003.

[91] K. Takuma, P. Phuagphong, E. Lee, K. Mori, A. Baba, and T. Matsuda, "Anti-apoptotic effect of cGMP in cultured astrocytes: inhibition by cGMP-dependent protein kinase of mitochondrial permeable transition pore," *Journal of Biological Chemistry*, vol. 276, no. 51, pp. 48093–48099, 2001.

[92] V. Borutaite, R. Morkuniene, O. Arandarcikaite, A. Jekabsone, J. Barauskaite, and G. C. Brown, "Nitric oxide protects the heart from ischemia-induced apoptosis and mitochondrial damage via protein kinase G mediated blockage of permeability transition and cytochrome c release," *Journal of Biomedical Science*, vol. 16, no. 1, article 70, 2009.

[93] M. Wyss, J. Smeitink, R. A. Wevers, and T. Wallimann, "Mitochondrial creatine kinase: a key enzyme of aerobic energy metabolism," *Biochimica et Biophysica Acta*, vol. 1102, no. 2, pp. 119–166, 1992.

[94] E. O'Gorman, G. Beutner, M. Dolder, A. P. Koretsky, D. Brdiczka, and T. Wallimann, "The role of creatine kinase in inhibition of mitochondrial permeability transition," *FEBS Letters*, vol. 414, no. 2, pp. 253–257, 1997.

[95] G. Beutner, A. Rück, B. Riede, and D. Brdiczka, "Complexes between porin, hexokinase, mitochondrial creatine kinase and adenylate translocator display properties of the permeability transition pore. Implication for regulation of permeability transition by the kinases," *Biochimica et Biophysica Acta*, vol. 1368, no. 1, pp. 7–18, 1998.

[96] T. Miura, M. Nishihara, and T. Miki, "Drug development targeting the glycogen synthase kinase-3β (GSK-3β)-mediated signal transduction pathway: role of GSK-3β in myocardial protection against ischemia/reperfusion injury," *Journal of Pharmacological Sciences*, vol. 109, no. 2, pp. 162–167, 2009.

[97] G. Fajardo, M. Zhao, G. Berry, L.-J. Wong, D. Mochly-Rosen, and D. Bernstein, "β2-adrenergic receptors mediate cardioprotection through crosstalk with mitochondrial cell death pathways," *Journal of Molecular and Cellular Cardiology*, vol. 51, no. 5, pp. 781–789, 2011.

[98] J. M. Adams and S. Cory, "Life-or-death decisions by the Bcl-2 protein family," *Trends in Biochemical Sciences*, vol. 26, no. 1, pp. 61–66, 2001.

[99] T. N. Chonghaile, K. A. Sarosiek, T.-T. Vo et al., "Pretreatment mitochondrial priming correlates with clinical response to cytotoxic chemotherapy," *Science*, vol. 334, no. 6059, pp. 1129–1133, 2011.

[100] F. Llambi, T. Moldoveanu, S. Tait et al., "A unified model of mammalian BCL-2 protein family interactions at the mitochondria," *Molecular Cell*, vol. 44, no. 4, pp. 517–531, 2011.

[101] I. Marzo, C. Brenner, N. Zamzami et al., "The permeability transition pore complex: a target for apoptosis regulation by caspases and Bcl-2-related proteins," *Journal of Experimental Medicine*, vol. 187, no. 8, pp. 1261–1271, 1998.

[102] S. Shimizu, M. Narita, and Y. Tsujimoto, "Bcl-2 family proteins regulate the release of apoptogenic cytochrome c by the mitochondrial channel VDAC," *Nature*, vol. 399, no. 6735, pp. 483–487, 1999.

[103] Z. X. Chen and S. Pervaiz, "Bcl-2 induces pro-oxidant state by engaging mitochondrial respiration in tumor cells," *Cell Death and Differentiation*, vol. 14, no. 9, pp. 1617–1627, 2007.

[104] K. N. Alavian, H. Li, L. Collis et al., "Bcl-x$_L$ regulates metabolic efficiency of neurons through interaction with the mitochondrial F 1 F O ATP synthase," *Nature Cell Biology*, vol. 13, no. 10, pp. 1224–1233, 2011.

[105] I. Marzo, C. Brenner, N. Zamzami et al., "Bax and adenine nucleotide translocator cooperate in the mitochondrial control of apoptosis," *Science*, vol. 281, no. 5385, pp. 2027–2031, 1998.

[106] C. Brenner, H. Cadiou, H. L. A. Vieira et al., "Bcl-2 and Bax regulate the channel activity of the mitochondrial adenine nucleotide translocator," *Oncogene*, vol. 19, no. 3, pp. 329–336, 2000.

[107] A. S. Belzacq, H. L. A. Vieira, F. Verrier et al., "Bcl-2 and Bax modulate adenine nucleotide translocase activity," *Cancer Research*, vol. 63, no. 2, pp. 541–546, 2003.

[108] S. Shimizu, A. Konishi, T. Kodama, and Y. Tsujimoto, "BH4 domain of antiapoptotic Bcl-2 family members closes voltage-dependent anion channel and inhibits apoptotic mitochondrial changes and cell death," *Proceedings of the National Academy of Sciences of the United States of America*, vol. 97, no. 7, pp. 3100–3105, 2000.

[109] H. L. A. Vieira, P. Boya, I. Cohen et al., "Cell permeable BH3-peptides overcome the cytoprotective effect of Bcl-2 and Bcl-XL," *Oncogene*, vol. 21, no. 13, pp. 1963–1977, 2002.

[110] R. Sugioka, S. Shimizu, T. Funatsu et al., "BH4-domain peptide from Bcl-xL exerts anti-apoptotic activity in vivo,"

Oncogene, vol. 22, no. 52, pp. 8432–8440, 2003.

[111] M. Ono, Y. Sawa, M. Ryugo et al., "BH4 peptide derivative from Bcl-xL attenuates ischemia/reperfusion injury thorough anti-apoptotic mechanism in rat hearts," *European Journal of Cardio-thoracic Surgery*, vol. 27, no. 1, pp. 117–121, 2005.

[112] D. Morin, R. Assaly, S. Paradis, and A. Berdeaux, "Inhibition of mitochondrial membrane permeability as a putative pharmacological target for cardioprotection," *Current Medicinal Chemistry*, vol. 16, no. 33, pp. 4382–4398, 2009.

[113] C. Plin, J. P. Tillement, A. Berdeaux, and D. Morin, "Resveratrol protects against cold ischemia-warm reoxygenation-induced damages to mitochondria and cells in rat liver," *European Journal of Pharmacology*, vol. 528, no. 1–3, pp. 162–168, 2005.

[114] H. Ligeret, S. Barthelemy, R. Zini, J. P. Tillement, S. Labidalle, and D. Morin, "Effects of curcumin and curcumin derivatives on mitochondrial permeability transition pore," *Free Radical Biology and Medicine*, vol. 36, no. 7, pp. 919–929, 2004.

[115] Z. Ungvari, W. E. Sonntag, R. De Cabo, J. A. Baur, and A. Csiszar, "Mitochondrial protection by resveratrol," *Exercise and Sport Sciences Reviews*, vol. 39, no. 3, pp. 128–132, 2011.

[116] S. Rimbaud, M. Ruiz, J. Piquereau et al., "Resveratrol improves survival, hemodynamics and energetics in a rat model of hypertension leading to heart failure," *PLoS One*, vol. 6, no. 10, article e26391, 2011.

[117] M. Izem-Meziane, B. Djerdjouri, S. Rimbaud et al., "Catecholamine-induced cardiac mitochondrial dysfunction and mPTP opening: protective effect of curcumin," *American Journal of Physiology*, vol. 302, no. 3, pp. H665–H674, 2012.

[118] V. Tanwar, J. Sachdeva, M. Golechha, S. Kumari, and D. S. Arya, "Curcumin protects rat myocardium against isoproterenol-induced ischemic injury: attenuation of ventricular dysfunction through increased expression of hsp27 alongwith strengthening antioxidant defense system," *Journal of Cardiovascular Pharmacology*, vol. 55, no. 4, pp. 377–384, 2010.

[119] P. Manikandan, M. Sumitra, S. Aishwarya, B. M. Manohar, B. Lokanadam, and R. Puvanakrishnan, "Curcumin modulates free radical quenching in myocardial ischaemia in rats," *The International Journal of Biochemistry & Cell Biology*, vol. 36, no. 10, pp. 1967–1980, 2004.

[120] G. F. Kelso, C. M. Porteous, C. V. Coulter et al., "Selective targeting of a redox-active ubiquinone to mitochondria within cells: antioxidant and antiapoptotic properties," *Journal of Biological Chemistry*, vol. 276, no. 7, pp. 4588–4596, 2001.

[121] D. Graham, N. N. Huynh, C. A. Hamilton et al., "Mitochondria-targeted antioxidant mitoq10 improves endothelial function and attenuates cardiac hypertrophy," *Hypertension*, vol. 54, no. 2, pp. 322–328, 2009.

[122] A. Vergeade, P. Mulder, C. Vendeville-Dehaudt et al., "Mitochondrial impairment contributes to cocaine-induced cardiac dysfunction: prevention by the targeted antioxidant MitoQ," *Free Radical Biology and Medicine*, vol. 49, no. 5, pp. 748–756, 2010.

[123] L. Plecitá-Hlavatá, J. Ježek, and P. Ježek, "Pro-oxidant mitochondrial matrix-targeted ubiquinone MitoQ10 acts as antioxidant at retarded electron transport or proton pumping within Complex I," *International Journal of Biochemistry and Cell Biology*, vol. 41, no. 8-9, pp. 1697–1707, 2009.

[124] J. M. Núñez-Córdoba and M. A. Martínez-González, "Antioxidant vitamins and cardiovascular disease," *Current Topics in Medicinal Chemistry*, vol. 11, no. 14, pp. 1861–1869, 2011.

[125] J. Q. Chen, P. R. Cammarata, C. P. Baines, and J. D. Yager, "Regulation of mitochondrial respiratory chain biogenesis by estrogens/estrogen receptors and physiological, pathological and pharmacological implications," *Biochimica et Biophysica Acta*, vol. 1793, no. 10, pp. 1540–1570, 2009.

[126] V. Regitz-Zagrosek, S. Oertelt-Prigione, U. Seeland, and R. Hetzer, "Sex and gender differences in myocardial hypertrophy and heart failure," *Wiener Medizinische Wochenschrift*, vol. 161, pp. 109–116, 2011.

[127] A. H. E. M. Maas, Y. T. Van Der Schouw, V. Regitz-Zagrosek et al., "Red alert for womens heart: the urgent need for more research and knowledge on cardiovascular disease in women," *European Heart Journal*, vol. 32, no. 11, pp. 1362–1368, 2011.

[128] S. P. Duckles and D. N. Krause, "Mechanisms of cerebrovascular protection: oestrogen, inflammation and mitochondria," *Acta Physiologica*, vol. 203, no. 1, pp. 149–154, 2011.

[129] C. M. Klinge, "Estrogenic control of mitochondrial function and biogenesis," *Journal of Cellular Biochemistry*, vol. 105, no. 6, pp. 1342–1351, 2008.

[130] K. A. Mattingly, M. M. Ivanova, K. A. Riggs, N. S. Wickramasinghe, M. J. Barch, and C. M. Klinge, "Estradiol stimulates transcription of nuclear respiratory factor-1 and increases mitochondrial biogenesis," *Molecular Endocrinology*, vol. 22, no. 3, pp. 609–622, 2008.

[131] R. Morkuniene, O. Arandarcikaite, L. Ivanoviene, and V. Borutaite, "Estradiol-induced protection against ischemia-induced heart mitochondrial damage and caspase activation is mediated by protein kinase G," *Biochimica et Biophysica Acta*, vol. 1797, no. 6-7, pp. 1012–1017, 2010.

[132] C. Borrás, J. Gambini, R. López-Grueso, F. V. Pallardó, and J. Viña, "Direct antioxidant and protective effect of estradiol on isolated mitochondria," *Biochimica et Biophysica Acta*, vol. 1802, no. 1, pp. 205–211, 2010.

[133] R. Morkuniene, A. Jekabsone, and V. Borutaite, "Estrogens prevent calcium-induced release of cytochrome c from heart mitochondria," *FEBS Letters*, vol. 521, no. 1-3, pp. 53–56, 2002.

[134] J. R. Muñoz-Castañeda, P. Montilla, M. C. Muñoz, I. Bujalance, J. Muntané, and I. Túnez, "Effect of 17-β-estradiol administration during adriamycin-induced cardiomyopathy in ovariectomized rat," *European Journal of Pharmacology*, vol. 523, no. 1–3, pp. 86–92, 2005.

[135] R. Ventura-Clapier, B. Mettauer, and X. Bigard, "Beneficial effects of endurance training on cardiac and skeletal muscle energy metabolism in heart failure," *Cardiovascular Research*, vol. 73, no. 1, pp. 10–18, 2007.

[136] A. Garnier, D. Fortin, J. Zoll et al., "Coordinated changes in mitochondrial function and biogenesis in healthy and diseased human skeletal muscle," *FASEB Journal*, vol. 19, no. 1, pp. 43–52, 2005.

[137] A. Ascensão, R. Ferreira, and J. Magalhães, "Exercise-induced cardioprotection—biochemical, morphological and functional evidence in whole tissue and isolated mitochondria," *International Journal of Cardiology*, vol. 117, no. 1, pp. 16–30, 2007.

[138] M. Ciminelli, A. Ascah, K. Bourduas, and Y. Burelle, "Short term training attenuates opening of the mitochondrial permeability transition pore without affecting myocardial function following ischemia-reperfusion," *Molecular and Cellular Biochemistry*, vol. 291, no. 1-2, pp. 39–47, 2006.

[139] M. Marcil, K. Bourduas, A. Ascah, and Y. Burelle, "Exercise training induces respiratory substrate-specific decrease in Ca^{2+}-induced permeability transition pore opening in heart

mitochondria," *American Journal of Physiology*, vol. 290, no. 4, pp. H1549–H1557, 2006.

[140] A. Ascensão, J. Lumini-Oliveira, N. G. Machado et al., "Acute exercise protects against calcium-induced cardiac mitochondrial permeability transition pore opening in doxorubicin-treated rats," *Clinical Science*, vol. 120, no. 1, pp. 37–49, 2011.

7

Mechanisms of Alcohol-Induced Endoplasmic Reticulum Stress and Organ Injuries

Cheng Ji

Southern California Research Center for ALPD and Cirrhosis, USC Research Center for Liver Disease, Department of Medicine, Keck School of Medicine, University of Southern California, Los Angeles, CA 90089, USA

Correspondence should be addressed to Cheng Ji, chengji@usc.edu

Academic Editor: Huiping Zhou

Alcohol is readily distributed throughout the body in the blood stream and crosses biological membranes, which affect virtually all biological processes inside the cell. Excessive alcohol consumption induces numerous pathological stress responses, part of which is endoplasmic reticulum (ER) stress response. ER stress, a condition under which unfolded/misfolded protein accumulates in the ER, contributes to alcoholic disorders of major organs such as liver, pancreas, heart, and brain. Potential mechanisms that trigger the alcoholic ER stress response are directly or indirectly related to alcohol metabolism, which includes toxic acetaldehyde and homocysteine, oxidative stress, perturbations of calcium or iron homeostasis, alterations of S-adenosylmethionine to S-adenosylhomocysteine ratio, and abnormal epigenetic modifications. Interruption of the ER stress triggers is anticipated to have therapeutic benefits for alcoholic disorders.

1. Introduction

Alcohol is the most socially accepted addictive drug. Alcohol abuse and dependence causes social problems such as domestic violence and loss of productivity in work place as well as traffic accident-related injuries and chronic organ disorders. Excessive alcohol use is the third leading cause of preventable death in the United States and is responsible for 3.8% of deaths worldwide [1–3]. Alcohol-related medical problems can be improved upon a good understanding of pathogenesis of alcohol-induced injuries. After its consumption, alcohol is readily distributed throughout the body in the blood stream and crosses biological membranes which affect virtually all organs and biological processes in the body. Most of the alcohol that enters the body is first oxidized to toxic acetaldehyde, which is catalyzed by the cytosolic alcohol dehydrogenase (ADH) (Figure 1). Acetaldehyde is then converted by acetaldehyde dehydrogenase (ALDH) to acetic acid, which occurs primarily in the liver [4]. Alcohol can also be oxidized to acetaldehyde by cytochrome P450IIE1 (CYP2E1) which generates hydrogen peroxide. Alcohol-related medical illness results directly or indirectly from the toxic alcohol metabolites in cells and tissues. Alcoholic injuries can be found in most organs including brain, gastrointestinal tract, immune system, kidney, lung, heart, pancreas, and most frequently liver (reviewed in [1, 5–13]). Alcohol-induced liver disease (ALD) is better characterized than in other organs. The progression of ALD includes a spectrum of liver diseases, ranging from steatosis, steatohepatitis, fibrosis, to cirrhosis and even cancer [1, 7, 13]. However, the underlying molecular mechanisms of ALD are not completely understood. Both primary factors and cofactors are involved in the pathogenesis of ALD. Primary factors include but are not limited to increased oxidative stress mainly from mitochondrial malfunction and CYP2E1, increased endotoxin production and TNF signaling, impaired innate and adaptive immunity, hypoxia, impaired methionine metabolism, and epigenetic modifications [7, 9, 10, 13–18]. Cofactors may include malnutrition or complications with diabetes, obesity, smoking, or HCV/HIV infections [1, 9, 10, 13]. Alcohol-induced perturbations of homeostasis in the endoplasmic reticulum (ER) have evolved as an important factor contributing to fatty liver disease, which has been reviewed by a few comprehensive reviews [19–22]. Evidence for the involvement of ER in the pathogenesis of alcoholic injury is now

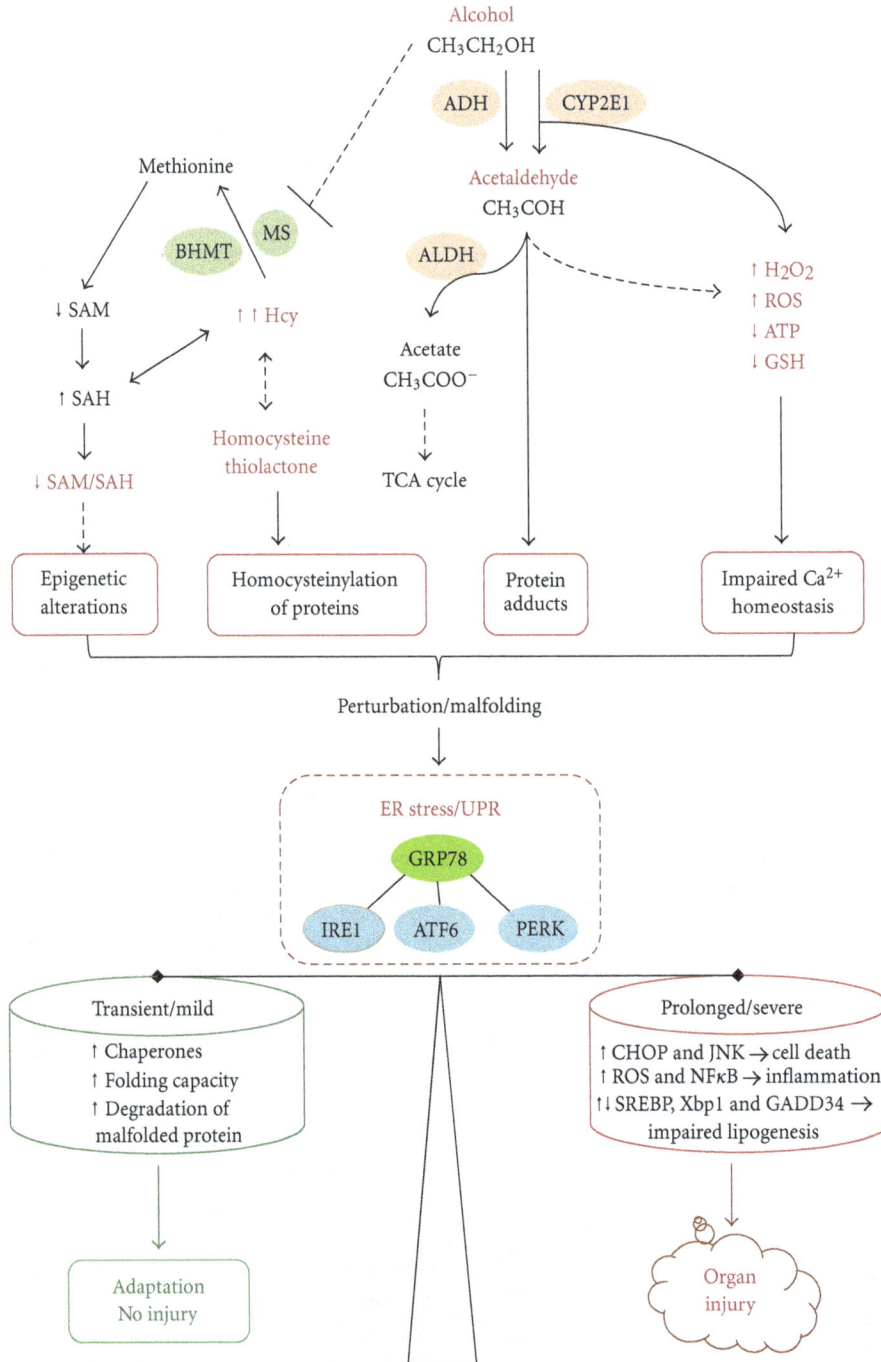

FIGURE 1: Mechanisms of alcohol-induced endoplasmic reticulum (ER) stress and organ injuries. ADH: alcohol dehydrogenase; ALDH: acetaldehyde dehydrogenase; CYP2E1: cytochrome P450 2E1; ROS: reactive oxidative stress; GSH: glutathione; BHMT: betaine-homocysteine methyltransferase; MS: methionine synthase; Hcy, homocysteine; SAM: S-adenosylmethionine, SAH: S-adenosylhomo-cysteine; TCA: tricarboxylic acid; UPR: unfolded protein response; GRP78: glucose-regulated protein 78; IRE1: inositol requiring enzyme; ATF6:activating transcription factor 6; PERK: protein kinase ds RNA-dependent-like ER kinase; CHOP: C/EBP-homologous protein; JNK, c-jun-N-terminal kinase; NFκB, nuclear factor κB; SREBP: sterol regulatory element binding protein; Xbp-1: X box binding protein 1; GADD34: growth arrest and DNA damage-inducible protein. See the context for details.

accumulating beyond the liver. The purpose of this review is to highlight phenomenological evidence for alcohol-induced ER stress in select organ disorders and to discuss potential molecular mechanisms causing alcoholic ER stress.

2. ER Stress and the Unfolded Protein Response (UPR)

The ER is an essential organelle for protein synthesis and modifications, for storing and releasing Ca^{2+}, for the biosynthesis of lipids and sterols, and for detoxification of certain drugs. ER stress is a condition under which unfolded or malfolded proteins accumulate in the ER (reviewed in [18–21]). ER stress results from perturbations in ER homeostasis such as calcium depletion, inhibition of glycosylation, alterations of the redox state, or lipid overloading. ER stress triggers the unfolded protein response (UPR), which constitutes a series of ER-to-nucleus signaling mediated by three ER resident transmembrane sensor proteins, inositol requiring protein 1 (IRE1), ds-RNA-activated protein kinase (PKR) like ER kinase (PERK), and activating transcription factor 6 (ATF6) (Figure 1). The three sensors are activated upon dissociation from their inhibitory binding with the chaperone GRP78/BiP. IRE1, which has kinase and endoribonuclease activities, is activated by transautophosphorylation. The activated IRE1 processes the transcription factor X-box binding protein-1 (XBP1) mRNA via the unconventional splicing to form transcriptionally active spliced XBP1 (sXBP1). sXBP1 activates UPR target genes, including chaperones and ER-associated degradation (ERAD) pathway genes. The second sensor PERK phosphorylates the eukaryotic initiation factor 2α-subunit (eIF2α), leading to an inhibition of the initiation of translation and a global attenuation in protein translation. Phosphorylation of eIF2α selectively activates activating transcription factor 4 (ATF4), which regulates ER chaperone genes, ERAD pathway genes, amino acid metabolism genes, and the transcription factor C/EBP homologous protein (CHOP) [19–21]. The third sensor ATF6 is cleaved in the Golgi to form a transcriptionally active fragment that traffics to the nucleus to activate UPR target genes. In general the UPR results in reduced synthesis of nascent proteins, increased unloading of unfolded proteins, and increased capacity of folding, which lead to restoration of ER homeostasis.

However, prolonged or severe UPR provokes a complex network of interacting and parallel responses contributing to pathological consequences such as apoptosis, inflammation, and fat accumulation [19–24]. The ER stress-induced apoptosis is mediated by a few factors. CHOP regulates growth arrest and DNA damage-inducible protein (GADD34). GADD34 binds protein phosphatase-1 and enhances eIF2α dephosphorylation, leading to premature restoration of translation and enhanced ER stress. CHOP can also regulate expression of the TRAIL receptor DR5, pro- and antiapoptotic Bcl-2 family protein Bim, Bax and Bcl-2 modulating cell death [19–21]. Sustained activation of IRE1 recruits the adaptor protein TRAF2 and activates JNK and NF-κB, both of which mediate apoptosis [23]. In addition, alterations in ER calcium homeostasis, upregulation of ER oxidase 1 (ERO1) by CHOP, activation of caspase 12, and activation of GSK3β by tribbles 3 (TRB3) and AKT are other mechanisms underlying ER stress-induced inflammation and apoptosis [21, 23, 25]. Lipid accumulation is also a main pathological feature of prolonged ER stress, and each of the three ER sensor pathways has direct molecular effects on lipid synthesis. The IRE1α-XBP1 branch regulates C/EBPα and C/EBPβ that control directly the expression of genes involved in *de novo* fatty acid biosynthesis [26]. The ATF6 branch is involved in phospholipid biosynthesis as well as

in fatty acid oxidation and lipoprotein secretion [27, 28]. The PERK-eIF2α branch influences expression of C/EBP family and PPARγ transcription factors via the eIF2α-specific phosphatase GADD34 and regulates SREBP1-related *de novo* lipid synthesis and accumulation [18–24, 29, 30].

3. ER Stress in Alcoholic Organ Injuries

3.1. Liver. Alcohol is mainly metabolized in the liver, and liver cells are rich in ER which assumes synthesis of a large amount of secretory and membrane proteins [19, 20, 29]. Partial role of ER in alcohol metabolism was initially realized decades ago as NADH from the hepatic oxidation of ethanol to acetaldehyde by ADH was found to support also microsomal ethanol oxidations [14, 15]. The inducible microsomal ethanol oxidizing system (MEOS) is associated with proliferation of the ER and a concomitant induction of cytochrome P4502E1 (CYP2E1) in rats and in humans. Free radical release as a consequence of CYP2E1 function in the ER and subsequent oxidative stress and lipid peroxidation generally contribute to ALD [14, 15]. However, alcohol-induced ER stress response was not recognized until recently. Molecular evidence for an impaired UPR was first found in the intragastric alcohol-fed mice using microarray gene expression profiling [18]. The alterations of selected ER stress markers were associated with severe steatosis, scattered apoptosis, and necroinflammatory foci. Moderate upregulation of expression of SREBP-1c and SREBP-2 and their responsive genes was detected by immuoblotting [18]. SREBP-1c knockout mice were protected against triglyceride accumulation [30–32]. Knocking out CHOP resulted in minimal alcohol-induced apoptosis in mouse liver [32–34]. In a setting of alcohol infusion and moderate obesity, there are synergistic effects of accentuated ER and mitochondrial stress, nitrosative stress mediated by M1 macrophage activation, and adiponectin resistance on hepatic necroinflammation and steatohepatitis [35]. In micropigs fed alcohol, liver steatosis and apoptosis were shown to be accompanied by increased mRNA levels of CYP2E1, GRP78 and SREBP-1c, and protein levels of CYP2E1, GRP78, activated SREBP and caspase 12 [36]. In addition, the ER stress response was correlated with elevated transcripts of lipogenic enzymes such as fatty acid synthase (FAS), acetyl-CoA carboxylase (ACC), and stearoyl-CoA desaturase (SCD). Further, alcohol-induced lipopolysaccharide (LPS) is linked to impaired UPR and advanced hepatic injury [37–39]. In cirrhotic rat livers, only eIF2α was activated in the basal state. After LPS challenge, full UPR as indicated by activation of IRE1α, ATF-6, and eIF2α was detected [37]. However, LPS-induced accumulation of NF-κB-dependent antiapoptotic proteins was not observed, suggesting that the UPR sensitized the cirrhotic livers to LPS/TNFα-mediated apoptosis. Alcohol-induced hepatic ER stress response not only occurs in rodents but also in livers of baboon and human patients [40, 41]. In baboon fed alcohol orally, upregulation of calpain 2, calpain p94, and ERD21 and downregulation of eIF2α were among the genes of altered expression that was revealed by using cDNA array analysis

[41]. Gene expression profiling of cirrhotic liver samples from human alcoholics also revealed alterations of calpain and calreticulin that are indicative of ER malfunction.

3.2. Pancreas. The pancreas is one of the important digestive organs adversely affected by alcohol abuse. Pancreatitis is among the most common alcohol-related hospital diagnosis in USA [11]. The underlying mechanisms for alcohol-induced pancreatitis are not well understood. Similar to the liver, the pancreas has the capacity to metabolize alcohol via both the oxidative and nonoxidative pathways yielding toxic metabolites such as acetaldehyde and lipid esters. Fatty acid ethyl and cholesteryl esters are known to accumulate in the acinar cell after chronic alcohol consumption which decreases the stability of the membranes of zymogen granules and lysosomes [42, 43], which cause a premature activation of intracellular digestive enzyme and may predispose the gland to autodigestive inflammation and injury. In respect to the role of organelles in alcoholic pancreatic injury, the ER has been considered as the acinar cell has the highest rate of protein synthesis among all tissues in adult organism. In fact, perturbations of ER homeostasis are found in acute pancreatitis [44, 45], and all the three ER stress/UPR transducers (i.e., IRE1, ATF6, and PERK) and their downstream pathways are activated. However, chronic alcohol feeding alone causes minimal pancreatic tissue injury in animal models [45, 46]. Further studies demonstrate that alcohol feeding activates the UPR in pancreas with upregulation of the transcription factor XBP1 in the intragastric alcohol infusion model [47, 48]. This suggests that alcohol induces a physiologic adaptive UPR that may prevent pathophysiologic pancreatitis responses. Indeed, heterozygous deletion of the XBP1 gene prevents XBP1 upregulation and results in pathologic changes including extensive dilation of the ER with occasional dense luminal inclusions, hallmarks of ER stress, and significant accumulation of autophagic vacuoles in acinar cells [48]. Thus, impaired UPR in the pancreas can potentiate alcohol-induced toxicity and aggravate pancreatic damages.

3.3. Brain. Alcohol exposure during development has devastating effects on the loss of neurons in selected brain areas, which leads to profound damages to the central nervous system (CNS). Alcohol consumption during pregnancy causes fetal alcohol spectrum disorders (FASDs) [1, 49]. Microcephaly, abnormal cortical thickness, reduced cerebral white matter volume, ventriculomegaly, and cerebellar hypoplasia are the prominent CNS abnormalities in FASDs. Children with (FASD) have a variety of cognitive, behavioral, and neurological impairments [49]. What cause ethanol-induced neurodegeneration are not clear. Considering that ER stress plays a role in the pathogenesis of several popular neurological diseases such as Huntington's disease, brain ischemia, Alzheimer's disease, and Parkinson's disease [50–53], an involvement of ER stress in alcohol-induced neuron toxicity has been hypothesized [54]. Recent evidence from both *in vitro* and *in vivo* tests appears to support the assumption. Exposure of SH-SY5Y neuroblastoma cells or primary cerebellar granule neurons to ethanol alone had little effect on the expression of ER stress markers [54]; however, ethanol markedly increased the expression of GRP78, CHOP, ATF4, ATF6, and phosphorylated PERK and eIF2α in the presence of tunicamycin or thapsigargin, which was accompanied with increased cell death. Acute exposure of seven-day-old mice to ethanol by subcutaneous injection at a dose of 5 g/kg significantly increased ER stress response. Increase of ATF6, CHOP, GRP78, and mesencephalic astrocyte-derived neurotrophic factor as well as the phosphorylation of IRE1, eIF2α, PERK, and PKR were detected within 24 hours after the ethanol exposure. Further, the ethanol-induced increase in phosphorylated eIF2α, caspase-12 and CHOP was distributed in neurons of specific areas of the cerebral cortex, hippocampus, and thalamus. Since the age of the animals used in this experiment is equivalent to the third trimester of pregnancy in humans, the above evidence suggests that ethanol directly induce ER stress in the developing brain.

3.4. Heart. It is well documented that chronic heavy alcohol drinking is a risk factor for cardiovascular disorders including cardiac hypertrophy, myofibrillar disruption, reduced contractility, and decreased ejection fraction [55]. Alcohol may change the circulatory hemodynamics resulting in stress on the heart. The stressed heart demands more cardiac output which leads to compensatory hypertrophic responses such as neurohormonal activation and increased growth factors and cytokines, resulting in enlarged cardiomyocytes and increased sarcomere assembly. ER stress may play a critical role in regulating protein synthesis in cardiac myocytes, and thereby produce cell enlargement and cardiac hypertrophy. Chronic alcohol consumption by FVB (Friend virus-B type) albino mice at 4% of diet for 12 weeks resulted in increased heart weight and heart-to-body weight ratio [56]. In the myocardium of the FVB mice chronically fed alcohol, GRP78, CHOP, and IRE1a protein expression levels were increased, indicative of the UPR. Class I alcohol dehydrogenase efficiently oxidizes alcohol resulting in increased production of acetaldehyde. Overexpressing alcohol dehydrogenase in the FVB mice during chronic ethanol treatment resulted in a greater UPR upregulation [56]. The finding indicates that acetaldehyde from alcohol metabolism may induce ER stress. Furthermore, overexpressing of the antioxidant protein metallothionein in FVB mice significantly reduced peak shortening and maximal shortening velocity of cardiac myocytes by LPS, which is often elevated in alcoholics [13–15, 39, 40]. In parallel, the transgenic FVB mice displayed decreased protein levels of GRP78, CHOP, PERK, and IRE1 whereas the wild type FVB displayed a significant increase in the protein levels of PERK, phospho-JNK, and phospho-p38 in the myocardium in response to LPS [56, 57].

4. Mechanisms of Alcohol-Induced ER Stress

4.1. Acetaldehyde Adducts and ER Stress. Alcohol-derived acetaldehyde is highly reactive [58–62]. At physiological temperature and pH, acetaldehyde reacts with nucleophilic groups in proteins, such as α-amino groups of internal lysine residues and the ε-amino group on the N-terminal amino acid of unblocked proteins forming unstable Schiff base acetaldehyde adducts. In addition, ethanol abuse may also lead to the formation of other types of protein adducts, such as malondialdehyde-acetaldehyde hybrids and α-hydroxy-ethyl protein-adducts. The acetaldehyde adducts initiate immunogenic reactions, cause conformational changes and inactivation of the adducted targets, or trigger aberrant protein degradation, which contribute to the development of alcoholic organ diseases (Figure 1). Malondialdehyde–acetaldehyde adduct is found to be the dominant epitope after malondialdehyde modification of proteins in atherosclerosis [63]. Antibodies to the aldehyde adducts have been detected in the serum of patients with atherosclerotic lesions and correlate with the progression of atherosclerosis. It is known that atherosclerosis develops as a result of protein unfolding and modification of protein and/or macromolecular complex function at the cellular level [63]. In supporting this, evidence for ER stress response was found in transgenic mice with cardiac overexpression of ADH that increased acetaldehyde exposure [56, 57]. The ADH transgene increased induction of IRE1, eIF-2α, GRP78, and CHOP and exacerbated chronic alcohol ingestion-induced myocardial dysfunction and hypertrophy. Further, in a mouse model of acute ethanol intoxication, inhibition of ADH causes downregulation of GRP78 mRNA levels [64]. This suggests a causal relationship between ethanol metabolism and ER stress response. Acetaldehyde adducts also affect ER Ca^{2+} handling in rat ventricular myocytes [65, 66], which may disturb ER calcium homeostasis playing a critical role in stress-mediated cellular injury [67]. In response to alcohol dosing *in vivo*, the actin in Type I and Type II fibre predominant muscles of rats was found to form stable covalent adducts with acetaldehyde [68]. Histochemical analysis showed that unreduced-acetaldehyde-protein adducts were located within the sarcolemmal (i.e., muscle membrane) and subsarcolemmal regions, which perturbed the membranes and increased protein and enzyme activity of sarcoplasmic-ER Ca^{2+}-ATPase, resulting in muscle cell death and alcoholic myopathy. In addition, acetaldehyde adducts are found in the central nervous system which may be responsible for alcoholic ER stress response. In the brain of a heavy drinker who had died suddenly while drinking continuously, acetaldehyde adducts were immunologically identified [69]. In a mouse model administered with the Lieber-DeCarli liquid diet and alcohol, acetaldehyde adducts were readily detected in degenerated neurons in the cerebral cortex [70]. The neural region that alcoholic ER stress response occurred colocalized with the acetaldehyde adducts. In young mice, ethanol-induced increase in ER stress protein markers was found to be distributed in the immature neurons of specific areas of the cerebral cortex, hippocampus and thalamus [54]. Thus, while most organs of the body can be affected by alcohol-derived acetaldehyde, cardiac and skeletal muscle cells and neurons appear to be particularly susceptible to acetaldehyde adducts that cause ER stress and injury.

4.2. Homocysteine Toxicity and ER Stress. Homocysteine (Hcy) is a normal intermediate involved in the metabolism of the essential amino acid-methionine (Figure 1). Excessive Hcy is toxic to cells. An abnormally elevated level of Hcy in the blood, a medical condition termed hyperhomocysteinemia (HHcy), is an independent risk factor in cardiovascular, neurodegenerative diseases, diabetes, obesity, and hepatic steatosis [32, 71–73]. It is generally accepted that aminoacyl thioester homocysteine thiolactone (HTL) derived from Hcy editing during protein synthesis contributes to the most of Hcy toxicity [74, 75]. HTL undergoes not only nucleophilic, which can be facilitated in the presence of acetaldehyde, but also electrophilic reactions to form protein adducts or homocysteinylation of protein lysine side chains and/or other free amine groups [75]. These reactions cause malfolding of proteins and trigger ER stress response. Evidence linking HHcy to ER stress and alcoholic liver injury has well been established in cell and animal models [16, 18–20, 32]. The intragastric alcohol feeding exhibited a greater than 5-fold increase in mouse plasma Hcy [18, 34, 35]. Hcy is metabolized normally by remethylation to methionine which is catalyzed by methionine synthase (MS) using folate as a methyl donor and by betaine-homocysteine methyltransferase (BHMT) using betaine as a methyl donor. Chronic alcohol-induced disturbance of methionine metabolism appears to contribute to the alcoholic HHcy. Alcohol inhibits enzyme activity of MS in mice and rats and reduces mRNA expression of BHMT and MS in mice [16, 17, 34, 76–79]. Simultaneous betaine feeding in the intragastric alcohol-fed mice decreased alcoholic HHcy and abrogated ER stress response in parallel with decreased ALT and amelioration of alcohol-induced necroinflammation, apoptosis, and fatty liver [18]. In cultured HepG2 cells, BHMT overexpression inhibited Hcy-induced ER stress response, lipid accumulation, and cell death [77]. In primary mouse hepatocytes, suppression of BHMT by RNA interference potentiated Hcy-induced but not tunicamycin-induced ER stress response and cell injury [77]. Transgenic mice expressing human BHMT in organs peripheral to the liver are resistant to alcohol or a high methionine and low folate diet induced HHcy and fatty liver [78]. In intragastric alcohol-fed rats, BHMT is induced, which minimizes the effect of inhibited MS on Hcy levels and subsequent ER stress response and injury [79]. In a survey using 14 mouse strains, Ivan Rusyn has found that the alcoholic HHcy is correlated with alcohol-induced liver jury (personal communication, 2011). Therefore, the above several lines of evidence support Hcy toxicity as a pathogenic factor contributing to alcohol-induced disorders.

4.3. SAM/SAH Ratio, Epigenetic Alterations and ER Stress. There are two types of important epigenetic regulations of gene expression: DNA methylation of cytosines within CpG dinucleotides and histone modifications [80, 81]. Aberrant epigenetic changes are involved in the etiology of a growing number of disorders such as alcohol dependence. Both global hypomethylation of DNA in liver and hypermethylation of DNA from peripheral blood cells have been reported in animal models and in human subjects with alcohol dependence [82–86]. This is because DNA methylation in general depends on the methyl donor S-adenosylmethionine (SAM) and is inhibited by S-adenosylhomocysteine (SAH). Both SAM and SAH are involved in methionine metabolism [87, 88]. Inside the cell, SAM is demethylated to SAH, which is readily converted to Hcy which is remethylated to methionine. Plasma Hcy is not metabolized and represents the cumulative export of Hcy from liver and other tissues. Alcohol consumption decreases levels of SAM and increases levels of SAH and/or Hcy resulting in a decrease in SAM to SAH ratio (Figure 1) [76, 78, 87–92]. Thus, alcohol has a marked impact on the hepatic methylation capacity. Evidence demonstrating epigenetic effects on alcoholic ER stress is emerging [17, 82]. In 66 male alcoholic patients with alcohol dependence, chronically elevated Hcy levels are associated with increased DNA methylation in the promoter region of homocysteine-inducible ER protein (HERP) and decreased expression of HERP mRNA in the blood [93, 94]. The decrease in HERP levels is followed by a lethal ER stress, mitochondrial dysfunction, and cell death in neurons of the developing and adult brain [94]. Thus it is conceivable that alcoholic Hcy regulates HERP and causes ER stress and injury through an epigenetic mechanism. In respect to the epigenetic modifications of histone, it is reported that alcohol causes a dose- and time-dependent selective acetylation of histone H3-K9 in cultured hepatocytes [95, 96]. Intragastric administration of ethanol increases the levels of acetylated H3-K9 by 2-3 folds in the liver of mice after 12 h [97]. Further analysis indicates that the increased acetylation is tissue specific as it is noted in liver, lung, and spleen but not in tissues from other organs tested. Thus, while other stress pathways such as the MAPK signaling may be involved, the alcoholic epigenetic effects on the ER stress pathways can be more relevant. For instance, in both cystathionine beta synthase heterozygous (CBS$^{+/-}$) and wild type (WT) mice fed ethanol diets by intragastric infusion for 4 weeks, steato-hepatitis, reduction in liver SAM, elevation in liver SAH, and reduction in the SAM/SAH ratio were observed [17]. Hepatic ER stress markers including GRP78, ATF4, CHOP, caspase 12, and SREBP-1c were upregulated and negative correlated with the SAM/SAH ratio in response to alcohol. Further, trimethylated histone H3 lysine-9 (3meH3K9) protein levels in centrilobular regions revealed by immunohistochemistry were reduced in ethanol-fed mice. The levels of 3meH3K9 in the promoter regions of GRP78, SREBP-1c, and CHOP revealed specifically by a chromatin immunoprecipitation assay were decreased only in CBS$^{+/-}$ mice fed alcohol. Since CBS is involved in transsulfuration of Hcy, the findings imply that interactions of CBS ablation and alcohol feeding impair methionine metabolism, which leads to epigenetic modifications of ER stress signaling pathways. In addition, the key modulator of UPR, sXBP1 has recently been found to be a nonhistone protein target of acetylation mediated by p300 and deacetylation mediated by the NAD$^+$-dependent class III deacetylase SIRT1 (sirtuin 1) [98, 99]. SIRT1 is demonstrated to be one of the major targets of alcohol action which influences TNF-α production in macrophages and alters glucose and lipid metabolism in the liver leading to hepatic steatosis and inflammation [100–102]. SIRT1 may

also play a role in alcohol-induced ER stress response and injury through an epigenetic mechanism.

4.4. Oxidative Stress and Disruption of Ca^{2+} or Iron Home-ostasis and ER Stress. In the ER, proteins undergo oxidative protein folding. PDI is a critical oxoreductase that catalyzes disulfide bond formation with consequent generation of reactive oxygen species (ROS) during the oxidative protein folding [19, 103]. ROS is normally under control due to cellular glutathione that sustains PDI ability to regenerate and form disulfide bridges repeatedly [103–105]. However, chronic ethanol consumption increases the production of a variety of ROS, including superoxide, H$_2$O$_2$, lipid per-oxides, and peroxynitrite [1, 13–15]. Alcoholic ROS reduce glutathione level and increase oxidized glutathione, which breaks the redox status of the ER (Figure 1). This loss of redox homeostasis perturbs the oxidative folding and makes PDI ineffective in the catalytic redox cycles leading to more utilization of reduced glutathione. Depletion of glutathione generates excessive ROS which triggers ER stress. Antioxidant treatment, CHOP deletion, or translation attenuation has been shown to reduce oxidative stress and preserve ER function [19–23]. Ethanol rapidly caused oxidative stress in cultured neuronal cells and antioxidants blocked alcoholic potentiation of ER stress and cell death [54]. An association of ER stress response with increased oxidized glutathione was found in the pancreatic acinar cell of the ethanol-fed rats [47]. In HepG2 cells, acetaldehyde impaired mitochondrial glutathione transport and stimulated mitochondrial choles-terol content, the latter of which was preceded by increased levels of CHOP and SREBP1 [106]. Chronic exposure of ani-mals to alcohol or overexpression of cytochrome CYP2E1 in hepatocytes increases the expression of superoxide dismutase (SOD) and activates nuclear factor erythroid 2-related factor 2 (Nrf2), which is an ER stress responsive factor [14, 107–109]. These lines of evidence suggest an intimate relation-ship between ER stress and ROS production. Furthermore, alcoholic oxidative stress plays a critical role in possible interplays between ER stress and mitochondrial stress, which can be mediated either by intracellular calcium or iron. Alcohol or Hcy induces alterations of lipid composition in the ER and affected ratio of phosphatidylcholine (PC) to phosphatidylethanolamine (PE) [20, 78]. Alterations of the PC/PE ratio disrupt ER calcium homeostasis causing ER stress [110]. Under ER stress, abnormal release of intracellular Ca^{2+} from the ER via inositol 1,4,5-triphosphate receptor (IP3R) channels leads to excessive mitochondrial Ca^{2+} uptake, which in turn promotes ROS production and apoptosis via multiple effects on the mitochondria [67, 111, 112]. Elevated serum iron indices (transferrin saturation, ferritin) and hepatic iron overloading are often observed in patients with alcoholic liver disease [113–117]. Excessive iron damages mitochondrial iron–sulfur clusters that generate defects in heme-containing cytochrome c and cytochrome oxidase leading to excess mitochondrial ROS [118]. Iron homeostasis is regulated by hepcidin, a circulatory antimi-crobial peptide synthesized in hepatocytes [119]. Critical-ly, ER stress response can regulate expression of hepcidin

[19, 29, 120]. Thus a vicious cycle exists: alcoholic ROS and/or ER stress damage mitochondria through iron, which in return augments ROS and stresses the ER further, all of which probably act synergistically to cause severe alcoholic injury.

4.5. Synergistic ER Stress by Alcohol, Drugs, Viral Infection and Environments. Acute alcohol or chronic alcohol at moderate concentrations may not induce readily detectable ER stress response in some cell and animal models [29, 47]. This does not rule out the doomed potential of alcohol to induce ER stress. Indeed, ER stress can be synergistically induced by alcohol in the presence of environmental factors, genetic predispositions, drugs, or virus infection. First, it is recently noted that an accelerated development of pancreatitis in alcoholic patients who smoke may result from an additive or multiplicative effect that is mediated by ER stress response [47]. Second, in a mouse model with liver-specific deletion of Grp78, low-level oral alcohol feeding did not induce HHcy that is often seen in mice fed high doses of alcohol [29]. However, the low alcohol feeding activated SREBP1 and unconventional splicing of Xbp1 (sXbp1) and decreased Insig 1 and ATF6 and its downstream targets such as ERp57 and Derl3 in the liver GRP78 knockouts, leading to aggravated lipid accumulation in the liver. Thus, compared to the aforementioned Hcy-ER stress mechanism, Grp78 deletion represents a genetic predisposition that unmasks a distinct mechanism by which alcohol induces ER stress, one that normally is largely obscured by compensatory changes in normal animals or presumably in the majority of human population who have low-to-moderate drinking. Similarly, certain drugs potentiate alcoholic ER stress response. For instance, some HIV protease inhibitors (HIV PIs) used in anti-HIV therapeutics can cause adverse side effects such as dyslipidemia and liver injury [29, 121, 122]. Portion of HIV-infected patients often concomitantly consume or abuse alcohol leading to more severe liver injury. One of the underlying mechanisms is severe ER stress responses that are caused by both alcohol and the HIV drugs. It has been demonstrated that single gavage dosing for alcohol alone or ritonavir and lopinavir combined did not induce detectable liver injury in wild type [29]. However, the gavage treatment with alcohol plus the two HIV drugs caused significant increase in plasma ALT as well as activation of CHOP, ATF4, and sXbp1. Thus, alcohol exacerbates some HIV drug-induced ER stress and subsequent injury. Third, it is known that both alcoholic activation of the ER stress sensor-IRE1α and alcohol-induced accumulation of proinflammatory cytokines such as TNFα, IL-6, and MCP-1 activate JNK and/or NF-κB pathways that mediate tissue/organ injuries [9, 10, 23, 29, 37–39]. This pathway overlap may be a result of interactions between ER stress and inflammation. The likely scenario is that mild ER stress under moderate alcohol dosing has a negative impact on ER function, which makes cells more susceptible to inflammatory signals, which subsequently augments ER stress response and injury via the JNK pathway. Fourth, alcohol may sensitize virus-infected cells to ER stress and apoptosis. It is reported that hepatitis C (HCV) infection causes ER stress in cell and animal models as well as in patients with chronic HCV [123–125]. HCV directly induces steatosis and development of hepatocellular carcinoma (HCC), which is correlated with a state of oxidative stress in mice transgenic for the HCV core protein [126, 127]. There is clinical evidence indicating that alcohol metabolism increases HCV replication and modulates the host response to HCV [128, 129]. The HCV nonstructural protein 5A (NS5A) localizes to the ER and is part of the HCV replication complex that forms altered cytoplasmic membrane structures. The membrane structure triggers ER stress and the UPR, leading to a release of ER Ca^{2+} stores and subsequent oxidative stress [124]. In addition, interactions between HCV core and destabilization of the mitochondrial electron transport chain result in increased production of ROS [130, 131]. Since alcohol alone perturbs Ca^{2+} homeostasis and stimulates ROS generation, it is conceivable that ROS mediates the synergistic interactions between alcohol consumption and HCV infection.

5. Concluding Remarks

While a large number of different stress responses and pathological pathways have been implicated in ethanol-induced injury [1, 7, 13–15], the occurrence of ER stress in the major organs including liver, brain, pancreas, and heart firmly supports its contributing role to alcoholic disorders. Alcohol causes alterations in many specific steps involved in the ER stress and UPR. The potential causes for alcohol-induced ER stress are directly or indirectly related to alcohol metabolism, which include but may not be limited to toxic acetaldehyde and homocysteine modifying proteins, oxidative stress from impaired CYP2E1 function and perturbations of calcium or iron homeostasis, alterations of SAM to SAH ratio and subsequent biochemical or epigenetic modifications, and, most importantly, interactions between these factors. Each of the factors may contribute more or less to the induction of the ER stress depending on tissues/organs or experimental models, dosage and duration of alcohol exposure, and presence of other environmental factors. Current investigations and conclusions on alcoholic ER stress appear depending on positive identifications of selective molecular markers of ER stress response, conclusions from which can be misleading sometimes. For instance, the ER stress-induced UPR is dynamic. It can be protective when most of the ER markers are positively detected or detrimental when most markers are latent or disappearing. The timing and quantity of the protection cannot be defined currently. Thus, circumstantially negative observations of the ER stress markers may not necessary rule out an existence of alcoholic ER stress. Future research should be directed at developing sensitive markers, particularly epigenetic markers, for identifying the alcoholic ER stress, and at defining timing and dynamics of the alcoholic ER stress and injuries using both acute and chronic models. Another point is that the ER is a cytosolic network that communicates readily with other cellular loci such as mitochondria, lysosome, cytoplasm, and nucleus. Simultaneous appearance of alcoholic dysfunctions of the other loci such as ATP depletion, abnormal degradation of the inside materials, oxidative stress, and numerous other stress responses could overshadow the role of ER stress in alcoholic diseases. Thus, the role of alcoholic ER stress in organ disorders can be defined precisely by studying complex interplays among the organelles and loci in disease pathogenesis, which could provide better therapeutic strategies targeting the ER. Finally, with respect to the therapeutic interventions at alcoholic ER stress, possible approaches include lowering homocysteine and raising SAM by nutritional support with betaine or folate [16, 20, 32], improving protein folding by using chemical chaperone PBA (sodium 4-phenylbutyrate) and TUDCA [19, 20, 29], blocking eIF2α dephosphorylation by using salubrinal [132], and ameliorating ROS production from the oxidative protein folding by using antioxidants. However, results of clinical trials are not available. Each of the individual approaches alone may not be a simple or universal cure as alcohol-induced pathogenesis is very complex. It is anticipated that properly combined treatments with all the beneficial agents can be effective.

Acknowledgments

This work has been supported by NIH Grants R01AA018846, R01AA018612, and R01AA014428 and by the USC Research Center for Liver Disease (P30 DK48522) and the Southern California Research Center for ALPD and Cirrhosis (P50 AA11999). The author thanks Dr. N. Kaplowitz and the graduate students and fellows who contributed to the studies.

References

[1] L. Gunzerath, B. G. Hewitt, T. K. Li, and K. R. Warren, "Alcohol research: past, present, and future," *Annals of the New York Academy of Sciences*, vol. 1216, no. 6648, pp. 1–23, 2011.

[2] M. P. Heron, D. L. Hoyert, S. L. Murphy, J. Q. Xu, K. D. Kochanek, and B. Tejada-Vera, "Deaths: final data for 2006," *National Vital Statistics Reports*, vol. 57, no. 14, pp. 1–134, 2009.

[3] J. Rehm, C. Mathers, S. Popova, M. Thavorncharoensap, Y. Teerawattananon, and J. Patra, "Global burden of disease and injury and economic cost attributable to alcohol use and alcohol-use disorders," *The Lancet*, vol. 373, no. 9682, pp. 2223–2233, 2009.

[4] S. Zakhari, "Overview: how is alcohol metabolized by the body?" *Alcohol Research and Health*, vol. 29, no. 4, pp. 245–254, 2006.

[5] J. Neiman, "Alcohol as a risk factor for brain damage: neurologic aspects," *Alcoholism: Clinical and Experimental Research*, vol. 22, no. 7, pp. 346–351, 1998.

[6] M. Epstein, "Alcohol's impact on kidney function," *Alcohol Health and Research World*, vol. 21, no. 1, pp. 84–92, 1997.

[7] S. W. French, "The mechanism of organ injury in alcoholics: implications for therapy," *Alcohol and Alcoholism Supplement*, vol. 1, pp. 57–63, 1991.

[8] C. Bode and J. C. Bode, "Alcohol's role in gastrointestinal tract disorders," *Alcohol Health and Research World*, vol. 21, no. 1, pp. 76–83, 1997.

[9] G. Szabo, J. R. Wands, A. Eken et al., "Alcohol and hepatitis C virus-interactions in immune dysfunctions and liver damage," *Alcoholism: Clinical and Experimental Research*, vol. 34, no. 10, pp. 1675–1686, 2010.

[10] B. Gao, E. Seki, D. A. Brenner et al., "Innate immunity in alcoholic liver disease," *American Journal of Physiology—Gastrointestinal and Liver Physiology*, vol. 300, no. 4, pp. 516–525, 2011.

[11] A. L. Yang, S. Vadhavkar, G. Singh, and M. B. Omary, "Epidemiology of alcohol-related liver and pancreatic disease in the United States," *Archives of Internal Medicine*, vol. 168, no. 6, pp. 649–656, 2008.

[12] A. George and V. M. Figueredo, "Alcohol and arrhythmias: a comprehensive review," *Journal of Cardiovascular Medicine*, vol. 11, no. 4, pp. 221–228, 2010.

[13] T. H. Frazier, A. M. Stocker, N. A. Kershner, L. S. Marsano, and C. J. McClain, "Treatment of alcoholic liver disease," *Therapeutic Advances in Gastroenterology*, vol. 4, no. 1, pp. 63–81, 2011.

[14] A. I. Cederbaum, Y. Lu, and D. Wu, "Role of oxidative stress in alcohol-induced liver injury," *Archives of Toxicology*, vol. 83, no. 6, pp. 519–548, 2009.

[15] R. G. Thurman, S. Ji, and J. J. Lemasters, "Alcohol-induced liver injury. The role of oxygen," *Recent Developments in Alcoholism*, vol. 2, pp. 103–117, 1984.

[16] C. H. Halsted, "Nutrition and alcoholic liver disease," *Seminars in Liver Disease*, vol. 24, no. 3, pp. 289–304, 2004.

[17] F. Esfandiari, V. Medici, D. H. Wong et al., "Epigenetic regulation of hepatic endoplasmic reticulum stress pathways in the ethanol-fed cystathionine β synthase-deficient mouse," *Hepatology*, vol. 51, no. 3, pp. 932–941, 2010.

[18] C. Ji and N. Kaplowitz, "Betaine decreases hyperhomocysteinemia, endoplasmic reticulum stress, and liver injury in alcohol-fed mice," *Gastroenterology*, vol. 124, no. 5, pp. 1488–1499, 2003.

[19] H. Malhi and R. J. Kaufman, "Endoplasmic reticulum stress in liver disease," *Journal of Hepatology*, vol. 54, no. 4, pp. 795–809, 2011.

[20] C. Ji, "Dissection of endoplasmic reticulum stress signaling in alcoholic and non-alcoholic liver injury," *Journal of Gastroenterology and Hepatology*, vol. 23, no. 1, pp. S16–S24, 2008.

[21] D. Ron and S. R. Hubbard, "How IRE1 Reacts to ER Stress," *Cell*, vol. 132, no. 1, pp. 24–26, 2008.

[22] S. M. Colgan, A. A. Hashimi, and R. C. Austi, "Endoplasmic reticulum stressand lipid dysregulation," *Expert Reviews in Molecular Medicine*, vol. 13, p. e4, 2011.

[23] M. Kitamura, "Control of NF-κB and inflammation by the unfolded protein response," *International Reviews of Immunology*, vol. 30, no. 1, pp. 4–15, 2011.

[24] A. H. Lee, E. F. Scapa, D. E. Cohen, and L. H. Glimcher, "Regulation of hepatic lipogenesis by the transcription factor XBP1," *Science*, vol. 320, no. 5882, pp. 1492–1496, 2008.

[25] H. Bommiasamy, S. H. Back, P. Fagone et al., "ATF6α induces XBP1-independent expansion of the endoplasmic reticulum," *Journal of Cell Science*, vol. 122, no. 10, pp. 1626–1636, 2009.

[26] D. T. Rutkowski, J. Wu, S. H. Back et al., "UPR pathways combine to prevent hepatic steatosis caused by ER stress-mediated suppression of transcriptional master regulators," *Developmental Cell*, vol. 15, no. 6, pp. 829–840, 2008.

[27] S. Oyadomari, H. P. Harding, Y. Zhang, M. Oyadomari, and D. Ron, "Dephosphorylation of translation initiation factor 2α enhances glucose tolerance and attenuates hepatosteatosis in mice," *Cell Metabolism*, vol. 7, no. 6, pp. 520–532, 2008.

[28] E. Bobrovnikova-Marjon, G. Hatzivassiliou, C. Grigoriadou et al., "PERK-dependent regulation of lipogenesis during mouse mammary gland development and adipocyte differentiation," *Proceedings of the National Academy of Sciences of the United States of America*, vol. 105, no. 42, pp. 16314–16319, 2008.

[29] C. Ji, N. Kaplowitz, M. Y. Lau, E. Kao, L. M. Petrovic, and A. S. Lee, "Liver-specific loss of glucose-regulated protein 78 perturbs the unfolded protein response andexacerbates a spectrum of liver diseases in mice," *Hepatology*, vol. 54, no. 1, pp. 229–239, 2011.

[30] C. Ji, C. Chan, and N. Kaplowitz, "Predominant role of sterol response element binding proteins (SREBP) lipogenic pathways in hepatic steatosis in the murine intragastric ethanol feeding model," *Journal of Hepatology*, vol. 45, no. 5, pp. 717–724, 2006.

[31] C. Ji and N. Kaplowitz, "ER stress: can the liver cope?" *Journal of Hepatology*, vol. 45, no. 2, pp. 321–333, 2006.

[32] L. Dara, C. Ji, and N. Kaplowitz, "The contribution of endoplasmic reticulum stress to liver diseases," *Hepatology*, vol. 53, no. 5, pp. 1752–1763, 2011.

[33] C. Ji, R. Mehrian-Shai, C. Chan, Y. H. Hsu, and N. Kaplowitz, "Role of CHOP in hepatic apoptosis in the murine model of intragastric ethanol feeding," *Alcoholism: Clinical and Experimental Research*, vol. 29, no. 8, pp. 1496–1503, 2005.

[34] C. Ji, Q. Deng, and N. Kaplowitz, "Role of TNF-α in ethanol-induced hyperhomocysteinemia and murine alcoholic liver injury," *Hepatology*, vol. 40, no. 2, pp. 442–451, 2004.

[35] J. Xu, K. K. Lai, A. Verlinsky et al., "Synergistic steatohepatitis by moderate obesity and alcohol in mice despite increased adiponectin and p-AMPK," *Journal of Hepatology*, vol. 55, no. 3, pp. 673–682, 2011.

[36] F. Esfandiari, J. A. Villanueva, D. H. Wong, S. W. French, and C. H. Halsted, "Chronic ethanol feeding and folate deficiency activate hepatic endoplasmic reticulum stress pathway in micropigs," *American Journal of Physiology—Gastrointestinal and Liver Physiology*, vol. 289, no. 1, pp. G54–G63, 2005.

[37] K. A. Tazi, I. Bièche, V. Paradis et al., "In vivo altered unfolded protein response and apoptosis in livers from lipopolysaccharide-challenged cirrhotic rats," *Journal of Hepatology*, vol. 46, no. 6, pp. 1075–1088, 2007.

[38] H. A. Järveläinen, T. Oinonen, and K. O. Lindros, "Alcohol-induced expression of the CD14 endotoxin receptor protein in rat Kupffer cells," *Alcoholism: Clinical and Experimental Research*, vol. 21, no. 8, pp. 1547–1551, 1997.

[39] G. L. Su, A. Rahemtulla, P. Thomas, R. D. Klein, S. C. Wang, and A. A. Nanji, "CD14 and lipopolysaccharide binding protein expression in a rat model of alcoholic liver disease," *American Journal of Pathology*, vol. 152, no. 3, pp. 841–849, 1998.

[40] L. He, J. C. Marecki, G. Serrero, F. A. Simmen, M. J. Ronis, and T. M. Badger, "Dose-dependent effects of alcohol on insulin signaling: partial explanation for biphasic alcohol impact on human health," *Molecular Endocrinology*, vol. 21, no. 10, pp. 2541–2550, 2007.

[41] D. Seth, M. A. Leo, P. H. McGuinness et al., "Gene expression profiling of alcoholic liver disease in the baboon (Papiohamadryas) and human liver," *American Journal of Pathology*, vol. 163, no. 6, pp. 2303–2317, 2003.

[42] S. J. Pandol, S. Periskic, I. Gukovsky et al., "Ethanol diet increases the sensitivity of rats to pancreatitis induced by cholecystokinin octapeptide," *Gastroenterology*, vol. 117, no. 3, pp. 706–716, 1999.

[43] A. S. Gukovskaya, O. A. Mareninova, I. V. Odinokova et al., "Cell death in pancreatitis: effects of alcohol," *Journal of Gastroenterology and Hepatology*, vol. 21, no. 3, pp. S10–S13, 2006.

[44] C. H. Kubisch, M. D. Sans, T. Arumugam, S. A. Ernst, J. A. Williams, and C. D. Logsdon, "Early activation of endoplasmic reticulum stress is associated with arginine-induced acute pancrea-titis," *American Journal of Physiology—Gastrointestinal and Liver Physiology*, vol. 291, no. 2, pp. G238–G245, 2006.

[45] I. Gukovsky, A. Lugea, M. Shahsahebi et al., "A rat model reproducing key pathological responses of alcoholic chronic pancreatitis," *American Journal of Physiology—Gastrointestinal and Liver Physiology*, vol. 294, no. 1, pp. G68–G79, 2007.

[46] M. Singh, M. M. LaSure, and D. E. Bockman, "Pancreatic acinar cell function and morphology in rats chronically fed on ethanol diet," *Gastroenterology*, vol. 82, no. 3, pp. 425–434, 1982.

[47] S. J. Pandol, F. S. Gorelick, A. Gerloff, and A. Lugea, "Alcohol abuse, endoplasmic reticulum stress and pancreatitis," *Digestive Diseases*, vol. 28, no. 6, pp. 776–782, 2010.

[48] A. Lugea, D. Tischler, J. Nguyen et al., "Adaptive unfolded protein response attenuates alcohol-induced pancreatic damage," *Gastroenterology*, vol. 140, no. 3, pp. 987–997, 2011.

[49] E. P. Riley, M. A. Infante, and K. R. Warren, "Fetal alcohol spectrum disorders: an overview," *Neuropsychology Review*, vol. 21, no. 2, pp. 73–80, 2011.

[50] S. Tajiri, S. Oyadomari, S. Yano et al., "Ischemia-induced neuronal cell death is mediated by the endoplasmic reticulum stress pathway involving CHOP," *Cell Death and Differentiation*, vol. 11, no. 4, pp. 403–415, 2004.

[51] T. Katayama, K. Imaizumi, T. Manabe, J. Hitomi, T. Kudo, and M. Tohyama, "Induction of neuronal death by ER stress in Alzheimer's disease," *Journal of Chemical Neuroanatomy*, vol. 28, no. 1-2, pp. 67–78, 2004.

[52] R. M. Silva, V. Ries, T. F. Oo et al., "CHOP/GADD153 is a mediator of apoptotic death in substantia nigra dopamine neurons in an in vivo neurotoxin model of parkinsonism," *Journal of Neurochemistry*, vol. 95, no. 4, pp. 974–986, 2005.

[53] W. Scheper and J. J. Hoozemans, "Endoplasmic reticulum protein quality control in neurodegenerative disease: the good, the bad and the therapy," *Current Medicinal Chemistry*, vol. 16, no. 5, pp. 615–626, 2009.

[54] Z. Ke, X. Wang, Y. Liu et al., "Ethanol induces endoplasmic reticulum stress in the developing brain," *Alcoholism: Clinical and Experimental Research*, vol. 35, no. 9, pp. 1574–1583, 2011.

[55] C. D. Spies, M. Sander, K. Stangl et al., "Effects of alcohol on the heart," *Current Opinion in Critical Care*, vol. 7, no. 5, pp. 337–343, 2001.

[56] S. Y. Li and J. Ren, "Cardiac overexpression of alcohol dehydrogenase exacerbates chronic ethanol ingestion-induced myocardial dysfunction and hypertrophy: role of insulin signaling and ER stress," *Journal of Molecular and Cellular Cardiology*, vol. 44, no. 6, pp. 992–1001, 2008.

[57] S. Y. Li, S. A. Gilbert, Q. Li, and J. Ren, "Aldehyde dehydrogenase-2 (ALDH2) ameliorates chronic alcohol ingestion-induced myocardial insulin resistance and endoplasmic reticulum stress," *Journal of Molecular and Cellular Cardiology*, vol. 47, no. 2, pp. 247–255, 2009.

[58] M. Setshedi, J. R. Wands, and S. M. Monte, "Acetaldehyde adducts in alcoholic liver disease," *Oxidative Medicine and Cellular Longevity*, vol. 3, no. 3, pp. 178–185, 2010.

[59] T. M. Donohue, D. J. Tuma, and M. F. Sorrell, "Binding of metabolically derived acetaldehyde to hepatic proteins in vitro," *Laboratory Investigation*, vol. 49, no. 2, pp. 226–231, 1983.

[60] V. J. Stevens, W. J. Fanil, C. B. Newman, R. V. Sims, A. Cerami, and C. M. Peterson, "Acetaldehyde adduct with hemoglobin," *Journal of Clinical Investigation*, vol. 67, no. 2, pp. 361–369, 1981.

[61] M. Hoerner, U. J. Behrens, T. Worner, and C. S. Lieber, "Humoral immune response to acetaldehyde adducts in alcoholic patients," *Research Communications in Chemical Pathology and Pharmacology*, vol. 54, no. 1, pp. 3–12, 1986.

[62] Y. Israel, E. Hurwitz, O. Niemela, and R. Arnon, "Monoclonal and polyclonal antibodies against acetaldehyde-containing epitopes in acetaldehyde-protein adducts," *Proceedings of the National Academy of Sciences of the United States of America*, vol. 83, no. 20, pp. 7923–7927, 1986.

[63] F. Ursini, K. J. Davies, M. Maiorino, T. Parasassi, and A. Sevanian, "Atherosclerosis: another protein misfolding disease?" *Trends in Molecular Medicine*, vol. 8, no. 8, pp. 370–374, 2002.

[64] Y. Nishitani and H. Matsumoto, "Ethanol rapidly causes activation of JNK associated with ER stress under inhibition of ADH," *FEBS Letters*, vol. 580, no. 1, pp. 9–14, 2006.

[65] K. R. Mills, K. Ward, F. Martin, and T. J. Peters, "Peripheral neuropathy and myopathy in chronic alcoholism," *Alcohol and Alcoholism*, vol. 21, no. 4, pp. 357–362, 1986.

[66] Y. H. Sun, Y. Q. Li, S. L. Feng et al., "Calcium-sensing receptor activation contributed to apoptosis stimulates TRPC6 channel in rat neonatal ventricular myocytes," *Biochemical and Biophysical Research Communications*, vol. 394, no. 4, pp. 955–961, 2010.

[67] D. Mekahli, G. Bultynck, J. B. Parys, H. de Smedt, and L. Missiaen, "Endoplasmic-reticulum calcium depletion and disease," *Cold Spring Harbor Perspectives in Biology*, vol. 3, no. 6, 2011.

[68] S. Worrall, O. Niemela, S. Parkkila, T. J. Peters, and V. R. Preedy, "Protein adducts in type I and type II fibre predominant muscles of the ethanol-fed rat: preferential localisation in the sarcolemmal and subsarcolemmal region," *European Journal of Clinical Investigation*, vol. 31, no. 8, pp. 723–730, 2001.

[69] K. Nakamura, K. Iwahashi, A. Furukawa et al., "Acetaldehyde adducts in the brain of alcoholics," *Archives of Toxicology*, vol. 77, no. 10, pp. 591–593, 2003.

[70] K. Nakamura, K. Iwahashi, M. Itoh et al., "Immunohistochemical study on acetaldehyde adducts in alcohol-fed mice," *Alcoholism: Clinical and Experimental Research*, vol. 24, no. 4, pp. 93–96, 2000.

[71] S. R. Lentz, "Mechanisms of homocysteine-induced atherothrombosis," *Journal of Thrombosis and Haemostasis*, vol. 3, no. 8, pp. 1646–1654, 2005.

[72] S. Seshadri, A. Beiser, J. Selhub et al., "Plasma homocysteine as a risk factor for dementia and Alzheimer's disease," *New England Journal of Medicine*, vol. 346, no. 7, pp. 476–483, 2002.

[73] C. Ji and N. Kaplowitz, "Hyperhomocysteinemia, endoplasmic reticulum stress, and alcoholic liver injury," *World Journal of Gastroenterology*, vol. 10, no. 12, pp. 1699–1708, 2004.

[74] H. Jakubowski, "Mechanism of the condensation of homocysteine thiolactone with aldehydes," *Chemistry*, vol. 12, no. 31, pp. 8039–8043, 2006.

[75] H. Jakubowski, "Molecular basis of homocysteine toxicity in humans," *Cellular and Molecular Life Sciences*, vol. 61, no. 4, pp. 470–487, 2004.

[76] S. H. Kenyon, A. Nicolaou, and W. A. Gibbons, "The effect of ethanol and its metabolites upon methionine synthase activity in vitro," *Alcohol*, vol. 15, no. 4, pp. 305–309, 1998.

[77] C. Ji, M. Shinohara, J. Kuhlenkamp, C. Chan, and N. Kaplowitz, "Mechanisms of protection by the betaine-homocysteine methyltransferase/betaine system in HepG2 cells and primary mouse hepatocytes," *Hepatology*, vol. 46, no. 5, pp. 1586–1596, 2007.

[78] C. Ji, M. Shinohara, D. Vance et al., "Effect of transgenic extrahepatic expression of betaine-homocysteine methyltransferase on alcohol or homocysteine-induced fatty liver," *Alcoholism: Clinical and Experimental Research*, vol. 32, no. 6, pp. 1049–1058, 2008.

[79] M. Shinohara, C. Ji, and N. Kaplowitz, "Differences in betaine-homocysteine methyltransferase expression, endoplasmic reticulum stress response, and liver injury between alcohol-fed mice and rats," *Hepatology*, vol. 51, no. 3, pp. 796–805, 2010.

[80] R. Brown and G. Strathdee, "Epigenomics and epigenetic therapy of cancer," *Trends in Molecular Medicine*, vol. 8, no. 4, pp. S43–S48, 2002.

[81] W. Fischle, Y. Wang, and C. D. Allis, "Histone and chromatin cross-talk," *Current Opinion in Cell Biology*, vol. 15, no. 2, pp. 172–178, 2003.

[82] J. Oliva, J. Dedes, J. Li, S. W. French, and F. Bardag-Gorce, "Epigenetics of proteasome inhibition in the liver of rats fed ethanol chronically," *World Journal of Gastroenterology*, vol. 15, no. 6, pp. 705–712, 2009.

[83] A. J. Garro, N. Espina, D. McBeth, S. L. Wang, and C. Y. Wu-Wang, "Effects of alcohol consumption on DNA methylation reactions and gene expression: implications for increased cancer risk," *European Journal of Cancer Prevention*, vol. 1, no. 3, pp. 19–23, 1992.

[84] A. J. Garro, D. L. McBeth, V. Lima, and C. S. Lieber, "Ethanol consumption inhibits fetal DNA methylation in mice: implications for the fetal alcohol syndrome," *Alcoholism: Clinical and Experimental Research*, vol. 15, no. 3, pp. 395–398, 1991.

[85] S. W. Choi, F. Stickel, H. W. Baik, Y. I. Kim, H. K. Seitz, and J. B. Mason, "Chronic alcohol consumption induces genomic but not p53-specific DNA hypomethylation in rat colon," *Journal of Nutrition*, vol. 129, no. 11, pp. 1945–1950, 1999.

[86] S. Bleich, B. Lenz, M. Ziegenbein et al., "Epigenetic DNA hypermethylation of the HERP gene promoter induces down-regulation of its mRNA expression in patients with alcohol dependence," *Alcoholism: Clinical and Experimental Research*, vol. 30, no. 4, pp. 587–591, 2006.

[87] S. C. Lu and J. M. Mato, "Role of methionine adenosyltransferase and S-adenosylmethionine in alcohol-associated liver cancer," *Alcohol*, vol. 35, no. 3, pp. 227–234, 2005.

[88] J. D. Finkelstein, "Methionine metabolism in mammals," *Journal of Nutritional Biochemistry*, vol. 1, no. 5, pp. 228–237, 1990.

[89] K. K. Kharbanda, D. D. Rogers II, M. E. Mailliard et al., "A comparison of the effects of betaine and S-adenosylmethionine on ethanol-induced changes in methionine metabolism and steatosis in rat hepatocytes," *Journal of Nutrition*, vol. 135, no. 3, pp. 519–524, 2005.

[90] A. J. Barak, H. C. Beckenhauer, M. E. Mailliard, K. K. Kharbanda, and D. J. Tuma, "Betaine lowers elevated

S-adenosylhomocysteine levels in hepatocytes from ethanol-fed rats," *Journal of Nutrition*, vol. 133, no. 9, pp. 2845–2848, 2003.

[91] K. C. Trimble, A. M. Molloy, J. M. Scott, and D. G. Weir, "The effect of ethanol on one-carbon metabolism: increased methionine catabolism and lipotrope methyl-group wastage," *Hepatology*, vol. 18, no. 4, pp. 984–989, 1993.

[92] E. Giovannucci, J. Chen, S. A. Smith-Warner et al., "Methylenetetrahydrofolate reductase, alcohol dehydrogenase, diet, and risk of colorectal adenomas," *Cancer Epidemiology Biomarkers and Prevention*, vol. 12, no. 10, pp. 970–979, 2003.

[93] B. Lenz, S. Bleich, S. Beutler et al., "Homocysteine regulates expression of herp by DNA methylation involving the AARE and CREB binding sites," *Experimental Cell Research*, vol. 312, no. 20, pp. 4049–4055, 2006.

[94] Y. Ma and L. M. Hendershot, "Herp is dually regulated by both the endoplasmic reticulum stress-specific branch of the unfolded protein response and a branch that is shared with other cellular stress pathways," *Journal of Biological Chemistry*, vol. 279, no. 14, pp. 13792–13799, 2004.

[95] P. H. Park, R. Miller, and S. D. Shukla, "Acetylation of histone H3 at lysine 9 by ethanol in rat hepatocytes," *Biochemical and Biophysical Research Communications*, vol. 306, no. 2, pp. 501–504, 2003.

[96] P. H. Park, R. W. Lim, and S. D. Shukla, "Involvement of histone acetyltransferase (HAT) in ethanol-induced acetylation of histone H3 in hepatocytes: potential mechanism for gene expression," *American Journal of Physiology—Gastrointestinal and Liver Physiology*, vol. 289, no. 6, pp. G1124–G1136, 2005.

[97] J. S. Kim and S. D. Shukla, "Acute in vivo effect of ethanol (binge drinking) on histone H3 modifications in rat tissues," *Alcohol and Alcoholism*, vol. 41, no. 2, pp. 126–132, 2006.

[98] F. M. Wang, Y. J. Chen, and H. J. Ouyang, "Regulation of unfolded protein response modulator XBP1s by acetylation and deacetylation," *Biochemical Journal*, vol. 433, no. 1, pp. 245–252, 2011.

[99] Y. Li, S. Xu, A. Giles et al., "Hepatic overexpression of SIRT1 in mice attenuates endoplasmic reticulum stress and insulin resistance in the liver," *FASEB Journal*, vol. 25, no. 5, pp. 1664–1679, 2011.

[100] C. S. Lieber, M. A. Leo, X. Wang, and L. M. DeCarli, "Effect of chronic alcohol consumption on Hepatic SIRT1 and PGC-1α in rats," *Biochemical and Biophysical Research Communications*, vol. 370, no. 1, pp. 44–48, 2008.

[101] Z. Shen, J. M. Ajmo, C. Q. Rogers et al., "Role of SIRT1 in regulation of LPS- or two ethanol metabolites-induced TNF-α production in cultured macrophage cell lines," *American Journal of Physiology—Gastrointestinal and Liver Physiology*, vol. 296, no. 5, pp. G1047–G1053, 2009.

[102] A. Purushotham, T. T. Schug, Q. Xu, S. Surapureddi, X. Guo, and X. Li, "Hepatocyte-specific deletion of SIRT1 alters fatty acid metabolism and results in hepatic steatosis and inflammation," *Cell Metabolism*, vol. 9, no. 4, pp. 327–338, 2009.

[103] Y. Shimizu and L. M. Hendershot, "Oxidative folding: cellular strategies for dealing with the resultant equimolar production of reactive oxygen species," *Antioxidants and Redox Signaling*, vol. 11, no. 9, pp. 2317–2331, 2009.

[104] J. D. Malhotra and R. J. Kaufman, "Endoplasmic reticulum stress and oxidative stress: a vicious cycle or a double-edged sword?" *Antioxidants and Redox Signaling*, vol. 9, no. 12, pp. 2277–2293, 2007.

[105] A. Görlach, P. Klappa, and T. Kietzmann, "The endoplasmic reticulum: folding, calcium homeostasis, signaling, and redox control," *Antioxidants and Redox Signaling*, vol. 8, no. 9-10, pp. 1391–1418, 2006.

[106] J. M. Lluis, A. Colell, C. García-Ruiz, N. Kaplowitz, and J. C. Fernández-Checa, "Acetaldehyde impairs mitochondrial glutathione transport in HepG2 cells through endoplasmic reticulum stress," *Gastroenterology*, vol. 124, no. 3, pp. 708–724, 2003.

[107] S. H. Bae, S. H. Sung, E. J. Cho et al., "Concerted action of sulfiredoxin and peroxiredoxin I protects against alcohol-induced oxidative injury in mouse liver," *Hepatology*, vol. 53, no. 3, pp. 945–953, 2011.

[108] O. R. Koch, G. Pani, S. Borrello et al., "Oxidative stress and antioxidant defenses in ethanol-induced cell injury," *Molecular Aspects of Medicine*, vol. 25, no. 1-2, pp. 191–198, 2004.

[109] P. Gong and A. I. Cederbaum, "Nrf2 is increased by CYP2E1 in rodent liver and HepG2 cells and protects against oxidative stress caused by CYP2E1," *Hepatology*, vol. 43, no. 1, pp. 144–153, 2006.

[110] S. Fu, L. Yang, P. Li et al., "Aberrant lipid metabolism disrupts calcium homeostasis causing liver endoplasmic reticulum stress in obesity," *Nature*, vol. 473, no. 7348, pp. 528–531, 2011.

[111] A. Deniaud, O. S. El Dein, E. Maillier et al., "Endoplasmic reticulum stress induces calcium-dependent permeability-transition, mitochondrial outer membranepermeabilization and apoptosis," *Oncogene*, vol. 27, no. 3, pp. 285–299, 2008.

[112] T. I. Peng and M. J. Jou, "Oxidative stress caused by mitochondrial calcium overload," *Annals of the New York Academy of Sciences*, vol. 1201, pp. 183–188, 2010.

[113] R. W. Chapman, M. Y. Morgan, M. Laulicht, A. V. Hoffbrand, and S. Sherlock, "Hepatic iron stores and markers of iron overload in alcoholics and patients with idiopathic hemochromatosis," *Digestive Diseases and Sciences*, vol. 27, no. 10, pp. 909–916, 1982.

[114] B. J. Potter, R. W. Chapman, R. M. Nunes, D. Sorrentino, and S. Sherlock, "Transferrin metabolism in alcoholic liver disease," *Hepatology*, vol. 5, no. 5, pp. 714–721, 1985.

[115] M. G. Irving, J. W. Halliday, and L. W. Powell, "Association between alcoholism and increased hepatic iron stores," *Alcoholism: Clinical and Experimental Research*, vol. 12, no. 1, pp. 7–13, 1988.

[116] T. M. de Feo, S. Fargion, L. Duca et al., "Non-transferrin-bound iron in alcohol abusers," *Alcoholism: Clinical and Experimental Research*, vol. 25, no. 10, pp. 1494–1499, 2001.

[117] J. B. Whitfield, G. Zhu, A. C. Heath, L. W. Powell, and N. G. Martin, "Effects of alcohol consumption on indices of iron stores and of iron stores on alcohol intake markers," *Alcoholism: Clinical and Experimental Research*, vol. 25, no. 7, pp. 1037–1045, 2001.

[118] E. Napoli, F. Taroni, and G. A. Cortopassi, "Frataxin, iron-sulfur clusters, heme, ROS, and aging," *Antioxidants and Redox Signaling*, vol. 8, no. 3-4, pp. 506–516, 2006.

[119] C. Pigeon, G. Ilyin, B. Courselaud, P. Leroyer, B. Turlin, and P. Brissot, "A new mouse liver-specific gene, encoding a protein homologous to humanantimicrobial peptide hepcidin, is overexpressed during iron overload," *Journal of Biological Chemistry*, vol. 276, no. 11, pp. 7811–7819, 2001.

[120] C. Vecchi, G. Montosi, K. Zhang et al., "ER stress controls iron metabolism through induction of hepcidin," *Science*, vol. 325, no. 5942, pp. 877–880, 2009.

[121] N. Bertholet, D. M. Cheng, J. H. Samet, E. Quinn, and R. Saitz, "Alcohol consumption patterns in HIV-infected adults with alcohol problems," *Drug and Alcohol Dependence*, vol. 112, no. 1-2, pp. 160–163, 2010.

[122] F. H. Galvan, E. G. Bing, J. A. Fleishman et al., "The prevalence of alcohol consumption and heavy drinking among people with HIV in the United States: results from the HIV cost and services utilization study," *Journal of Studies on Alcohol*, vol. 63, no. 2, pp. 179–186, 2002.

[123] T. Asselah, I. Bièche, A. Mansouri et al., "In vivo hepatic endoplasmic reticulum stress in patients with chronic hepatitis C," *Journal of Pathology*, vol. 221, no. 3, pp. 264–274, 2010.

[124] M. A. Joyce, K. A. Walters, S. E. Lamb et al., "HCV induces oxidative and ER stress, and sensitizes infected cells to apoptosis in SCID/Alb-uPA mice," *PLoS Pathogens*, vol. 5, no. 2, p. e1000291, v2009.

[125] K. D. Tardif, G. Waris, and A. Siddiqui, "Hepatitis C virus, ER stress, and oxidative stress," *Trends in Microbiology*, vol. 13, no. 4, pp. 159–163, 2005.

[126] H. Lerat, H. L. Kammoun, I. Hainault et al., "Hepatitis C virus proteins induce lipogenesis and defective triglyceride secretion in transgenic mice," *Journal of Biological Chemistry*, vol. 284, no. 48, pp. 33466–33474, 2009.

[127] H. Lerat, M. Honda, M. R. Beard et al., "Steatosis and liver cancer in transgenic mice expressing the structural and nonstructural proteins of hepatitis C virus," *Gastroenterology*, vol. 122, no. 2, pp. 352–365, 2002.

[128] K. Safdar and E. R. Schiff, "Alcohol and hepatitis C," *Seminars in Liver Disease*, vol. 24, no. 3, pp. 305–315, 2004.

[129] F. Pessione, F. Degos, P. Marcellin et al., "Effect of alcohol consumption on serum hepatitis C virus RNA and histological lesions in chronic hepatitis C," *Hepatology*, vol. 27, no. 6, pp. 1717–1722, 1998.

[130] K. Otani, M. Korenaga, M. R. Beard et al., "Hepatitis C virus core protein, cytochrome P450 2E1, and alcohol produce combined mitochondrial injury and cytotoxicity in hepatoma cells," *Gastroenterology*, vol. 128, no. 1, pp. 96–107, 2005.

[131] M. Korenaga, M. Okuda, K. Otani, T. Wang, Y. Li, and S. A. Weinman, "Mitochondrial dysfunction in hepatitis C," *Journal of Clinical Gastroenterology*, vol. 39, no. 4, pp. S162–S166, 2005.

[132] M. Boyce, K. F. Bryant, C. Jousse et al., "A selective inhibitor of eIF2α dephosphorylation protects cells from ER stress," *Science*, vol. 307, no. 5711, pp. 935–939, 2005.

8

Ubiquitin-Mediated Regulation of Endocytosis by Proteins of the Arrestin Family

Michel Becuwe,[1] Antonio Herrador,[2] Rosine Haguenauer-Tsapis,[1]
Olivier Vincent,[2] and Sébastien Léon[1]

[1] Institut Jacques Monod, Centre National de la Recherche Scientifique, UMR 7592, Université Paris Diderot, Sorbonne Paris Cité, 75205 Paris, France
[2] Instituto de Investigaciones Biomédicas, CSIC-UAM, Arturo Duperier, 4, 28029 Madrid, Spain

Correspondence should be addressed to Olivier Vincent, ovincent@iib.uam.es
and Sébastien Léon, leon.sebastien@ijm.univ-paris-diderot.fr

Academic Editor: Dmitry Karpov

In metazoans, proteins of the arrestin family are key players of G-protein-coupled receptors (GPCRS) signaling and trafficking. Following stimulation, activated receptors are phosphorylated, thus allowing the binding of arrestins and hence an "arrest" of receptor signaling. Arrestins act by uncoupling receptors from G proteins and contribute to the recruitment of endocytic proteins, such as clathrin, to direct receptor trafficking into the endocytic pathway. Arrestins also serve as adaptor proteins by promoting the recruitment of ubiquitin ligases and participate in the agonist-induced ubiquitylation of receptors, known to have impact on their subcellular localization and stability. Recently, the arrestin family has expanded following the discovery of arrestin-related proteins in other eukaryotes such as yeasts or fungi. Surprisingly, most of these proteins are also involved in the ubiquitylation and endocytosis of plasma membrane proteins, thus suggesting that the role of arrestins as ubiquitin ligase adaptors is at the core of these proteins' functions. Importantly, arrestins are themselves ubiquitylated, and this modification is crucial for their function. In this paper, we discuss recent data on the intricate connections between arrestins and the ubiquitin pathway in the control of endocytosis.

1. Introduction

The name of "arrestin" was initially given to a 48-kDa protein that was essential to "arrest" the signal following the photoexcitation of rhodopsin, a photoreceptor of the G-protein-coupled receptors (GPCRS) family expressed in rod and cone cells of the retina [1, 2]. A second isoform involved in the same process has later been identified; both of these proteins are now designated visual arrestins (or arrestin-1 and -4) (for review see [3]). A similar regulatory system was described for another GPCR, the β2-adrenergic receptor (β2-AR), which involves two other arrestins, named β-arrestin-1 and -2 (or arrestin-2 and -3, resp.) [4–6]. β-arrestins are ubiquitously expressed and were later found to regulate a large number of receptors in addition to β2-AR.

2. Arrestin-Mediated Regulation of GPCRs

2.1. Arrestin-Dependent Uncoupling of GPCRs from G-Proteins. Arrestins are key players in the regulation of GPCR signaling activity. Upon agonist stimulation, GPCRs undergo conformational changes leading to their association to heterotrimeric G proteins and subsequent activation, thereby triggering appropriate signal transduction pathways. Receptor desensitization is initiated after ligand binding through the phosphorylation of residues within their cytosolic loops by G-protein-coupled receptor kinases (GRKs). This modification allows arrestin docking to the GPCR, which in turn favors the uncoupling between the receptor and the G protein. Indeed, β-arrestins are cytosolic proteins that, in response to receptor stimulation, relocalize rapidly to the plasma membrane [7]. Structural and structure-function studies of visual arrestins identified a phosphate-sensor domain in the polar core of the protein [8]. Intramolecular interactions between the C-terminal tail and the phosphate sensor region maintain the arrestin in an inactive state, and this interaction is disrupted upon binding to the phosphorylated receptor. This interaction is followed by a conformational change of the arrestin molecule and leads to a high-affinity receptor-binding state. Arrestin recruitment onto the phosphorylated receptor hinders its interaction with G protein, and consequently silences the activation of the GPCR-G protein-signaling module.

2.2. Arrestins and GPCR Endocytosis. Another crucial component of GPCR regulation operates at the level of their localization [9]. Endocytosis plays a major role in the modulation of GPCR signaling activity, and, again, this regulation involves β-arrestins [10]. Indeed, β-arrestins act as endocytic adaptor proteins that recruit components of the endocytosis machinery to promote GPCR internalization and/or degradation. β-arrestins interact with clathrin through a clathrin-binding motif [11–13] to promote GPCR association to clathrin-coated pits (CCPs). Deletion of the clathrin-binding site abrogates arrestin-promoted trafficking of the β2-AR [14]. Additionally, β-arrestins interact with the clathrin adaptor complex AP-2 upon GPCR binding [14–16], to promote clathrin-coat assembly and receptor targeting to CCPs [17]. In addition to clathrin, β-arrestins also bind to other components of the endocytic machinery such as the N-ethylmaleimide-sensitive fusion protein (NSF), the small G protein ARF6, and the phosphatidylinositol 4-phosphate 5 kinase PIP5 K Iα [18–20].

2.3. Arrestins as Signaling Scaffolds. Besides their functions in GPCR desensitization and trafficking, β-arrestins are also capable of generating their own signals by scaffolding signaling molecules, such as non-receptor tyrosine kinases of the Src family, or MAP (mitogen-activated protein) kinases (ERK1/2, c-Jun N-terminal kinase 3 JNK3) (reviewed in [21]). β-arrestins therefore mediate a second wave of signaling distinct from G-protein-dependent signaling.

2.4. Arrestins and Ubiquitin. β-arrestins were also shown to regulate the final fate of the receptor, by acting on the balance between receptor recycling to the plasma membrane, or its lysosomal degradation. The posttranslational modification of plasma membrane proteins, including receptors, by ubiquitin is known to affect their sorting along the endocytic pathway [22, 23]. β-arrestins have the ability to recruit ubiquitin ligases and promote receptor ubiquitylation, therefore acting as "adaptor" proteins [24]. Interestingly, a phylogenetic study has revealed that proteins of the arrestin family are present in all eukaryotes, except plants [25, 26]. A body of evidence (detailed later in this review, [27]) indicates that these arrestin-related proteins are also involved in the regulation of plasma membrane proteins trafficking by acting as ubiquitin ligase adaptors. Therefore,

this function seems to be one of the most conserved features within the arrestin family [28, 29].

Both arrestins and arrestin-related proteins are themselves targets of ubiquitylation. This was discovered very early on for β-Arr2 in response to agonist stimulation [30]. Likewise, the fungal arrestin-related protein PalF was shown to be ubiquitylated in response to alkaline ambient pH, in a signal- and receptor- (PalH) dependent manner [31]. This ubiquitylation appeared crucial for the proper function of arrestins [24, 32–35], but the precise role of this modification is poorly understood. An additional layer of complexity has recently been added following the observations that β-arrestins interact with deubiquitylating enzymes that regulate their ubiquitylation status as well as receptor ubiquitylation and, consequently, their fate [35–37].

In this review, we will focus on the connections between arrestins and ubiquitin. We will detail the function of arrestins and arrestin-related proteins as ubiquitin ligase adaptor and discuss how arrestin functions could be regulated by ubiquitylation.

3. Arrestins as Ubiquitin Ligase Adaptors

3.1. Ubiquitin and Endocytic Protein Sorting. Studies in the last decades have shown that ubiquitin is a master regulator of endocytosis in eukaryotes. Early work performed in the yeast *Saccharomyces cerevisiae* demonstrated that ubiquitin is involved in the endocytosis of plasma membrane proteins, such as ABC (ATP-binding cassette) transporters [38], receptor [39, 40], or permeases [41]. The ubiquitylation of plasma membrane proteins appears to trigger their internalization and targeting to endosomes [42], although the existence of an ubiquitin-independent internalization mechanism is also documented [43]. In mammalian cells, the situation is more complex, as several internalization pathways exist in the cell with only some of them regulated by ubiquitin [44].

Initially, it has been proposed that ubiquitylated cargoes are recognized in yeast and mammals by the ubiquitin-binding motifs of various proteins involved in endocytosis, such as Eps15 (Ede1 in yeast) and Epsin (Eps15 interacting; Ent1 and Ent2 in yeast) which display UIM (ubiquitin-interacting motif) or UBA (ubiquitin-associated) domains [45, 46]. In addition, these endocytic proteins can also interact with phosphoinositides and clathrin, making them ideal candidates to coordinate ubiquitin recognition and cargo internalization. While such a function appears to be established in mammalian cells [47–49], recent data in yeast favor a more complex model, where ubiquitin-binding domains would play a more general role in protein interactions and the assembly of the endocytic network [50]. Noteworthy, in mammalian cells, endocytic adaptors are often ubiquitylated in response to extracellular stimuli, and this contributes greatly to the ubiquitin-based signaling triggered upon cell stimulation [23, 51–53].

A second major ubiquitin-dependent step in the endocytic pathway occurs at multivesicular bodies (MVBs) and is required for cargo delivery into lysosomes [54]. Cargo ubiquitylation provides the crucial signal for entering into this pathway. A series of protein complexes, collectively named ESCRT (endosomal sorting complex required for transport) carry ubiquitin-binding domains and act in concert to allow the recognition and sorting of ubiquitylated cargoes into luminal vesicles of MVBs [55]. Therefore, lack of cargo ubiquitylation at this stage leads to a defective targeting to the lysosome, and, eventually, recycling [56].

In mammalian cells, initial studies showed that the ubiquitin conjugation system is important for the downregulation of the growth hormone receptor (GHR) [57]. Also, the study of the amiloride-sensitive epithelial sodium channel ENaC clearly established that its ubiquitylation regulates the channel's stability [58]. Subsequent work on ENaC, GHR, and many other receptors (such as EGFR, PDGFR, c-Met, TGF-βR, β2-AR) confirmed the critical function of ubiquitin in endocytosis in mammals [23, 59, 60]. However, where this ubiquitylation occurs in the cell (plasma membrane or endosomal compartments), and how ubiquitylation impacts on the target receptor's fate (internalization, progression through the endocytic pathway, or degradation) are still a matter of debate and seem to vary upon the receptor and the physiological situation considered [61]. Also, it should be noted that while ubiquitin-mediated endocytosis appears as the main pathway in yeast, ubiquitin-independent endocytosis is more represented in higher eukaryotes [44, 62].

3.2. The "Classic" β-Arr2/β2-AR Couple.
A first evidence for the role of arrestins in receptor ubiquitylation came from a study by Shenoy and colleagues who observed that the β2-AR is ubiquitylated within 15 min of isoproterenol stimulation, ultimately leading to receptor degradation [30]. β2-AR ubiquitylation requires β-Arr2 and the ubiquitin ligase MDM2, which turned out to ubiquitylate β-Arr2 rather than β2-AR (see below) [30]. A mutant β2-AR lacking the ubiquitylation sites ($\beta2 - AR^{KO}$) is normally internalized, but not degraded [30]. In contrast, a translational fusion of ubiquitin to the β2-AR, which mimics its constitutive ubiquitylation, is internalized similarly as the wild-type β2-AR, but is degraded more efficiently [24].

Therefore, ubiquitylation is a critical signal for β2-AR degradation upon stimulation. A similar implication of GPCR ubiquitylation in its degradation, but not in its internalization, was also reported in the case of CXCR4 [63]. The identity of the ubiquitin ligase responsible for β2-AR ubiquitylation was revealed more recently. Indeed, β-Arr2 was found to interact with the HECT-type (homologous to E6-AP C-terminus) ubiquitin ligase Nedd4 (discussed below in this paper). Nedd4 (neural precursor cell expressed developmentally downregulated protein 4) promotes β2-AR ubiquitylation at endosomes, leading to its lysosomal targeting [24].

Once internalized, GPCRs can also escape degradation and recycle back to the plasma membrane in a functional state to mediate further signaling. Because GPCR ubiquitylation appears to trigger its degradation, deubiquitylation could regulate GPCR recycling to the plasma membrane.

Indeed, two deubiquitylating enzymes named USP33 and USP20 regulate β2-AR deubiquitylation, recycling and resensitization [35, 37]. USP33 was first identified as a β-arrestin interactant, thus suggesting that β-Arr2 could be involved in USP33 recruitment to β2-AR [35]. However, USP33 was found to interact with β2-AR even before agonist stimulation, that is, when β-Arr2 is not yet translocated to the plasma membrane [37]. In fact, USP33 appears to be transferred from agonist-activated β2-AR to β-Arr2, thus triggering its deubiquitylation and dissociation from the receptor, once internalized. Reassociation of USP33 with β2-AR in endosomal compartments would regulate its deubiquitylation and recycling to the plasma membrane. Thus, the association and dissociation of β-Arr2 from β2-AR may coordinate the ubiquitin conjugating/deconjugating activities towards β2-AR to tune the balance between receptor degradation and recycling. This positions the ubiquitin ligase adaptor function of β-Arr2 as a key regulator of GPCR signaling.

3.3. β-Arrestins as Ubiquitin Ligase Adaptors: Other Examples.
The function of β-arrestins as ubiquitin ligase adaptors is not restricted to β2-AR. Additional studies identified β-arrestins as ubiquitin ligase adaptors for non-GPCR proteins: β-Arr1, as β-Arr2, acts as an adaptor for ubiquitin ligases of the Nedd4 family such as Itch/AIP4 (Atrophin-1-interacting protein 4) for ubiquitylation of the TRPV4 (transient receptor potential) channel [64], and Nedd4 for that of the Na$^+$/H$^+$ exchanger 1 (NHE1) [65]. In the latter case, however, and in contrast to the situation described for the β2-AR, cargo ubiquitylation is required for its internalization. Because β-arrestins interact with both ubiquitin ligase and clathrin (see above), they may then act at two levels: first, for cargo ubiquitylation, which could recruit Eps15/Epsin endocytic adaptors, and, second, to assist the latter in the recruitment of a clathrin coat.

The contribution of β-arrestins to the trafficking of another classical GPCR, the chemokine receptor, CXCR4, was also studied. Early reports had shown that the ligand-induced ubiquitylation of CXCR4 by the Nedd4-like ubiquitin ligase AIP4 is required for its lysosomal sorting [63, 66]. β-Arr1 interacts with AIP4 at endosomes, and knockdown experiments revealed that β-Arr1 is an important player in CXCR4 degradation but, surprisingly, is not required for its ubiquitylation [67]. Instead, CXCR4 is phosphorylated at the plasma membrane after ligand binding, which allows the direct recruitment of the ubiquitin ligase AIP4 via its WW domains, and hence CXCR4 ubiquitylation [68]. β-Arr1 was later found to interact with the ESCRT-0 complex and to direct the ubiquitylation of one of its components, HRS (hepatocyte growth factor-regulated tyrosine kinase substrate) in a CXCR4-dependent manner [69].

Interestingly, β-arrestins appear to act primarily as adaptors for ubiquitin ligase of the Nedd4 family. These enzymes display WW domains that can interact with specific proline-rich motifs (usually, a [L/P]PxY sequence). Although this motif is sometimes present on the targeted substrates, as in the case of ENaC [70], in most cases this interaction motif is present on an adaptor protein in charge of substrate recognition [71]. However, no PPxY motif has been found in β-arrestins, and polyproline regions are not involved in Nedd4 interaction [24]. In addition, Nedd4 recruitment to β2-AR was not affected by mutations in Nedd4 WW domains. This indicates that this interaction involves a noncanonical binding of β-Arr2 to Nedd4, for which the molecular determinants remain to be addressed.

In some cases, β-arrestins act as adaptors for ubiquitin ligase which do not belong to the Nedd4 family. β-Arr1 was proposed to act as an adaptor for the RING (Really interesting new gene) ubiquitin ligase Mdm2 to mediate insulin-like growth factor I (IGF-1) receptor ubiquitylation and downregulation [72, 73]. A similar role was appointed to β-Arr2 for the ubiquitylation of the androgen receptor [74]. Again, the molecular basis of this interaction awaits further investigations.

4. Arrestin-Related Proteins: New Players in the Field

Visual and β-arrestins share a similar structure, with an arrestin fold in their N-terminal domains and a C-terminal tail [75–80]. It was proposed that visual and β-arrestins actually originate from an ancestral arrestin family from which they diverged relatively recently [25]. This ancestral family would also have given rise to proteins whose expression is not limited to metazoans: members of the Vps26 family, which display an arrestin-like fold [81, 82], as well as arrestin-related proteins (also coined α-arrestins) [25]. Indeed, proteins displaying sequence homologies to arrestins were first identified in the filamentous fungus *Aspergillus nidulans*, named CreD [83] and PalF [31], and more recent work in yeast allowed to identify additional members of this protein family renamed "ART" (arrestin-related trafficking adaptors) [34, 84, 85] that will be discussed later in this paper. In human, the ART family is composed of six members, named arrestin-domain containing 1–5 (ARRDC1–5) and TXNIP (Thioredoxin-interacting protein) (Figure 1). Therefore, arrestin-related proteins are expressed in all eukaryotes, except plants, which interestingly do not harbor Nedd4-like genes either [25].

4.1. Arrestin-Related Proteins as Endocytic Adaptors. A main difference between visual/β-arrestins and arrestin-related proteins is that the latter possess PPxY motifs (Figure 1). In agreement with the reported function of these motifs (see above), many studies have documented the ability of yeast arrestin-related proteins to interact with the only Nedd4-like ubiquitin ligase in *S. cerevisiae*, named Rsp5 [33, 34, 85–90]. Rsp5 is critical for ubiquitin-dependent intracellular trafficking pathways, such as endocytosis and MVB sorting [91]. However, most of the transporters lack PPxY motifs, and until recently, the molecular basis for the interaction between Rsp5 and transporters was unknown. It has become clear that yeast arrestin-related proteins fulfill this function, by acting as Rsp5 adaptors to mediate ubiquitylation and subsequent endocytosis of transporters

[33, 34, 84, 85, 90]. Using a chemical-genetic screen, Emr and colleagues have identified Ldb19/Art1 as a regulator of the endocytosis of Can1, an arginine transporter [34]. The function of Ldb19/Art1 was also extended to the endocytosis of other amino acid transporters. In a parallel study, Nikko et al. showed that two other arrestin-related proteins, named Ecm21/Art2 and Csr2/Art8, are specifically involved in the downregulation of the manganese transporter Smf1 [85]. Altogether, around 10 arrestin-related proteins were identified in yeast [34, 84, 85], and gathered in a family referred to as "ART" (arrestin-related trafficking adaptors).

Contrary to the situation in mammalian cells, where β-arrestins mainly act at a late step in cargo sorting, studies in yeast suggested a role for arrestin-related proteins in cargo internalization at the plasma membrane [27]. Indeed, several yeast ARTs are involved in the signal-induced internalization of transporters in response to specific environmental signals [33, 34, 84, 85, 90]. In addition, Art1 relocalizes to the plasma membrane in response to the signal that induces amino acid transporter endocytosis [34, 92].

However, as for β-arrestins, the situation is probably more complex, and the role of ARTs in endocytosis may not be restricted to the plasma membrane. In the course of their study of the high-affinity iron-uptake protein complex Fet3/Ftr1 in yeast, Burd and colleagues documented an example of ubiquitin-independent internalization [43]. The results show that a nonubiquitylatable form of Fet3/Ftr1 can still be internalized but is constitutively recycled back to the plasma membrane, leading to an apparent defect in internalization. Although the involvement of an arrestin remains to be determined, it strongly suggests that in this system, cargo ubiquitylation by Rsp5 is required at endosomal compartments, rather than at the plasma membrane. In addition, two yeast arrestin-related proteins, Aly1/Art6 and Aly2/Art3, have been shown to localize to intracellular compartments and to control the trafficking of the general amino-acids transporter Gap1 between trans-Golgi and endosomes [93]. Consistent with these findings, Aly1 and Aly2 interact with both clathrin and Golgi-specific clathrin adaptor complex AP-1, thus suggesting that arrestin related proteins, as β-arrestins, promote clathrin-coat assembly and cargo targeting to clathrin-coated vesicles. Therefore, futures studies will be necessary to precise where yeast arrestin-related proteins act on cargo trafficking. Regarding their intracellular localizations, we can already hypothesize several modes of action within the ART family of proteins.

Like their yeast homologs, several human ARRDC proteins are able to interact with ubiquitin ligases of the Nedd4 family [94–97]. Among those, ARRDC3 was isolated in a screen designed to identify proteins involved in β2-AR ubiquitylation and degradation after agonist treatment [95]. ARRDC3, as β-Arr2 [24], was shown to bridge the interaction between Nedd4 and β2-AR, leading to the intriguing possibility that arrestin-related proteins might coordinate, together with β-arrestins, receptor ubiquitylation and degradation. Because both classes of arrestins have the ability to dimerize [94, 98], this raises the possibility of potential heterooligomers between arrestin and arrestin-related proteins that

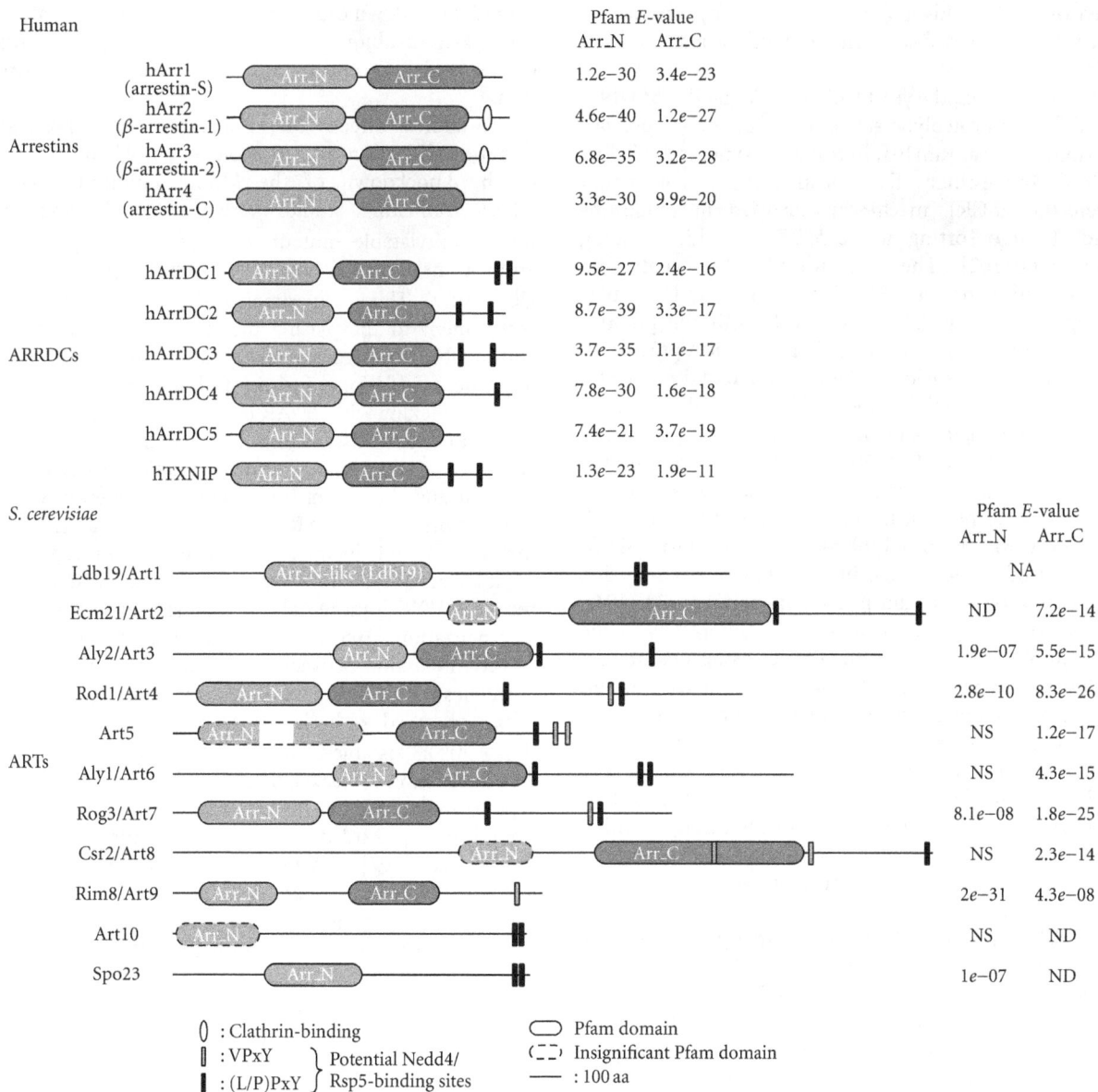

FIGURE 1: Schematic representation of the domain organization of human arrestins and arrestin-domain containing (ARRDC) proteins, and yeast ARTs (arrestin-related trafficking adaptors). Domains detected by Pfam 26.0 (http://pfam.sanger.ac.uk/) are shown and correspond to the following Pfam-A accessions: Arr_N: PF00339, Arr_C: PF02752, Arr_N-like (Ldb19): PF13002, along with the corresponding E-values for each domain (NA: not applicable; ND: not detected; NS: not significant). A putative Arr_N domain in Ecm21/Art2 was identified by alignment with the presumed Arr_N domain of Csr2/Art8. Clathrin-binding sites are depicted on β-arrestins; potential binding sites for ubiquitin ligases of the Nedd4 family are also indicated on arrestin-related proteins.

could reveal a complementary role between arrestin classes.

The basic function of arrestin-related proteins as ubiquitin ligase adaptor therefore seems strongly conserved. Of note, a role for ARRDC3 in the degradation of a cell surface adhesion molecule, integrin β4, was also pointed out, but its role as ubiquitin ligase adaptor in this context has not been investigated [99]. Future studies will indubitably unravel new connections between ubiquitin and arrestin-related proteins.

4.2. Other Functions of Arrestin-Related Proteins: An ESCRT Connection. As previously mentioned, arrestin-related proteins were initially identified in A. nidulans and named CreD [83] and PalF [31]. Interestingly, in both cases, a connection with the ubiquitin pathway was established. CreD was shown to interact physically with the Nedd4 homologue in A. nidulans, HulA, whereas PalF was found to be ubiquitylated in vivo. PalF, a protein involved in the ambient pH signaling in fungi, binds to the seven-transmembrane and putative pH sensor, PalH. This pointed out to many similarities between mammalian β-arrestins and this arrestin-related protein [31].

As in *A. nidulans*, the yeast PalF homologue Rim8/Art9 is essential for the proteolytic activation of the pH-responsive transcription factor, Rim101, in response to neutral-alkaline pH [100]. Interestingly, there is an intricate connection between the ESCRT machinery, involved in ubiquitin-dependent cargo sorting at the MVB, and this signaling pathway [100–102]. The ART Rim8/Art9 is central to the coordination of the ESCRT machinery and the pH-signaling pathway, as it interacts with both the putative pH sensor Rim21 and the ESCRT-I subunit Vps23 [89]. ESCRT appears to provide a platform for recruitment of a protein complex containing the ESCRT-III binding protein and ALIX homologue Rim20, that enables the proteolytic activation of the Rim101 transcription factor in response to the pH signal. Although initial studies suggested that this process takes place at the endosomal membrane [103], subsequent work supported the idea that arrestin-mediated recruitment of ESCRT in the fungal ambient pH signaling pathway may occur at the plasma membrane [89, 104]. Similarly, some of the human ARRDCs interact with the Vps23 homologue TSG101 or the ESCRT-associated protein ALIX [90, 96]. In particular, ARRDC1-mediated recruitment of ESCRT appears to drive the formation of microvesicles at the plasma membrane that may be involved in intercellular communication [94, 105]. Interestingly, this situation is reminiscent of the budding step of different enveloped RNA viruses, which recruit ESCRT components through similar interactions to promote membrane scission and subsequent viral particle release [106]. Accordingly, overexpression of several ARRDCs inhibits murine leukemia virus (MLV) viral particle release in a PPxY-specific way [96]. Therefore, ARRDCs may act as adaptors between Nedd4-like enzymes and the ESCRT machinery, in viral budding.

Finally, the connection between arrestin and the ESCRT machinery may not be restricted to arrestin-related proteins. Indeed, and as previously mentioned, β-Arr1 was found to interact with STAM-1 (signal-transducing adaptor molecule), a component of ESCRT-0, to regulate endosomal sorting of CXCR4 [69].

5. Regulation of Arrestin Function by Ubiquitin

5.1. Regulation of β-Arrestins by Ubiquitylation. Arrestins are specifically recruited to the cargoes following agonist stimulation (receptor) or in response to the presence of the substrate (transporter, channel), thus suggesting that they are regulated to mediate an adapted response of the cell to extracellular changes.

As mentioned previously, Shenoy and colleagues showed in a seminal article that β-Arr2 is itself ubiquitylated in response to agonist treatment [30]. Ubiquitylation of β-Arr2, in contrast to that of β2-AR, does not require Nedd4, but an ubiquitin ligase of the RING family, Mdm2. As this modification occurred upon stimulation, this suggested a role for β-Arr2 ubiquitylation in β-AR trafficking. Indeed,

Mdm2 knockdown caused a defect in β2-AR internalization. Thus, β-Arr2 ubiquitylation appears to play a key role in GPCR trafficking, and several lines of evidences support this idea.

To address directly the importance of this posttranslational modification and to avoid potential indirect effects of the knockdown of the Mdm2 ubiquitin ligase on β2-AR trafficking, studies were performed using both a nonubiquitylatable mutant form of β-Arr2 ($\beta - Arr2^{0K}$) and a translational fusion of ubiquitin to β-Arr2 (β-Arr2-Ub) [107]. These experiments showed that $\beta - Arr2^{0K}$ recruitment to the plasma membrane was only transient and unable to trigger internalization of β2-AR. On the opposite, translational fusion of ubiquitin to β-Arr2 led to its co-trafficking with β2-AR into endosomal compartments [107]. Previous observations had classified GPCRs in two classes (A and B), based on the interaction pattern between receptor and β-arrestin. Interaction of β-arrestin with class A receptors (e.g., β2-AR) only takes place at the plasma membrane, while its interaction with class B receptors (e.g., angiotensin II type 1a receptor: AT1aR, or vasopressin V2 receptor: V2R) is more stable and persists even after receptor internalization [108]. Interestingly, the increased stability of the interaction between class B receptors and β-arrestin correlates with a sustained β-arrestin ubiquitylation, which is not observed with class A receptors [36]. Indeed, even if $\beta - Arr2^{K0}$ is able to interact with the receptor *in vitro*, this interaction is weaker than that displayed with the wild type form *in vivo*. On the opposite, translational fusion of ubiquitin to β-Arr2 displays a stronger binding than wild type β-Arr2 [107]. β-Arr2 ubiquitylation therefore appears to reinforce the interaction with β2-AR.

Because β-Arr2 is capable of interacting with the endocytic machinery, such as clathrin or clathrin adaptors, the failure of $\beta - Arr2^{K0}$ to promote β2-AR internalization could originate from an impaired interaction with these components. Indeed, $\beta - Arr2^{K0}$ exhibits a weaker interaction with clathrin than the wild-type form [107]. While clathrin is not known to interact with ubiquitin, β-Arr2 ubiquitylation might stabilize the interaction with clathrin through other components of the endocytic machinery such as Eps15/epsin proteins that are able to bind both ubiquitin and clathrin.

β-Arr2 ubiquitylation was also shown to affect its scaffolding function for signaling proteins. The amplitude of β-arrestin-mediated activation of ERK (extracellular signal-regulated kinase) correlates with the β-Arr2 ubiquitylation status. Although β-Arr2 ubiquitylation was not required for its interaction with MAP kinases (such as c-Raf and ERK), translational fusion of ubiquitin to β-Arr2 led to an increased level of ERK activity in endosome localized receptor complexes (signalosomes). Consistent with these findings, β-arrestin ubiquitylation promote its association with membrane. Again, ubiquitylation appears to function in stabilizing the β-arrestin-mediated interaction between the receptor and signaling proteins [107].

Finally, arrestins undergo conformational changes upon binding to activated receptors [109]. Ubiquitin modification could therefore contribute to the proper rearrangement of the β-arrestin structure, leading to optimal interactions with its partners, and this awaits further investigations.

5.2. Regulation of Arrestin-Related Protein by Ubiquitylation.

Many arrestin-related proteins have also been reported as substrates for ubiquitylation, both in fungi and human [31, 33, 34, 89, 90, 94, 96, 97, 110]. Ubiquitylation of these proteins, in contrast to that of β-arrestins, is triggered by ubiquitin ligases of the Nedd4 family. Therefore, arrestin-related proteins are adaptors as well as targets of the same ubiquitin ligases.

The yeast arrestin-related protein Ldb19/Art1 is required for the endocytosis of amino acid permeases, such as the arginine transporter, Can1. Failure to endocytose Can1 leads to sensitivity of the cells to canavanine, a toxic analog of arginine. A nonubiquitylatable mutant of Ldb19/Art1 cannot grow on this drug, suggesting that Can1 remains at the plasma and therefore that Art1 is not functional [34]. The importance of ubiquitylation for ART function was also demonstrated for Rod1/Art4, involved in the glucose-induced endocytosis of carbon sources transporters [33, 84]. Rod1/Art4 is ubiquitylated in response to glucose exposure and a nonubiquitylatable mutant is unable to promote the endocytosis of the lactate transporter, Jen1, following glucose treatment [33]. Altogether, these data indicate that ART ubiquitylation is crucial for their function in endocytosis. Human arrestin-related protein ARRDC3 was isolated in a screen designed to identify genes involved in β2-AR degradation, and acts as a Nedd4 adaptor for β2-AR ubiquitylation [95]. While ARRDC3 ubiquitylation has not yet been observed, ARRDC1 and TXNIP were shown to be ubiquitylated by ubiquitin ligases of the Nedd4 family [90, 96, 97]. Thus, it is tempting to speculate that the same regulation applies in fungi and human.

The ubiquitylation of the arrestin-related protein PalF in *A. nidulans* is triggered in a signal-(alkaline pH) and receptor-(PalH) dependent manner [31]. PalF ubiquitylation appears as a major determinant of its activity, since the translational fusion of ubiquitin to PalF leads to a constitutive activation of the pathway [110]. The yeast PalF homologue Rim8/Art9 was shown to be monoubiquitylated [89]. Monoubiquitylation of Rim8/Art9 occurs on a lysine residue in its C-terminus and, as for all other ARTs described to date in yeast, is performed by Rsp5, which binds to a PxY motif near the ubiquitylation site. This monoubiquitylated residue, together with a SxP motif, contributes to the interaction of Rim8/Art9 with the ESCRT-I subunit Vps23 via its ubiquitin-binding domain, UEV (ubiquitin E2 variant) [89]. Interestingly, Vps23 binding appears to control the levels of monoubiquitylated Rim8/Art9, thus suggesting that this interaction either promotes Rim8 ubiquitylation or prevents its further polyubiquitylation and possibly its degradation. Interaction of human ARRDC1 with the Vps23 homologue Tsg101 was shown to be mediated by a PSAP motif which, like the SxP motif in Rim8, is located at the protein C-terminus [94, 96]. In addition, it was proposed that the ubiquitylation of ARRDC1 is important for its function [94].

However, this is based on results obtained upon depletion of the corresponding Nedd4-like ligase WWP1, which in principle could also impair a potential adaptor function and may have off-target effects. Therefore, the identification and mutation of the ubiquitylation sites will be critical to address this question.

5.3. Dynamic Regulation of Arrestin Ubiquitylation. Phosphorylation-Dependent Ubiquitylation?

Because ubiquitin ligases target a large number of proteins in the cell, their activity toward a given substrate is usually indirectly regulated through substrate accessibility, either by the use of adaptor proteins, or by post-translational modification of the substrate, such as phosphorylation [111].

Interestingly, cytosolic β-arrestins are constitutively phosphorylated, and undergo dephosphorylation upon binding to the activated receptor. β-Arr1 dephosphorylation is required for β2-AR internalization, but not for its desensitization [112]. Indeed, a β-Arr1 mutant mimicking constitutive phosphorylation displays a weaker interaction with clathrin but an unaltered β2-AR binding [112]. Similar data were reported for β-Arr2, and the phosphorylation site was localized near the clathrin and AP-2 binding motifs, thus providing an explanation as to why β-Arr2 phosphorylation regulates the interaction with clathrin/AP-2 [113]. Additionally, the phosphorylation of the major visual arrestin in *Drosophila* (Arr2) also regulates its interaction with clathrin [114].

Importantly, once the receptor is internalized, β-Arr1 is rephosphorylated. These dynamic phosphorylation/dephosphorylation events suggest the involvement of kinases and phosphatases whose activation is coordinated in response to agonist exposure. Interestingly, β-Arr1 is phosphorylated *in vitro* by ERK kinases and accordingly, the modulation of ERK activity *in vivo* affects β-Arr1 phosphorylation, thus providing an inhibitory feedback control of its function [115].

Although β-arrestins are both dephosphorylated and ubiquitylated upon receptor binding, an eventual relationship between these two modifications remained to be addressed. Such a link has been described for the yeast arrestin-related protein Rod1/Art4, involved in the glucose-induced endocytosis of carbon sources transporters [33]. As for β-arrestins, Rod1/Art4 dephosphorylation and ubiquitylation occurs in response to an external signal-in this case, glucose. The yeast homologue of AMPK (5′-AMP-activated protein kinase), Snf1, and its counteracting phosphatase PP1 (protein phosphatase 1) control the phosphorylation status of Art4/Rod1 in response to glucose availability. Therefore, in the absence of glucose, Art4/Rod1 is phosphorylated and endocytosis is inhibited. This inhibitory effect results from the ability of phosphorylated Art4/Rod1 to bind 14-3-3 proteins, thereby hindering its ubiquitylation by Rsp5 and hence preventing its activation [33].

Interestingly, phosphorylation of another yeast arrestin-related protein, Ldb19/Art1, also regulates its function. A recent study indicated that Ldb19/Art1 is subject to phosphoinhibition through the action of the TOR (target of rapamycin) effector and protein kinase Npr1, thus allowing cells to regulate amino acid transporter endocytosis in response to the nitrogen status [92]. While the overall phospho-inhibition mechanism recalls that of Rod1/Art4, Ldb19/Art1 ubiquitylation is uncoupled from its phosphorylation, suggesting a different regulatory mechanism [34]. The identification of Npr1-dependent phosphorylation sites on Ldb19/Art1 allowed generating a nonphosphorylatable mutant form of the protein. Interestingly, this mutant fails to be translocated at the plasma membrane upon stimulation, which likely explains why transporters endocytosis is impaired [92]. Further work will be needed to understand the molecular mechanism of this phosphorylation-dependent inhibition.

Other examples of arrestin-related proteins that are subjected to phosphorylation which include *A. nidulans* PalF, involved in ambient pH sensing, although the regulatory mechanism appears to be different since PalF undergoes phosphorylation instead of dephosphorylation, in response to the ambient pH signal [31]. The same result has been observed in the pathogenic yeast *Candida albicans*, where the PalF homologue Rim8 is also phosphorylated in response to neutral-alkaline pH [116]. Although phosphorylation of Rim8/Art9 in baker's yeast has not been reported, its ubiquitylation does not appear to be regulated by ambient pH, in contrast to that of PalF, which is induced by alkaline pH [89]. The apparent lack of regulation of Rim8/Art9 ubiquitylation is consistent with its role in Vps23 binding, which appears to occur even in nonstimulated conditions. Thus, the pH-dependent regulation of PalF ubiquitylation in *A. nidulans* may reflect an additional level of regulation in this organism.

From these studies, it emerges that arrestin-related proteins are often modified posttranslationally in response to stimulation, either by phosphorylation, ubiquitylation, or both. A crosstalk between phosphorylation and ubiquitylation has been evidenced. How these modifications operate to coordinate arrestin function is unknown, and this provides new avenues for research in this field.

Ubiquitylation and deubiquitylation. Ubiquitylation, akin to phosphorylation, is a reversible process. Therefore, deubiquitylation appeared as a possible mechanism for regulation of arrestin function. In support of this idea, the transient β-Arr2 ubiquitylation associated to class A receptor suggested that deubiquitylation occurs rapidly after agonist stimulation. Indeed, the ubiquitin-specific protease 33 (USP33) was shown to deubiquitylate β-Arr2 following β2-AR binding [35]. USP33 knock-down led to an increase in β-Arr2 ubiquitylation. This was accompanied by a stronger interaction with the receptor, and a prolonged β-Arr2-dependent MAP kinase signaling. These findings are consistent with the phenotypes of cells expressing a translational fusion of ubiquitin to β-Arr2 (see above) and provide an additional mechanism for the regulation of arrestin function. Strikingly, recruitment of USP33 to β-Arr2 depends on the receptor class, in agreement with previous finding that receptor class determines the kinetics of β-Arr2 deubiquitylation [36]. In addition, receptors belonging to the same class can target different lysine residues on β-Arr2 for ubiquitylation [32]. Therefore, binding of β-Arr2 to different receptors may trigger distinct conformational change that could modulate both ubiquitylation sites accessibility and association with deubiquitylating enzymes, hence leading to a different functional output—in full support of the concept of an "ubiquitylation code."

6. Concluding Remarks

In this paper, we emphasized the many relationships between ubiquitin metabolism and arrestin biology. Many proteins of the arrestin family act as ubiquitin ligase adaptors and are required for ubiquitylation of endocytic cargo. In particular, recent data obtained on arrestin-related proteins have pointed out several features shared with β-arrestins, such as the intimate connection existing between these proteins and ubiquitin ligases of the Nedd4 family. In addition, ubiquitin also regulates arrestin function in a yet undefined way, which is now critical to understand. However, the emergence of new regulatory mechanisms involving a now expanded family of arrestin proteins, combined with the multiplicity of model organisms, is likely to favor the rapid evolution of concepts in arrestin biology.

Abbreviations

AIP4:	Atrophin-interacting protein 4
AP-1/AP-2:	Clathrin adaptor protein
ARRDC:	Arrestin-domain Containing
ART:	Arrestin-related trafficking adaptors
β-Arr:	β-arrestin
β2-AR:	β2-adrenergic receptor
ERK:	Extracellular signal-regulated Kinase
ESCRT:	Endosomal sorting complex required for transport
GPCRS:	G-protein-coupled receptors.
MAP:	Mitogen-activated protein
MVBs:	Multivesicular bodies
Nedd4:	Neural precursor cell expressed developmentally downregulated protein 4
RING:	Really interesting new gene
TXNIP:	Thioredoxin-interacting protein.

Acknowledgments

This work was supported by the CNRS and by a Grant from the Fondation ARC pour la recherche sur le cancer (SFI20101201844) to S. Léon, and by a Grant from the Spanish CICYT (BFU2008-02005) to O. Vincent. M. Becuwe and A. Herrador are recipients of Ph.D. fellowships from the French Ministère de l'Enseignement Supérieur et de la Recherche and from CSIC-JAE, respectively.

References

[1] U. Wilden, S. W. Hall, and H. Kuhn, "Phosphodiesterase activation by photoexcited rhodopsin is quenched when rhodopsin is phosphorylated and binds the intrinsic 48-kDa

protein of rod outer segments," *Proceedings of the National Academy of Sciences of the United States of America*, vol. 83, no. 5, pp. 1174–1178, 1986.

[2] R. Zuckerman and J. E. Cheasty, "A 48 kDa protein arrests cGMP phosphodiesterase activation in retinal rod disk membranes," *FEBS Letters*, vol. 207, no. 1, pp. 35–41, 1986.

[3] V. V. Gurevich, S. M. Hanson, X. Song, S. A. Vishnivetskiy, and E. V. Gurevich, "The functional cycle of visual arrestins in photoreceptor cells," *Progress in Retinal and Eye Research*, vol. 30, no. 6, pp. 405–430, 2011.

[4] J. L. Benovic, H. Kühn, I. Weyand, J. Codina, M. G. Caron, and R. J. Lefkowitz, "Functional desensitization of the isolated β-adrenergic receptor by the beta-adrenergic receptor kinase: potential role of an analog of the retinal protein arrestin (48-kDa protein)," *Proceedings of the National Academy of Sciences of the United States of America*, vol. 84, no. 24, pp. 8879–8882, 1987.

[5] M. J. Lohse, J. L. Benovic, J. Codina, M. G. Cargon, and R. J. Lefkowitz, "β-arrestin: a protein that regulates β-adrenergic receptor function," *Science*, vol. 248, no. 4962, pp. 1547–1550, 1990.

[6] H. Attramadal, J. L. Arriza, C. Aoki et al., "β-arrestin2, a novel member of the arrestin/β-arrestin gene family," *Journal of Biological Chemistry*, vol. 267, no. 25, pp. 17882–17890, 1992.

[7] S. M. DeWire, S. Ahn, R. J. Lefkowitz, and S. K. Shenoy, "β-arrestins and cell signaling," *Annual Review of Physiology*, vol. 69, pp. 483–510, 2007.

[8] V. V. Gurevich and E. V. Gurevich, "The molecular acrobatics of arrestin activation," *Trends in Pharmacological Sciences*, vol. 25, no. 2, pp. 105–111, 2004.

[9] A. C. Hanyaloglu and M. von Zastrow, "Regulation of GPCRs by endocytic membrane trafficking and its potential implications," *Annual Review of Pharmacology and Toxicology*, vol. 48, pp. 537–568, 2008.

[10] C. A. C. Moore, S. K. Milano, and J. L. Benovic, "Regulation of receptor trafficking by GRKs and arrestins," *Annual Review of Physiology*, vol. 69, pp. 451–482, 2007.

[11] J. G. Krupnick, O. B. Goodman Jr., J. H. Keen, and J. L. Benovic, "Arrestin/clathrin interaction. Localization of the clathrin binding domain of nonvisual arrestins to the carboxyl terminus," *Journal of Biological Chemistry*, vol. 272, no. 23, pp. 15011–15016, 1997.

[12] O. B. Goodman Jr., J. G. Krupnick, F. Santini et al., "β-arrestin acts as a clathrin adaptor in endocytosis of the β_2-adrenergic receptor," *Nature*, vol. 383, no. 6599, pp. 447–450, 1996.

[13] O. B. Goodman Jr., J. G. Krupnick, V. V. Gurevich, J. L. Benovic, and J. H. Keen, "Arrestin/clathrin interaction. Localization of the arrestin binding locus to the clathrin terminal domain," *Journal of Biological Chemistry*, vol. 272, no. 23, pp. 15017–15022, 1997.

[14] Y. M. Kim and J. L. Benovic, "Differential roles of arrestin-2 interaction with clathrin and adaptor protein 2 in G protein-coupled receptor trafficking," *Journal of Biological Chemistry*, vol. 277, no. 34, pp. 30760–30768, 2002.

[15] S. A. Laporte, W. E. Miller, K. M. Kim, and M. G. Caron, "β-arrestin/AP-2 interaction in G protein-coupled receptor internalization. Identification of a β-arrestin binding site β_2-adaptin," *Journal of Biological Chemistry*, vol. 277, no. 11, pp. 9247–9254, 2002.

[16] S. A. Laporte, R. H. Oakley, J. Zhang et al., "The β_2-adrenergic receptor/βarrestin complex recruits the clathrin adaptor AP-2 during endocytosis," *Proceedings of the National*

Academy of Sciences of the United States of America, vol. 96, no. 7, pp. 3712–3717, 1999.

[17] M. A. Edeling, S. K. Mishra, P. A. Keyel et al., "Molecular switches involving the AP-2 β_2 appendage regulate endocytic cargo selection and clathrin coat assembly," *Developmental Cell*, vol. 10, no. 3, pp. 329–342, 2006.

[18] A. Claing, W. Chen, W. E. Miller et al., "β-arrestin-mediated ADP-ribosylation factor 6 activation and β_2-Adrenergic receptor endocytosis," *Journal of Biological Chemistry*, vol. 276, no. 45, pp. 42509–42513, 2001.

[19] P. H. McDonald, N. L. Cote, F. T. Lin, R. T. Premont, J. A. Pitcher, and R. J. Lefkowitz, "Identification of NSF as a β-arrestin1-binding protein: implications for β_2-adrenergic receptor regulation," *Journal of Biological Chemistry*, vol. 274, no. 16, pp. 10677–10680, 1999.

[20] C. D. Nelson, J. J. Kovacs, K. N. Nobles, E. J. Whalen, and R. J. Lefkowitz, "β-arrestin scaffolding of phosphatidylinositol 4-phosphate 5-kinase Iα promotes agonist-stimulated sequestration of the β_2-adrenergic receptor," *Journal of Biological Chemistry*, vol. 283, no. 30, pp. 21093–21101, 2008.

[21] L. M. Luttrell and D. Gesty-Palmer, "Beyond desensitization: physiological relevance of arrestin-dependent signaling," *Pharmacological Reviews*, vol. 62, no. 2, pp. 305–330, 2010.

[22] F. Acconcia, S. Sigismund, and S. Polo, "Ubiquitin in trafficking: the network at work," *Experimental Cell Research*, vol. 315, no. 9, pp. 1610–1618, 2009.

[23] S. Polo, "Signaling-mediated control of ubiquitin ligases in endocytosis," *BMC Biology*, vol. 10, article 25, 2012.

[24] S. K. Shenoy, K. Xiao, V. Venkataramanan, P. M. Snyder, N. J. Freedman, and A. M. Weissman, "Nedd4 mediates agonist-dependent ubiquitination, lysosomal targeting, and degradation of the β_2-adrenergic receptor," *Journal of Biological Chemistry*, vol. 283, no. 32, pp. 22166–22176, 2008.

[25] C. E. Alvarez, "On the origins of arrestin and rhodopsin," *BMC Evolutionary Biology*, vol. 8, no. 1, article 222, 2008.

[26] L. Aubry, D. Guetta, and G. Klein, "The arrestin fold: variations on a theme," *Current Genomics*, vol. 10, no. 2, pp. 133–142, 2009.

[27] S. Polo and P. P. di Fiore, "Finding the right partner: science or ART?" *Cell*, vol. 135, no. 4, pp. 590–592, 2008.

[28] A. K. Shukla, K. Xiao, and R. J. Lefkowitz, "Emerging paradigms of β-arrestin-dependent seven transmembrane receptor signaling," *Trends in Biochemical Sciences*, vol. 36, no. 9, pp. 457–469, 2011.

[29] S. K. Shenoy and R. J. Lefkowitz, "β-arrestin-mediated receptor trafficking and signal transduction," *Trends in Pharmacological Sciences*, vol. 32, no. 9, pp. 521–533, 2011.

[30] S. K. Shenoy, P. H. McDonald, T. A. Kohout, and R. J. Lefkowitz, "Regulation of receptor fate by ubiquitination of activated β_2-adrenergic receptor and β-arrestin," *Science*, vol. 294, no. 5545, pp. 1307–1313, 2001.

[31] S. Herranz, J. M. Rodríguez, H. J. Bussink et al., "Arrestin-related proteins mediate pH signaling in fungi," *Proceedings of the National Academy of Sciences of the United States of America*, vol. 102, no. 34, pp. 12141–12146, 2005.

[32] S. K. Shenoy and R. J. Lefkowitz, "Receptor-specific ubiquitination of β-arrestin directs assembly and targeting of seven-transmembrane receptor signalosomes," *Journal of Biological Chemistry*, vol. 280, no. 15, pp. 15315–15324, 2005.

[33] M. Becuwe, N. Vieira, D. Lara et al., "A molecular switch on an arrestin-like protein relays glucose signaling to transporter endocytosis," *Journal of Cell Biology*, vol. 196, no. 2, pp. 247–259, 2012.

[34] C. H. Lin, J. A. MacGurn, T. Chu, C. J. Stefan, and S. D. Emr, "Arrestin-related ubiquitin-ligase adaptors regulate endocytosis and protein turnover at the cell surface," *Cell*, vol. 135, no. 4, pp. 714–725, 2008.

[35] S. K. Shenoy, A. S. Modi, A. K. Shukla et al., "β-arrestin-dependent signaling and trafficking of 7-transmembrane receptors is reciprocally regulated by the deubiquitinase USP$_{33}$ and the E$_3$ ligase Mdm2," *Proceedings of the National Academy of Sciences of the United States of America*, vol. 106, no. 16, pp. 6650–6655, 2009.

[36] S. K. Shenoy and R. J. Lefkowitz, "Trafficking patterns of β-arrestin and G protein-coupled receptors determined by the kinetics of β-arrestin deubiquitination," *Journal of Biological Chemistry*, vol. 278, no. 16, pp. 14498–14506, 2003.

[37] M. Berthouze, V. Venkataramanan, Y. Li, and S. K. Shenoy, "The deubiquitinases USP33 and USP20 coordinate β2 adrenergic receptor recycling and resensitization," *The EMBO Journal*, vol. 28, no. 12, pp. 1684–1696, 2009.

[38] R. Kolling and C. P. Hollenberg, "The ABC-transporter Ste6 accumulates in the plasma membrane in a ubiquitinated form in endocytosis mutants," *The EMBO Journal*, vol. 13, no. 14, pp. 3261–3271, 1994.

[39] L. Hicke and H. Riezman, "Ubiquitination of a yeast plasma membrane receptor signals its ligand-stimulated endocytosis," *Cell*, vol. 84, no. 2, pp. 277–287, 1996.

[40] A. F. Roth and N. G. Davis, "Ubiquitination of the yeast a-factor receptor," *Journal of Cell Biology*, vol. 134, no. 3, pp. 661–674, 1996.

[41] J. M. Galan, V. Moreau, B. Andre, C. Volland, and R. Haguenauer-Tsapis, "Ubiquitination mediated by the Npi1p/Rsp5p ubiquitin-protein ligase is required for endocytosis of the yeast uracil permease," *Journal of Biological Chemistry*, vol. 271, no. 18, pp. 10946–10952, 1996.

[42] E. Lauwers, Z. Erpapazoglou, R. Haguenauer-Tsapis, and B. André, "The ubiquitin code of yeast permease trafficking," *Trends in Cell Biology*, vol. 20, no. 4, pp. 196–204, 2010.

[43] T. I. Strochlic, B. C. Schmiedekamp, J. Lee, D. J. Katzmann, and C. G. Burd, "Opposing activities of the snx3-retromer complex and ESCRT proteins mediate regulated cargo sorting at a common endosome," *Molecular Biology of the Cell*, vol. 19, no. 11, pp. 4694–4706, 2008.

[44] L. M. Traub, "Tickets to ride: selecting cargo for clathrin-regulated internalization," *Nature Reviews Molecular Cell Biology*, vol. 10, no. 9, pp. 583–596, 2009.

[45] S. C. Shih, D. J. Katzmann, J. D. Schnell, M. Sutanto, S. D. Emr, and L. Hicke, "Epsins and Vps27p/Hrs contain ubiquitin-binding domains that function in receptor endocytosis," *Nature Cell Biology*, vol. 4, no. 5, pp. 389–393, 2002.

[46] R. C. Aguilar, H. A. Watson, and B. Wendland, "The yeast epsin Ent1 is recruited to membranes through multiple independent interactions," *Journal of Biological Chemistry*, vol. 278, no. 12, pp. 10737–10743, 2003.

[47] H. Barriere, C. Nemes, D. Lechardeur, M. Khan-Mohammad, K. Fruh, and G. L. Lukacs, "Molecular basis of oligoubiquitin-dependent internalization of membrane proteins in mammalian cells," *Traffic*, vol. 7, no. 3, pp. 282–297, 2006.

[48] M. Kazazic, V. Bertelsen, K. W. Pedersen et al., "Epsin 1 is involved in recruitment of ubiquitinated EGF receptors into clathrin-coated pits," *Traffic*, vol. 10, no. 2, pp. 235–245, 2009.

[49] S. Sigismund, T. Woelk, C. Puri et al., "Clathrin-independent endocytosis of ubiquitinated cargos," *Proceedings of the*

[50] M. R. Dores, J. D. Schnell, L. Maldonado-Baez, B. Wendland, and L. Hicke, "The function of yeast epsin and Ede1 ubiquitin-binding domains during receptor internalization," *Traffic*, vol. 11, no. 1, pp. 151–160, 2010.

[51] K. Haglund, N. Shimokawa, I. Szymkiewicz, and I. Dikic, "Cbl-directed monoubiquitination of CIN85 is involved in regulation of ligand-induced degradation of EGF receptors," *Proceedings of the National Academy of Sciences of the United States of America*, vol. 99, no. 19, pp. 12191–12196, 2002.

[52] M. Katz, K. Shtiegman, P. Tal-Or et al., "Ligand-independent degradation of epidermal growth factor receptor involves receptor ubiquitylation and Hgs, an adaptor whose ubiquitin-interacting motif targets ubiquitylation by Nedd4," *Traffic*, vol. 3, no. 10, pp. 740–751, 2002.

[53] S. Polo, S. Sigismund, M. Faretta et al., "A single motif responsible for ubiquitin recognition and monoubiquitination in endocytic proteins," *Nature*, vol. 416, no. 6879, pp. 451–455, 2002.

[54] S. B. Shields and R. C. Piper, "How Ubiquitin Functions with ESCRTs," *Traffic*, vol. 12, no. 10, pp. 1306–1317, 2011.

[55] W. M. Henne, N. J. Buchkovich, and S. D. Emr, "The ESCRT pathway," *Developmental Cell*, vol. 21, no. 1, pp. 77–91, 2011.

[56] D. J. Katzmann, G. Odorizzi, and S. D. Emr, "Receptor downregulation and multivesicular-body sorting," *Nature Reviews Molecular Cell Biology*, vol. 3, no. 12, pp. 893–905, 2002.

[57] G. J. Strous, P. van Kerkhof, R. Govers, A. Ciechanover, and A. L. Schwartz, "The ubiquitin conjugation system is required for ligand-induced endocytosis and degradation of the growth hormone receptor," *The EMBO Journal*, vol. 15, no. 15, pp. 3806–3812, 1996.

[58] O. Staub, I. Gautschi, T. Ishikawa et al., "Regulation of stability and function of the epithelial Na$^+$ channel (ENaC) by ubiquitination," *The EMBO Journal*, vol. 16, no. 21, pp. 6325–6336, 1997.

[59] L. Hicke and R. Dunn, "Regulation of membrane protein transport by ubiquitin and ubiquitin-binding proteins," *Annual Review of Cell and Developmental Biology*, vol. 19, pp. 141–172, 2003.

[60] D. Mukhopadhyay and H. Riezman, "Proteasome-independent functions of ubiquitin in endocytosis and signaling," *Science*, vol. 315, no. 5809, pp. 201–205, 2007.

[61] J. N. Hislop and M. von Zastrow, "Role of ubiquitination in endocytic trafficking of G-protein-coupled receptors," *Traffic*, vol. 12, no. 2, pp. 137–148, 2011.

[62] A. M. Hommelgaard, K. Roepstorff, F. Vilhardt, M. L. Torgersen, K. Sandvig, and B. van Deurs, "Caveolae: stable membrane domains with a potential for internalization," *Traffic*, vol. 6, no. 9, pp. 720–724, 2005.

[63] A. Marchese and J. L. Benovic, "Agonist-promoted ubiquitination of the G protein-coupled receptor CXCR4 mediates lysosomal sorting," *Journal of Biological Chemistry*, vol. 276, no. 49, pp. 45509–45512, 2001.

[64] A. K. Shukla, J. Kim, S. Ahn et al., "Arresting a transient receptor potential (TRP) channel: β-arrestin 1 mediates ubiquitination and functional down-regulation of TRPV4," *Journal of Biological Chemistry*, vol. 285, no. 39, pp. 30115–30125, 2010.

[65] A. Simonin and D. Fuster, "Nedd4-1 and β-arrestin-1 are key regulators of Na$^+$/H$^+$ exchanger 1 ubiquitylation,

National Academy of Sciences of the United States of America, vol. 102, no. 8, pp. 2760–2765, 2005.

endocytosis, and function," *Journal of Biological Chemistry*, vol. 285, no. 49, pp. 38293–38303, 2010.

[66] A. Marchese, C. Raiborg, F. Santini, J. H. Keen, H. Stenmark, and J. L. Benovic, "The E3 ubiquitin ligase AIP4 mediates ubiquitination and sorting of the G protein-coupled receptor CXCR4," *Developmental Cell*, vol. 5, no. 5, pp. 709–722, 2003.

[67] D. Bhandari, J. Trejo, J. L. Benovic, and A. Marchese, "Arrestin-2 interacts with the ubiquitin-protein isopeptide ligase atrophin-interacting protein 4 and mediates endosomal sorting of the chemokine receptor CXCR4," *Journal of Biological Chemistry*, vol. 282, no. 51, pp. 36971–36979, 2007.

[68] D. Bhandari, S. L. Robia, and A. Marchese, "The E3 ubiquitin ligase atrophin interacting protein 4 binds directly to the chemokine receptor CXCR4 via a novel WW domain-mediated interaction," *Molecular Biology of the Cell*, vol. 20, no. 5, pp. 1324–1339, 2009.

[69] R. Malik and A. Marchese, "Arrestin-2 interacts with the endosomal sorting complex required for transport machinery to modulate endosomal sorting of CXCR4," *Molecular Biology of the Cell*, vol. 21, no. 14, pp. 2529–2541, 2010.

[70] O. Staub, S. Dho, P. C. Henry et al., "WW domains of Nedd4 bind to the proline-rich PY motifs in the epithelial Na$^+$ channel deleted in Liddle's syndrome," *The EMBO Journal*, vol. 15, no. 10, pp. 2371–2380, 1996.

[71] S. Léon and R. Haguenauer-Tsapis, "Ubiquitin ligase adaptors: regulators of ubiquitylation and endocytosis of plasma membrane proteins," *Experimental Cell Research*, vol. 315, no. 9, pp. 1574–1583, 2009.

[72] F. T. Lin, Y. Daaka, and R. J. Lefkowitz, "β-arrestins regulate mitogenic signaling and clathrin-mediated endocytosis of the insulin-like growth factor I receptor," *Journal of Biological Chemistry*, vol. 273, no. 48, pp. 31640–31643, 1998.

[73] L. Girnita, S. K. Shenoy, B. Sehat et al., "β-arrestin is crucial for ubiquitination and down-regulation of the insulin-like growth factor-1 receptor by acting as adaptor for the MDM2 E3 ligase," *Journal of Biological Chemistry*, vol. 280, no. 26, pp. 24412–24419, 2005.

[74] V. Lakshmikanthan, L. Zou, J. I. Kim et al., "Identification of βArrestin2 as a corepressor of androgen receptor signaling in prostate cancer," *Proceedings of the National Academy of Sciences of the United States of America*, vol. 106, no. 23, pp. 9379–9384, 2009.

[75] J. Granzin, U. Wilden, H. W. Choe, J. Labahn, B. Krafft, and G. Buldt, "X-ray crystal structure of arrestin from bovine rod outer segments," *Nature*, vol. 391, no. 6670, pp. 918–921, 1998.

[76] M. Han, V. V. Gurevich, S. A. Vishnivetskiy, P. B. Sigler, and C. Schubert, "Crystal structure of β-arrestin at 1.9 Å: possible mechanism of receptor binding and membrane translocation," *Structure*, vol. 9, no. 9, pp. 869–880, 2001.

[77] J. A. Hirsch, C. Schubert, V. V. Gurevich, and P. B. Sigler, "The 2.8 Å crystal structure of visual arrestin: a model for arrestin's regulation," *Cell*, vol. 97, no. 2, pp. 257–269, 1999.

[78] S. K. Milano, H. C. Pace, Y. M. Kim, C. Brenner, and J. L. Benovic, "Scaffolding functions of arrestin-2 revealed by crystal structure and mutagenesis," *Biochemistry*, vol. 41, no. 10, pp. 3321–3328, 2002.

[79] S. K. Milano, Y. M. Kim, F. P. Stefano, J. L. Benovic, and C. Brenner, "Nonvisual arrestin oligomerization and cellular localization are regulated by inositol hexakisphosphate binding," *Journal of Biological Chemistry*, vol. 281, no. 14, pp. 9812–9823, 2006.

[80] R. B. Sutton, S. A. Vishnivetskiy, J. Robert et al., "Crystal structure of cone arrestin at 2.3 Å: evolution of receptor specificity," *Journal of Molecular Biology*, vol. 354, no. 5, pp. 1069–1080, 2005.

[81] H. Shi, R. Rojas, J. S. Bonifacino, and J. H. Hurley, "The retromer subunit Vps26 has an arrestin fold and binds Vps35 through its C-terminal domain," *Nature Structural and Molecular Biology*, vol. 13, no. 6, pp. 540–548, 2006.

[82] B. M. Collins, S. J. Norwood, M. C. Kerr et al., "Structure of Vps26B and mapping of its interaction with the retromer protein complex," *Traffic*, vol. 9, no. 3, pp. 366–379, 2008.

[83] N. A. Boase and J. M. Kelly, "A role for creD, a carbon catabolite repression gene from Aspergillus nidulans, in ubiquitination," *Molecular Microbiology*, vol. 53, no. 3, pp. 929–940, 2004.

[84] E. Nikko and H. R. B. Pelham, "Arrestin-mediated endocytosis of yeast plasma membrane transporters," *Traffic*, vol. 10, no. 12, pp. 1856–1867, 2009.

[85] E. Nikko, J. A. Sullivan, and H. R. B. Pelham, "Arrestin-like proteins mediate ubiquitination and endocytosis of the yeast metal transporter Smf1," *EMBO Reports*, vol. 9, no. 12, pp. 1216–1221, 2008.

[86] T. Andoh, Y. Hirata, and A. Kikuchi, "PY motifs of Rod1 are required for binding to Rsp5 and for drug resistance," *FEBS Letters*, vol. 525, no. 1–3, pp. 131–134, 2002.

[87] Y. Kee, W. Muñoz, N. Lyon, and J. M. Huibregtse, "The deubiquitinating enzyme Ubp2 modulates Rsp5-dependent Lys^{63}-linked polyubiquitin conjugates in *Saccharomyces cerevisiae*," *Journal of Biological Chemistry*, vol. 281, no. 48, pp. 36724–36731, 2006.

[88] R. Gupta, B. Kus, C. Fladd et al., "Ubiquitination screen using protein microarrays for comprehensive identification of Rsp5 substrates in yeast," *Molecular Systems Biology*, vol. 3, article 116, 2007.

[89] A. Herrador, S. Herranz, D. Lara, and O. Vincent, "Recruitment of the ESCRT machinery to a putative seven-transmembrane-domain receptor is mediated by an arrestin-related protein," *Molecular and Cellular Biology*, vol. 30, no. 4, pp. 897–907, 2010.

[90] R. Hatakeyama, M. Kamiya, T. Takahara, and T. Maeda, "Endocytosis of the aspartic acid/glutamic acid transporter Dip5 is triggered by substrate-dependent recruitment of the Rsp5 ubiquitin ligase via the arrestin-like protein Aly2," *Molecular and Cellular Biology*, vol. 30, no. 24, pp. 5598–5607, 2010.

[91] N. Belgareh-Touzé, S. Léon, Z. Erpapazoglou, M. Stawiecka-Mirota, D. Urban-Grimal, and R. Haguenauer-Tsapis, "Versatile role of the yeast ubiquitin ligase Rsp5p in intracellular trafficking," *Biochemical Society Transactions*, vol. 36, part 5, pp. 791–796, 2008.

[92] J. A. MacGurn, P.-C. Hsu, M. B. Smolka, and S. D. Emr, "TORC1 regulates endocytosis via npr1-mediated phospho-inhibition of a ubiquitin ligase adaptor," *Cell*, vol. 147, no. 5, pp. 1104–1117, 2011.

[93] A. F. O'Donnell, A. Apffel, R. G. Gardner, and M. S. Cyert, "α-arrestins Aly1 and Aly2 regulate intracellular trafficking in response to nutrient signaling," *Molecular Biology of the Cell*, vol. 21, no. 20, pp. 3552–3566, 2010.

[94] J. F. Nabhan, R. Hu, R. S. Oh, S. N. Cohen, and Q. Lu, "Formation and release of arrestin domain-containing protein 1-mediated microvesicles (ARMMs) at plasma membrane by recruitment of TSG101 protein," *Proceedings of the National*

Academy of Sciences of the United States of America, vol. 109, no. 11, pp. 4146–4151, 2012.

[95] J. F. Nabhan, H. Pan, and Q. Lu, "Arrestin domain-containing protein 3 recruits the NEDD4 E3 ligase to mediate ubiquitination of the β_2-adrenergic receptor," *EMBO Reports*, vol. 11, no. 8, pp. 605–611, 2010.

[96] S. Rauch and J. Martin-Serrano, "Multiple interactions between the ESCRT machinery and arrestin-related proteins: implications for PPXY-dependent budding," *Journal of Virology*, vol. 85, no. 7, pp. 3546–3556, 2011.

[97] P. Zhang, C. Wang, K. Gao et al., "The ubiquitin ligase itch regulates apoptosis by targeting thioredoxin-interacting protein for ubiquitin-dependent degradation," *Journal of Biological Chemistry*, vol. 285, no. 12, pp. 8869–8879, 2010.

[98] H. Storez, M. G. H. Scott, H. Issafras et al., "Homo- and hetero-oligomerization of β-arrestins in living cells," *Journal of Biological Chemistry*, vol. 280, no. 48, pp. 40210–40215, 2005.

[99] K. M. Draheim, H. B. Chen, Q. Tao, N. Moore, M. Roche, and S. Lyle, "ARRDC3 suppresses breast cancer progression by negatively regulating integrin $\beta4$," *Oncogene*, vol. 29, no. 36, pp. 5032–5047, 2010.

[100] M. A. Peñalva, J. Tilburn, E. Bignell, and H. N. Arst Jr., "Ambient pH gene regulation in fungi: making connections," *Trends in Microbiology*, vol. 16, no. 6, pp. 291–300, 2008.

[101] M. Hayashi, T. Fukuzawa, H. Sorimachi, and T. Maeda, "Constitutive activation of the pH-responsive Rim101 pathway in yeast mutants defective in late steps of the MVB/ESCRT pathway," *Molecular and Cellular Biology*, vol. 25, no. 21, pp. 9478–9490, 2005.

[102] W. Xu, F. J. Smith Jr., R. Subaran, and A. P. Mitchell, "Multivesicular body-ESCRT components function in pH response regulation in *Saccharomyces cerevisiae* and *Candida albicans*," *Molecular Biology of the Cell*, vol. 15, no. 12, pp. 5528–5537, 2004.

[103] J. H. Boysen and A. P. Mitchell, "Control of Bro1-domain protein Rim20 localization by external pH, ESCRT machinery, and the *Saccharomyces cerevisiae* Rim101 pathway," *Molecular Biology of the Cell*, vol. 17, no. 3, pp. 1344–1353, 2006.

[104] A. Galindo, A. M. Calcagno-Pizarelli, H. N. Arst Jr., and M. Á. Peñalva, "An ordered pathway for the assembly of fungal ESCRT-containing ambient pH signalling complexes at the plasma membrane," *Journal of Cell Science*, vol. 125, no. 7, pp. 1784–1795, 2012.

[105] L. Kuo and E. O. Freed, "ARRDC1 as a mediator of microvesicle budding," *Proceedings of the National Academy of Sciences of the United States of America*, vol. 109, no. 11, pp. 4025–4026, 2012.

[106] E. Morita, "Differential requirements of mammalian ESCRTs in multivesicular body formation, virus budding and cell division," *The FEBS Journal*, vol. 279, no. 8, pp. 1399–1406, 2012.

[107] S. K. Shenoy, L. S. Barak, K. Xiao et al., "Ubiquitination of β-arrestin links seven-transmembrane receptor endocytosis and ERK activation," *Journal of Biological Chemistry*, vol. 282, no. 40, pp. 29549–29562, 2007.

[108] R. H. Oakley, S. A. Laporte, J. A. Holt, M. G. Caron, and L. S. Barak, "Differential affinities of visual arrestin, βarrestin1, and βarrestin2 for G protein-coupled receptors delineate two major classes of receptors," *Journal of Biological Chemistry*, vol. 275, no. 22, pp. 17201–17210, 2000.

[109] E. V. Gurevich and V. V. Gurevich, "Arrestins: ubiquitous regulators of cellular signaling pathways," *Genome Biology*, vol. 7, no. 9, article 236, 2006.

[110] A. Hervás-Aguilar, A. Galindo, and M. A. Peñalva, "Receptor-independent ambient pH signaling by ubiquitin attachment to fungal arrestin-like PalF," *Journal of Biological Chemistry*, vol. 285, no. 23, pp. 18095–18102, 2010.

[111] T. Hunter, "The age of crosstalk: phosphorylation, ubiquitination, and beyond," *Molecular Cell*, vol. 28, no. 5, pp. 730–738, 2007.

[112] F. T. Lin, K. M. Krueger, H. E. Kendall et al., "Clathrin-mediated endocytosis of the β-adrenergic receptor is regulated by phosphorylation/dephosphorylation of β-arrestin1," *Journal of Biological Chemistry*, vol. 272, no. 49, pp. 31051–31057, 1997.

[113] F. T. Lin, W. Chen, S. Shenoy, M. Cong, S. T. Exum, and R. J. Lefkowitz, "Phosphorylation of β-arrestin2 regulates its function in internalization of β_2-adrenergic receptors," *Biochemistry*, vol. 41, no. 34, pp. 10692–10699, 2002.

[114] A. Kiselev, M. Socolich, J. Vinos, R. W. Hardy, C. S. Zuker, and R. Ranganathan, "A molecular pathway for light-dependent photoreceptor apoptosis in Drosophila," *Neuron*, vol. 28, no. 1, pp. 139–152, 2000.

[115] F. T. Lin, W. E. Miller, L. M. Luttrell, and R. J. Lefkowitz, "Feedback regulation of β-arrestin1 function by extracellular signal- regulated kinases," *Journal of Biological Chemistry*, vol. 274, no. 23, pp. 15971–15974, 1999.

[116] J. Gomez-Raja and D. A. Davis, "The β-arrestin-like protein Rim8 is hyperphosphorylated and complexes with Rim21 and Rim101 to promote adaptation to neutral-alkaline pH," *Eukaryotic Cell*, vol. 11, no. 5, pp. 683–693, 2012.

9

Divalent Metal Ion Transport across Large Biological Ion Channels and Their Effect on Conductance and Selectivity

Elena García-Giménez, Antonio Alcaraz, and Vicente M. Aguilella

Laboratory of Molecular Biophysics, Department of Physics, Universitat Jaume I, 12071 Castellón, Spain

Correspondence should be addressed to Vicente M. Aguilella, aguilell@uji.es

Academic Editor: Vladimir Uversky

Electrophysiological characterization of large protein channels, usually displaying multi-ionic transport and weak ion selectivity, is commonly performed at physiological conditions (moderate gradients of KCl solutions at decimolar concentrations buffered at neutral pH). We extend here the characterization of the OmpF porin, a wide channel of the outer membrane of *E. coli,* by studying the effect of salts of divalent cations on the transport properties of the channel. The regulation of divalent cations concentration is essential in cell metabolism and understanding their effects is of key importance, not only in the channels specifically designed to control their passage but also in other multiionic channels. In particular, in porin channels like OmpF, divalent cations modulate the efficiency of molecules having antimicrobial activity. Taking advantage of the fact that the OmpF channel atomic structure has been resolved both in water and in MgCl$_2$ aqueous solutions, we analyze the single channel conductance and the channel selectivity inversion aiming to separate the role of the electrolyte itself, and the counterion accumulation induced by the protein channel charges and other factors (binding, steric effects, etc.) that being of minor importance in salts of monovalent cations become crucial in the case of divalent cations.

1. Introduction

The lipid membrane of the cells forms an insulating barrier to the passage of ions, metabolites, and other larger molecules [1]. However, the selective transport of charged solutes and large molecules across the cell membrane is a physiological function necessary for the survival of the cells and hence of the living organisms. That specialized physiological function is carried out by ion channels, a large family of specialized proteins present in all living organisms that open pores of nanometer dimensions in the cell membranes [2–4]. The actual size of the pore determines mostly the specific function of each channel [5]. Thus, narrow channels can efficiently discriminate between different charged species while other processes that require the rapid transport of many ions across the cell membrane are more easily achieved by wider pores also known as mesoscopic channels [6]. This paper focuses on the transport properties of wide channels, in particular on the channel conductance and the ion selectivity. The latter refers here to the ability to favor the passage of certain kind of ions against others when both species are present in solution at the same time (e.g., cations and anions). The fact that the transport through wide channels is passive and multi-ionic makes them suitable for regulating the influx of nutrients and to extrude waste products, necessary in cell metabolism. Their weak selectivity for low-molecular-weight inorganic ions is relevant for several reasons. First, because its comprehension is a first, necessary step to interpret adequately the highly sophisticated mechanisms responsible for the specific selectivity in narrow channels [5]. Second, because the study and understanding of their function have contributed to develop a variety of biotechnological, analytical, and medical applications [7–9].

Mesoscopic channels usually discriminate ions by their charge, that is, the channel is cation selective or anion selective, depending on whether cation or anion transport is favored by the protein [5, 10–12]. The selectivity of these wide channels is mainly regulated by electrostatic interactions between protein ionizable residues and the permeating ions [2]. However, additional factors such as diffusion and even short range interactions may play a role in certain specific cases [6, 13, 14]. Examples of mesoscopic channels extensively studied are bacterial porins like OmpF from *E. coli* [15–17], the voltage-gated anion channel (VDAC) of the outer mitochondrial membrane [18], pore-opening toxins like the alpha-hemolysin channel secreted by *S. Aureus* [19, 20], and antibiotic peptides like alamethicin [21–23]. A common feature of porins is their beta barrel structure. Their hydrophilic environment provides a water-filled pore through which hydrated positive and negative small ions, metabolites like ATP [24] or antibiotic molecules [25], are able to pass. Because of this, porins are also called general diffusion pores [16, 17] and they are weakly selective for small ions. One of them, the bacterial porin OmpF, has been chosen as a model system representative of wide channels in numerous studies. The main reasons are that it is easily genetically modified, overexpressed [26], and crystallized [27]. Besides, its well-known structure obtained at atomic resolution [15] allows establishing a relationship between the channel structure and its function by means of Continuum theories (e.g., Poisson-Nernst-Planck) [28–30] and computational approaches like molecular dynamics (MD) and Brownian dynamics (BD) [28, 31–36]. Furthermore, in this study in particular, where the effects of divalent cations are analyzed in detail, it is especially relevant the fact that the OmpF atomic structure has been reported not only in the presence of salts of monovalent [37] but also in divalent [38] cations.

Crystallographic Structures of the Ompf Porin. OmpF (outer membrane protein F) is a general diffusion porin. It is a homotrimeric protein which forms wide, water-filled, voltage-gated pores in the outer membrane of *E. coli*. Each subunit of the channel has an asymmetric structure, with relatively large entrances of ~4 nm of diameter and a narrow region with diameter ~1 nm at approximately half of the channel length. One of the first crystal structures of the OmpF porin was obtained in 1995 with a resolution of 2.4 Å from X-ray analysis of crystals grown in absence of salt. It is available in the Protein Data Bank (PDB) with 2OMF code [15]. Subsequently, a variety of structures of the OmpF channel and mutants were solved [15, 33, 37–45]. The OmpF structure with PDB code 2ZFG [38] is especially relevant to the present study. It was obtained in 2008 with a resolution of 1.59 Å from crystals grown in a 1 M MgCl$_2$ aqueous solution. It displays a Mg^{2+} cation located between the two acid residues of the loop L3 of the structure. Figure 1 shows the location of the Mg^{2+} cation according to the 2ZFG structure.

The knowledge of the 3D atomic structure of the OmpF porin is a great advantage to establish a relationship between the channel structure and its functional properties. A complete channel characterization requires the combination of different theoretical approaches. Sophisticated methods like MD and BD simulations can provide significant microscopic details such as the crucial effect on the ion transport of the residues present in the narrow constriction of the channel,

where a strong electric field transversal to the pore axis is generated separating the pathways for cations and anions along the channel [28, 34]. Using all-atom MD simulations it is also possible to obtain the channel conductance and analyze the behavior of a single protein residue. Alternatively, mean field theories based on the Poison and Nernst-Planck (PNP) equations and the Teorell-Meyer-Sievers model can be used for estimating the conductance and selectivity of the channel under different conditions (salt concentration, solution pH, etc.) and for their comparison with experiments [29, 30, 46, 47].

2. Materials and Methods

2.1. Channel Reconstitution in Planar Bilayers. A technique widely used for measuring selectivity and conductance in the OmpF channel is its reconstitution in planar lipid membranes. This technique, introduced by Mueller et al. [48] and later improved by Montal and Mueller [49], consists of forming a lipid membrane by the apposition of two lipid monolayers made from a 1% solution of diphytanoyl phosphatidylcholine (DPhPC) (neutral lipid provided by Avanti Polar Lipids, Inc., Alabaster, AL, USA). The lipid bilayer is formed in a micrometric hole (around 80 μm) made on a 15 μm thick Teflon film separating two solutions [50]. The membrane is formed by raising the level of the buffer solution where a small amount of lipid dissolved in an organic solvent has been deposited. Previously, the orifice is pretreated with a 1% solution of hexadecane in pentane to allow adherence.

The capacitance of the bilayer membrane formed depends on the actual location of the orifices in the film and its size but is always around 80–120 pF. Single-channel insertion is achieved by adding 0.1–0.3 μL of a 1 μg/mL solution of OmpF in a buffer that contains 1 M KCl and 1% (v/v) of Octyl POE (Alexis, Switzerland) to a 2 mL aqueous phase only on the cis side of the membrane. The membrane potential is applied using Ag/AgCl electrodes in 2 M KCl, 1.5% agarose bridges assembled within standard 250 μL pipette tips [50]. Potential is defined as positive when it is greater on the side of protein addition (the *cis* side of the membrane cell). Figure 2 shows a schematic representation of the experimental setup for the reconstitution of the protein channel on a lipid bilayer.

The membrane chamber and the headstage are isolated from external noise sources with a double metal screen (Amuneal Manufacturing Corp., Philadelphia, PA, USA). To measure the current and apply the potential an Axopatch 200B amplifier (Molecular Devices Corp., Sunnyvale, CA, USA) is used in the voltage-clamp mode. Data are saved directly into the computer memory. Data treatment is done using the PClamp software.

2.2. Reversal Potential Measurements and Correction by the Liquid Junction Potential. Channel selectivity is commonly evaluated by measuring the reversal potential (RP). RP is defined as the applied transmembrane voltage that yields zero electric current when there is a concentration gradient

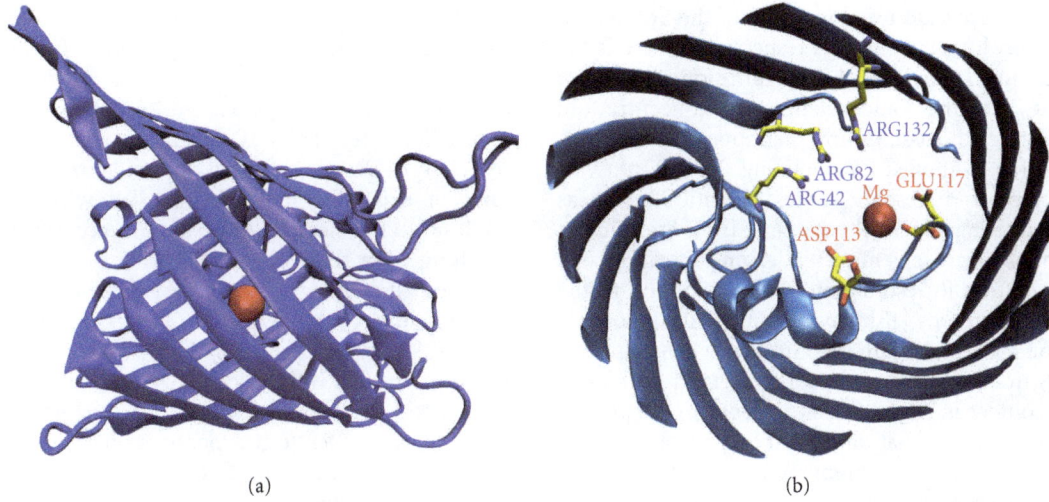

FIGURE 1: Location of the magnesium ion in the crystal structure of the OmpF channel (PDB code 2ZFG) resolved in 1 M MgCl$_2$ solution [38].

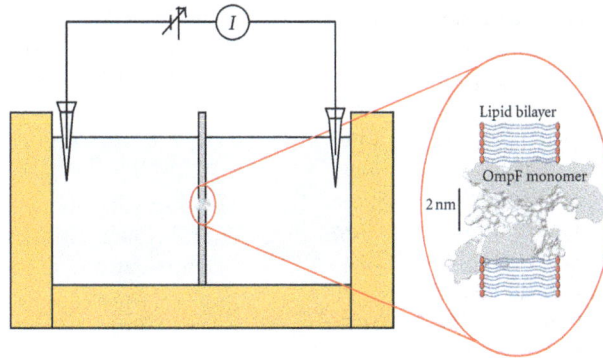

FIGURE 2: Schematic representation of the experimental setup and reconstitution of the OmpF channel on a lipid bilayer.

across the channel [2]. In the experiments reported here, the RP is obtained as follows. Once a lipid membrane is formed at a given salt concentration gradient, a single OmpF channel is inserted without any applied potential. Next, the channel conductance is checked by applying +50 mV (-50 mV in divalent salts) and later switching the potential polarity. Afterwards, the ionic current through the channel is manually set to zero by adjusting the applied potential. The experimental method used for determining the RP described above introduces two contributions to the electric potential measured (V_{exp}) [13]: On one hand, the potential difference across each electrode-bridge/solution interface, known as liquid junction potential (LJP); On the other hand, the potential drop across the channel itself, the RP. Hence, the overall LJP has to be subtracted from the raw zero current potential measurement, V_{exp},

$$RP = V_{exp} - LJP. \tag{1}$$

Usually, salt bridges made of a KCl-concentrated solution are used to measure the channel selectivity following pioneer studies performed under physiological conditions with KCl used as electrolyte [51, 52]. Under those conditions the LJP

is small (~ 1 mV). This is a negligible and usually disregarded quantity as it is comparable to the experimental error of electrophysiology experiments. That is why the protocol is sometimes erroneously extended to altogether different conditions without taking into account that in experiments with other salts (NaCl, LiCl, CaCl$_2$, MgCl$_2$, etc.) the LJP contribution becomes significant and may be comparable to the actual RP [13]. This repeated oversight, although long noted [53], has led to some inconsistencies in the selectivity data [29, 30, 54–57]. Since direct measurements of LJP are difficult [52, 53], it is necessary to rely on LJP theoretical estimates to determine the actual RP. Assuming ideal electrolyte solutions and linear ion concentration profiles in the junction of the two solutions of salt bridge/and cell compartment, many authors use Henderson's equation [58, 59] to calculate the LJP between two solutions (left (L) and right (R)):

$$LJP \equiv \phi_L - \phi_R$$
$$= -\left(\frac{k_B T}{e}\right) \frac{\sum_i z_i D_i [C_{i,L} - C_{i,R}]}{\sum_i z_i^2 D_i [C_{i,L} - C_{i,R}]} \ln \frac{\sum_i z_i^2 D_i C_{i,L}}{\sum_i z_i^2 D_i C_{i,R}}, \tag{2}$$

where k_B and T have their usual meaning of the Boltzmann constant and absolute temperature, respectively, and e is the elementary charge. D_i denote the ionic diffusion coefficients and z_i and C_i are the ionic valence and concentration, respectively. Under the assumption mentioned above, Henderson's equation is applied for obtaining the LJP in a vast majority of cases of interest and it yields identical results as those obtained using the PNP equations [60]. Specifically, for ion channel measurements, Henderson's equation is a good approximation for estimating LJP contribution to the selectivity measurements [13]. Apart from what has already been said, one has to take into account that when solutions cannot be regarded as ideal or the ionic strength of the two solutions in contact is very different, Henderson's equation becomes a poor approximation, and then LJP calculations demand a proper estimation of ionic activity coefficients and ion mobilities as a function of concentration [61].

2.3. Numerical Procedure. We have used the PNP model for computing the conductance and selectivity of the OmpF channel in different salts of monovalent and divalent cations. This model is used here in its one-dimensional version starting from the channel effective fixed charge concentration calculated along the pore. This fixed charge concentration profile is calculated from the 3D electric potential distribution created by the protein-charged residues and it is then averaged along the channel. For calculating this potential it is necessary to know the dissociation constant K_a (or its equivalent in the logarithmic scale, the pK_a) of each titratable residue inside the protein once the interaction with the protein permanent charges (due to the different electronegativities of the atoms in the molecule) and the other titratable residues are considered. These pK_a values, known as the apparent pK_a or effective pK_a ($pK_{a\,eff}$), are calculated according the procedure described by Aguilella-Arzo et al. [30] using the UHBD code [62] for two crystallographic structures of the OmpF channel (PDB codes 2OMF and 2ZFG).

3. Effects of Divalent Cations on the Transport Properties of the Channel

3.1. Channel Selectivity in Salts of Monovalent and Divalent Cations. The characterization of ion channel selectivity is crucial for understanding the molecular basis of ion transport and establishing a relationship between the structure and the function of the channel [29, 30]. OmpF, like other wide multi-ionic channels, is too large to be specific to a certain ion (as happens in Na or K channels) but still has a clear influence on the permeability. Several experiments aimed to assess ion selectivity [16, 17, 63], as well as MD and BD simulations and continuum electrodiffusion models [28, 34] reveal that the transport of small inorganic ions (K^+, Na^+, Cl^-, etc.) across the channel is principally regulated by the electrostatic interactions between the permeating ions and the channel ionizable residues, in particular (although not solely) by the acidic and basic residues of the channel constriction [28, 31–33].

In many cases of interest, RP measurements are used to assess selectivity because the sign of RP provides a quick estimation of the effective charge of the channel: anionic selectivity is associated to a positive net charge and cationic selectivity is directly connected to a negative net charge. Despite being useful as a first estimation, this kind of reasoning must be handled with care [13]. In addition to the electrostatic exclusion/accumulation of ions resulting from their interaction with the protein ionizable residues, another important factor is the difference between the mobilities of the permeating ions themselves within the channel [64]. These diffusional effects as well as other short-range or specific interactions that may take place between the protein residues and mobile ions definitely play a role in the measured RP so it is necessary to design experiments enabling to separate the different contributions as far as possible. The comparison of a number of experiments done with chloride salts of different cations in a variety of conditions and several laboratories allows us to discuss the various sources of ion selectivity in large channels as follows.

(a) A salt whose cation and anion have the same valence, similar size, and consequently similar bulk mobilities is a suitable electrolyte to study the interaction between ionizable residues and permeating ions because the diffusion effects are negligible. In that respect, KCl is an ideal candidate [29, 30].

(b) Other salts whose cation and anion present different bulk ion mobilities are appropriate to analyze the contribution of the diffusion potential to the RP measurement. These effects are expected to become more important in salts with ions of different valence.

In view of the OmpF crystal structure obtained from a concentrated solution of $MgCl_2$ where a Mg^{2+} ion appears between the two acidic residues (Glu117 and Asp113) of the channel constriction (see Figure 1), experiments done in salts of divalent cations seem appropriate to study possible specific interactions between protein residues and permeating ions that may contribute to the channel RP. Figure 3 shows RP measurements as a function of the solution pH for KCl, $CaCl_2$, and $MgCl_2$.

Given that the bulk mobilities of K^+ and Cl^- are very similar, the diffusion potential arising from a 0.1/1 M gradient of KCl can be considered negligible and the RP measurements provide almost direct information about the channel interaction with the permeating ions. Thus, the pH sensitivity seen in the experiments with KCl can be rationalized in terms of the ionization equilibria of all protein ionizable residues (having different pK_a's) [30]. The overall effect is a different protein net effective charge at each pH as a consequence of the protonation and deprotonation of some protein residues. Therefore, at pH low enough to titrate the acidic residues, the channel is anion selective. That results in a positive net charge of the channel. At pH higher than 3.7 the channel is cation selective, which is consistent with the negative net charge of the channel. This channel pH sensitivity in KCl has previously been analyzed [29, 30, 65, 66] and it is almost independent on the absolute salt concentration.

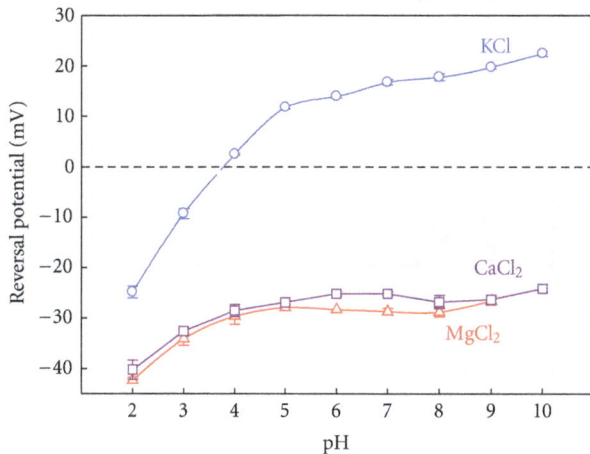

FIGURE 3: OmpF reversal potential measurements in 0.1/1 M KCl, CaCl$_2$, and MgCl$_2$ solutions. Adapted with permission from [13].

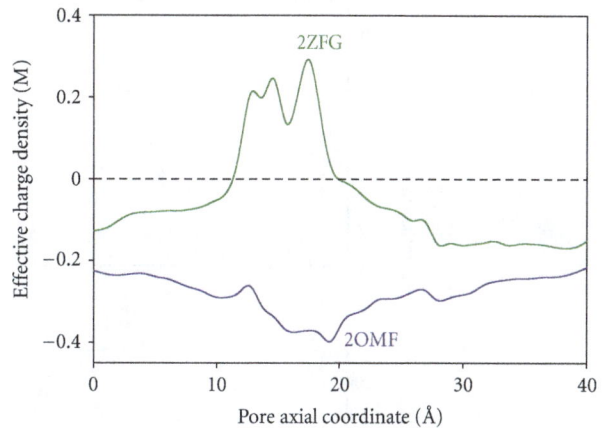

FIGURE 4: Average fixed charge concentration along the OmpF channel calculated from two OmpF structures: 2OMF (bottom curve) and 2ZFG (top curve). Reprinted with permission from [67].

The RP measurements performed in salts of divalent cations paint a quite different picture [67, 68]. Focusing around neutral pH, the cationic selectivity displayed by the channel in KCl solutions turns into anionic in MgCl$_2$ and CaCl$_2$. That is why one can think that a charge inversion effect could be taking place on the channel (this occurs when interfacial charges attract counterions in excess of their own nominal charge [68]). In addition, the clear sensitivity to pH shown in KCl disappears both in MgCl$_2$ as in CaCl$_2$. This change can be explained in terms of a binding process [67]. The presence of divalent cations hinders the protonation of the acidic groups in such a way that an abnormally high amount of protons (and then a lower effective pK_a) is required to neutralize the site. This means that the sensitivity to pH is not lost, but is shifted to lower pH. Interestingly, the pH sensitivity can be restored by lowering the absolute concentration of MgCl$_2$ or CaCl$_2$ [67]. This result indicates that when the concentration is low enough, the binding of cations is unlikely and has a limited effect on the residue protonation. This is not an exclusive feature of MgCl$_2$ and CaCl$_2$ salts. The effect of the divalent cations on the sensitivity of the channel to the pH variations and the apparent charge inversion is also observed in a variety of salts of divalent cations such as BaCl$_2$ and NiCl$_2$ [67].

To get further insight into these RP measurements, we can analyze the connection between charge and selectivity on the basis of the channel 3D structure. The pK_a calculation leads to a charge concentration profile along the protein for a particular pH [30]. Figure 4 shows the average fixed charge concentration calculated for pH 6 using the two OmpF above mentioned structures: the 3D structure resolved from crystals grown in the absence of salt (2OMF) [15] and the 3D structure obtained from crystals grown in 1 M MgCl$_2$ (2ZFG) [38].

The negative charge obtained from the 2OMF structure denoted by the bottom line in Figure 4 is consistent with the cation selectivity of the protein at pH 6 (see Figure 3). The effective charge profile calculated for the 2ZFG structure is very different. The presence of a divalent cation in the narrow

part of the channel has a significant effect on the titration of the adjacent residues. As far as this region regulates the selective permeation of ions through the pore, this positive effective charge around the central constriction could be the explanation to the observed anionic selectivity of the channel. However, note that this kind of reasoning is only valid when salts with similar anion and cation mobilities are used and diffusion effects can be negligible. In other salts, the difference between cation and anion diffusivities generates a diffusion potential that necessarily contributes to the measured RP. The diffusion potential would be the electric potential drop across a neutral, ideal pore, devoid of any electrostatic interaction, connecting two solutions at different salt concentrations. Furthermore, in a channel with negative net charge immersed in chloride salts of divalent cations, the diffusion potential and the interfacial Donnan potential may have opposite signs [46]. According to this, the anionic selectivity of the OmpF channel in salts of divalent cations may be simply a consequence of the counterbalance between the diffusional contribution and the channel electrostatic preference for cations.

To get further insight on this selectivity inversion and the role of the diffusion of divalent cations in the channel selectivity, we present a comparison between experiments done in KCl and MgCl$_2$ and the 1D PNP model predictions from the effective charge of 2OMF and 2ZFG structures. Note that this original approach based on the structure means an increase in the level of complexity in relation to previous studies using purely phenomenological approaches [13, 14].

Figure 5(a) shows the theoretical RP calculated from the 1D PNP model by using the two effective charge profiles displayed in Figure 4 and omitting the difference in the ion mobilities (i.e., considering only the electrostatic exclusion) over a wide range of concentration ratios. The upper plot shows the diffusion potential of MgCl$_2$ calculated in a neutral pore using the ion bulk diffusion coefficients. The comparison between Figures 5(a) and 5(b) shows three significant features.

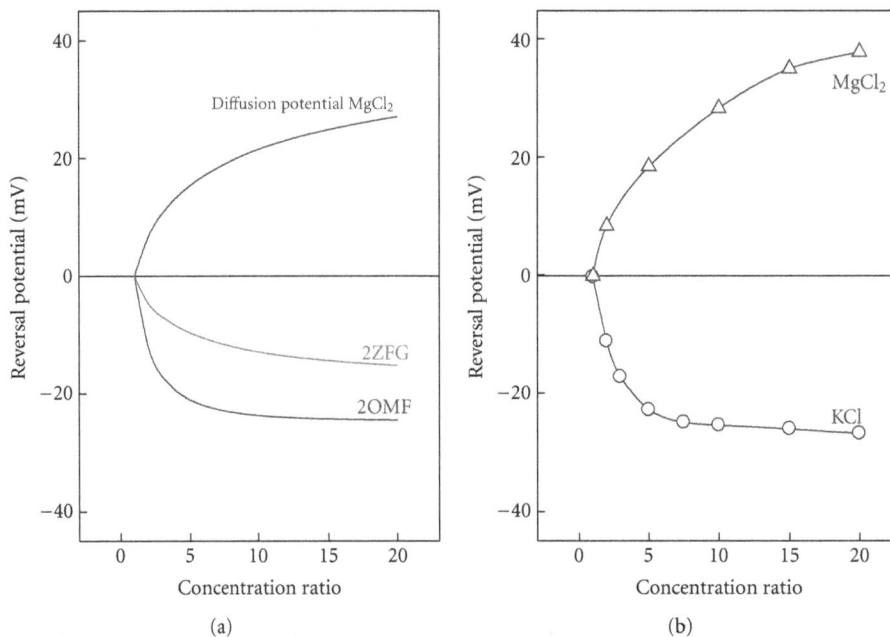

FIGURE 5: (a) Theoretical predictions of 1D PNP model for the electrostatic exclusion using 2OMF and 2ZFG effective charge profiles from Figure 4 and for the diffusion potential of $MgCl_2$. (b) OmpF channel reversal potential measurements in KCl and $MgCl_2$ versus concentration ratio (C_{cis}/C_{trans}). In all experiments the concentration on trans side is kept constant at 0.1 M and the concentration on cis side is varied. Measurements were done at pH 6.

(a) The model calculations (1D PNP) using the 2OMF effective charge (blue line in Figure 5(a)) correlate very well with the RP measurements in KCl.

(b) According to the 2ZFG structure the channel net charge is still negative (Figure 4). Thus, the model predictions considering only the electrostatic exclusion effects (green line in Figure 5(a)) do not account for the RP measurements in $MgCl_2$, but even give the opposite sign. This indicates that diffusion potentials (brown line in Figure 5(a)) are very significant in the present case and mostly determine the total RP.

(c) The RP measurements in $MgCl_2$ are a little greater than the bulk diffusion potential shown in the figure. This fact has two alternative interpretations, although they do not mutually exclude each other completely.

    (1) The negative effective charge of the channel is overcompensated by the divalent cations generating a "charge inversion" phenomenon in the channel [14].

    (2) The negative charge of the channel is compensated or almost compensated and the measured RP scales with an effective electric diffusion potential, slightly different from the bulk diffusion potential because of the divalent cations binding effect.

3.2. Charge Inversion and Selectivity Inversion. The experiments with OmpF channels reported so far make clear that the connection between selectivity and charge is not as obvious as one might think. Thus, the inversion of selectivity found in salts of multivalent cations does not necessarily imply that interfacial charges attract counterions in excess of their own nominal charge, but it can be alternatively caused by a complex interplay of several factors like exclusion, diffusion, and binding. Site-directed mutagenesis has proved to be a powerful tool for understanding the role of certain residues on channel selectivity. For example, it has been reported that the proper mutation of the residues located in the constriction of the OmpF channel can turn it into highly selective to Ca^{2+} ions, with similar transport properties to the Ca^{2+} channel [55, 69]. Having in mind the OmpF crystal structure solved in a 1 M $MgCl_2$ where a Mg^{2+} ion appears between the two acidic residues Glu117 and Asp113, we have investigated if the negative charge of those residues is essential to cause an inversion of selectivity. To this end, we compared the reversal potential of the wild-type (OmpF-WT) channel and two mutants in which the above residues had been replaced either by two neutral cysteines (OmpF-CC) or by two positively charged arginines (OmpF-RR). Previous studies using these mutants showed that the dimensions of the narrowest part of the channel are not significantly changed by chemical modification [57], so that additional steric or entropic effects are unlikely. Figure 6 shows a sketch of the cross-section of the OmpF eyelet in the three cases mentioned. Control experiments of selectivity in monovalent KCl solutions were also carried out. Table 1 shows the RP measurements in tenfold concentration gradients (1 M cis/0.1 M trans) at pH 6.

Interestingly, the replacement of the two negative residues Asp113 and Glu117 by two neutral ones (see OmpF-CC) does not have a critical effect. The cationic selectivity in salts of KCl is preserved and the selectivity inversion

TABLE 1: Ion selectivity of OmpF (WT and mutants) in KCl and $CaCl_2$.

OmpF channel	Δq^*	RP (mV) 1/0.1 M KCl	RP (mV) 1/0.1 M $CaCl_2$	Selectivity inversion
WT	0	-25.4 ± 0.8	22.1 ± 0.7	Yes
CC	+2	-23.8 ± 0.8	30.1 ± 1.1	Yes
RR	+4	31.9 ± 1.0	35.4 ± 1.7	No

$^*\Delta q$: effective charge compared to WT OmpF.

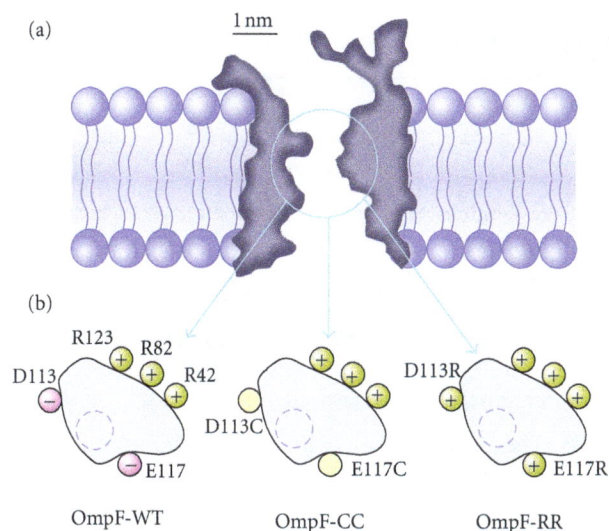

FIGURE 6: Sketch of the OmpF channel. (a) Longitudinal cross-section. (b) Idealized cross-section of the OmpF channel eyelet for the wild-type (OmpF-WT) protein channel and the mutants with residues Asp113 and Glu117 replaced with cysteines (OmpF-CC) or arginines (OmpF-RR). The dashed contour line represents the hypothetical binding site for a divalent cation based on the atomic structure of OmpF-WT in 1 M $MgCl_2$ [38].

produced by Ca^{2+} ions in OmpF-WT is not removed. Indeed, the anionic selectivity of OmpF-CC is even 50% higher than OmpF-WT. The substitution of Asp113 and Glu117 by two arginines (OmpF-RR) makes the channel anion selective in both salts of monovalent cations (KCl) and divalent cations ($CaCl_2$). Therefore, we cannot speak of selectivity inversion in this case. The comparison between OmpF-WT, OmpF-CC, and OmpF-RR suggests that the observed channel preference for anions in salts of divalent cations is not a pure surface effect based on the charge. Rather than that it is probably a joint effect of the long range coulombic interactions between protein and mobile charges and the short-range interaction involving particular functional groups in a precise arrangement [14].

A study using full atomic MD simulations [70] suggested that the inversion of selectivity of OmpF channel can be originated by electrostatic correlations typical of multivalent ions [71]. Thus, the binding of counterions would be an interfacial analogue of the Bjerrum correlations between ions in bulk electrolyte. According to the simulations, the existence of correlations does not require a highly charged

interface and depends strongly on the nature (the structure and charge distribution) of the chemical groups present in the interface.

3.3. Channel Conductance in Salts of Monovalent and Divalent Cations. The conductance measurements can provide an alternative way of studying the effect of divalent cations on channel transport properties. An initial, necessary step, involves separating channel and electrolyte effects. Otherwise, intrinsic properties of the salt could be incorrectly attributed to the channel action. It is wellknown that in non-ideal solutions conductivity increases with the ionic activity rather than with the ion concentration [72]. In addition, the change of the activity coefficient with concentration may be totally different in monovalent cations and divalent cations as is shown in Figure 7(a), where the tabulated values from the literature [73] have been translated from molal to molar scale and later interpolated [73–75].

Figure 7(a) shows marked differences in the activity coefficient between salts of monovalent and divalent cations, especially above 1 M. Whereas in KCl and NaCl it is almost insensitive to concentration, in $CaCl_2$ and $MgCl_2$ it slightly decreases in the low concentration range, then reaches a minimum and finally shows a steep increase. Figure 7(b) shows the measured conductivity as a function of the electrolyte activity in solution. In the case of salts of divalent cations, the importance of studying the intrinsic properties of the electrolyte becomes apparent. The solution conductivity of $CaCl_2$ and $MgCl_2$ saturates with increasing activity. The origin of this saturation is likely to be a strong reduction of the divalent ion diffusion coefficient with the concentration [73]. One might ask about the use of ionic activity here, since ion selectivity has been discussed in terms of ion concentration in the previous section. Note that the study of ion selectivity is done in terms of concentration ratio, not in terms of absolute concentration. Since RP measurements normally involve moderate concentration gradients, the concentration ratio (C_{cis}/C_{trans}) [29, 46, 66, 76] is practically equal to the activity ratio (a_{cis}/a_{trans}) [72]. Hence, there was no need to invoke the subtle distinction between activity and concentration. However, when the channel conductance is studied as a function of absolute salt concentration, it is necessary to take into account that the activity coefficient may change considerably depending on the salt concentration.

Channel conductance is obtained from single channel current measurements under an applied potential of +100 mV in symmetrical salt solutions. It is defined as the

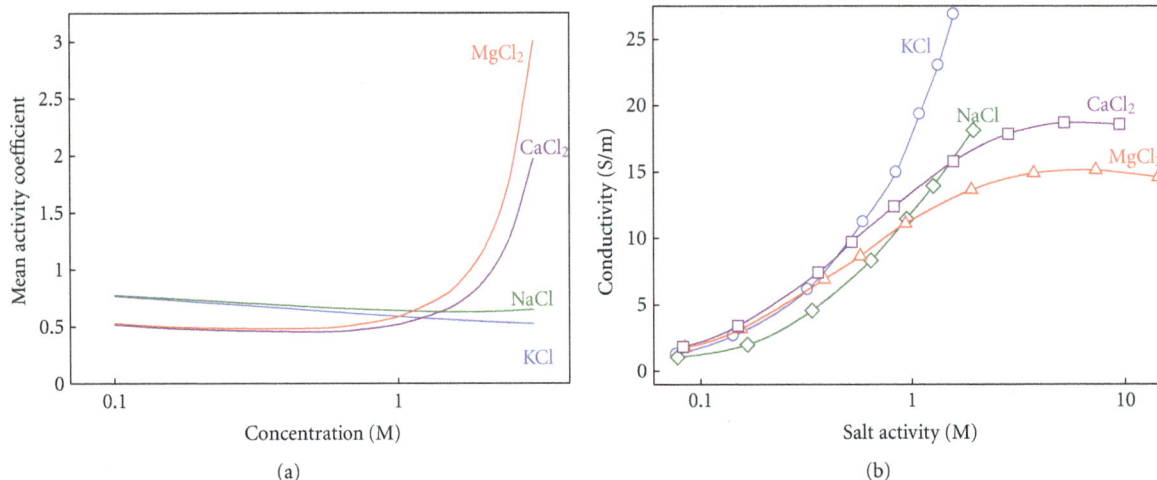

FIGURE 7: (a) Mean activity coefficient in molar reference as a function of the solution concentration for the electrolytes used in the experiments. (b) Measured conductivity as a function of the electrolyte activity in solution.

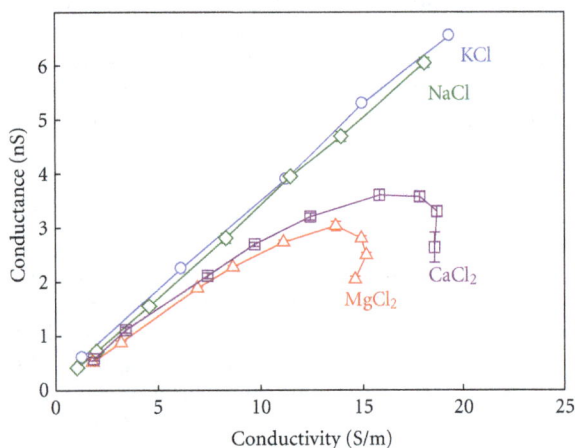

FIGURE 8: OmpF single channel conductance versus bulk solution conductivity at pH 6. Conductivity is measured in solutions of different salts of concentrations ranging 0.1–3 M. Reprinted with permission from [77].

current per voltage unit ($G = I/V$). Note that in conductance measurements, there is no correction for the LJP because both electrodes are in contact with solutions with the same concentration. Given that the bulk properties of the electrolytes under study are known (Figure 7(b)), any interaction between the permeating ions and the protein should be seen in the relationship between channel conductance and the respective solution conductivity. In Figure 8 OmpF channel conductance measurements are plotted versus the bulk solution conductivity in a wide range of salt concentration (up to 3 M).

From Figure 8 some conclusions can be drawn:

(a) A linear correlation is seen between channel conductance and electrolyte bulk conductivity in KCl and NaCl. This would be the expected outcome according to the principle of independent movement of ions

applied to the permeation through a channel: the more conductive a solution is the higher channel conductance is measured.

(b) The conductance-conductivity linear correlation is almost identical in both salts of monovalent cations. This is consistent with a large number of previous studies [13, 14, 30, 35, 54, 66, 76] where it is emphasized that the OmpF channel does not favor the permeation of any particular monovalent ion, which rules out any chemical specificity.

(c) Two regimes are observed in the relationship between the channel conductance and the solution conductivity in salts of divalent cations: at low and moderate salt concentrations (up to 1 M) conductance scales with bulk conductivity as happens in salts of monovalent cations. This suggests that, in this range, the ion transport is regulated by the electrolyte properties. Above 1 M, the channel conductance as well as the solution conductivity decreases as the concentration increases, breaking the linearity. This indicates that a specific interaction between the divalent cations with the channel can be taking place and shows that current saturation and blocking are not exclusive properties of narrow (single-file) ion channels but may be observed in large, multi-ionic channels like bacterial porins as well.

Modeling Ion Conductance. Mean field theories as the 1D PNP model have been used to model the ion transport across the OmpF channel in 1 : 1 salts by using the crystal 3D structure of the porin [30, 47]. In these studies, the tabulated diffusion coefficients for bulk solutions were used and a satisfactory agreement between theory and experiment was found. However, the modeling of channel conductance in 2 : 1 salts cannot ignore the electrolyte intrinsic properties shown in Figure 7. The strong dependence of the electrolyte conductivity on the ion activity can be translated into

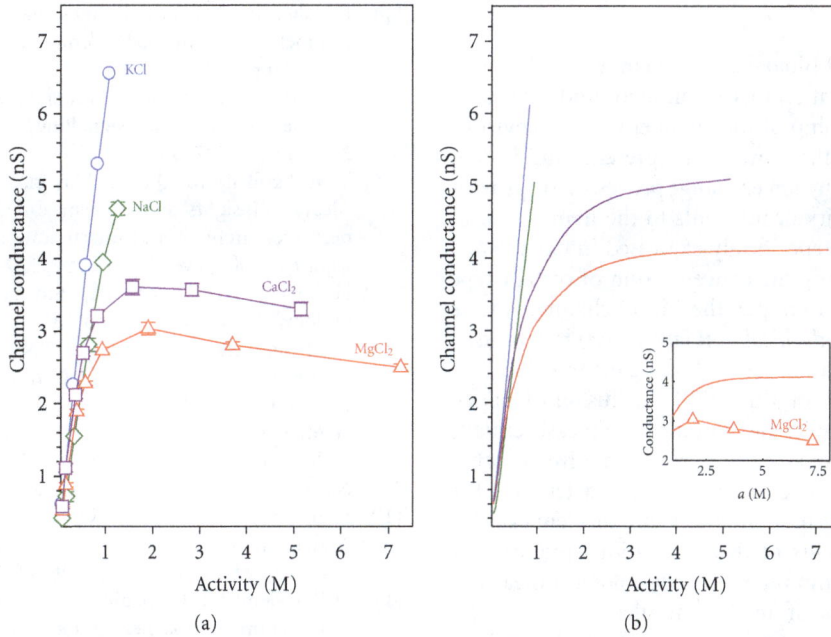

FIGURE 9: (a) OmpF channel conductance measurements over a wide range of concentrations of several salt solutions. (b) Conductance calculated using the PNP electrodiffusion model and the crystal structure of the OmpF channel as an input. The inset shows a comparison of measured and calculated conductance in highly concentrated $MgCl_2$ solutions.

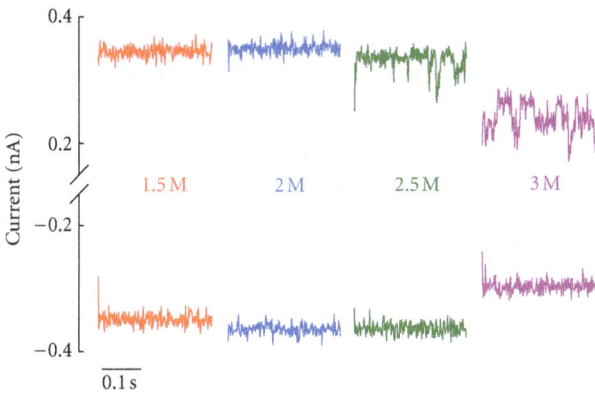

FIGURE 10: OmpF single-channel current recordings in concentrated solutions of $CaCl_2$ for applied voltages of both polarities (+100 mV and −100 mV). Reprinted with permission from [77].

effective salt-dependent diffusion coefficients as shown in [77]. Figure 9 shows a qualitative comparison between measurements and model calculations of OmpF conductance in different 1 : 1 and 2 : 1 salts.

From the comparison between Figures 9(a) and 9(b) it follows that:

(a) the measured conductance in KCl and NaCl agree quantitatively, and qualitatively with PNP calculations revealing that at low and moderate concentrations the pore conductance is controlled mainly by the electrolyte properties;

(b) the saturation in the conductance measurements with the $CaCl_2$ and $MgCl_2$ concentration is predicted

satisfactorily by the PNP model by using the effective salt-dependent diffusion coefficients. This shows that the saturation should be attributed to the electrolyte properties and not to the channel as could be mistakenly thought;

(c) the decreasing trend of the measured conductance at high enough concentrations in $CaCl_2$ and $MgCl_2$ (inset of Figure 9(b)) contrasts with the plateau anticipated by the theoretical model. This suggests that a close interaction between the channel residues and the permeating ions not considered in the PNP framework might be involved.

Conductance and Binding Site of Divalent Cations. Having in mind the OmpF structure obtained in 1 M $MgCl_2$ aqueous solution (2ZFG) [38] showing a Mg^{2+} ion bound to the protein and the experimental evidence of channel selectivity inversion in salts of divalent and trivalent cations [13, 14], one could wonder whether such binding could be behind the conductance decrease observed at high salt concentration of $CaCl_2$ or $MgCl_2$. Current traces of Figure 10 recorded at several salt concentrations and taken at high sampling frequency can be a clue of whether binding of divalent cations to the protein is modulating the channel conductance. In the lack of further evidence, the traces at high salt concentration point to the existence of substates of lower conductance as one of the causes of the conductance decrease.

Similar observations have been reported for another multi-ionic porin, the lysenin channel, which also displays discrete current changes upon Ca^{2+} addition [78].

4. Concluding Remarks

The characterization of biological ion channels in the presence of salts with divalent cations is crucial for understanding the functional relationship of the channel with its environment where several of these metals are present. Multivalent ions are involved in many ion exchange processes for providing an adequate quantity of nutrients to the living cells. In fact, many proteins are specifically expressed in determined conditions for controlling the concentration of certain type of ions in the cell. For example, the MnoP channel (in the outer membrane of *Bradyrhizobium japonicum*) is expressed under manganese limitation for facilitating the translocation of Mn^{2+}, but not Co^{2+} or Cu^{2+} [79]. In this short paper, we have analyzed selectivity and conductance measurements in the OmpF porin, considered as representative of other similar large multi-ionic channels, for characterizing the channel function in salts of monovalent and divalent cations. We have shown that salts of divalent cations induce new effects not found in common electrophysiological measurements performed in salts of small monovalent ions. A careful dissection of the different contributions to channel selectivity is needed for a proper characterization of the cationic or anionic preference of the channel because of the specific interactions between divalent cations and protein residues and the significant diffusion potential often involved. Once the diffusional contribution to selectivity (coming from the different ionic mobilities) is set aside, the experiments with OmpF mutants suggest that certain protein residues are responsible for the specific interaction of divalent cations with the protein. The binding of divalent cations to large channels can be of importance for the translocation of molecules with antimicrobial properties across bacterial porins [80]. In the case of OmpF channel, such a binding has been reported to favor antibiotic permeation [81, 82]. Other channels like lysenin have a binding site for divalent cations too [78], which allows using the lysenin channel as a biosensor for multivalent cations [83]. We have also shown that single channel measurements conductance over a wide range of salt concentrations make it possible to separate the intrinsic properties of the 2 : 1 electrolyte itself and other short range interactions of divalent cations with the protein.

Acknowledgments

Support from the Spanish Ministry of Science and Innovation (MICINN Project FIS2010-19810), Generalitat Valenciana (Project Prometeu 2012/069), and Fundació Caixa Castelló-Bancaixa (Project no. P1-1A2009-13) is acknowledged.

References

[1] A. Parsegian, "Energy of an ion crossing a low dielectric membrane: solutions to four relevant electrostatic problems," *Nature*, vol. 221, no. 5183, pp. 844–846, 1969.

[2] B. Hille, *Ion Channels of Excitable Membranes*, Sinauer Associates, Sunderland, Mass, USA, 2001.

[3] D. J. Aidley and P. R. Stanfield, *Ion Channels: Molecules in Action*, Cambridge University Press, New York, NY, USA, 1996.

[4] F. Ashcroft, D. Benos, F. Bezanilla et al., "The state of ion channel research in 2004," *Nature Reviews Drug Discovery*, vol. 3, no. 3, pp. 239–278, 2004.

[5] B. Corry, "Understanding ion channel selectivity and gating and their role in cellular signalling," *Molecular Biosystems*, vol. 2, no. 11, pp. 527–535, 2006.

[6] V. M. Aguilella, M. Queralt-Martín, M. Aguilella-Arzo, and A. Alcaraz, "Insights on the permeability of wide protein channels: measurement and interpretation of ion selectivity," *Integrative Biology*, vol. 3, no. 3, pp. 159–172, 2011.

[7] H. Bayley and P. S. Cremer, "Stochastic sensors inspired by biology," *Nature*, vol. 413, no. 6852, pp. 226–230, 2001.

[8] V. M. Aguilella and A. Alcaraz, "Nanobiotechnology: a fluid approach to simple circuits," *Nature Nanotechnology*, vol. 4, no. 7, pp. 403–404, 2009.

[9] G. Maglia, A. J. Heron, W. L. Hwang et al., "Droplet networks with incorporated protein diodes show collective properties," *Nature Nanotechnology*, vol. 4, no. 7, pp. 437–440, 2009.

[10] S. B. Laughlin, R. R. de Ruyter van Steveninck, and J. C. Anderson, "The metabolic cost of neural information," *Nature Neuroscience*, vol. 1, no. 1, pp. 36–41, 1998.

[11] D. P. Tieleman, P. C. Biggin, G. R. Smith, and M. S. P. Sansom, "Simulation approaches to ion channel structure-function relationships," *Quarterly Reviews of Biophysics*, vol. 34, no. 4, pp. 473–561, 2001.

[12] B. Roux, T. Allen, S. Bernèche, and W. Im, "Theoretical and computational models of biological ion channels," *Quarterly Reviews of Biophysics*, vol. 37, no. 1, pp. 15–103, 2004.

[13] A. Alcaraz, E. M. Nestorovich, M. L. López, E. García-Giménez, S. M. Bezrukov, and V. M. Aguilella, "Diffusion, exclusion, and specific binding in a large channel: a study of OmpF selectivity inversion," *Biophysical Journal*, vol. 96, no. 1, pp. 56–66, 2009.

[14] E. García-Giménez, A. Alcaraz, and V. M. Aguilella, "Overcharging below the nanoscale: multivalent cations reverse the ion selectivity of a biological channel," *Physical Review E*, vol. 81, no. 2, Article ID 021912, 2010.

[15] S. W. Cowan, R. M. Garavito, J. N. Jansonius et al., "The structure of OmpF porin in a tetragonal crystal form," *Structure*, vol. 3, no. 10, pp. 1041–1050, 1995.

[16] A. H. Delcour, "Solute uptake through general porins," *Frontiers in Bioscience*, vol. 8, pp. d1055–d1071, 2003.

[17] H. Nikaido, "Molecular basis of bacterial outer membrane permeability revisited," *Microbiology and Molecular Biology Reviews*, vol. 67, no. 4, pp. 593–656, 2003.

[18] M. Colombini, E. Blachly-Dyson, and M. Forte, "VDAC, a channel in the outer mitochondrial membrane," *Ion Channels*, vol. 4, pp. 169–202, 1996.

[19] E. Gouaux, "α-Hemolysin from *Staphylococcus aureus*: an archetype of β-barrel, channel-forming toxins," *Journal of Structural Biology*, vol. 121, no. 2, pp. 110–122, 1998.

[20] L. Song, M. R. Hobaugh, C. Shustak, S. Cheley, H. Bayley, and J. E. Gouaux, "Structure of staphylococcal α-hemolysin, a heptameric transmembrane pore," *Science*, vol. 274, no. 5294, pp. 1859–1866, 1996.

[21] V. M. Aguilella and S. M. Bezrukov, "Alamethicin channel conductance modified by lipid charge," *European Biophysics Journal*, vol. 30, no. 4, pp. 233–241, 2001.

[22] D. S. Cafiso, "Alamethicin: a peptide model for voltage gating and protein-membrane interactions," *Annual Review of Biophysics and Biomolecular Structure*, vol. 23, pp. 141–165, 1994.

[23] J. E. Hall, I. Vodyanoy, T. M. Balasubramanian, and G. R. Marshall, "Alamethicin. A rich model for channel behavior," *Biophysical Journal*, vol. 45, no. 1, pp. 233–247, 1984.

[24] M. Colombini, "VDAC: the channel at the interface between mitochondria and the cytosol," *Molecular and Cellular Biochemistry*, vol. 256-257, no. 1-2, pp. 107–115, 2004.

[25] E. M. Nestorovich, C. Danelon, M. Winterhalter, and S. M. Bezrukov, "Designed to penetrate: time-resolved interaction of single antibiotic molecules with bacterial pores," *Proceedings of the National Academy of Sciences of the United States of America*, vol. 99, no. 15, pp. 9789–9794, 2002.

[26] A. Prilipov, P. S. Phale, P. Van Gelder, J. P. Rosenbusch, and R. Koebnik, "Coupling site-directed mutagenesis with high-level expression: large scale production of mutant porins from *E. coli*," *FEMS Microbiology Letters*, vol. 163, no. 1, pp. 65–72, 1998.

[27] R. A. Pauptit, H. Zhang, G. Rummel, T. Schirmer, J. N. Jansonius, and J. P. Rosenbusch, "Trigonal crystals of porin from *Escherichia coli*," *Journal of Molecular Biology*, vol. 218, no. 3, pp. 505–507, 1991.

[28] W. Im and B. Roux, "Ions and counterions in a biological channel: a molecular dynamics simulation of ompf porin from *Escherichia coli* in an explicit membrane with 1 M KCl aqueous salt solution," *Journal of Molecular Biology*, vol. 319, no. 5, pp. 1177–1197, 2002.

[29] A. Alcaraz, E. M. Nestorovich, M. Aguilella-Arzo, V. M. Aguilella, and S. M. Bezrukov, "Salting out the ionic selectivity of a wide channel: the asymmetry of OmpF," *Biophysical Journal*, vol. 87, no. 2, pp. 943–957, 2004.

[30] M. Aguilella-Arzo, J. J. García-Celma, J. Cervera, A. Alcaraz, and V. M. Aguilella, "Electrostatic properties and macroscopic electrodiffusion in OmpF porin and mutants," *Bioelectrochemistry*, vol. 70, no. 2, pp. 320–327, 2007.

[31] D. P. Tieleman and H. J. C. Berendsen, "A molecular dynamics study of the pores formed by *Escherichia coli* OmpF porin in a fully hydrated palmitoyloleoylphosphatidylcholine bilayer," *Biophysical Journal*, vol. 74, no. 6, pp. 2786–2801, 1998.

[32] T. Schirmer and P. S. Phale, "Brownian dynamics simulation of ion flow through porin channels," *Journal of Molecular Biology*, vol. 294, no. 5, pp. 1159–1167, 1999.

[33] P. S. Phale, A. Philippsen, C. Widmer, V. P. Phale, J. P. Rosenbusch, and T. Schirmer, "Role of charged residues at the OmpF porin channel constriction probed by mutagenesis and simulation," *Biochemistry*, vol. 40, no. 21, pp. 6319–6325, 2001.

[34] W. Im and B. Roux, "Ion permeation and selectivity of OmpF porin: a theoretical study based on molecular dynamics, brownian dynamics, and continuum electrodiffusion theory," *Journal of Molecular Biology*, vol. 322, no. 4, pp. 851–869, 2002.

[35] C. Chimerel, L. Movileanu, S. Pezeshki, M. Winterhalter, and U. Kleinekathöfer, "Transport at the nanoscale: temperature dependence of ion conductance," *European Biophysics Journal*, vol. 38, no. 1, pp. 121–125, 2008.

[36] J. Faraudo, C. Calero, and M. Aguilella-Arzo, "Ionic partition and transport in multi-ionic channels: a molecular dynamics simulation study of the OmpF bacterial porin," *Biophysical Journal*, vol. 99, no. 7, pp. 2107–2115, 2010.

[37] B. Dhakshnamoorthy, S. Raychaudhury, L. Blachowicz, and B. Roux, "Cation-selective pathway of OmpF porin revealed by anomalous X-ray diffraction," *Journal of Molecular Biology*, vol. 396, no. 2, pp. 293–300, 2010.

[38] E. Yamashita, M. V. Zhalnina, S. D. Zakharov, O. Sharma, and W. A. Cramer, "Crystal structures of the OmpF porin: function in a colicin translocon," *EMBO Journal*, vol. 27, no. 15, pp. 2171–2180, 2008.

[39] K. L. Lout, N. Saint, A. Prilipov et al., "Structural and functional characterization of OmpF porin mutants selected for larger pore size. I. Crystallographic analysis," *Journal of Biological Chemistry*, vol. 271, no. 34, pp. 20669–20675, 1996.

[40] I. R. Vetter, M. W. Parker, A. D. Tucker, J. H. Lakey, F. Pattus, and D. Tsernoglou, "Crystal structure of a colicin N fragment suggests a model for toxicity," *Structure*, vol. 6, no. 7, pp. 863–874, 1998.

[41] P. S. Phale, A. Philippsen, T. Kiefhaber et al., "Stability of trimeric OmpF porin: the contributions of the latching loop L2," *Biochemistry*, vol. 37, no. 45, pp. 15663–15670, 1998.

[42] R. Dutzler, G. Rummel, S. Alberti et al., "Crystal structure and functional characterization of OmpK36, the osmoporin of Klebsiella pneumoniae," *Structure*, vol. 7, no. 4, pp. 425–434, 1999.

[43] S. Reitz, M. Cebi, P. Reiß et al., "On the function and structure of synthetically modified porins," *Angewandte Chemie*, vol. 48, no. 26, pp. 4853–4857, 2009.

[44] G. Kefala, C. Ahn, M. Krupa et al., "Structures of the OmpF porin crystallized in the presence of foscholine-12," *Protein Science*, vol. 19, no. 5, pp. 1117–1125, 2010.

[45] N. G. Housden, J. A. Wojdyla, J. Korczynska et al., "Directed epitope delivery across the *Escherichia coli* outer membrane through the porin OmpF," *Proceedings of the National Academy of Sciences of the United States of America*, vol. 107, no. 50, pp. 21412–21417, 2010.

[46] M. L. López, M. Aguilella-Arzo, V. M. Aguilella, and A. Alcaraz, "Ion selectivity of a biological channel at high concentration ratio: insights on small ion diffusion and binding," *Journal of Physical Chemistry B*, vol. 113, no. 25, pp. 8745–8751, 2009.

[47] M. L. López, E. García-Giménez, V. M. Aguilella, and A. Alcaraz, "Critical assessment of OmpF channel selectivity: merging information from different experimental protocols," *Journal of Physics Condensed Matter*, vol. 22, no. 45, Article ID 454106, 2010.

[48] P. Mueller, D. O. Rudin, H. T. Tien, and W. C. Wescott, "Reconstitution of excitable cell membrane structure in vitro," *Circulation*, vol. 26, pp. 1167–1171, 1962.

[49] M. Montal and P. Mueller, "Formation of bimolecular membranes from lipid monolayers and a study of their electrical properties," *Proceedings of the National Academy of Sciences of the United States of America*, vol. 69, no. 12, pp. 3561–3566, 1972.

[50] S. M. Beznukov and I. Vodyanoy, "Probing alamethicin channels with water-soluble polymers. Effect on conductance of channel states," *Biophysical Journal*, vol. 64, no. 1, pp. 16–25, 1993.

[51] E. Neher, "Correction for liquid junction potentials in patch clamp experiments," *Methods in Enzymology*, vol. 207, pp. 123–131, 1992.

[52] P. H. Barry and J. M. Diamond, "Junction potentials, electrode standard potentials, and other problems in interpreting electrical properties of membranes," *Journal of Membrane Biology*, vol. 3, no. 1, pp. 93–122, 1970.

[53] P. H. Barry and J. W. Lynch, "Liquid junction potentials and small cell effects in patch-clamp analysis," *Journal of Membrane Biology*, vol. 121, no. 2, pp. 101–117, 1991.

[54] C. Danelon, A. Suenaga, M. Winterhalter, and I. Yamato, "Molecular origin of the cation selectivity in OmpF porin: single channel conductances vs. free energy calculation," *Biophysical Chemistry*, vol. 104, no. 3, pp. 591–603, 2003.

[55] H. Miedema, A. Meter-Arkema, J. Wierenga et al., "Permeation properties of an engineered bacterial OmpF porin containing the EEEE-locus of Ca^{2+} channels," *Biophysical Journal*, vol. 87, no. 5, pp. 3137–3147, 2004.

[56] N. Saint, K. L. Lou, C. Widmer, M. Luckey, T. Schirmer, and J. P. Rosenbusch, "Structural and functional characterization of OmpF porin mutants selected for larger pore size. II. Functional characterization," *Journal of Biological Chemistry*, vol. 271, no. 34, pp. 20676–20680, 1996.

[57] M. Vrouenraets, J. Wierenga, W. Meijberg, and H. Miedema, "Chemical modification of the bacterial porin OmpF: gain of selectivity by volume reduction," *Biophysical Journal*, vol. 90, no. 4, pp. 1202–1211, 2006.

[58] D. A. MacInnes, *The Principles of Electrochemistry*, Dover Publications, New York, NY, USA, 1961.

[59] J. W. Perram and P. J. Stiles, "On the nature of liquid junction and membrane potentials," *Physical Chemistry Chemical Physics*, vol. 8, no. 36, pp. 4200–4213, 2006.

[60] T. Sokalski, P. Lingelfelter, and A. Lewenstam, "Numerical solution of the coupled Nernst-Planck and Poisson equations for liquid junction and ion selective membrane potentials," *Journal of Physical Chemistry B*, vol. 107, no. 11, pp. 2443–2452, 2003.

[61] H. W. Harper, "Calculation of liquid junction potentials," *Journal of Physical Chemistry*, vol. 89, no. 9, pp. 1659–1664, 1985.

[62] J. D. Madura, J. M. Briggs, R. C. Wade et al., "Electrostatics and diffusion of molecules in solution: simulations with the University of Houston Brownian Dynamics program," *Computer Physics Communications*, vol. 91, no. 1–3, pp. 57–95, 1995.

[63] R. Benz, A. Schmid, and R. E. W. Hancock, "Ion selectivity of gram-negative bacterial porins," *Journal of Bacteriology*, vol. 162, no. 2, pp. 722–727, 1985.

[64] D. Gillespie and R. S. Eisenberg, "Physical descriptions of experimental selectivity measurements in ion channels," *European Biophysics Journal*, vol. 31, no. 6, pp. 454–466, 2002.

[65] E. M. Nestorovich, T. K. Rostovtseva, and S. M. Bezrukov, "Residue ionization and ion transport through OmpF channels," *Biophysical Journal*, vol. 85, no. 6, pp. 3718–3729, 2003.

[66] A. Alcaraz, P. Ramírez, E. García-Giménez, M. L. López, A. Andrio, and V. M. Aguilella, "A pH-tunable nanofluidic diode: electrochemical rectification in a reconstituted single ion channel," *Journal of Physical Chemistry B*, vol. 110, no. 42, pp. 21205–21209, 2006.

[67] M. Queralt-Martín, E. García-Giménez, S. Mafé, and A. Alcaraz, "Divalent cations reduce the pH sensitivity of OmpF channel inducing the pK(a) shift of key acidic residues," *Physical Chemistry Chemical Physics*, vol. 13, no. 2, pp. 563–569, 2011.

[68] J. Lyklema, "Overcharging, charge reversal: chemistry or physics?" *Colloids and Surfaces A*, vol. 291, no. 1–3, pp. 3–12, 2006.

[69] H. Miedema, M. Vrouenraets, J. Wierenga et al., "Ca^{2+} selectivity of a chemically modified OmpF with reduced pore volume," *Biophysical Journal*, vol. 91, no. 12, pp. 4392–4400, 2006.

[70] M. Aguilella-Arzo, C. Calero, and J. Faraudo, "Simulation of electrokinetics at the nanoscale: inversion of selectivity in a bio-nanochannel," *Soft Matter*, vol. 6, no. 24, pp. 6079–6082, 2010.

[71] J. Faraudo and A. Travesset, "The many origins of charge inversion in electrolyte solutions: effects of discrete interfacial charges," *Journal of Physical Chemistry C*, vol. 111, no. 2, pp. 987–994, 2007.

[72] F. Helfferich, *Ion Exchange*, McGraw-Hill, New York, NY, USA, 1962.

[73] R. A. Robinson and R. H. Stokes, *Electrolyte Solutions*, Dover Publications, New York, NY, USA, 2002.

[74] V. M. M. Lobo and J. L. Quaresma, *Handbook of Electrolyte Solutions*, Elsevier, Amsterdam, The Netherlands, 1989.

[75] H. S. Harned and B. B. Owen, *The Physical Chemistry of Electrolyte Solutions*, Reinhold Publishing, New York, NY, USA, 1967.

[76] E. García-Giménez, A. Alcaraz, V. M. Aguilella, and P. Ramírez, "Directional ion selectivity in a biological nanopore with bipolar structure," *Journal of Membrane Science*, vol. 331, no. 1-2, pp. 137–142, 2009.

[77] E. García-Giménez, M. L. López, V. M. Aguilella, and A. Alcaraz, "Linearity, saturation and blocking in a large multiionic channel: divalent cation modulation of the OmpF porin conductance," *Biochemical and Biophysical Research Communications*, vol. 404, no. 1, pp. 330–334, 2011.

[78] D. Fologea, E. Krueger, R. Al Faori et al., "Multivalent ions control the transport through lysenin channels," *Biophysical Chemistry*, vol. 152, no. 1–3, pp. 40–45, 2010.

[79] T. H. Hohle, W. L. Franck, G. Stacey, and M. R. O'Brian, "Bacterial outer membrane channel for divalent metal ion acquisition," *Proceedings of the Academy of Sciences of the United States of America*, vol. 108, no. 37, pp. 15390–15395, 2011.

[80] A. J. H. Marshall and L. J. V. Piddock, "Interaction of divalent cations, quinolones and bacteria," *Journal of Antimicrobial Chemotherapy*, vol. 34, no. 4, pp. 465–483, 1994.

[81] P. R. Singh, M. Ceccarelli, M. Lovelle, M. Winterhalter, and K. R. Mahendran, "Antibiotic permeation across the OmpF channel: modulation of the affinity site in the presence of magnesium," *Journal of Physical Chemistry B*, vol. 116, no. 15, pp. 4433–4438, 2012.

[82] A. Brauser, I. Schroeder, T. Gutsmann et al., "Modulation of enrofloxacin binding in OmpF by Mg^{2+} as revealed by the analysis of fast flickering single-porin current," *The Journal of General Physiology*, vol. 140, no. 1, pp. 69–82, 2012.

[83] D. Fologea, R. Al Faori, E. Krueger et al., "Potential analytical applications of lysenin channels for detection of multivalent ions," *Analytical and Bioanalytical Chemistry*, vol. 401, no. 6, pp. 1871–1879, 2011.

10

Endoplasmic Reticulum Stress-Associated Lipid Droplet Formation and Type II Diabetes

Xuebao Zhang[1] and Kezhong Zhang[1, 2, 3]

[1] Center for Molecular Medicine and Genetics, The Wayne State University School of Medicine, 540 East Canfield Avenue, Detroit, MI 48201, USA
[2] Department of Immunology and Microbiology, The Wayne State University School of Medicine, Detroit, MI 48201, USA
[3] Karmanos Cancer Institute, The Wayne State University School of Medicine, Detroit, MI 48201, USA

Correspondence should be addressed to Kezhong Zhang, kzhang@med.wayne.edu

Academic Editor: Huiping Zhou

Diabetes mellitus (DM), a metabolic disorder characterized by hyperglycemia, is caused by insufficient insulin production due to excessive loss of pancreatic β cells (type I diabetes) or impaired insulin signaling due to peripheral insulin resistance (type II diabetes). Pancreatic β cell is the only insulin-secreting cell type that has highly developed endoplasmic reticulum (ER) to cope with high demands of insulin synthesis and secretion. Therefore, ER homeostasis is crucial to the proper function of insulin signaling. Accumulating evidence suggests that deleterious ER stress and excessive intracellular lipids in nonadipose tissues, such as myocyte, cardiomyocyte, and hepatocyte, cause pancreatic β-cell dysfunction and peripheral insulin resistance, leading to type II diabetes. The excessive deposition of lipid droplets (LDs) in specialized cell types, such as adipocytes, hepatocytes, and macrophages, has been found as a hallmark in ER stress-associated metabolic diseases, including obesity, diabetes, fatty liver disease, and atherosclerosis. However, much work remains to be done in understanding the mechanism by which ER stress response regulates LD formation and the pathophysiologic role of ER stress-associated LD in metabolic disease. This paper briefly summarizes the recent advances in ER stress-associated LD formation and its involvement in type II diabetes.

1. Introduction to ER Stress

ER is an intracellular organelle where dynamic protein folding and assembly, storing cellular calcium, and lipid biosynthesis occur. A variety of biochemical or pathophysiological stimuli can interrupt protein folding process in the ER by disrupting protein glycosylation, disulfide bond formation, or ER calcium pool. These disruptions can cause the accumulation of unfolded or misfolded proteins in the ER lumen, a condition termed as "ER stress" [1, 2]. To protect cells from proteotoxicity caused by ER stress, the unfolded protein response (UPR) is activated through attenuating general protein translation, increasing in protein folding capacity, and expediting degradation of misfolded proteins. Three major ER stress sensors or transducers have been found: inositol-requiring 1α (IRE1α), double-stranded RNA-dependent protein kinase- (PKR-) like ER kinase (PERK), and activating transcription factor 6 (ATF6), which have been comprehen-sively reviewed [2, 3]. The UPR signaling, mediated through ER stress sensors, modulates transcriptional and translation programs in cells under ER stress. As a double-edged sword, the UPR provides survival signals at the initial phase of stress response, leading to cell adaption to ER stress [1, 2, 4]. When ER stress gets prolonged, the UPR can induce cell death programs to kill the stressed cells. In recent years, the scope and consequence of ER stress and UPR have been significantly expanded. Many pathophysiologic stimuli, such as oxidative stress, proinflammatory stimuli, fatty acids, and energy fluctuations, can directly or indirectly cause ER stress and the UPR activation in specialized cell types, such as macrophages, hepatocytes, and pancreatic β cells [2, 5]. The UPR signaling is fundamental to the initiation and progress of a variety of diseases, including metabolic disease, cancer, cardiovascular disease, and neurodegenerative disease [2, 6, 7].

2. LD Formation

LD, also known as adiposome or fat body, has been found ubiquitously present in lipid-overloaded cells from yeast to mammals [8, 9]. For a long time, LD was thought simply as an inert lipid storage reservoir since its earliest description in 19th century. The discovery of perilipin, an LD-associated protein that coats LD in adipocytes, makes researchers to challenge the understanding of LD as lipid storage [10]. LD is now recognized as a dynamic organelle composed of a monolayer phospholipid, embedded with numerous proteins without transmembrane spanning domains, and a hydrophobic core that contains triacylglycerols (TGs) and sterol esters [11, 12]. TGs are the key neutral lipid required for LDs formation in adipocytes. Deletion of genes encoding enzymes responsible for neutral lipid synthesis eliminated LDs formation [13]. Evidence showed that, without DGAT enzymes, LDs cannot form in adipocytes. Therefore, by segregation of extra TG or hydrophobic molecules into LDs, cells are protected from lipotoxicity. These features make LD a regulatory organelle in lipid homeostasis. The biogenesis and assembly of LD are still largely unknown. It has been suggested that ER is the site where LD is synthesized and assembled. Over ninety percent of LDs were found in close apposition to the ER [14]. ER budding model, Bicelle model, and vesicular budding model have been suggested to explain how LD is formed in ER [15]. Perhaps, the most accepted model is ER budding model in which LD originated between the two leaflets of ER bilayer buds into the cytosol. Newly formed LD can increase its size (0.2 μm–20 μm in diameter) by homotypic fusion that depends on microtubule system, most likely motor protein dynein. Under this mechanism, the growth of LD may proceed without ongoing biosynthesis of TGs and sterol esters [16, 17].

3. ER Stress and LD Formation

LD formation has been proposed as an exit model in the removal of unfolded or misfolded proteins or some ubiquitinated proteins from the ER [18, 19]. LD may serve as a transient depot to sequester unfolded or misfolded as well as excessive proteins to alleviate ER stress (Figure 1). Diverse groups of LD-associated proteins were found in yeast S. cerevisiae, Drosophila embryos, and human hepatocyte cell line Huh7 [20–22]. Some of the LD-associated proteins, such as Acl-CoA synthetases, lanosterol synthetase, and GAPDH, are conserved from yeast to human. The proteins detected in LD seem to be specific, since the organelle-specific proteins, including lactate dehydrogenase (LDH) (cytosolic marker), integrin (plasma membrane marker), calnexin (ER marker), and GS28 (Golgi marker), were hardly detected in LD fractions [22]. Interestingly, a number of proteins which were thought to be organelle-specific, including histones (nucleus), caveolins (plasma membrane), HSP70 (cytosol), ApoB (ER), and Nir2 (Golgi), were detected in LD fraction [23]. Furthermore, LD dynamically interacts with ER, peroxisomes, mitochondria, and plasma membrane [15]. LD can be transported along microtubules, following the same way that the ER, Golgi, and mitochondria were positioned

and delivered [24]. It was proposed that the dynamical interactions between LD and the other compartments facilitate the exchange of proteins and lipids in cells. The LD is functionally and structurally similar to the extracellular counterpart of lipoprotein particles [15, 21]. This notion was supported by the finding that LD provides a platform for degradation of excessive ApoB protein by converging ubiquitin-proteasomal and autophagy-lysosomal pathways, thereby preventing cytotoxicity resulted from aggregation of excessive proteins [25]. Previous studies have shown that disruption of ER functions leads to the accumulation of intracellular lipids [26–28]. Disrupted protein glycosylation or ER-associated protein degradation by ER stress-inducing reagents, such as tunicamycin and brefeldin, has been demonstrated to increase LD accumulation in budding yeast Saccharomyces cerevisiae or mammalian cells [28, 29]. Previously, it is known that intracellular LD formation is through the lipogenic program activated by sterol regulatory element-binding proteins (SREBPs). Recent study suggested that more ER-localized, stress-responsive protein factors, such as hepatocyte-specific cAMP responsive element-binding protein (CREBH), can also regulate lipogenic programs to promote LD formation under metabolic stress signals, such as insulin and saturated fatty acids [30]. Moreover, ER stress response may directly facilitate LD synthesis and assembly as a mechanism to defend intracellular stress [29, 31] (Figure 1). This is consistent with the observations that lipids can be recruited to the stressed cells to sequester misfolded proteins in the ER at the early stage of ER stress and that the ER is expanded significantly to alleviated ER stress independent of the UPR [23, 32].

4. LD Formation and Type II Diabetes

Previous studies demonstrated that excessive accumulation of lipids in peripheral tissues is closely associated with insulin resistance in type II diabetes [33, 34]. Although ER stress and UPR pathways in metabolic disease have been extensively reviewed, ER stress-associated LD formation, which is independent of UPR pathway, did not draw much attention. The interaction between LD and mitochondrial might affect the peripheral tissue insulin resistance [35, 36] (Figure 1). Recent studies indicated that insulin resistance is not simply associated with the amount of intracellular lipids. Despite elevated lipids content in skeletal muscle of the trained enduring athletes, the insulin-signal in these individuals is still markedly sensitive [36]. The combination of weight loss and physical activity in obesity improves insulin sensitivity and reduces the size of LD, but not the overall intramyocellular lipid [37]. One possible explanation for these phenomena is that increased mitochondrial oxidative activity for lipid oxidation may decrease insulin resistance. This is supported by the facts that lower oxidative capacity is found in insulin resistant skeletal muscle and that exercise can improve the capacity for lipid oxidation [36]. Several mitochondrial proteins including prohibitin, a subunit of ATP synthase, and pyruvate carboxylase were identified in LD fractions by proteomic analysis [35]. In addition, numerous lipid metabolic enzymes, such as hormone-sensitive

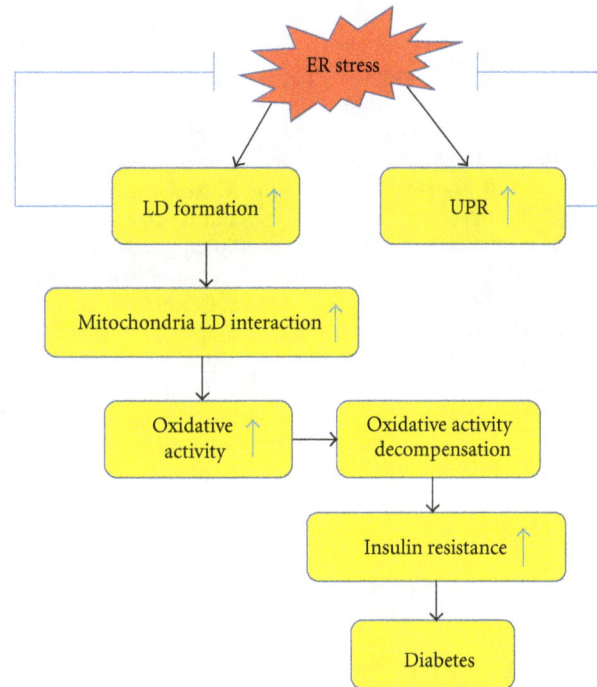

FIGURE 1: Interactions between ER stress, oxidative stress, and lipid droplets in type II diabetes. LD, lipid droplet; UPR, unfolded protein response.

lipase, lanosterol synthase, and acyl-CoA synthetase, were also found to be associated with LD complex, and the overall LD protein composition can be changed in response to lipolysis stimulation [35, 38]. Despite these observations, further study is required to explore how mitochondria communicate and interact with LD in metabolic processes.

Fat-specific protein 27 (Fsp27) is a member of cell death-inducing DNA fragmentation factor family proteins that is localized to LD. Fsp27 plays an important role in lipid storage and mitochondrial activity in adipocytes [39–41]. Genetic depletion of Fsp27 in mice is characterized by increased glucose uptake, improved insulin sensitivity, and significantly increased mitochondrial metabolism [39, 40]. Small sizes of LDs and increased mitochondrial activity were found in Fsp27-deficient white adipocytes, suggesting that ectopic LD formation represents an imbalance between lipid supply and lipid oxidation in peripheral tissue. Likely, LD-associated proteins and the interactions between LD and the other intracellular organelles may play direct roles in the pathogenesis of diabetes [42]. Type II diabetes is often correlated with increased serum levels of proinflammatory cytokines secreted by ER stress-activated macrophage. Previous research demonstrated that the proinflammatory cytokine TNFα blunts the insulin signaling pathway therefore causing insulin resistance by activating the JNK1/2 signaling pathway which is involved in serine phosphorylation of IRS1 (insulin receptor substrate 1) [43, 44]. However, a new study by Ranjit found that proinflammatory cytokines, such as TNFα, IL1β, and INFγ, act on lipolysis by decreasing the expression of FSP27 and the size of LD in adipocytes [45]. Since decreased FSP27 is evidenced to improve insulin

resistance and LDs, it is likely that the proinflammatory cytokines play double-edged roles in type II diabetes.

5. Conclusion

Accumulating evidence demonstrated a strong link between ER stress, LD formation, and type II diabetes. It is important to note that ER stress response is a fundamental stress signaling underlying many life styles, such as air pollution, chronic alcohol consumption, and smoking, which may be associated with the development of metabolic disease [46–48]. Therefore, for the future research, it is important to delineate ER mechanisms in LD formation that is associated with the development of type II diabetes. Key questions include what is the mechanism by which ER stress regulates LD formation? Is there any ER chaperones or UPR targets present in the LD complex? Does ER stress-associated LD formation provide survival or devastating pathways in the progression of type II diabetes? Is it possible to modulate LD formation by targeting ER stress signaling? Answering these questions will benefit and direct the future understanding and treatment of type II diabetes and the other types of metabolic disease.

Acknowledgments

Portion of the research work in the Zhang laboratory is supported by American Heart Association Grants 09GRNT-2280479; National Institutes of Health Grants DK090313 and ES017829 (K. Zhang)

References

[1] K. Zhang and R. J. Kaufman, "Signaling the unfolded protein response from the endoplasmic reticulum," *Journal of Biological Chemistry*, vol. 279, no. 25, pp. 25935–25938, 2004.

[2] D. Ron and P. Walter, "Signal integration in the endoplasmic reticulum unfolded protein response," *Nature Reviews Molecular Cell Biology*, vol. 8, no. 7, pp. 519–529, 2007.

[3] M. Schroder and R. J. Kaufman, "The mammalian unfolded protein response," *Annual Review of Biochemistry*, vol. 74, pp. 739–789, 2005.

[4] C. Rubio, D. Pincus, A. Korennykh, S. Schuck, H. El-Samad, and P. Walter, "Homeostatic adaptation to endoplasmic reticulum stress depends on Ire1 kinase activity," *Journal of Cell Biology*, vol. 193, no. 1, pp. 171–184, 2011.

[5] K. Zhang and R. J. Kaufman, "From endoplasmic-reticulum stress to the inflammatory response," *Nature*, vol. 454, no. 7203, pp. 455–462, 2008.

[6] H. Yoshida, "ER stress and diseases," *FEBS Journal*, vol. 274, no. 3, pp. 630–658, 2007.

[7] R. J. Kaufman, "Orchestrating the unfolded protein response in health and disease," *Journal of Clinical Investigation*, vol. 110, no. 10, pp. 1389–1398, 2002.

[8] M. K. Clausen, K. Christiansen, P. K. Jensen, and O. Behnke, "Isolation of lipid particles from baker's yeast," *FEBS Letters*, vol. 43, no. 2, pp. 176–179, 1974.

[9] K. Christiansen and P. K. Jensen, "Membrane-bound lipid particles from beef heart chemical composition and structure," *Biochimica et Biophysica Acta*, vol. 260, no. 3, pp. 449–459, 1972.

[10] A. S. Greenberg, J. J. Egan, S. A. Wek, N. B. Garty, E. J. Blanchette-Mackie, and C. Londos, "Perilipin, a major hormonally regulated adipocyte-specific phosphoprotein associated with the periphery of lipid storage droplets," *Journal of Biological Chemistry*, vol. 266, no. 17, pp. 11341–11346, 1991.

[11] D. A. Brown, "Lipid droplets: proteins floating on a pool of fat," *Current Biology*, vol. 11, no. 11, pp. R446–R449, 2001.

[12] P. T. Bozza and J. P. B. Viola, "Lipid droplets in inflammation and cancer," *Prostaglandins Leukotrienes and Essential Fatty Acids*, vol. 82, no. 4–6, pp. 243–250, 2010.

[13] C. A. Harris, J. T. Haas, R. S. Streeper et al., "DGAT enzymes are required for triacylglycerol synthesis and lipid droplets in adipocytes," *Journal of Lipid Research*, vol. 52, no. 4, pp. 657–667, 2011.

[14] K. M. Szymanski, D. Binns, R. Bartz et al., "The lipodystrophy protein seipin is found at endoplasmic reticulum lipid droplet junctions and is important for droplet morphology," *Proceedings of the National Academy of Sciences of the United States of America*, vol. 104, no. 52, pp. 20890–20895, 2007.

[15] Y. Guo, K. R. Cordes, R. V. Farese Jr., and T. C. Walther, "Lipid droplets at a glance," *Journal of Cell Science*, vol. 122, no. 6, pp. 749–752, 2009.

[16] P. Bostrom, L. Andersson, L. Li et al., "The assembly of lipid droplets and its relation to cellular insulin sensitivity," *Biochemical Society Transactions*, vol. 37, no. 5, pp. 981–985, 2009.

[17] P. Bostrom, M. Rutberg, J. Ericsson et al., "Cytosolic lipid droplets increase in size by microtubule-dependent complex formation," *Arteriosclerosis, Thrombosis, and Vascular Biology*, vol. 25, pp. 1945–1951, 2005.

[18] H. L. Ploegh, "A lipid-based model for the creation of an escape hatch from the endoplasmic reticulum," *Nature*, vol. 448, no. 7152, pp. 435–438, 2007.

[19] I. Z. Hartman, P. S. Liu, J. K. Zehmer et al., "Sterol-induced dislocation of 3-Hydroxy-3-methylglutaryl coenzyme a reductase from endoplasmic reticulum membranes into the cytosol through a subcellular compartment resembling lipid droplets," *Journal of Biological Chemistry*, vol. 285, no. 25, pp. 19288–19298, 2010.

[20] K. Athenstaedt, D. Zweytick, A. Jandrositz, S. D. Kohlwein, and G. Daum, "Identification and characterization of major lipid particle proteins of the yeast Saccharomyces cerevisiae," *Journal of Bacteriology*, vol. 181, no. 20, pp. 6441–6448, 1999.

[21] S. Cermelli, Y. Guo, S. P. Gross, and M. A. Welte, "The lipid-droplet proteome reveals that droplets are a protein-storage depot," *Current Biology*, vol. 16, no. 18, pp. 1783–1795, 2006.

[22] Y. Fujimoto, H. Itabe, J. Sakai et al., "Identification of major proteins in the lipid droplet-enriched fraction isolated from the human hepatocyte cell line HuH7," *Biochimica et Biophysica Acta*, vol. 1644, no. 1, pp. 47–59, 2004.

[23] M. A. Welte, "Proteins under new management: lipid droplets deliver," *Trends in Cell Biology*, vol. 17, no. 8, pp. 363–369, 2007.

[24] M. A. Welte, S. Cermelli, J. Griner et al., "Regulation of lipid-droplet transport by the perilipin homolog LSD2," *Current Biology*, vol. 15, no. 14, pp. 1266–1275, 2005.

[25] Y. Ohsaki, J. Cheng, A. Fujita, T. Tokumoto, and T. Fujimoto, "Cytoplasmic lipid droplets are sites of convergence of proteasomal and autophagic degradation of apolipoprotein B," *Molecular Biology of the Cell*, vol. 17, no. 6, pp. 2674–2683, 2006.

[26] A. J. Kim, Y. Shi, R. C. Austin, and G. H. Werstuck, "Valproate protects cells fom ER stress-induced lipid accumulation and apoptosis by inhibiting glycogen synthase kinase-3," *Journal of Cell Science*, vol. 118, no. 1, pp. 89–99, 2005.

[27] G. H. Werstuck, S. R. Lentz, S. Dayal et al., "Homocysteine-induced endoplasmic reticulum stress causes dysregulation of the cholesterol and triglyceride biosynthetic pathways," *Journal of Clinical Investigation*, vol. 107, no. 10, pp. 1263–1273, 2001.

[28] W. Fei, H. Wang, C. Bielby, and H. Yang, "Conditions of endoplasmic reticulum stress stimulate lipid droplet formation in Saccharomyces cerevisiae," *Biochemical Journal*, vol. 424, no. 1, pp. 61–67, 2009.

[29] J.-S. Lee, R. Mendez, H. H. Heng, Z. Yang, and K. Zhang, "Pharmacological ER stress promotes hepatic lipogenesis and lipid droplet formation," *American Journal of Translational Research*, vol. 4, pp. 102–113, 2012.

[30] C. Zhang, G. Wang, Z. Zheng et al., "Endoplasmic reticulum-tethered transcription factor cAMP responsive element-binding protein, hepatocyte specific, regulates hepatic lipogenesis, fatty acid oxidation, and lipolysis upon metabolic stress in mice," *Hepatology*. In press.

[31] K. Zhang, S. Wang, J. Malhotra et al., "The unfolded protein response transducer IRE1alpha prevents ER stress-induced hepatic steatosis," *The EMBO Journal*, vol. 30, pp. 1357–1375, 2011.

[32] S. Schuck, W. A. Prinz, K. S. Thorn, C. Voss, and P. Walter, "Membrane expansion alleviates endoplasmic reticulum stress independently of the unfolded protein response," *Journal of Cell Biology*, vol. 187, no. 4, pp. 525–536, 2009.

[33] S. E. Thomas, L. E. Dalton, M. L. Daly, E. Malzer, and S. J. Marciniak, "Diabetes as a disease of endoplasmic reticulum stress," *Diabetes/Metabolism Research and Reviews*, vol. 26, no. 8, pp. 611–621, 2010.

[34] D. Scheuner and R. J. Kaufman, "The unfolded protein response: a pathway that links insulin demand with β-cell

failure and diabetes," *Endocrine Reviews*, vol. 29, no. 3, pp. 317–333, 2008.

[35] D. L. Brasaemle, G. Dolios, L. Shapiro, and R. Wang, "Proteomic analysis of proteins associated with lipid droplets of basal and lipolytically stimulated 3T3-L1 adipocytes," *Journal of Biological Chemistry*, vol. 279, no. 5, pp. 46835–46842, 2004.

[36] B. H. Goodpaster, J. He, S. Watkins, and D. E. Kelley, "Skeletal muscle lipid content and insulin resistance: evidence for a paradox in endurance-trained athletes," *Journal of Clinical Endocrinology and Metabolism*, vol. 86, no. 12, pp. 5755–5761, 2001.

[37] J. He, B. H. Goodpaster, and D. E. Kelley, "Effects of weight loss and physical activity on muscle lipid content and droplet size," *Obesity Research*, vol. 12, no. 5, pp. 761–769, 2004.

[38] J. G. Granneman, H. P. Moore, R. L. Granneman, A. S. Greenberg, M. S. Obin, and Z. Zhu, "Analysis of lipolytic protein trafficking and interactions in adipocytes," *Journal of Biological Chemistry*, vol. 282, no. 8, pp. 5726–5735, 2007.

[39] Z. Nian, Z. Sun, L. Yu, S. Y. Toh, J. Sang, and P. Li, "Fat-specific protein 27 undergoes ubiquitin-dependent degradation regulated by triacylglycerol synthesis and lipid droplet formation," *Journal of Biological Chemistry*, vol. 285, no. 13, pp. 9604–9615, 2010.

[40] S. Y. Toh, J. Gong, G. Du et al., "Up-regulation of mitochondrial activity and acquirement of brown adipose tissue-like property in the white adipose tissue of Fsp27 deficient mice," *PLoS One*, vol. 3, no. 8, Article ID e2890, 2008.

[41] N. Nishino, Y. Tamori, S. Tateya et al., "FSP27 contributes to efficient energy storage in murine white adipocytes by promoting the formation of unilocular lipid droplets," *Journal of Clinical Investigation*, vol. 118, no. 8, pp. 2808–2821, 2008.

[42] M. A. Abdul-Ghani and R. A. Defronzo, "Pathogenesis of insulin resistance in skeletal muscle," *Journal of Biomedicine and Biotechnology*, vol. 2010, Article ID 476279, 19 pages, 2010.

[43] H. A. Tuttle, G. Davis-Gorman, S. Goldman, J. G. Copeland, and P. F. McDonagh, "Proinflammatory cytokines are increased in type 2 diabetic women with cardiovascular disease," *Journal of Diabetes and Its Complications*, vol. 18, no. 6, pp. 343–351, 2004.

[44] S. Fernandez-Veledo, R. Vila-Bedmar, I. Nieto-Vazquez, and M. Lorenzo, "C-Jun N-terminal kinase 1/2 activation by tumor necrosis factor-α induces insulin resistance in human visceral but not subcutaneous adipocytes: reversal by liver X receptor agonists," *Journal of Clinical Endocrinology and Metabolism*, vol. 94, no. 9, pp. 3583–3593, 2009.

[45] S. Ranjit, E. Boutet, P. Gandhi et al., "Regulation of fat specific protein 27 by isoproterenol and TNF-α to control lipolysis in murine adipocytes," *Journal of Lipid Research*, vol. 52, no. 2, pp. 221–236, 2011.

[46] S. Laing, G. Wang, T. Briazova et al., "Airborne particulate matter selectively activates endoplasmic reticulum stress response in the lung and liver tissues," *American Journal of Physiology*, vol. 299, no. 4, pp. C736–C749, 2010.

[47] S. J. Pandol, F. S. Gorelick, A. Gerloff, and A. Lugea, "Alcohol abuse, endoplasmic reticulum stress and pancreatitis," *Digestive Diseases*, vol. 28, no. 6, pp. 776–782, 2010.

[48] E. Jorgensen, A. Stinson, L. Shan, J. Yang, D. Gietl, and A. P. Albino, "Cigarette smoke induces endoplasmic reticulum stress and the unfolded protein response in normal and malignant human lung cells," *BMC Cancer*, vol. 8, article 229, 2008.

11

Sphingolipid and Ceramide Homeostasis: Potential Therapeutic Targets

Simon A. Young,[1] John G. Mina,[2] Paul W. Denny,[2, 3] and Terry K. Smith[1]

[1] School of Biology and Chemistry, Biomedical Sciences Research Complex, University of St Andrews, North Haugh, KY16 9ST, UK
[2] Biophysical Sciences Institute, School of Biological and Biomedical Sciences and Department of Chemistry, University of Durham University Science Laboratories, South Road, Durham DH1 3LE, UK
[3] School of Medicine and Health, Durham University, Queen's Campus, Stockton-on-Tees TS17 6BH, UK

Correspondence should be addressed to Paul W. Denny, p.w.denny@durham.ac.uk and Terry K. Smith, tks1@st-andrews.ac.uk

Academic Editor: Todd B. Reynolds

Sphingolipids are ubiquitous in eukaryotic cells where they have been attributed a plethora of functions from the formation of structural domains to polarized cellular trafficking and signal transduction. Recent research has identified and characterised many of the key enzymes involved in sphingolipid metabolism and this has led to a heightened interest in the possibility of targeting these processes for therapies against cancers, Alzheimer's disease, and numerous important human pathogens. In this paper we outline the major pathways in eukaryotic sphingolipid metabolism and discuss these in relation to disease and therapy for both chronic and infectious conditions.

1. Introduction

Sphingolipids are a class of natural products that were first characterised by the German-born chemist and clinician Johann L. W. Thudichum in 1884. They consist of an sphingoid base backbone, for example sphingosine, that can be N-acylated with fatty acids forming ceramides. To these lipid anchors is attached a variety of charged, neutral, phosphorylated and/or glycosylated moieties forming complex sphingolipids, for example phosphorylcholine to make the most abundant mammalian sphingolipid, sphingomyelin. These moieties result in both polar and nonpolar regions giving the molecules an amphipathic character which accounts for their tendency to aggregate into membranous structures. Furthermore, the divergence encountered in their chemical structures allows them to play distinctive roles within cellular metabolism (Figure 1) [1, 2].

Sphingolipids are ubiquitous and essential structural components of eukaryotic membranes [3] as well as some prokaryotic organisms and viruses [2]. They are found predominantly in the outer leaflet of the plasma membrane [4],

the lumen of intracellular organelles [5], and lipoproteins [2]. Sphingolipids (most notably ceramide) are also bioactive signalling molecules that control a plethora of cellular events including signal transduction, cell growth, differentiation, and apoptosis [6–10]. In addition, their role in protein kinase C regulation has more recently been elucidated [11]. It is noteworthy that some sphingolipid metabolites can exhibit both structural and signalling functionalities. For example, glycosphingolipids have been reported to be involved in cellular recognition complexes, for example, blood group antigens, cell adhesion, and the regulation of cell growth [4].

Over the last decade, there has been an exponential increase in the study of sphingolipids. However, the investigation and deciphering of the functions of each specific sphingolipid remains challenging due to the complexity in sphingolipid metabolic interconnection, their varied biophysical properties (neutral or charged), the hydrophobic nature of the enzymes involved, and the presence of multiple pathways that can operate in parallel [12]. The interaction of sphingolipid biosynthesis with other cellular metabolic pathways, for example glycerolipid metabolism, introduces another

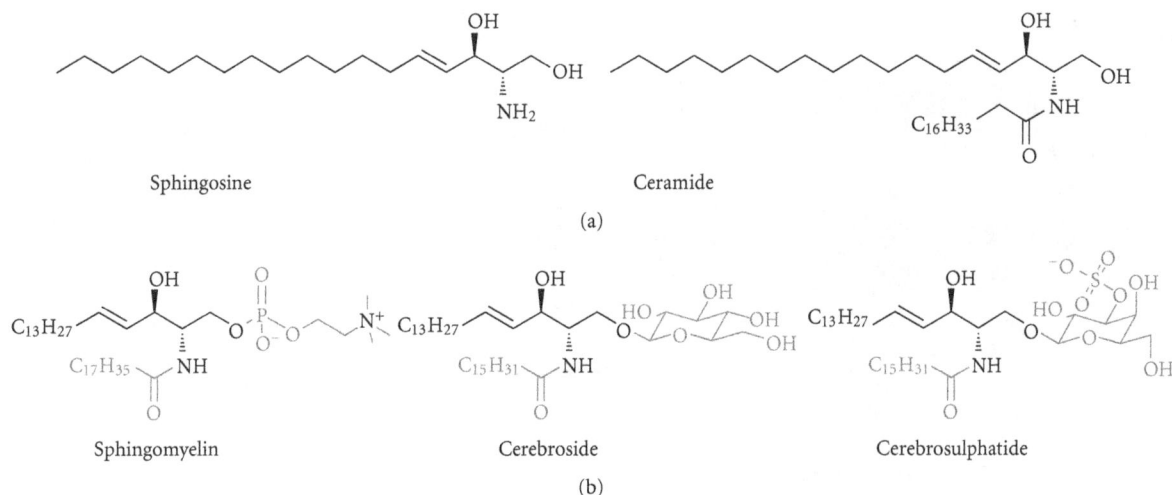

FIGURE 1: (a) The chemical structures of sphingosine and C_{18}-ceramide; (b) the lipids isolated by Thudichum.

layer of complexity and the cellular role of an individual sphingolipid could be defined as a multidimensional in terms of subcellular localisation, regulation and mechanism of action(s) [12].

Whilst the scientific literature has been enriched by articles focused on structural diversity [13–18] and cellular metabolism [2, 8–10, 12, 19–21], this paper focuses on the key enzymes involved in the regulation of ceramide, a central sphingolipid and a key bioactive molecule [12]. To this end we discuss the roles of these enzymes in the regulation of biosynthesis, and in the recycling, salvage, and degradation of complex sphingolipids, in mammalian, fungal and protozoan systems. In addition, we relate these observations to disease and potential therapies.

2. Sphingolipid Metabolism

Sphingolipid metabolism is a critical cell process [22] constituting a highly complex network of interconnected pathways, with ceramide (and to a lesser degree dihydroceramide) occupying a central position in both biosynthesis and catabolism. Therefore, this simple but highly bioactive sphingolipid represents a metabolic hub [12]. In terms of ceramide, the routes of formation can be grouped into either *de novo* synthesis; or recycling, salvage, and degradation (Figure 2). Sphingolipid metabolism has been extensively studied in mammalian and fungal systems, where many of the enzymes involved have been identified and characterised. Consequently, the mammalian pathways will be used as the reference model in the following discussion.

2.1. De Novo Synthesis. The first step in the *de novo* biosynthesis of sphingolipids is the condensation of serine and palmitoyl CoA, a reaction catalysed by the normally rate-limiting serine palmitoyltransferase (SPT, EC 2.3.1.50) to produce 3-ketodihydrosphingosine [23]. SPTs are members of the pyridoxal $5'$-phosphate-dependent α-oxoamine synthase family who share a conserved motif

(T[FL][GTS]**K**[SAG][FLV]G) around the PLP-binding lysine (in bold). The mammalian SPTs [23] (and those of other eukaryotes [24, 25]) are membrane bound in the endoplasmic reticulum as a heterodimer of subunits LCB1 and LCB2 (~53 and ~63 kDa); these are both type I integral membrane proteins sharing ~20% identity. The bacterial SPT is ~30% identical to both mammalian LCB1 and LCB2 at the amino acid level and has the conserved lysine residue in the PLP-binding motif, however the soluble 45 kDa protein forms active homodimers [26]. Palmitoyl-CoA functions as the best substrate of mammalian SPT *in vitro*, while it is also the dominant acyl-CoA *in vivo*, and thus the sphingoid bases from mammalian cells are predominantly C16 [23]. In contrast, the enzyme from the bacteria *Sphingomonas wittichii* utilises stearoyl-CoA most efficiently [27]. A third subunit increasing enzyme activity has been identified in *Saccharomyces cerevisiae* [28] and more recently 2 nonhomologous but functionally related proteins have been characterised in a mammalian system [29]. Furthermore, these additional subunits confer distinct acyl-CoA substrate specificities to the mammalian SPT thus explaining the diversity of long chain bases found in mammals [30]. As the "gatekeeper" of sphingolipid biosynthesis, loss of SPT has a catastrophic effect on mammalian cell viability with a partial loss of SPT function seen in the inherited progressive disorder, Hereditary Sensory Neuropathy type I (HSN1) [23]. The molecular basis of this condition is discussed later in this paper.

Following sphinganine (dihydrosphingosine) formation, metabolic differences are encountered. Whilst in fungi and higher plants sphinganine is hydroxylated to phytosphingosine then acylated to produce phytoceramide, in animal cells sphinganine is acylated to dihydroceramide which is later desaturated to form ceramide [31]. Ceramide (or phytoceramide), a central sphingolipid, is then transported from the ER to the Golgi apparatus where further synthesis of complex sphingolipids takes place [7, 19, 20, 32]. Ceramide can be phosphorylated by ceramide kinase [33], glycosylated by glucosyl or galactosyl ceramide synthases [34], or acquire a variety of neutral or charged head groups to form various

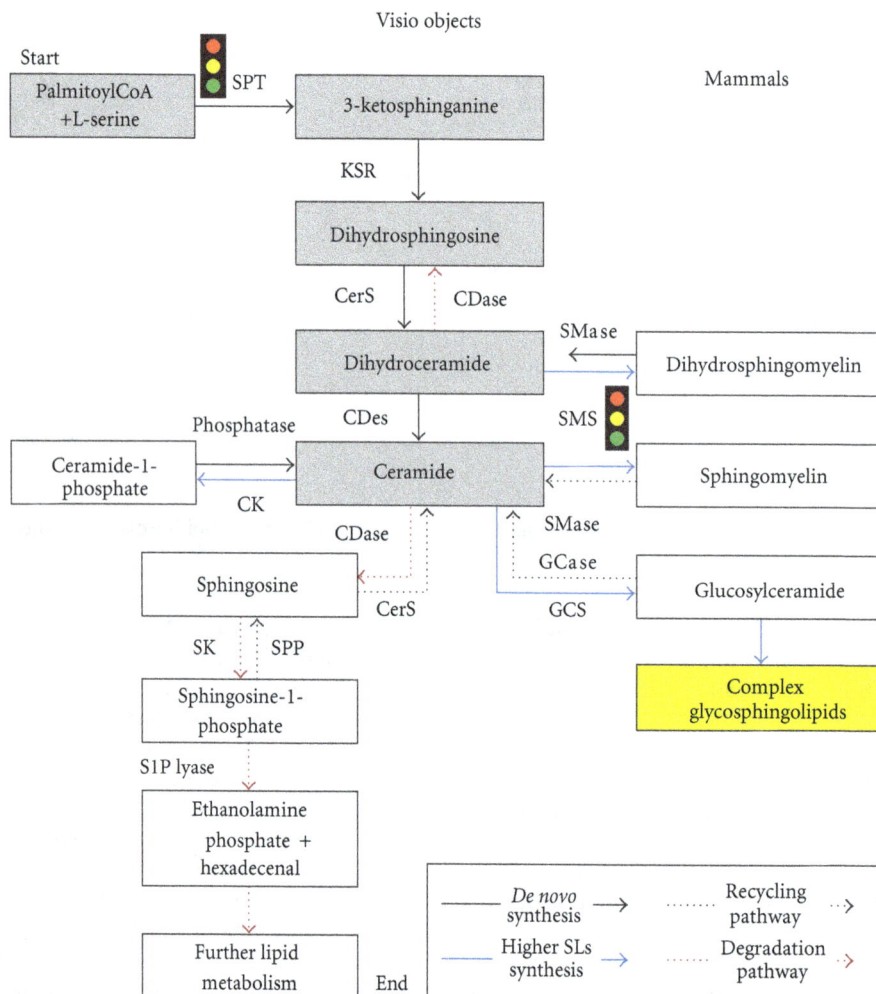

FIGURE 2: A simplified diagram of sphingolipid metabolism in mammals. The key regulatory synthetic steps are indicated by the "traffic light" symbols. SPT: Serine Palmitoyltransferase; 3-KSR: 3-Ketosphingosine Reductase; CerS: Ceramide Synthase; CDase: Ceramidase; CDes: Ceramide Desaturase; SMS: Sphingomyelin Synthase; SMase: Sphingomyelinase; CK: Ceramide Kinase; GCS: Glucosylceramide Synthase; GCase: Glycosidases; SK: Sphingosine Kinase; SPP: Sphingosine-1-Phosphate Phosphatase; S1P: Sphingosine-1-Phosphate.

complex phosphosphingolipids depending on the host organism. For example, in animal cells ceramide is a substrate for sphingomyelin (SM) synthase to produce SM [35]. In contrast, fungi and higher plants utilise phytoceramide to produce inositol phosphorylceramide (IPC) as their principal phosphosphingolipid, a reaction catalysed by IPC synthase [36, 37]. In these organisms IPC is later glycosylated to produce more complex phosphosphingolipids, for example, mannose-IPC (MIPC), in yeast [38, 39]. Finally, the protozoa (exemplified by the Kinetoplastidae) represent a distinct third group in which ceramide [21] acquires a phosphorylinositol head group from phosphatidylinositol (PI) to produce IPC via IPC synthase [40] (Figure 3).

The synthesis of complex sphingolipids such as SM are key regulatory synthetic steps, as the rate of synthesis not only decreases the amount of ceramide, but also indirectly increases the total amount of ceramide-containing molecules that potentially could be degraded/catabolised to form ceramide.

Importantly, in such biosynthetic steps the evolutionarily divergent SM and IPC synthases are central in controlling the delicate balance of glycerolipids (PI/PC in and diacylglycerol-DAG out) on one hand, and sphingolipids (phytoceramide/ceramide in and IPC/SM out) on the other. Therefore, these enzymes have an important role as regulators of proapoptotic ceramide and promitogenic DAG [41]. In addition to a mitogenic role, DAG has also been attributed to play a role in several enzyme activation and regulatory functions [1, 2]. Notably, IPC synthase inhibitors are acutely fungicidal, with the accumulation of ceramide proposed to induce apoptosis [42]. Thus this enzyme represents an attractive target for antifungals and more recently this has been extended to the kinetoplastid protozoa [43, 44].

2.2. Recycling, Salvage, and Degradation. In addition to the de novo synthesis, the recycling, salvage, and degradation pathways modulate cellular levels of sphingolipids. These pathways operate in the direction of ceramide regeneration from

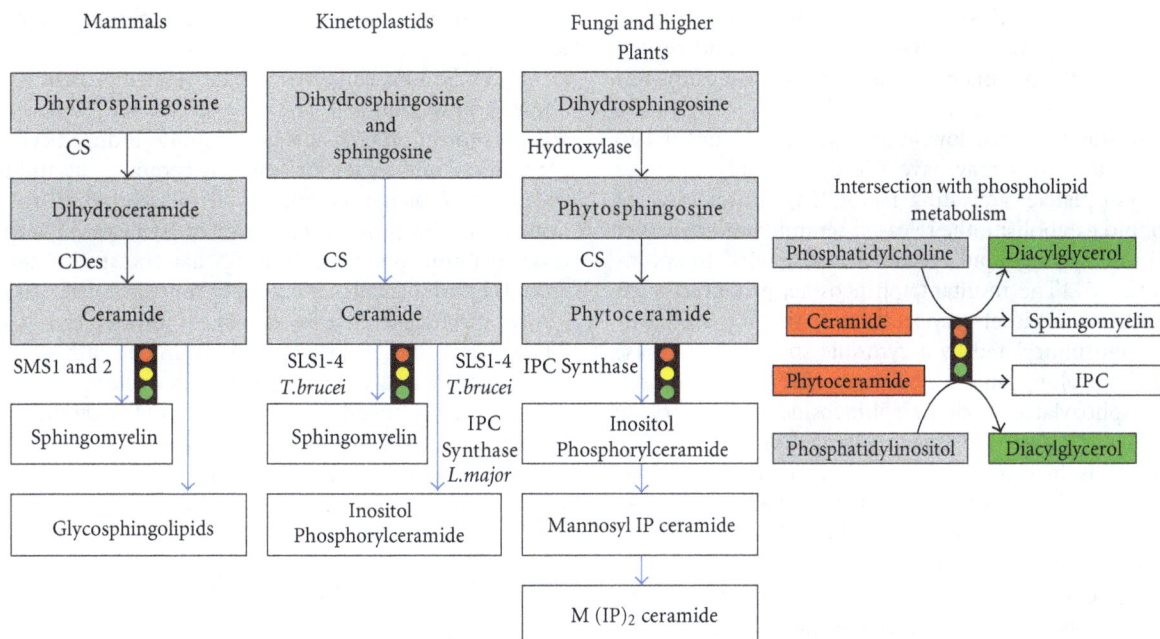

FIGURE 3: Divergence in the postceramide biosynthetic steps highlighting the conserved intersection with phospholipid metabolism as a regulator of the balance between promitogenic diacylglycerol (DAG) and proapoptotic ceramide. The key regulatory synthetic steps are indicated by the "traffic light" symbols. CS: Ceramide Synthase; CDes: Ceramide Desaturase; SMS: Sphingomyelin Synthase; SL: Sphingolipid; IPC: Inositol Phosphorylceramide; PC: Phosphatidylcholine.

complex sphingolipid reservoirs, for example glycosphingolipids (GSLs) and (SMs), through the action of specific hydrolases and phosphodiesterases.

Sphingolipid recycling can be categorised as either lysosomal or nonlysosomal degradation. In lysosomal degradation, catabolism of GSLs occurs as sugar residues are cleaved leading to the formation of glucosylceramide and galactosylceramide. In turn, specific β-glucosidases and galactosidases hydrolyse these lipids to form ceramide which can then be subsequently deacylated by an acid ceramidases to form sphingosine [12, 45], which can then be salvaged to form ceramide by reacylation. Defects in the function of these enzymes lead to a variety of lysosomal storage disorders such as Gaucher, Sandhoff, and Tay-Sachs diseases, resulting from the impairment of membrane degradation [46]. This will be discussed in more detail later.

Degradation of sphingolipids is a necessary part of maintaining lipid homeostasis, thus SM levels are maintained by catabolic action of sphingomyelinases (SMases), either neutral or acidic, releasing ceramide and the corresponding headgroup, phosphorylcholine in the case of SM. Acid sphingomyelinase (aSMase) was the first cloned human SMase and was initially assumed to be not much more than a housekeeping gene with a prominent role in the turnover of sphingomyelin in the lysosome. However, unusually aSMase has revealed itself to encode two unique enzymes through the differential trafficking of a single-protein precursor [47]. In addition to the commonly studied lysosomal aSMase, an alternative form is secreted extracellularly and may have a role in the nonlysosomal hydrolysis of SM both in the outer leaflet of the plasma membrane and in lipoproteins in the bloodstream [48]. These studies indicate that ceramide production by sphingolipid hydrolysis in different cellular or extracellular locations may provide different metabolic effects and biological impacts. Indeed, the nonlysosomal degradation of SM is catalysed by neutral and alkaline SMases in a variety of intracellular and extracellular locations. The least studied of these SMases, the alkaline SMase (Alk-SMase), is highly tissue specific with trypsin resistance and bile salt dependency and has a key role in the dietary acquisition of ceramide by digesting SM in the gut [49]. Notably, animal studies have shown Alk-SMase is specifically down regulated in colon cancer, while membrane SM accumulates. The supplementation of dietary SM can prevent the promotion of further colonic tumors [50].

SMases are commonly activated by growth factors, cytokines, chemotherapeutic agents, irradiation and nutrient removal [51], and though they can differ in their subcellular localisation and tissue specificity, all are thought to regulate the local ceramide concentration and any corresponding stress-induced responses [52]. Ceramide and associated metabolites, such as sphingosine-1-phosphate, are known to function as second messengers, stimulating various biological activities in mammalian cells, including the activation of protein-kinases and/or protein-phosphatases 2A [53]. Increased levels of ceramides can exert antiproliferation effects, induce apoptosis, and play major roles in mitogenesis and endocytosis. There is a growing body of evidence that suggests Mg^{2+}-dependent neutral sphingomyelinases (nSMases) are the major source for stress-induced ceramide production [51]. nSMases are ubiquitously expressed in mammalian cells, predominately membrane bound on the outer leaflet of the plasma membrane where most of the SM is located [52]. Other mammalian nSMases localise to the ER, where the predicted low abundance of SM has led to speculation that they may have additional lipid substrates such as lyso-platelet-activating factor [54]. In all cases of sphingolipid catabolism, the released ceramide can be either recycled into sphingolipid synthesis or degraded to sphingosine [12, 45]. The resultant sphingosine, produced from either pathway, is either recycled into sphingolipid biosynthesis or phosphorylated by a cytosolic sphingosine kinase (SK) yielding sphingosine-1-phosphate (S1P). S1P can itself be dephosphorylated back to sphingosine or irreversibly degraded by S1P lyase into the nonsphingolipid species ethanolamine phosphate and hexadecenal, representing a unique exit point from the sphingolipid metabolic pathway [12, 45]. In fact this is the mechanism by which the kinetoplastid *Leishmania* obtain ethanolamine [21].

Another kinetoplastid, *Trypanosoma brucei*, has shown that an ER nSMase directly involved in sphingolipid catabolism is essential because its formation of ceramide is required for post-Golgi sorting and deposition of the essential glycosylphosphatidylinositol-anchored variant surface glycoprotein on the cell surface [55]. Similarly, the *Leishmania* nSMase is essential for virulence and, whilst able to catabolise inositol phosphorylceramide (IPC), demonstrated greater activity with SM [56].

The corresponding yeast nSMase homologue (Isc1) is also capable of IPC catabolism, generating ceramide. During early growth Isc1p resides in the ER, but in late logarithmic growth it is found in the outer leaflet of the mitochondria, where the resulting ceramide formation plays a crucial role in the reprogramming of mitochondrial gene expression during the transition from anaerobic to aerobic metabolism, coupled with a change in carbon source, that is, glucose to ethanol [57, 58].

3. Ceramide Homeostasis

As discussed, ceramides are central intermediates of sphingolipid metabolism. In addition to forming the basis of complex sphingolipids, ceramide is a bioactive molecule that regulates a myriad of cellular pathways including apoptosis, cell senescence, the cell cycle, and differentiation [59]. In addition, this lipid species is involved in the cell response to stress challenge. Notably, several anticancer drugs, for example, etoposide and daunorubicin, have been found to function by elevating the level of cellular ceramide triggering apoptosis [60–63]. The apoptotic role of ceramide [64, 65] contrasts with that of the mitogenic agonist DAG. Whilst the former stimulates signal transduction pathways associated with cell death or growth inhibition, DAG activates the various isoforms of protein kinase C associated with cell growth and survival. Thus, ceramide and DAG generation may regulate cellular homeostasis by inducing death and growth, respectively. Given that ceramide and DAG are a substrate and a byproduct, respectively, of SM and IPC synthases, these

enzymes are hypothesized to play a central role in homeostasis [9].

Recently, a human SM synthase-related protein has been shown to function as a ceramide sensor [66] with a crucial role in protecting cells against ceramide-induced cell death. Disruption of this sensor leads to ceramide accumulation in the endoplasmic reticulum and mitochondrial-mediated apoptosis. This process is suppressed by targeting a ceramidase to mitochondria indicating that transfer of ceramide from the endoplasmic reticulum to mitochondria, via an unknown mechanism, is a key step in committing cells to death. The presence of a mitochondrial ceramide synthase has also been reported and hypothesized to play a role in this apoptotic process [67]. Together, these findings provide mechanistic evidence for the proapoptotic accumulation of ceramide in the mitochondria and demonstrate that the regulation of ceramide homeostasis is a vital cellular function.

4. Defects in Sphingolipid Metabolism

Despite sphingolipids being minor components in some cells, their accumulation in certain cells and tissues forms the basis of many human diseases.

4.1. Sphingolipidoses. Defects in sphingolipid catabolism, that is, lipid hydrolases, form the basis of a wide variety of human diseases. These diseases, collectively known as sphingolipidoses (Figure 4), belong to the lysosomal storage diseases and are inherited disorders characterised by accumulations in specific lipids in certain tissues and/or organs.

The most common is Gaucher disease, in which glucosylceramide accumulates due to a deficiency of glucosylceramide-β-glucosidase, causing changes in the specialised membrane microdomains termed lipid rafts. This in turn seems to impair lipid and protein sorting and consequently causes the pathology characterising this disorder. For example, lipid rafts are necessary for correct insulin signalling, and a perturbed lipid raft composition impairs insulin signalling leading to the insulin-resistance observed in patients with Gaucher disease [68, 69].

Fabry disease is an X-chromosomal-linked inherited deficiency of lysosomal α-galactosidase A, causing deposition of globotriaosylceramide in the lysosomes of endothelial, perithelial and smooth-muscle cells of blood vessels. This leads to renal, cardiac and/or cerebral complications and, most commonly, death before the age of 50 [70, 71].

Tay-Sachs disease or GM2-gangliosidosis comes in various forms, the most extreme, infantile form being caused by defects in β-hexosaminidase A and has a high heterozygote frequency (1 : 27) among Ashkenazi Jews. This condition leads to death between the second and fourth years of life [72, 73].

Sandhoff disease is characterized by storage of negatively charged glycolipids and elevation of uncharged glycolipids. This disease has various clinical forms, infantile, juvenile, and adult, all with varying pathological manifestations, including a chronic variant similar to Tay-Sachs disease [74, 75].

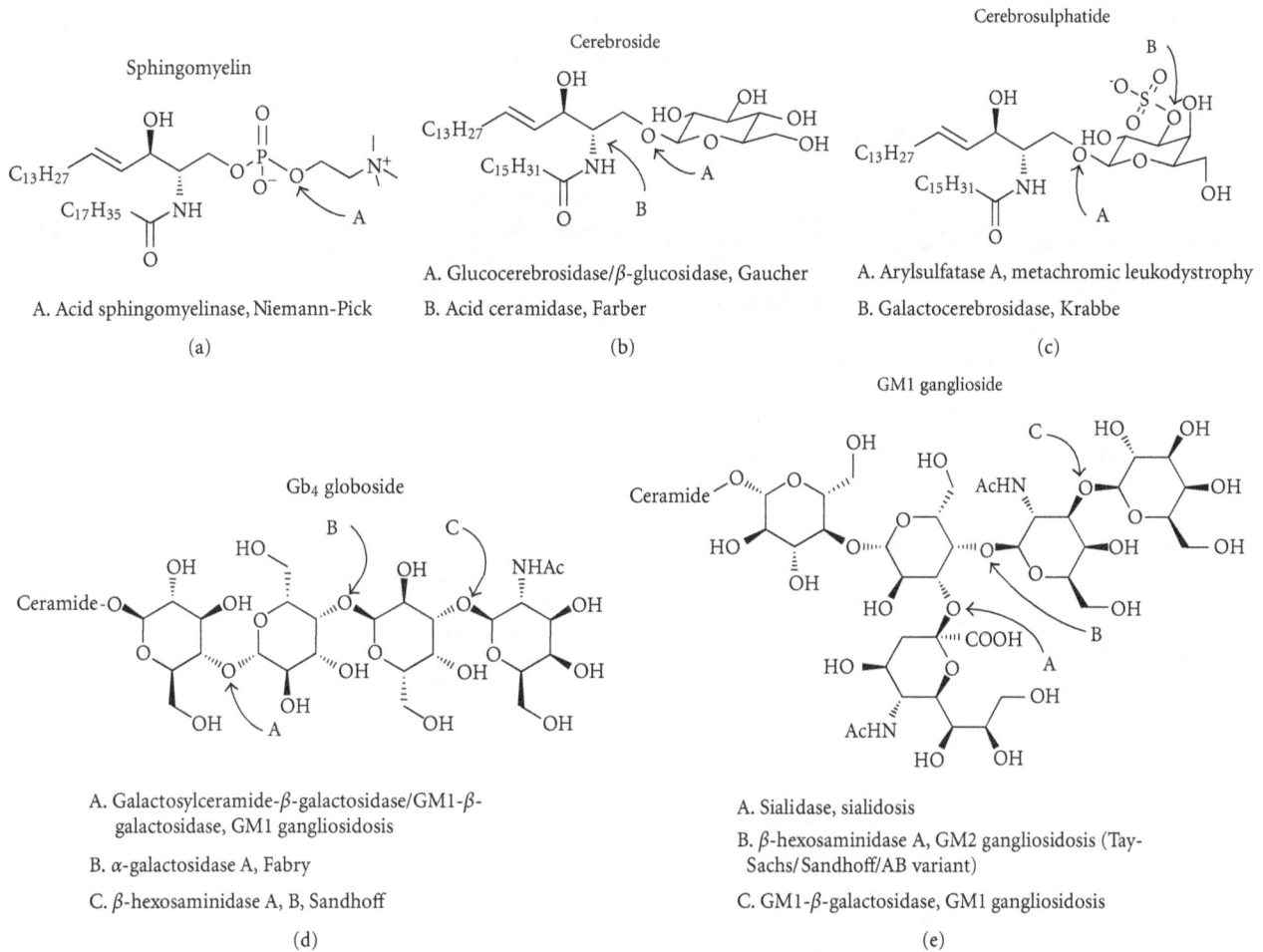

FIGURE 4: Defects in mammalian sphingolipid catabolism and their associated sphingolipidoses. The enzymes responsible are indicated at the glycosidic linkages on which they act.

The inherited Niemann-Pick disease (types A and B) is characterised by a deficiency in the lysosomal acidic SMase, causing an accumulation not only of sphingomyelin but also of glycosphingolipids, sphingosine and others in multilamellar storage bodies [76].

There are many other related and associated genetic diseases, Metachromatic Leukodystrophy caused by a deficiency of arylsulfatase A, Krabbe disease or globoid cell leukodystrophy caused by a deficiency of galactosylceramide-β-galactosidase and Farber disease caused by a deficiency of lysosomal acid ceramidase causing storage of excess ceramide in the lysosomes. For further details of these and other sphingolipid metabolic diseases, refer to an excellent review by Kolter and Sandhoff [77].

4.2. Hereditary Sensory Neuropathy.
Clinical disorders are also associated with alterations in SPT activity, although a complete lack of SPT activity is predicted to be embryonically lethal. The inherited disease hereditary sensory neuropathy type I (HSN1) is a progressive degeneration of lower limb sensory and autonomic neurons and has been associated with mutations in the human *LCB1* gene [78, 79]. LCB1

(SPTLC1) mutations confer dominant negative effects on SPT, thus substantially reducing SPT activity and hence the rate of *de novo* synthesis of sphingolipids [80]. Recently, point mutations have been found in the catalytic SPT subunit (LCB2 or SPTLC2) in patients suffering from HSN1. These were confirmed to affect SPT activity using an *in vitro* system. No mutations were observed in the third (SPTLC3) subunit in these patients [81].

Point mutations in the SPT complex also affect SPT activity in terms of substrate specificity and alanine can be used instead of serine in the condensation with palmitoyl-CoA, resulting in the formation and accumulation of 1-deoxy-sphinganine in the serum of these patients [82].

4.3. Alzheimer's Disease.
A relatively recent discovery was the highly altered sphingolipid metabolism in brain cells in Alzheimer's disease [83]. The latest study has highlighted the key role of an nSMase in the disease by promoting the damaging effects of fibrillar amyloid-β 1–42 peptide-activated astroglia through ceramide production and thus apoptosis in neuronal cells [84].

In addition, SPT has been shown to be downregulated by the amyloid precursor protein [85]. This novel physiological function of the amyloid precursor protein suggests that SPT and sphingolipid metabolism is involved in Alzheimer's disease pathology.

4.4. Other Diseases. Obesity and its established association with insulin resistance, type 2 diabetes, and cardiovascular disease are directly and/or indirectly involved in the overaccumulation of long-chain fatty acids. The resulting surplus to the storage capacity of adipose tissue results in deposition in nonadipose tissues, such as the liver, muscle, heart, and pancreatic islets. This leads to deleterious effects, not only as atherosclerosis, but excess lipids are also forced into alternative nonoxidative pathways resulting in the formation of reactive lipid moieties, such as sphingolipids, that promote metabolically relevant cellular dysfunction (lipotoxicity) and programmed cell death (lipoapoptosis) [86–88].

5. Sphingolipid Biosynthesis: An Attractive Drug Target

Due to the complexities of sphingolipid metabolism and associated defects in a variety of tissue types and cell compartments, there is a significant challenge in the understanding, diagnosis, and treatment of genetic diseases such as those discussed above. However, clinical manipulation sphingolipid metabolism will prove key in the treatment of these conditions; in addition it is becoming clear that by inducing the accumulation of proapoptotic ceramide, therapies for cancer and infectious disease may be developed.

5.1. Sphingolipidoses. The general strategy to treating the inherited human diseases involving sphingolipid metabolism, the sphingolipidoses is the restoration of the defective lysosomal degradation. These include enzyme replacement, heterologous bone marrow transplantation as a form of cell-mediated therapy, gene therapy and the use of chemical chaperones for enzyme-enhancement therapy. An alternative strategy is the reduction of substrate influx into the lysosomes using substrate reduction (substrate deprivation) therapy ([77] and references therein).

5.2. Cancer. The roles of ceramide in diverse cellular responses to stress, particularly apoptosis, has been discussed above and as mentioned elevated ceramide often result from treatment with anticancer drugs and also irradiation. Therefore the manipulation of sphingolipid biosynthesis and homeostasis to elevate ceramide levels and induce programmed cell death is a viable strategy for anticancer therapies [89]. Conversely, dysregulation of ceramide metabolism affects the cellular response to chemotherapy or other anticancer regimens by rendering the cells more resistant to killing; in these cases therapeutic manipulation of ceramide metabolism could overcome this resistance [90, 91]. Further developments in the manipulation of sphingolipid metabolism as an anticancer strategy will undoubtedly follow with the breakthrough discovery that FTY720, a water-soluble sphingosine

analogue effective in many cancer models which acts by downregulating nutrient transporter proteins in cancer cells at least partially via ceramide generation [92]. The resulting starvation induces a homeostatic autophagy selectively in cancer cells sensitive to nutrient limitation, while normal cells have the ability to adapt and survive by becoming quiescent. Notably AAL-149, an FTY720 analogue, similarly kills patient-derived leukaemic cells, but not cells of healthy donors, without the dose-limiting toxicity of FTY720. Thus, by targeting the sphingolipid metabolism of cancer cells rather than any specific oncogenic defect, such compounds should have potent activity against a range of tumours, particularly if applied in combination with inhibitors of autophagy.

5.3. Pathogens. The essential functions of sphingolipids, coupled with the divergence of the biosynthetic pathway between mammals and eukaryotic pathogens have resulted in the investigation of the biosynthetic enzymes as possible drug targets for antifungal and antiprotozoals. Consequently, inhibitors of many of the steps in sphingolipid biosynthesis have been described [2, 43]. Before the synthesis of ceramide/ phytoceramide the sphingolipid biosynthetic pathway is largely conserved across evolution. At least in part because of this, all the inhibitors identified as targeting fungal enzymes in this part of the biosynthetic pathway are nonselective and inhibit the mammalian orthologues [43]. This has curtailed their clinical application as anti-fungal agents, for example fumonisin B_1 which inhibits the fungal phytoceramide synthase demonstrated mammalian toxicity [93]. In contrast the post-ceramide divergence represented by the absence of IPC synthase and inositol-based sphingolipids in mammalian cells, highlights the therapeutic potential of inhibitors targeting fungal IPC synthases. Such inhibitors could result in selective antifungal drugs with minimal host toxicity. Additionally, the identification and isolation of functional orthologues of the fungal enzyme in the kinetoplastid protozoan parasites (*Leishmania* spp., *Trypanosoma brucei* and *Trypanosoma cruzi*) indicated that IPC synthase is a valid target for antiprotozoal compounds [3]. One of the *T. brucei* sphingolipid synthases, a novel bifunctional enzyme catalysing the synthesis of both IPC and SM, is essential for parasite growth and can be inhibited *in vitro* by the antifungal aureobasidin A at low nanomolar concentrations [43]. As the causative agent of Chagas' disease, *Trypanosoma cruzi* has a complex lifecycle with an essential intracellular stage in vertebrate hosts, in addition to an extracellular existence in an insect vector. Necessary to persistence of the lifecycle is the synthesis of surface glycosylphosphatidylinositol (GPI) anchored glycoconjugates, meaning that the biosynthesis of GPI anchors is attractive target for new therapies against Chagas' disease [93]. As many *T. cruzi* GPI anchors contain IPC as the lipid portion, the sole IPC synthase is highlighted as a new therapeutic target for Chagas disease.

To date only the natural compounds—aureobasidin A [33, 94, 95], Khafrefungin [94], and Rustmicin [96, 97] have been reported as potent inhibitors of the fungal IPC synthase. As discussed, there is an inhibitory effect of aureobasidin A against the Kinetoplastid enzyme orthologues [43, 98, 99], although the specificity of this remains unclear [3, 100, 101].

Unfortunately, further development of all the three known inhibitors of IPC synthase has stalled either due to lack of physical properties required for an acceptable pharmacokinetic profile [42, 102], or because their highly complex structures render chemical synthesis challenging. Moreover, the few synthetic efforts to modify or synthesise analogues that have been reported, resulted in compounds with either reduced or no activity [103, 104]. What has proven more successful however is the development of substrate (ceramide) analogues, with targeted inhibition against the protozoal IPC synthases *in cellulo* [105].

In addition to sphingolipid synthesis as a therapeutic target against parasitic protozoa, degradation of sphingolipids similarly is an area providing new opportunities for antiprotozoal compounds. *Leishmania* spp. use sphingolipid biosynthesis to generate ethanolamine (Etn), essential for the survival and differentiation from procyclics to virulent metacyclics [21]. A likely starting point for Etn production is the degradation of IPC, and a putative neutral SMase and/or IPC hydrolase (IPCase), designated *ISCL* was identified in the *L. major* genome [106]. ISCL showed much greater activity against non-self SM over IPC, suggesting a role in host SL degradation confirmed by *ISCL* null mutants failing to induce lesions in susceptible BALB/c mice. Further investigation revealed that host SL catabolism by *Leishmania* was essential to resist the harsh acidic environment in the phagolysosomes of macrophages [107]. These findings reveal that SL catabolism, as well as anabolism, by *Leishmania* is necessary for proliferation of the parasite in the mammalian host, making the ISCL enzyme an equally attractive target for inhibition studies. *Trypanosoma cruzi* invades mammalian cells by attaching and mimicking injury to the host plasma membrane, inducing a repair process that involves the Ca^{2+}-dependent exocytosis of lysosomes [108]. As host acid sphingomyelinase (aSMase) is delivered by lysosomes to the plasma membrane, its ceramide-generating activity promotes rapid endocytosis to internalise the seemingly damaged membrane and the attached parasites. Consequently any inhibition or reduction of this lysosomal aSMase blocks *T. cruzi* invasion, though subsequent treatment with an extracellular sphingomyelinase can restore the infection to normal levels. In a similar approach it has been demonstrated that inhibition of host cell SPT, and so sphingolipid biosynthesis, by myriocin, suppresses hepatitis C virus replication [109]. In addition, it has emerged that ceramide induces activation of double-stranded RNA-dependent protein kinase-mediated antiviral response [110]. These recent findings suggest that manipulation of host sphingolipid metabolism may provide a new combined therapeutic strategy for treatment of both protozoal and viral infections.

6. Perspective

Recent studies of the genetic, biochemical and cell biology of sphingolipids have provided exciting new insights into their function, regulation and control, allowing the consideration of future manipulations to aid the fight against human diseases including cancer and major fungal and parasitic infections. In addition, this information may ultimately aid

the treatment of several rare genetic disorders (e.g. the sphingolipidoses) and, perhaps, Alzheimer's disease. However, in most cases these studies are at an early stage and further work is required to establish proof of concept. This will undoubtedly be achieved as progress towards a fuller understanding of the complex and multilayered metabolic pathways of sphingolipid metabolism is realized, and as inhibitors of the enzymes involved become available.

Abbreviations

SM:	Sphingomyelin
IPC:	Inositol phosphorylceramide
MIPC:	Mannose-IPC
PI:	Phosphatidylinositol
DAG:	Diacylglycerol
CS:	Ceramide synthase
CDes:	Ceramide desaturase
SMS:	Sphingomyelin synthase
SL:	Sphingolipid
PC:	Phosphatidylcholine
SMases:	Sphingomyelinases
aSMase:	Acid sphingomyelinase
Alk-SMase:	Alkaline SMase
SPT:	Serine palmitoyltransferase
3-KSR:	3-Ketosphingosine reductase
CerS:	Ceramide synthase
CDase:	Ceramidase
CDes:	Ceramide desaturase
SMS:	Sphingomyelin synthase
SMase:	Sphingomyelinase
CK:	Ceramide kinase
GCS:	Glucosylceramide synthase
GCase:	Glycosidases
SK:	Sphingosine kinase
SPP:	Sphingosine-1-phosphate phosphatase
S1P:	Sphingosine-1-phosphate
HSN1:	Hereditary sensory neuropathy type I.

Author Contribution

Both Simon A. Young and John G. Mina contributed equally to this paper.

Acknowledgments

Research in TKS's laboratory is supported in part by Wellcome Trust project Grants (086658 and 093228) and studentships from the Biotechnology and Biological Sciences Research Council and SUSLA. Research in PWD's laboratory is supported in part by Wellcome Trust project Grant (094759), a grant from the Open Lab Foundation, studentships from the Engineering and Physical Sciences and Medical Research Councils and the Wolfson Research Institute.

References

[1] D. E. Metzler, *Biochemistry; The Chemical Reactions of Living Cells*, vol. 1-2, Elsevier, 2nd edition, 2003.

[2] A. H. Merrill and K. Sandhoff, "Sphingolipids: metabolism and cell signalling," in *Biochemistry of Lipids, Lipoproteins and Membranes*, D. E. Vance and J. E. Vance, Eds., pp. 373–407, Elsevier, Amsterdam, The Netherlands, 2002.

[3] P. W. Denny, H. Shams-Eldin, H. P. Price, D. F. Smith, and R. T. Schwarz, "The protozoan inositol phosphorylceramide synthase: a novel drug target that defines a new class of sphingolipid synthase," *Journal of Biological Chemistry*, vol. 281, no. 38, pp. 28200–28209, 2006.

[4] M. I. Gurr, J. L. Harwood, and K. N. Frayn, *Lipid Biochemistry: An Introduction*, Blackwell Science, 5th edition, 2002.

[5] G. Van Meer and Q. Lisman, "Sphingolipid transport: rafts and translocators," *Journal of Biological Chemistry*, vol. 277, no. 29, pp. 25855–25858, 2002.

[6] A. Huwiler, T. Kolter, J. Pfeilschifter, and K. Sandhoff, "Physiology and pathophysiology of sphingolipid metabolism and signaling," *Biochimica et Biophysica Acta*, vol. 1485, no. 2-3, pp. 63–99, 2000.

[7] J. Ohanian and V. Ohanian, "Sphingolipids in mammalian cell signaling," *Cellular and Molecular Life Sciences*, vol. 58, no. 14, pp. 2053–2068, 2001.

[8] O. Cuvillier, "Sphingosine in apoptosis signaling," *Biochimica et Biophysica Acta*, vol. 1585, no. 2-3, pp. 153–162, 2002.

[9] B. J. Pettus, C. E. Chalfant, and Y. A. Hannun, "Ceramide in apoptosis: an overview and current perspectives," *Biochimica et Biophysica Acta*, vol. 1585, no. 2-3, pp. 114–125, 2002.

[10] R. Buccoliero and A. H. Futerman, "The roles of ceramide and complex sphingolipids in neuronal cell function," *Pharmacological Research*, vol. 47, no. 5, pp. 409–419, 2003.

[11] Y. A. Hannun, C. R. Loomis, A. H. Merrill, and R. M. Bell, "Sphingosine inhibition of protein kinase C activity and of phorbol dibutyrate binding in vitro and in human platelets," *Journal of Biological Chemistry*, vol. 261, no. 27, pp. 12604–12609, 1986.

[12] Y. A. Hannun and L. M. Obeid, "Principles of bioactive lipid signalling: lessons from sphingolipids," *Nature Reviews Molecular Cell Biology*, vol. 9, no. 2, pp. 139–150, 2008.

[13] S. T. Pruett, A. Bushnev, K. Hagedorn et al., "Thematic review series: sphingolipids—biodiversity of sphingoid bases ("sphingosines") and related amino alcohols," *Journal of Lipid Research*, vol. 49, no. 8, pp. 1621–1639, 2008.

[14] E. Fahy, S. Subramaniam, H. A. Brown et al., "A comprehensive classification system for lipids," *Journal of Lipid Research*, vol. 46, no. 5, pp. 839–861, 2005.

[15] E. Fahy, S. Subramaniam, R. C. Murphy et al., "Update of the LIPID MAPS comprehensive classification system for lipids," *Journal of Lipid Research*, vol. 50, supplement, pp. S9–S14, 2009.

[16] R. M. Bell, A. H. Merrill Jr., Y. A. Hannun et al., *Advances in Lipid Research: Sphingolipids, Part A: Functions and Breakdown Products*, vol. 25, 1993.

[17] R. M. Bell, A. H. Merrill Jr., Y. A. Hannun, and R. M. Bell, "Introduction: sphingolipids and their metabolites in cell regulation," *Advances in lipid research*, vol. 25, pp. 1–24, 1993.

[18] A. H. Merrill, E. M. Schmelz, D. L. Dillehay et al., "Sphingolipids—the enigmatic lipid class: biochemistry, physiology and pathophysiology," *Toxicology and Applied Pharmacology*, vol. 142, no. 1, pp. 208–225, 1997.

[19] M. O. Pata, Y. A. Hannun, and C. K.-Y. Ng, "Plant sphingolipids: decoding the enigma of the Sphinx," *New Phytologist*, vol. 185, no. 3, pp. 611–630, 2010.

[20] N. Bartke and Y. A. Hannun, "Bioactive sphingolipids: metabolism and function," *Journal of Lipid Research*, vol. 50, supplement, pp. S91–S96, 2009.

[21] K. Zhang, F. F. Hsu, D. A. Scott, R. Docampo, J. Turk, and S. M. Beverley, "Leishmania salvage and remodelling of host sphingolipids in amastigote survival and acidocalcisome biogenesis," *Molecular Microbiology*, vol. 55, no. 5, pp. 1566–1578, 2005.

[22] B. Ségui, N. Andrieu-Abadie, J. P. Jaffrézou, H. Benoist, and T. Levade, "Sphingolipids as modulators of cancer cell death: potential therapeutic targets," *Biochimica et Biophysica Acta*, vol. 1758, no. 12, pp. 2104–2120, 2006.

[23] K. Hanada, "Serine palmitoyltransferase, a key enzyme of sphingolipid metabolism," *Biochimica et Biophysica Acta*, vol. 1632, no. 1—3, pp. 16–30, 2003.

[24] P. W. Denny, D. Goulding, M. A. J. Ferguson, and D. F. Smith, "Sphingolipid-free Leishmania are defective in membrane trafficking, differentiation and infectivity," *Molecular Microbiology*, vol. 52, no. 2, pp. 313–327, 2004.

[25] M. Chen, G. Han, C. R. Dietrich, T. M. Dunn, and E. B. Cahoon, "The essential nature of sphingolipids in plants as revealed by the functional identification and characterization of the Arabidopsis LCB1 subunit of serine palmitoyltransferase," *Plant Cell*, vol. 18, no. 12, pp. 3576–3593, 2006.

[26] H. Ikushiro, H. Hayashi, and H. Kagamiyama, "Bacterial serine palmitoyltransferase: a water-soluble homodimeric prototype of the eukaryotic enzyme," *Biochimica et Biophysica Acta*, vol. 1647, no. 1-2, pp. 116–120, 2003.

[27] M. C. Raman, K. A. Johnson, D. J. Clarke, J. H. Naismith, and D. J. Campopiano, "The serine palmitoyltransferase from sphingomonas wittichii RW1: an interesting link to an unusual acyl carrier protein," *Biopolymers*, vol. 93, no. 9, pp. 811–822, 2010.

[28] K. Gable, H. Slife, D. Bacikova, E. Monaghan, and T. M. Dunn, "Tsc3p is an 80-amino acid protein associated with serine palmitoyltransferase and required for optimal enzyme activity," *Journal of Biological Chemistry*, vol. 275, no. 11, pp. 7597–7603, 2000.

[29] G. Han, S. D. Gupta, K. Gable et al., "Identification of small subunits of mammalian serine palmitoyltransferase that confer distinct acyl-CoA substrate specificities," *Proceedings of the National Academy of Sciences of the United States of America*, vol. 106, no. 20, pp. 8186–8191, 2009.

[30] T. Hornemann, A. Penno, M. F. Rütti et al., "The SPTLC3 subunit of serine palmitoyltransferase generates short chain sphingoid bases," *Journal of Biological Chemistry*, vol. 284, no. 39, pp. 26322–26330, 2009.

[31] Y. Sugimoto, H. Sakoh, and K. Yamada, "IPC synthase as a useful target for antifungal drugs," *Current Drug Targets*, vol. 4, no. 4, pp. 311–322, 2004.

[32] P. E. Bromley, Y. O. Li, S. M. Murphy, C. M. Sumner, and D. V. Lynch, "Complex sphingolipid synthesis in plants: characterization of inositolphosphorylceramide synthase activity in bean microsomes," *Archives of Biochemistry and Biophysics*, vol. 417, no. 2, pp. 219–226, 2003.

[33] P. Rovina, A. Schanzer, C. Graf, D. Mechtcheriakova, M. Jaritz, and F. Bornancin, "Subcellular localization of ceramide kinase and ceramide kinase-like protein requires interplay of their Pleckstrin Homology domain-containing N-terminal regions together with C-terminal domains," *Biochimica et Biophysica Acta*, vol. 1791, no. 10, pp. 1023–1030, 2009.

[34] A. Raas-Rothschild, I. Pankova-Kholmyansky, Y. Kacher, and A. H. Futerman, "Glycosphingolipidoses: beyond the enzymatic defect," *Glycoconjugate Journal*, vol. 21, no. 6, pp. 295–304, 2004.

[35] K. Huitema, J. Van Den Dikkenberg, J. F. H. M. Brouwers, and J. C. M. Holthuis, "Identification of a family of animal

sphingomyelin synthases," *EMBO Journal*, vol. 23, no. 1, pp. 33–44, 2004.

[36] M. M. Nagiec, E. E. Nagiec, J. A. Baltisberger, G. B. Wells, R. L. Lester, and R. C. Dickson, "Sphingolipid synthesis as a target for antifungal drugs," *Journal of Biological Chemistry*, vol. 272, no. 15, pp. 9809–9817, 1997.

[37] W. Wang, X. Yang, S. Tangchaiburana et al., "An inositolphosphorylceramide synthase is involved in regulation of plant programmed cell death associated with defense in arabidopsis," *Plant Cell*, vol. 20, no. 11, pp. 3163–3179, 2008.

[38] R. C. Dickson and R. L. Lester, "Yeast sphingolipids," *Biochimica et Biophysica Acta*, vol. 1426, no. 2, pp. 347–357, 1999.

[39] R. L. Lester and R. C. Dickson, "Sphingolipids with inositolphosphate-containing head groups," *Advances in Lipid Research*, vol. 26, pp. 253–274, 1993.

[40] J. G. Mina, J. A. Mosely, H. Z. Ali et al., "A plate-based assay system for analyses and screening of the Leishmania major inositol phosphorylceramide synthase," *International Journal of Biochemistry and Cell Biology*, vol. 42, no. 9, pp. 1553–1561, 2010.

[41] J. C. M. Holthuis, F. G. Tafesse, and P. Ternes, "The multigenic sphingomyelin synthase family," *Journal of Biological Chemistry*, vol. 281, no. 40, pp. 29421–29425, 2006.

[42] N. H. Georgopapadakou, "Antifungals targeted to sphingolipid synthesis: focus on inositol phosphorylceramide synthase," *Expert Opinion on Investigational Drugs*, vol. 9, no. 8, pp. 1787–1796, 2000.

[43] J. G. Mina, S. Y. Pan, N. K. Wansadhipathi et al., "The Trypanosoma brucei sphingolipid synthase, an essential enzyme and drug target," *Molecular and Biochemical Parasitology*, vol. 168, no. 1, pp. 16–23, 2009.

[44] S. S. Sutterwala, F. F. Hsu, E. S. Sevova et al., "Developmentally regulated sphingolipid synthesis in African trypanosomes," *Molecular Microbiology*, vol. 70, no. 2, pp. 281–296, 2008.

[45] T. Kolter and K. Sandhoff, "Sphingolipids—their metabolic pathways and the pathobiochemistry of neurodegenerative diseases," *Angewandte Chemie*, vol. 38, no. 11, pp. 1532–1568, 1999.

[46] H. Schulze and K. Sandhoff, "Lysosomal lipid storage diseases," *Cold Spring Harbor Perspectives in Biology*, vol. 3, no. 6, 2011.

[47] S. L. Schissel, E. H. Schuchman, K. J. Williams, and I. Tabas, "Zn^{2+}-stimulated sphingomyelinase is secreted by many cell types and is a product of the acid sphingomyelinase gene," *Journal of Biological Chemistry*, vol. 271, no. 31, pp. 18431–18436, 1996.

[48] R. W. Jenkins, D. Canals, and Y. A. Hannun, "Roles and regulation of secretory and lysosomal acid sphingomyelinase," *Cellular Signalling*, vol. 21, no. 6, pp. 836–846, 2009.

[49] R. D. Duan, T. Bergman, N. Xu et al., "Identification of human intestinal alkaline sphingomyelinase as a novel ectoenzyme related to the nucleotide phosphodiesterase family," *Journal of Biological Chemistry*, vol. 278, no. 40, pp. 38528–38536, 2003.

[50] Y. Zhang, Y. Cheng, G. H. Hansen et al., "Crucial role of alkaline sphingomyelinase in sphingomyelin digestion: a study on enzyme knockout mice," *Journal of Lipid Research*, vol. 52, no. 4, pp. 771–781, 2011.

[51] T. Yabu, S. Imamura, M. Yamashita, and T. Okazaki, "Identification of Mg^{2+}-dependent neutral sphingomyelinase 1 as a mediator of heat stress-induced ceramide generation and apoptosis," *Journal of Biological Chemistry*, vol. 283, no. 44, pp. 29971–29982, 2008.

[52] S. Tomiuk, K. Hofmann, M. Nix, M. Zumbansen, and W. Stoffel, "Cloned mammalian neutral sphingomyelinase: functions in sphingolipid signaling?" *Proceedings of the National Academy of Sciences of the United States of America*, vol. 95, no. 7, pp. 3638–3643, 1998.

[53] A. Huwiler, T. Kolter, J. Pfeilschifter, and K. Sandhoff, "Physiology and pathophysiology of sphingolipid metabolism and signaling," *Biochimica et Biophysica Acta*, vol. 1485, no. 2-3, pp. 63–99, 2000.

[54] H. Sawai, Y. Okamoto, C. Luberto et al., "Identification of ISC1 (YERO19w) as inositol phosphosphingolipid phospholipase C in Saccharomyces cerevisiae," *Journal of Biological Chemistry*, vol. 275, no. 50, pp. 39793–39798, 2000.

[55] S. A. Young and T. K. Smith, "The essential neutral sphingomyelinase is involved in the trafficking of the variant surface glycoprotein in the bloodstream form of Trypanosoma brucei," *Molecular Microbiology*, vol. 76, no. 6, pp. 1461–1482, 2010.

[56] O. Zhang, M. C. Wilson, W. Xu et al., "Degradation of host sphingomyelin is essential for Leishmania virulence," *Plos Pathogens*, vol. 5, no. 12, Article ID e1000692, 2009.

[57] S. Vaena de Avalos, Y. Okamoto, and Y. A. Hannun, "Activiation and localization of inositolphosphosphingolipid phospholipase C Isc1p, to the mitochondria during growth of Saccharomyces cerevisiae," *The Journal of Biological Chemistry*, vol. 279, no. 12, pp. 11507–11545, 1998.

[58] H. Kitagaki, L. A. Cowart, N. Matmati et al., "ISC1-dependent metabolic adaptation reveals an indispensable role for mitochondria in induction of nuclear genes during the diauxic shift in Saccharomyces cerevisiae," *Journal of Biological Chemistry*, vol. 284, no. 16, pp. 10818–10830, 2009.

[59] J. C. M. Holthuis, T. Pomorski, R. J. Raggers, H. Sprong, and G. Van Meer, "The organizing potential of sphingolipids in intracellular membrane transport," *Physiological Reviews*, vol. 81, no. 4, pp. 1689–1723, 2001.

[60] R. Bose, M. Verheij, A. Halmovitz-Friedman, K. Scotto, Z. Fuks, and R. Kolesnick, "Ceramide synthase mediates daunorubicin-induced apoptosis: an alternative mechanism for generating death signals," *Cell*, vol. 82, no. 3, pp. 405–414, 1995.

[61] C. E. Chalfant, K. Rathman, R. L. Pinkerman et al., "De novo ceramide regulates the alternative splicing of caspase 9 and Bcl-x in A549 lung adenocarcinoma cells. Dependence on protein phosphatase-1," *Journal of Biological Chemistry*, vol. 277, no. 15, pp. 12587–12595, 2002.

[62] D. K. Perry, J. Carton, A. K. Shah, F. Meredith, D. J. Uhlinger, and Y. A. Hannun, "Serine palmitoyltransferase regulates de novo ceramide generation during etoposide-induced apoptosis," *Journal of Biological Chemistry*, vol. 275, no. 12, pp. 9078–9084, 2000.

[63] H. T. Wang, B. J. Maurer, C. P. Patrick, and M. C. Cabot, "N-(4-hydroxyphenyl)retinamide elevates ceramide in neuroblastoma cell lines by coordinate activation of serine palmitoyltransferase and ceramide synthase," *Cancer Research*, vol. 61, no. 13, pp. 5102–5105, 2001.

[64] J. D. Fishbein, R. T. Dobrowsky, A. Bielawska, S. Garrett, and Y. A. Hannun, "Ceramide-mediated growth inhibition and CAPP are conserved in Saccharomyces cerevisiae," *Journal of Biological Chemistry*, vol. 268, no. 13, pp. 9255–9261, 1993.

[65] J. Quintans, J. Kilkus, C. L. McShan, A. R. Gottschalk, and G. Dawson, "Ceramide mediates the apoptotic response of WEHI 231 cells to anti- immunoglobulin, corticosteroids and irradiation," *Biochemical and Biophysical Research Communications*, vol. 202, no. 2, pp. 710–714, 1994.

[66] A. M. Vacaru, F. G. Tafesse, P. Ternes et al., "Sphingomyelin synthase-related protein SMSr controls ceramide homeostasis in the ER," *Journal of Cell Biology*, vol. 185, no. 6, pp. 1013–1027, 2009.

[67] S. A. Novgorodov, D. A. Chudakova, B. W. Wheeler et al., "Developmentally regulated ceramide synthase 6 increases mitochondrial Ca^{2+} loading capacity and promotes apoptosis," *Journal of Biological Chemistry*, vol. 286, no. 6, pp. 4644–4658, 2011.

[68] L. A. Boven, M. van Meurs, R. G. Boot et al., "Gaucher cells demonstrate a distinct macrophage phenotype and resemble alternatively activated macrophages," *American Journal of Clinical Pathology*, vol. 122, no. 3, pp. 359–369, 2004.

[69] A. H. Futerman, J. L. Sussman, M. Horowitz, I. Silman, and A. Zimran, "New directions in the treatment of Gaucher disease," *Trends in Pharmacological Sciences*, vol. 25, no. 3, pp. 147–151, 2004.

[70] P. Ashton-Prolla, B. Tong, J. Shabbeer, K. H. Astrin, C. M. Eng, and R. J. Desnick, "Fabry disease: twenty-two novel mutations in the α-galactosidase a gene and genotype/phenotype correlations in severely and mildly affected hemizygotes and heterozygotes," *Journal of Investigative Medicine*, vol. 48, no. 4, pp. 227–235, 2000.

[71] Y. Kacher and A. H. Futerman, "Genetic diseases of sphingolipid metabolism: pathological mechanisms and therapeutic options," *FEBS Letters*, vol. 580, no. 23, pp. 5510–5517, 2006.

[72] R. Myerowitz, D. Lawson, H. Mizukami, Y. Mi, C. J. Tifft, and R. L. Proia, "Molecular pathophysiology in Tay-Sachs and Sandhoff diseases as revealed by gene expression profiling," *Human Molecular Genetics*, vol. 11, no. 11, pp. 1343–1350, 2002.

[73] G. M. Petersen, J. I. Rotter, and R. M. Cantor, "The Tay-Sachs disease gene in North American Jewish populations: geographic variations and origin," *American Journal of Human Genetics*, vol. 35, no. 6, pp. 1258–1269, 1983.

[74] T. Yamashita, Y. P. Wu, R. Sandhoff et al., "Interruption of ganglioside synthesis produces central nervous system degeneration and altered axon-glial interactions," *Proceedings of the National Academy of Sciences of the United States of America*, vol. 102, no. 8, pp. 2725–2730, 2005.

[75] T. Yamashita, R. Wada, T. Sasaki et al., "A vital role for glycosphingolipid synthesis during development and differentiation," *Proceedings of the National Academy of Sciences of the United States of America*, vol. 96, no. 16, pp. 9142–9147, 1999.

[76] R. O. Brady, J. N. Kanfer, M. B. Mock, and D. S. Fredrickson, "The metabolism of sphingomyelin. II. Evidence of an enzymatic deficiency in Niemann-Pick diseae," *Proceedings of the National Academy of Sciences of the United States of America*, vol. 55, no. 2, pp. 366–369, 1966.

[77] T. Kolter and K. Sandhoff, "Sphingolipid metabolism diseases," *Biochimica et Biophysica Acta*, vol. 1758, no. 12, pp. 2057–2079, 2006.

[78] J. L. Dawkins, D. J. Hulme, S. B. Brahmbhatt, M. Auer-Grumbach, and G. A. Nicholson, "Mutations in SPTLC1, encoding serine palmitoyltransferase, long chain base subunit-1, cause hereditary sensory neuropathy type I," *Nature Genetics*, vol. 27, no. 3, pp. 309–312, 2001.

[79] K. Bejaoui, C. Wu, M. D. Scheffler et al., "SPTLC1 is mutated in hereditary sensory neuropathy, type 1," *Nature Genetics*, vol. 27, no. 3, pp. 261–262, 2001.

[80] K. Bejaoui, Y. Uchida, S. Yasuda et al., "Hereditary sensory neuropathy type 1 mutations confer dominant negative effects on serine palmitoyltransferase, critical for sphingolipid

[81] A. Rothier et al., "Mutations in the SPTLC2 subunit of serine pamitoyltransferase cause heredity sensory and automatic neuropathy type 1," *American Journal of Human Genetics*, vol. 87, pp. 513–522, 2010.

[82] K. Gable, S. D. Gupta, G. Han, S. Niranjanakumari, J. M. Harmon, and T. M. Dunn, "A disease-causing mutation in the active site of serine palmitoyltransferase causes catalytic promiscuity," *Journal of Biological Chemistry*, vol. 285, no. 30, pp. 22846–22852, 2010.

[83] N. J. Haughey, V. V. Bandaru, M. Bae, and M. P. Mattson, "Roles for dysfunctional sphingolipid metabolism in Alzheimer's disease neuropathogenesis," *Biochimica et Biophysica Acta*, vol. 1801, no. 8, pp. 878–886, 2010.

[84] A. Jana and K. Pahan, "Fibrillar amyloid-β-activated human astroglia kill primary human neurons via neutral sphingomyelinase: implications for Alzheimer's disease," *Journal of Neuroscience*, vol. 30, no. 38, pp. 12676–12689, 2010.

[85] M. O. W. Grimm, S. Grsgen, T. L. Rothhaar et al., "Intracellular APP domain regulates serine-palmitoyl-CoA transferase expression and is affected in alzheimer's disease," *International Journal of Alzheimer's Disease*, vol. 2011, Article ID 695413, 2011.

[86] C. M. Kusminski, S. Shetty, L. Orci, R. H. Unger, and P. E. Scherer, "Diabetes and apoptosis: lipotoxicity," *Apoptosis*, vol. 14, no. 12, pp. 1484–1495, 2009.

[87] X. C. Jiang, C. Yeang, Z. Li et al., "Sphingomyelin biosynthesis: its impact on lipid metabolism and atherosclerosis," *Clinical Lipidology*, vol. 4, no. 5, pp. 595–609, 2009.

[88] M. Fuller, "Sphingolipids: the nexus between Gaucher disease and insulin resistance," *Lipids in Health and Disease*, vol. 9, article 113, 2010.

[89] W. C. Huang, C.-L. Chen, Y.-S. Lin et al., "Apoptotic sphingolipid ceramide in cancer therapy," *Journal of Lipids*, vol. 2011, Article ID 565316, 15 pages, 2011.

[90] P. P. Ruvolo, "Ceramide regulates cellular homeostasis via diverse stress signaling pathways," *Leukemia*, vol. 15, no. 8, pp. 1153–1160, 2001.

[91] A.-M. Domijan and A. Y. Abramov, "Fumonisin B1 inhibits mitochondrial respiration and deregulates calcium homeostasis—implication to mechanism of cell toxicity," *International Journal of Biochemistry and Cell Biology*, vol. 43, no. 6, pp. 897–904, 2011.

[92] K. Romero Rosales, G. Singh, K. Wu et al., "Sphingolipid-based drugs selectively kill cancer cells by down-regulating nutrient transporter proteins," *Biochemical Journal*, vol. 439, no. 2, pp. 299–311, 2011.

[93] C. M. Koeller and N. Heise, "The sphingolipid biosynthetic pathway is a potential target for chemotherapy against chagas disease," *Enzyme Research*, vol. 2011, Article ID 648159, 13 pages, 2011.

[94] N. Awazu, K. Ikai, J. Yamamoto et al., "Structures and antifungal activities of new aureobasidins," *Journal of Antibiotics*, vol. 48, no. 6, pp. 525–527, 1995.

[95] K. Ikai, K. Takesako, K. Shiomi et al., "Structure of aureobasidin A," *Journal of Antibiotics*, vol. 44, no. 9, pp. 925–933, 1991.

[96] S. M. Mandala, R. A. Thornton, M. Rosenbach et al., "Khafrefungin, a novel inhibitor of sphingolipid synthesis," *Journal of Biological Chemistry*, vol. 272, no. 51, pp. 32709–32714, 1997.

[97] S. M. Mandala, R. A. Thornton, J. Milligan et al., "Rustmicin, a potent antifungal agent, inhibits sphingolipid synthesis

at inositol phosphoceramide synthase," *Journal of Biological Chemistry*, vol. 273, no. 24, pp. 14942–14949, 1998.

[98] G. H. Harris, A. Shafiee, M. A. Cabello et al., "Inhibition of fungal sphingolipid biosynthesis by rustmicin, galbonolide B and their new 21-hydroxy analogs," *Journal of Antibiotics*, vol. 51, no. 9, pp. 837–844, 1998.

[99] M. L. Salto, L. E. Bertello, M. Vieira, R. Docampo, S. N. J. Moreno, and R. M. De Lederkremer, "Formation and remodeling of inositolphosphoceramide during differentiation of Trypanosoma cruzi from trypomastigote to amastigote," *Eukaryotic Cell*, vol. 2, no. 4, pp. 756–768, 2003.

[100] A. K. Tanaka, V. B. Valero, H. K. Takahashi, and A. H. Straus, "Inhibition of Leishmania (Leishmania) amazonensis growth and infectivity by aureobasidin A," *Journal of Antimicrobial Chemotherapy*, vol. 59, no. 3, pp. 487–492, 2007.

[101] J. M. Figueiredo, W. B. Dias, L. Mendonça-Previato, J. O. Previato, and N. Heise, "Characterization of the inositol phosphorylceramide synthase activity from Trypanosoma cruzi," *Biochemical Journal*, vol. 387, no. 2, pp. 519–529, 2005.

[102] E. S. Sevova, M. A. Goren, K. J. Schwartz et al., "Cell-free synthesis and functional characterization of sphingolipid synthases from parasitic trypanosomatid protozoa," *Journal of Biological Chemistry*, vol. 285, no. 27, pp. 20580–20587, 2010.

[103] Y. Sugimoto, H. Sakoh, and K. Yamada, "IPC synthase as a useful target for antifungal drugs," *Current Drug Targets*, vol. 4, no. 4, pp. 311–322, 2004.

[104] P. A. Aeed, C. L. Young, M. M. Nagiec, and Å. P. Elhammer, "Inhibition of inositol phosphorylceramide synthase by the cyclic peptide aureobasidin A," *Antimicrobial Agents and Chemotherapy*, vol. 53, no. 2, pp. 496–504, 2009.

[105] J. G. Mina, J. A. Mosely, H. Z. Ali, P. W. Denny, and P. G. Steel, "Exploring Leishmania major Inositol Phosphorylceramide Synthase (LmjIPCS): Insights into the ceramide binding domain," *Organic and Biomolecular Chemistry*, vol. 9, no. 6, pp. 1823–1830, 2011.

[106] O. Zhang, M. C. Wilson, W. Xu et al., "Degradation of host sphingomyelin is essential for Leishmania virulence," *Plos Pathogens*, vol. 5, no. 12, Article ID e1000692, 2009.

[107] W. Xu, L. Xin, L. Soong, and K. Zhang, "Sphingolipid degradation by Leishmania major is required for its resistance to acidic pH in the mammalian host," *Infection and Immunity*, vol. 79, no. 8, pp. 3377–3387, 2011.

[108] M. C. Fernandes, M. Cortez, A. R. Flannery, C. Tam, R. A. Mortara, and N. W. Andrews, "Trypanosoma cruzi subverts the sphingomyelinase-mediated plasma membrane repair pathway for cell invasion," *Journal of Experimental Medicine*, vol. 208, no. 5, pp. 909–921, 2011.

[109] K. Tatematsu, Y. Tanaka, M. Sugiyama, M. Sudoh, and M. Mizokami, "Host sphingolipid biosynthesis is a promising therapeutic target for the inhibition of hepatitis B virus replication," *Journal of Medical Virology*, vol. 83, no. 4, pp. 587–593, 2011.

[110] S. D. Spassieva, T. D. Mullen, D. M. Townsend, and L. M. Obeid, "Disruption of ceramide synthesis by CerS2 downregulation leads to autophagy and the unfolded protein response," *Biochemical Journal*, vol. 424, no. 2, pp. 273–283, 2009.

12

Behaviour of Human Erythrocyte Aggregation in Presence of Autologous Lipoproteins

C. Saldanha,[1] J. Loureiro,[2] C. Moreira,[3] and J. Martins e Silva[4]

[1] *Instituto de Medicina Molecular, Unidade de Biologia Microvascular e Inflamação,*
 Instituto de Bioquímica Faculdade de Medicina da Universidade de Lisboa. Av Prof. Egas Moniz, 1649-028 Lisboa, Portugal
[2] *Departmento de Cardiologia, Hospital Fernando da Fonseca. IC19, 2720-276 Amadora, Portugal*
[3] *Departmento de Medicina I, Faculdade de Medicina da Universidade de Lisboa. Av Prof. Egas Moniz, 1649-028 Lisboa, Portugal*
[4] *Instituto de Biopatologia Química Faculdade de Medicina da Universidade de Lisboa. Av Prof. Egas Moniz, 1649-028 Lisboa, Portugal*

Correspondence should be addressed to C. Saldanha, carlotasaldanha@fm.ul.pt

Academic Editor: Terry K. Smith

The aim of this work was to evaluate *in vitro* the effect of autologous plasma lipoprotein subfractions on erythrocyte tendency to aggregate. Aliquots of human blood samples were enriched or not (control) with their own HDL-C, LDL-C, or VLDL-C fractions obtained from the same batch by density gradient ultracentrifugation. Plasma osmolality and erythrocyte aggregation index (EAI) were determined. Blood aliquots enriched with LDL-C and HDL-C showed significant higher EAI than untreated aliquots, whereas enrichment with VLDL-C does not induce significant EAI changes. For the same range of lipoprotein concentrations expressed as percentage of osmolality variation, the EAI variation was positive and higher in presence of HDL-C than upon enrichment with LDL-C ($P < 0.01$). Particle size, up to LDL diameter values, seems to reinforce erythrocyte tendency to aggregate at the same plasma osmolality (particle number) range of values.

1. Introduction

There is scientific agreement that a high serum level of low-density lipoproteins cholesterol (LDL-C) is a risk factor for atherosclerosis and cardiovascular diseases [1–3]. A linear relationship between LDL-C levels and the occurrence of coronary artery disease is well documented in two meta-analysis [4, 5]. Conversely, it has been shown that HDL-C when at normal or high serum levels acts as a vascular protector and consequently without contribution such as a risk factor for atherosclerosis [6]. However, if its antioxidant capacity is diminished in patients with systolic heart failure, it will predict a higher risk of incident long term for adverse cardiac events [7].

Several clinical studies evidenced associations between complex lipid macromolecules; for example, high LDL-C concentrations and blood rheological behaviour, like blood hyperviscosity, that are both referred to as cardiovascular risk factors [8–10]. Blood viscosity is dependent on macro-(hematocrit and plasma viscosity) and micro-(erythrocyte deformability and aggregation) hemorheological parameters. Disturbances in blood rheological behaviour, such as high values of the blood and plasma viscosity and increased erythrocyte aggregation tendency, have been described in patients with ischemic heart diseases [11]. Red blood cells (RBCs) participate in acute coronary occlusion, mainly under conditions of lower shear rate, for example, within the microcirculation in the peri-infarct domain of myocardium [12].

Under *in vitro* stasis conditions, RBCs in normal human blood form loose aggregates with a characteristic morphology, similar to a stack of coins. Such aggregation is frequently named as rouleaux formation [13]. After prolonged stases, individual rouleaux can cluster, thereby forming three-dimensional structures, [14, 15]. Under circulation, the attractive forces involved are relatively weak, and aggregates can be dispersed during flow by the shear rate [16]. RBCs aggregation increasing at low shear rate affects blood viscosity and microvascular flow dynamics being markedly enhanced in several clinical states [17–21].

Factors influencing RBCs aggregation can be divided into (i) extrinsic factors such as levels of plasma proteins (e.g., fibrinogen, lipoproteins, macroglobulins, or immunoglobulins), hematocrit, and shear rate, and (ii) intrinsic factors, for example, RBCs shape, deformability and membrane surface properties [22–32]. RBC membrane surface properties and structure, such as surface charge and the ability of macromolecules to penetrate the membrane glycocalyx, greatly affect aggregation for cells suspended in a defined medium [33, 34]. Different studies have shown that hyperlipoproteinemia is associated with erythrocyte hyperaggregation [35–37]. The inverse correlation of erythrocyte aggregation with HDL2-C subfraction was reported in hypercholesterolemia middle-aged male population without apparent symptoms of cardiovascular disease [38]. It was evidenced *in vitro* that LDL-C enhances the RBCs aggregation induced by fibrinogen according to two aggregation models [39]. Considering the particle-like nature of the lipoproteins we raise the hypothesis that increased amounts of lipoprotein particles may change plasma osmolality with repercussions in erythrocyte aggregation.

The aim of our work was to study *in vitro* the erythrocyte aggregation tendency in blood samples collected from healthy male adults and enriched with their own plasma lipoproteins subfractions.

2. Material and Methods

2.1. Blood Samples. On consecutive days, venous blood samples were obtained with previous consent from healthy fasting volunteers adult males ($n = 10$) after 15 min in the recumbent position and collected (for two plastic tubes) with anticoagulant (10 I.U. of heparin/mL or 0.1% EDTA).

2.2. Lipoprotein Fractions. Lipoproteins fractions were prepared by a discontinuous NaCl/KBr density gradient ultracentrifugation using an SW 50.1 rotor (Beckman) [40]. Lipoprotein fractions were characterised by electrophoresis (Electra HR Helena Laboratories) buffer tris-barbital-sodium buffer pH 8.8) in cellulose acetate by comparison with serum controls (Lipotrol, Helena Laboratories).

2.3. Erythrocyte Aggregation Index. Erythrocyte aggregation was determined using the MA1 aggregometer from Myrenne GMBH (Roetgen, Germany). The MA1 aggregometer consists of a rotating cone plate chamber which disperses the sample by high shear rate of $600\,s^{-1}$ and a photometer that determines the extent of aggregation. The intensity of light (emitted by a light emitting diode) is measured after transmission through the blood sample. The aggregation was determined in stasis for 10 seconds after dispersion of the blood sample [41].

2.4. Plasma Osmolality. Plasma osmolality was determined with the Osmomat 030 Cryoscopic Osmometer from Gonotec (Berlin, Germany).

2.5. Experimental Design. Blood samples from each donor were divided on aliquots, and after centrifugation and small volumes of plasma (0, $5\,\mu L$, $10\,\mu L$, $20\,\mu L$, and $40\,\mu L$) were discharged and replaced by equal values of their own previously enriched lipoprotein subfractions prepared a day before. With this procedure, no hematocrit variations were obtained. Blood aliquots were gently mixed by inversion, and erythrocyte aggregation was assessed. At the end of each assay, the aliquots were centrifuged at 12000 rpm for 1 minute in the Biofuge 15 centrifuge from Heraeus, and plasma osmolality was determined. HDL-C, LDL-C, and VLDL-C concentrations were expressed as percentage of osmolality variation values.

2.6. Statistical Analysis. The statistical evaluation performed utilized the "one-way" ANOVA with homogeneity test, cluster analysis, and average method.

3. Results

The major plasma lipoprotein subfractions were obtained by the discontinuous density gradient centrifugation between the density range of 1.006 and 1.300 g/mL. Each lipoprotein fraction was well banded with VLDL-C at the top, LDL-C in the upper middle and the HDL-C in the lower middle portion of the tube. After that, each fraction was pooled and submitted to electrophoresis where their obtained migration was confirmed by comparison with the serum control.

The volume of each lipoprotein sub-fraction added to the autologous blood samples aliquots caused variation of osmolality concentrations in relation to its absence. Using the cluster analysis and the beverage method, four classes of concentration range expressed as osmolality variation were grouped for each LDL-C; the VLDL-C and the HDL-C enriched blood aliquots, namely, Class I 0.005–0.025; Class II 0.030–0.035; Class III 0.045–0.055; Class IV 0.077–0.095 (Figure 1).

The erythrocyte aggregation variation for each enriched LDL, HDL, and VLDL blood aliquots in relation to the initial value is grouped by the different osmolality class variation values (Figure 2). The variation of erythrocyte aggregation in relation to the initial values depends on the plasma osmolality values, as well as of the type of lipoprotein subfraction. The enriched VLDL-C blood samples aliquots do not induce statistical significant variation on erythrocyte aggregation values (Figure 3). At variance in relation to the initial erythrocyte aggregation values, the enriched LDL-C blood samples presented significant statistical enhanced values at all range of percentage of osmolality variation (Figure 4). The same behaviour was verified in the enriched HDL-C blood aliquots, with exception for the higher values of percentages of osmolality variation, where a very significant ($P < 0.0001$) decrease was obtained for the erythrocyte aggregation (Figure 5). For the same range of osmolality variation, the two types of lipoproteins LDL-C and HDL-C induced different variation of erythrocyte aggregation ($P < 0.01$). When the values of osmolality variation were plotted against the respective values of aggregation variation

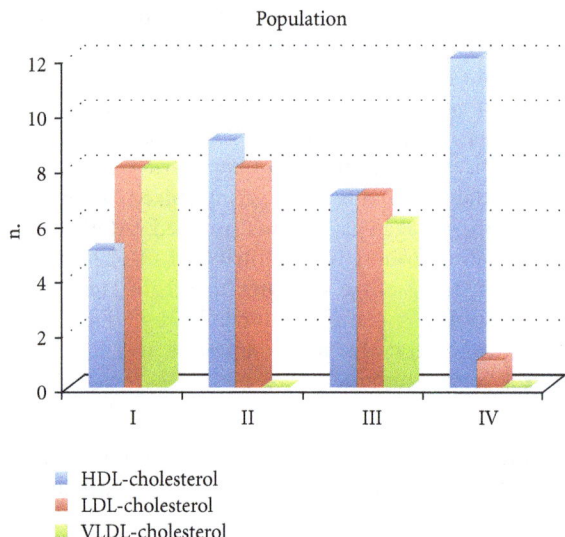

FIGURE 1: Histogram presenting the concentrations of enriched blood samples with LDL-C (red), HDL-C (blue), VLDL-C (green) distributed by four classes according the osmolality scale of values (class I 0.005–0.025; class II 0.030–0.035: class III 0.045–0.055; class IV 0.077–0.095).

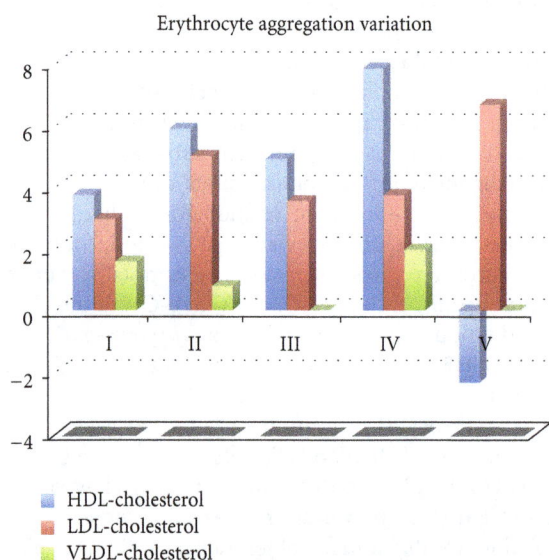

FIGURE 2: Histogram presenting the values of the erythrocyte aggregation variation obtained in enriched blood samples with LDL-C (red), HDL-C (blue), VLDL-C (green) in the four classes of concentration according the osmolality scale of values (I 0.005–0.025; II 0.030–0.035: III 0.045–0.055; IV 0.077–0.095).

obtained in each enriched HDL-C aliquot, a significant ($R^2 = 0.383$) inverse linear regression was obtained (Figure 6).

4. Discussion

In the present *in vitro* study, we investigated the induction of human erythrocyte tendency to aggregation by autologous lipoproteins sub-fractions. With the amount of lipoprotein

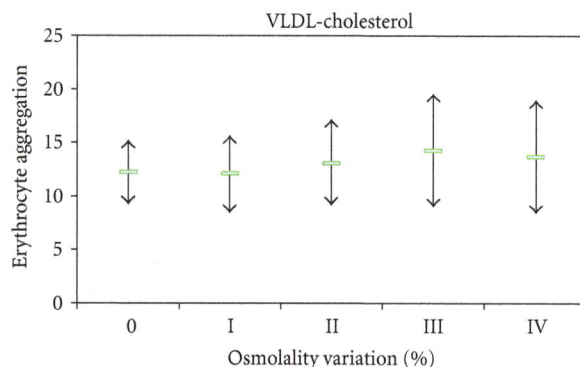

FIGURE 3: Values of erythrocyte aggregation index (mean +/− sd) obtained in enriched blood samples with HDL-C and without enrichment.

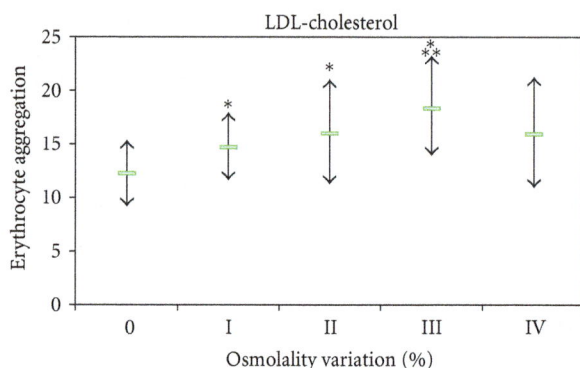

*Vs 0—$P < 0.01$
**Vs 5—$P < 0.001$

FIGURE 4: Values of erythrocyte aggregation index (mean +/− sd) obtained in enriched blood samples with HDL-C and without enrichment

sub-fractions added, changes in plasma osmolality were observed as a consequence of the increase number of particles. We used cluster method on the values of osmolality to define four classes (Figure 1).

In blood aliquots enriched with VLDL-C, no changes in erythrocyte aggregation index were verified (Figure 3). For the same amount of particles corresponding to the same class of osmolality variation, both HDL-C and LDL-C enrichment induce enhancement of erythrocyte aggregation (Figures 4 and 5). VLDL has higher diameter than the other two lipoprotein classes [42] and may either rest in the plasma bulk not interfering with erythrocyte aggregation tendency. LDL-C particles bind in a nonabsolute specific way with erythrocyte membrane, while 60% of membrane area can be occupied by HDL-C as has been described [43]. The occupancy of some areas of erythrocyte membrane by HDL-C or LDL-C, causing some interference in the promotion of EAI tendency, may be an explanation for our results. Significant ($P < 0.01$) higher values of erythrocyte aggregation were obtained in HDL-C-enriched aliquots more than for LDL-C-enriched ones at the same class of osmolality variation (under the same number of particles). The exception was

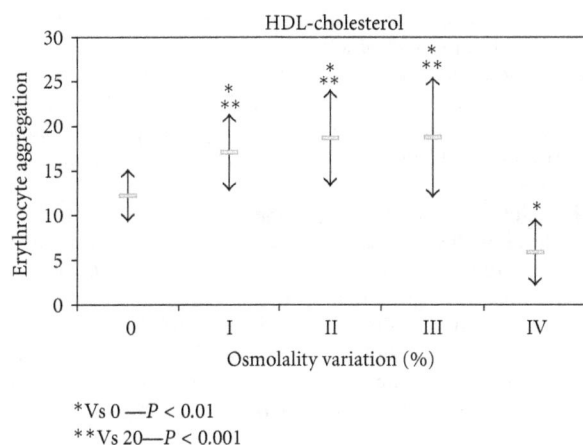

*Vs 0 —$P < 0.01$
**Vs 20—$P < 0.001$

FIGURE 5: Values of erythrocyte aggregation index (mean +/− sd) obtained in enriched blood samples with LDL-C and without enrichment.

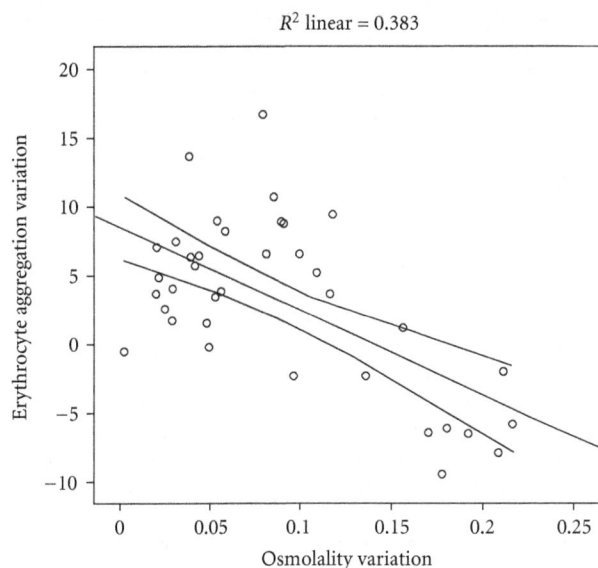

FIGURE 6: Presentation of the association obtained between the erythrocyte aggregation values and the osmolality values in enriched blood sample with HDL-C.

for the HDL-C-enriched aliquots with the highest percentage of osmolality variation which significantly ($P < 0.0001$) decreased EAI to lower values than the aliquot control obtained. We raise the hypothesis that the higher number of unbinding HDL-C particles may increase the ionic strength to a threshold and consequently pull away the erythrocytes decreasing their tendency to aggregate.

Our results suggest that in healthy human blood aliquots enriched with autologous HDL-C or LDL-C, (where fibrinogen is present at normal range), when submitted to shear rate to disaggregate RBCs, and stopped after that, EAI tendency increases. The association between EAI values and HDL-C obtained in our experimental design is in accordance with others studies [39], which is a decrease of fibrinogen-induced

RBCs aggregation in presence of the HDL2 subclass as not observed in a different experimental approach.

Recently [44], it was verified that the LDL particles number must be considered as an indicator for atherosclerosis risk factor, which has been previously remembered by others in the prevention of cardiovascular disease [45].

Two models, the cross-bridging and the depletion layer, were described to explain the reversible erythrocyte aggregation process at physiological conditions [46], but there are controversies and difficulties to adopt one or reject an other. The depletion model developed for polymer solutions of dextran demonstrated that there is an optimal molecular weight value to reach the greater erythrocyte aggregation tendency [47]. The hydrodynamic radius (Rh) determined for VLDL [48] is in the range between 15 nm and 40 nm which is much above of others belonging to macromolecules promoters of erythrocyte aggregation increase [49]. Particle size is an influent factor that may explain the absence of VLDL effect in the erythrocyte aggregation obtained. Regarding our results, they may also fit the cross-bridging model if, by an unknown mechanism, we assume that the particles numbers and size favour fibrinogen binding in a similar way as previously reported for immunoglobulin effects on fibrinogen-mediated erythrocyte aggregation [50, 51]. More studies are needed to explain in what model fit the effects of lipoproteins in erythrocyte aggregation. Our results may contribute to better understand the direct associations between high erythrocyte aggregation tendency and other cardiovascular risk factors such as hyperlipoproteinemia.

Acknowledgments

The authors are grateful to Emilia Alves for typewriting the paper. This work was supported by "Fundação para a Ciência e a Tecnologia."

References

[1] W. B. Kannel, W. P. Castelli, and T. Gordon, "Cholesterol in the prediction of atherosclerotic disease: new perspectives based on the Framingham study," *Annals of Internal Medicine*, vol. 90, no. 1, pp. 85–91, 1979.

[2] N. E. Miller, F. Hammett, S. Saltissi, H. Vanzeller, S. Coltart, and B. Lewis, "Relation of angiographically defined coronary artery disease to plasma lipoprotein subfractions and apolipoproteins," *The British Medical Journal*, vol. 282, no. 6278, pp. 1741–1744, 1981.

[3] A. M. Alshehri, "Metabolic syndrome and cardiovascular risk," *Journal of Family and Community Medicine*, vol. 17, pp. 73–78, 2010.

[4] C. Baigent, A. Keech, P. M. Kearney, L. Blackwell, and G. Buck, "Efficacy and safety of cholesterol-lowering treatment prospective meta-analysis of data from 90,056 participants in 14 randomised trials of statins," *The Lancet*, vol. 366, pp. 1267–1278, 2005.

[5] J. G. Robinson, B. Smith, N. Maheshwari, and H. Schrott, "Pleiotropic effects of statins: benefit beyond cholesterol reduction? A meta-regression analysis," *Journal of the American College of Cardiology*, vol. 46, no. 10, pp. 1855–1862, 2005.

[6] A. V. Khera, M. Cuchel, M. de la Llera-Moya et al., "Choles-terol efflux capacity, high-density lipoprotein function, and atherosclerosis," *The New England Journal of Medicine*, vol. 364, no. 2, pp. 127–135, 2011.

[7] W. H. Tang, Y. Wu, S. Mann et al., "Diminished antioxidant activity of high-density lipoprotein-associated proteins in systolic heart failure," *Circulation—Heart Failure*, vol. 4, no. 1, pp. 59–64, 2011.

[8] A. H. Seplowitz, S. Chien, and F. R. Smith, "Effects of lipoproteins on plasma viscosity," *Atherosclerosis*, vol. 38, no. 1-2, pp. 89–95, 1981.

[9] G. Lowe, "Blood viscosity, lipoproteins, and cardiovascular risk," *Circulation*, vol. 85, no. 6, pp. 2329–2331, 1992.

[10] W. Koenig, M. Sund, E. Ernst, W. Mraz, V. Hombach, and U. Keil, "Association between rheology and components of lipoproteins in human blood: results from the MONICA project," *Circulation*, vol. 85, no. 6, pp. 2197–2204, 1992.

[11] G. Kesmarky, K. Toth, L. Habon, G. Vajda, and I. Juricskay, "Hemorheological parameters in coronary artery disease," *Clinical Hemorheology and Microcirculation*, vol. 18, no. 4, pp. 245–251, 1998.

[12] J. Dormandy, E. Ernst, A. Matrai, and P. T. Flute, "Hemor-rheologic changes following acute myocardial infarction," *The American Heart Journal*, vol. 104, no. 6, pp. 1364–1367, 1982.

[13] A. M. Ehrly, *Therapeutic Hemorheology*, Springer, New York, NY, USA, 1991.

[14] G. D. O. Lowe, *Clinical Blood Rheology*, CRC Press, Boca Raton, Fla, USA, 1988.

[15] J. F. Stoltz, M. Singh, and P. Riha, *Hemorheology in Practice*, IOS Press, Amsterdam, The Netherlands, 1999.

[16] O. K. Baskurtand and H. J. Meiselman, "Cellular determinants of low-shear blood viscosity," *Biorheology*, vol. 34, no. 3, pp. 235–247, 1997.

[17] P. Sousa-Ramalho, R. Martins, C. Saldanha, L. Cardoso, D. Santos, and J. Martins e Silva, "Haemorheological changes in glaucoma," *New Trends in Ophthalmology*, vol. 1, pp. 45–51, 1986.

[18] S. Chien and L. A. Lang, "Physicochemical basis and clinical implications of red cell aggregation," *Clinical Hemorheology*, vol. 7, no. 1, pp. 71–91, 1987.

[19] P. Souza-Ramalho, H. Hormigo, R. Martins, C. Saldanha, and J. Martins e Silva, "Haematological changes in retinal vasculitis," *Eye*, vol. 2, no. 3, pp. 278–282, 1988.

[20] L. Sargento, L. Zabala, T. Gomes, C. Saldanha, J. Martins-Silva, and P. Souza-Ramalho, "The effect of angiography with disodium fluorecein on the hemorheologic parametes of diabetes," *Acta Médica Portuguesa*, vol. 9, pp. 303–307, 1996.

[21] E. Konstantinova, L. Ivanova, T. Tolstaya, and E. Mironova, "Rheological properties of blood and parameters of platelets aggregation in arterial hypertension," *Clinical Hemorheology and Microcirculation*, vol. 35, no. 1-2, pp. 135–138, 2006.

[22] T. Shiga, N. Maeda, and K. Kon, "Erythrocyte rheology," *Critical Reviews in Oncology/Hematology*, vol. 10, no. 1, pp. 9–48, 1990.

[23] N. Maeda, M. Seike, Y. Suzuki, and T. Shiga, "Effect of pH on the velocity of erythrocyte aggregation," *Biorheology*, vol. 25, no. 1-2, pp. 25–30, 1988.

[24] G. Yardin and H. J. Meiselman, "Effects of cellular morphology on the viscoelastic behavior of high hematocrit RBC suspen-sions," *Biorheology*, vol. 26, no. 2, pp. 153–175, 1989.

[25] C. Saldanha and J. Martins e Silva, "Effect of plasma osmolality on erythrocyte aggregation index. Hémorheologie et aggre-gation erythrocytaire," in *Théories et Applications Cliniques*, J. F. Stoltz, M. Donner, and A. L. Copley, Eds., vol. 3, pp. 142–145, Editions Medicales Internationales, 1991.

[26] R. Mesquita, M. I. Gonçalves, S. Dias, L. Sargento, C. Saldanha, and J. Martins e Silva, "Ethanol and erythrocyte membrane interaction: a hemorheologic perspective," *Clinical Hemorheology and Microcirculation*, vol. 21, no. 2, pp. 95–98, 1999.

[27] I. Gonçalves, C. Saldanha, and J. Martins e Silva, "Beta-estradiol effects on erythrocyte aggregation—a controlled in vitro study," *Clinical Hemorheology and Microcirculation*, vol. 25, no. 3-4, pp. 127–134, 2001.

[28] R. Mesquita, I. Pires, C. Saldanha, and J. Martins-Silva, "Effects of acetylcholine and spermineNONOate on erythro-cyte hemorheologic and oxygen carrying properties," *Clinical Hemorheology and Microcirculation*, vol. 25, no. 3-4, pp. 153–163, 2001.

[29] T. Santos, R. Mesquita, J. Martins e Silva, and C. Saldanha, "Effects of choline on hemorheological properties and NO metabolism of human erythrocytes," *Clinical Hemorheology and Microcirculation*, vol. 29, no. 1, pp. 41–51, 2003.

[30] S. Hilário, C. Saldanha, and J. Martins e Silva, "An in vitro study of adrenaline effect on human erythrocyte properties in both gender," *Clinical Hemorheology and Microcirculation*, vol. 28, no. 2, pp. 89–98, 2003.

[31] F. A. Carvalho, A. V. Maria, J. M. Braz Nogueira, J. Guerra, J. Martins-Silva, and C. Saldanha, "The relation between the erythrocyte nitric oxide and hemorheological parameters," *Clinical Hemorheology and Microcirculation*, vol. 35, no. 1-2, pp. 341–347, 2006.

[32] J. P. Almeida, F. A. Carvalho, T. Freitas, and C. Saldanha, "Modulation of hemorheological parameters by the erythro-cyte redox thiol status," *Clinical Hemorheology and Microcir-culation*, vol. 40, no. 2, pp. 99–111, 2008.

[33] M. W. Rampling, H. J. Meiselman, B. Neu, and O. K. Baskurt, "Influence of cell-specific factors on red blood cell aggregation," *Biorheology*, vol. 41, no. 2, pp. 91–112, 2004.

[34] V. Schechner, I. Shapira, S. Berliner et al., "Significant dominance of fibrinogen over immunoglobulins, C-reactive protein, cholesterol and triglycerides in maintaining increased red blood cell adhesiveness/aggregation in the peripheral venous blood: a model in hypercholesterolaemic patients," *The European Journal of Clinical Investigation*, vol. 33, no. 11, pp. 955–961, 2003.

[35] R. H. Jay, M. W. Rampling, and D. J. Betteridge, "Abnor-malities of blood rheology in familial hypercholesterolaemia: effects of treatment," *Atherosclerosis*, vol. 85, no. 2-3, pp. 249–256, 1990.

[36] S. Muller, O. Ziegler, M. Donner, P. Drouin, and J. F. Stoltz, "Rheological properties and membrane fluidity of red blood cells and platelets in primary hyperlipoproteinemia," *Atherosclerosis*, vol. 83, no. 2-3, pp. 231–237, 1990.

[37] A. Vayá, M. Martínez, R. Carmena, and J. Aznar, "The lipid composition of red blood cells and their hemorheological behaviour in patients with primary hyperlipoproteinemia," *Clinical Hemorheology and Microcirculation*, vol. 13, pp. 447–457, 1993.

[38] S. M. Razavian, V. Atger, P. Giral et al., "Influence of HDL sub-fractions on erythrocyte aggregation in hypercholesterolemic men," *Arteriosclerosis and Thrombosis*, vol. 14, no. 3, pp. 361–366, 1994.

[39] E. Simon, S. M. Razavian, J. L. Beyec et al., "Influence of lipoprotein subfractions on dextran- and fibrinogen-induced erythrocyte aggregation," *Clinical Hemorheology*, vol. 15, no. 4, pp. 667–676, 1995.

[40] B. H. Chung, T. Wilkinson, J. C. Geer, and J. P. Segrest, "Preparative and quantitative isolation of plasma lipoproteins: rapid, single discontinuous density gradient ultracentrifugation in a vertical rotor," *Journal of Lipid Research*, vol. 21, no. 3, pp. 284–291, 1980.

[41] H. Schmid-Schonbein, E. Volger, P. Teitel, H. Kieswetter, V. Dauer, and L. Heilman, "New hemorheological techniques for the routine laboratory," *Clinical Hemorheology*, vol. 2, no. 1-2, pp. 93–105, 1982.

[42] A. M. Gotto Jr., H. J. Pownall, and R. J. Havel, "Introduction to the plasma lipoproteins," *Methods in Enzymology*, vol. 128, pp. 3–41, 1986.

[43] D. Y. Hui, J. G. Noel, and J. A. Harmony, "Binding of plasma low density lipoproteins to erythrocytes," *Biochimica et Biophysica Acta*, vol. 664, no. 3, pp. 513–526, 1981.

[44] J. D. Otvos, S. Mora, I. Shalaurova, P. Greenland, R. H. MacKey, and D. C. Goff, "Clinical implications of discordance between low-density lipoprotein cholesterol and particle number," *Journal of Clinical Lipidology*, vol. 5, no. 2, pp. 105–113, 2011.

[45] W. C. Cromwell, J. D. Otvos, M. J. Keyes et al., "LDL particle number and risk of future cardiovascular disease in the Framingham Offspring Study—implications for LDL management," *Journal of Clinical Lipidology*, vol. 1, no. 6, pp. 583–592, 2007.

[46] H. J. Meiselman, B. Neu, M. W. Rampling, and O. K. Baskurt, "RBC aggregation: laboratory data and models," *The Indian Journal of Experimental Biology*, vol. 45, no. 1, pp. 9–17, 2007.

[47] B. Neu, R. Wenby, and H. J. Meiselman, "Effects of dextran molecular weight on red blood cell aggregation," *Biophysical Journal*, vol. 95, no. 6, pp. 3059–3065, 2008.

[48] D. C. Rambaldi, A. Zattoni, S. Casolari, P. Reschiglian, D. Roessner, and C. Johann, "An analytical method for size and shape characterization of blood lipoproteins," *Clinical Chemistry*, vol. 53, no. 11, pp. 2026–2029, 2007.

[49] K. J. Armstrong, R. B. Wenby, H. J. Meiselman, and T. C. Fisher, "The hydrodynamic radii of macromolecules and their effect on red blood cell aggregation," *Biophysical Journal*, vol. 87, no. 6, pp. 4259–4270, 2004.

[50] N. Maeda and T. Shiga, "Opposite effect of albumin on the erythrocyte aggregation induced by immunoglobulin G and fibrinogen," *Biochimica et Biophysica Acta*, vol. 855, no. 1, pp. 127–135, 1986.

[51] R. Ben-Ami, G. Barshtein, T. Mardi et al., "A synergistic effect of albumin and fibrinogen on immunoglobulin-induced red blood cell aggregation," *The American Journal of Physiology—Heart and Circulatory Physiology*, vol. 285, pp. H2663–H2669, 2003.

13

C-Lobe of Lactoferrin: The Whole Story of the Half-Molecule

Sujata Sharma, Mau Sinha, Sanket Kaushik, Punit Kaur, and Tej P. Singh

Department of Biophysics, All India Institute of Medical Sciences, New Delhi 110029, India

Correspondence should be addressed to Tej P. Singh; tpsingh.aiims@gmail.com

Academic Editor: Andrei Surguchov

Lactoferrin is an iron-binding diferric glycoprotein present in most of the exocrine secretions. The major role of lactoferrin, which is found abundantly in colostrum, is antimicrobial action for the defense of mammary gland and the neonates. Lactoferrin consists of two equal halves, designated as N-lobe and C-lobe, each of which contains one iron-binding site. While the N-lobe of lactoferrin has been extensively studied and is known for its enhanced antimicrobial effect, the C-lobe of lactoferrin mediates various therapeutic functions which are still being discovered. The potential of the C-lobe in the treatment of gastropathy, diabetes, and corneal wounds and injuries has been indicated. This review provides the details of the proteolytic preparation of C-lobe, and interspecies comparisons of its sequence and structure, as well as the scope of its therapeutic applications.

1. Lactoferrin: A Bilobal Protein

Lactoferrin is an iron-binding glycoprotein which is abundantly present in colostrum and is the major mediator of defence for the newborn [1–4]. The three-dimensional structures of lactoferrin from various species, such as human [5, 6], bovine [7, 8], buffalo [9, 10], equine [11, 12], and camel [13] have been solved. In all these cases, the overall structures of iron-saturated lactoferrin were found to have a similar folding. Lactoferrin is an 80 kDa single-chain glycoprotein which can be equally divided into two homologous halves, N-lobe and C-lobe, each of which is about 40 kDa in size and is connected to the other by a short helical segment (Figure 1). Each lobe is further subdivided into two domains, which are designated as N1 and N2 domains of N-lobe and C1 and C2 domains of C-lobe. The iron-binding site is situated inside the interdomain cleft in each lobe. The iron-binding site consists of four residues 2 tyrosines, 1 aspartate, and 1 histidine residues. The iron-binding residues in N-lobe are Asp 60, Tyr 92, Tyr 192, and His 253 while the corresponding iron-binding residues in C-lobe are Asp 395, Tyr 433, Tyr 526, and His 595. The iron-binding residues are coordinated to the ferric ion and a synergistic bidentate carbonate anion (Figure 2).

Though there has been a considerable amount of information on the structure of native lactoferrin, a similar knowledge on the structures of two independent lobes upon cleavage of lactoferrin has been scarce. Hence, there have been efforts since the discovery of lactoferrin to uncouple the two lobes using limited proteolysis and thereby study their structures and associated functions.

2. Preparation of Monoferric Lobes of Lactoferrin: Early Days to Present Times

The earliest reports of the production of C-lobe came in 1976 when both iron-saturated and apolactoferrin were digested with trypsin [14]. It was found that the iron-saturated lactoferrin was more resistant to proteolysis than apolactoferrin. However, instead of generating two equal 40 kDa lobes, five different fragments with molecular weights ranging from 25,000 to 52,700 were produced.

Later reports of proteolysis of lactoferrin with trypsin did not report success on the generation of two equal halves of lactoferrin. In fact, it was established that proteolysis of iron-saturated human lactoferrin with trypsin generates a 30 kDa N-terminal fragment and a 50 kDa C-terminal fragment. Upon further hydrolysis of N-lobe with a second tryptic digestion, the ND2 domain of 18 kDa was generated [15].

On the other hand, the first report on the human N-lobe produced from the cloned cDNA expressed in baby hamster kidney (BHK) cells appeared [16]. Iron binding and release studies on this recombinant N-lobe showed that the absence of stabilizing contacts between the N- and C-lobes led to faster and easier release of iron from the N-lobes as compared to that from native lactoferrin. Similarly, C-lobe of human lactoferrin was also cloned and expressed in due time [17].

The first report on the generation of equal 40 kDa N-lobe and C-lobe came in 1999, where it was reported that limited proteolysis of buffalo lactoferrin with proteinase K, a nonspecific serine protease, led to the production of two 40 kDa equal-sized lobes along with low molecular mass peptides (<14 kDa) [18]. The proteolysis of buffalo lactoferrin with trypsin and pepsin produced two major fragments of approximately 35 and 23 kDa together with small molecular mass peptides. Subtilisin, another enzyme, hydrolyzed lactoferrin to produce two fragments of 40 and 26 kDa along with low molecular mass peptides [18]. N-lobe and C-lobe, generated by proteolysis with proteinase K, were further purified using ion-exchange and gel filtration chromatography. Upon further hydrolysis, it was found that N-lobe was completely digested into low molecular mass peptides, while C-lobe remained intact for a long time indicating it to be significantly resistant to digestion by proteases. The resistance of C-lobe to proteolysis was attributed to a peptide found within the C-lobe, which bound with and hence inhibited proteinase K.

The secondary structure elements of both the purified lobes and the native lactoferrin were analyzed by circular dichroism studies. It was found that lower helical structures were present in the N- and C-lobes as compared to native lactoferrin [18]. Similarly, the iron saturations of the native lactoferrin was found to be higher than that of N- and C-lobes.

Another report on the cloning and expression of C-lobe followed where a *Rhodococcus erythropolis* expression system for the bovine lactoferrin C-lobe was constructed which yielded a native and a denatured form of C-lobe [19]. However, the denatured protein had to be refolded by stepwise dialysis against refolding buffers in order to generate a bioactive, folded C-lobe. After almost three decades of work on the generation of the N-lobe and C-lobe which included both the proteolysis as well as cloning efforts, the proteolysis of lactoferrin with proteinase K was the most effective way, as it also led to the crystallization of C-lobe and to its successful structure determination [20]. The structure of cloned C-lobe has not been determined to date.

3. Structure of C-Lobe within Native Lactoferrins

The structures of lactoferrin were solved in iron-saturated form from five species, equine [11] (PDB Code: 1B1X), human [5] (PDB Code: 1LFG), bovine [7, 8] (PDB Code: 1BLF), and buffalo [9, 10] (PDB Code: 1BIY), while the structures of iron-free (apo) forms of lactoferrin could be determined only from three species, equine [12] (PDB Code: 1B7U), human [6] (PDB Code: 1CB6), and camel [13] (PDB Code: 1DTZ). Within the structure of native lactoferrin in all species, the C-lobe was found to have a two-fold internal homology with

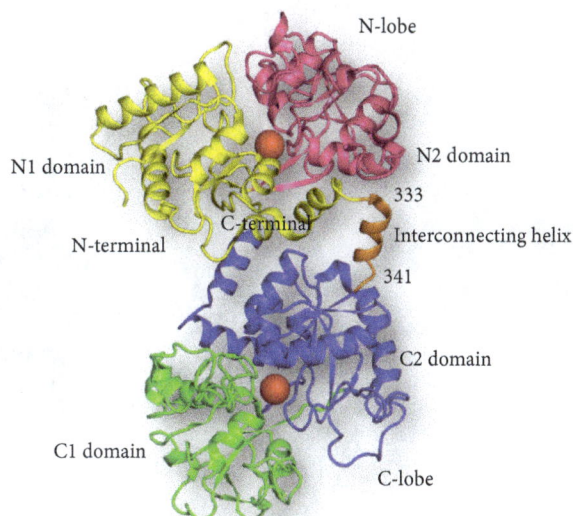

FIGURE 1: Schematic diagram of the bovine lactoferrin molecule (PDB code: 1BLF). The N1 and N2 domains are colored in yellow and pink, respectively, while the C1 and C2 domains are colored in green and blue, respectively. The interconnecting helix between the lobes is colored in orange. The two iron atoms are shown as red spheres.

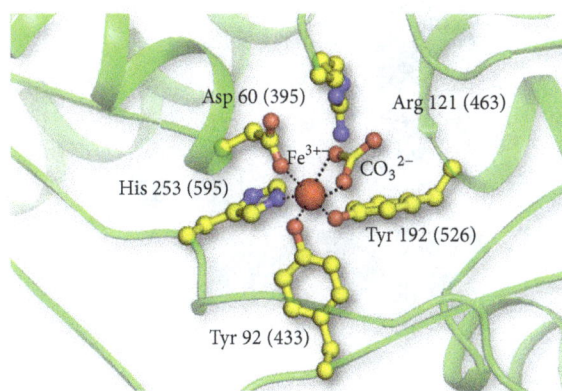

FIGURE 2: Schematic figure of the iron-binding site of lactoferrin. The iron atom is shown as a red sphere, while the interacting amino acid residues of lactoferrin are in yellow. The residue numbers correspond to N-lobe, while the corresponding residues of C-lobe are in brackets.

N-lobe, and both the lobes were connected by a short helical segment.

In the iron-saturated forms of lactoferrin, both the N-lobe and C-lobe adopt a closed conformation [5, 7–11], though there were differences in the degree of closure of the two domains of each lobe in each species. However, in the case of apolactoferrin, the N-lobe and C-lobe showed differential conformations depending on the species from which the lactoferrin originated. In case of apoequine lactoferrin (AELF), both the N- and C-lobes adopt the closed conformation [12] (Figure 3(a)). In case of human apolactoferrin (AHLF), the N-lobe adopts the open conformation while C-lobe remained in the closed form [6] (Figure 3(b)). However,

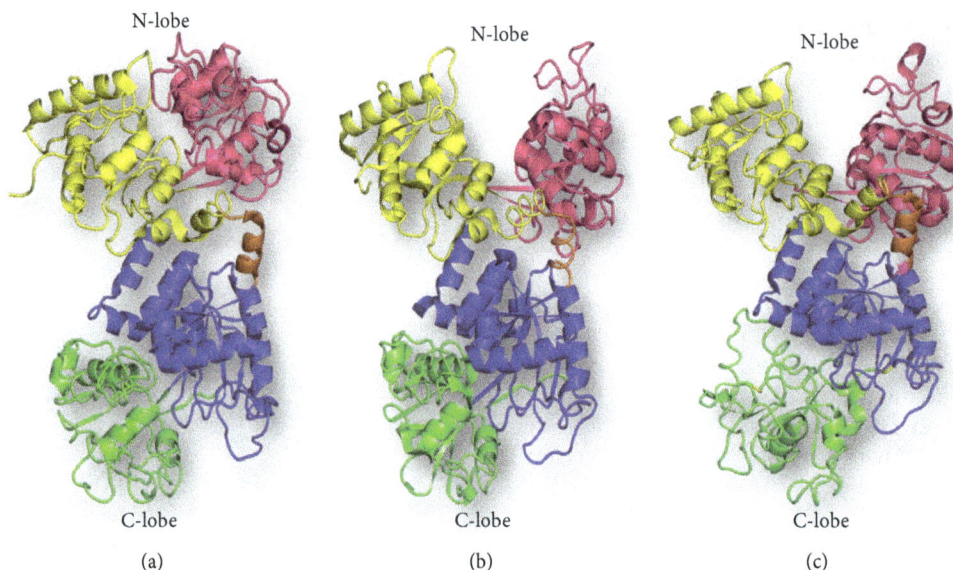

FIGURE 3: Schematic diagrams of apolactoferrins from three species showing variable behavior of the conformation of the domains. (a) Equine apolactoferrin (PDB code: 1B7U) shows both the lobes in closed conformation. (b) Human apolactoferrin (PDB code: 1CB6) shows open N-lobe and closed C-lobe. (c) Camel apolactoferrin (PDB code: 1DTZ) shows both the lobes in open conformation.

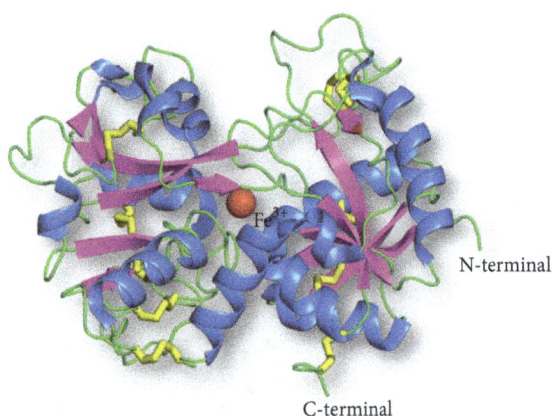

FIGURE 4: Schematic diagrams of C-lobe of bovine lactoferrin, produced using limited proteolysis with proteinase K (PDB code: 1NKX). α-helices are represented as blue helices, β-sheets are indicated by magenta arrows, and disulfide bonds are indicated as yellow sticks. The iron atom is shown in red.

in the case of apocamel lactoferrin (AULF), both the N-lobe and C-lobe were found in open conformations, showing wide distances between the iron-binding residues in the native iron-free form of lactoferrin [13] (Figure 3(c)).

The differences in the conformations of N-lobe of apolactoferrins from three different species were explained by the presence of four crucial interactions at the mouth of the iron-binding site in the N-lobe of equine lactoferrin, which were absent from other lactoferrins [11, 12]. The first interaction was a hydrogen bond between Ser12 in N1 domain and Ser185 in N2 domain which brought the two domains together. The second interaction which was responsible for the closure of N-lobe was a salt bridge between Lys301 and Glu216. Also, the

interactions between Arg249$\cdots$Gln89 and Asp211$\cdots$Gln83 are also responsible for the N-lobe to remain closed in the apostate in AELF [12].

While the C-lobes of both equine and human lactoferrins remain closed in the apostate, the C-lobe of camel lactoferrin is observed to be in the open conformation. This was explained by the absence of one interaction in camel apolactoferrin which was present in equine and human lactoferrins. This occurs due to differences in the sequence of camel lactoferrin, as compared to other lactoferrins. In camel lactoferrin, Gln418 is replaced by Pro418, and hence the crucial interaction between Gln418 with Arg587 (human lactoferrin) and Gln587 (equine lactoferrin) is missing.

It was also seen that the positioning and conformations of residues of the iron-binding site of N-lobe of camel apolactoferrin were similar to those of the N-lobe in human apolactoferrin. However, the corresponding residues in the C-lobe were similar to those in the C-lobes of duck and hen apoovotransferrins [13]. This indicated that in terms of structure and iron binding and release, while the N-lobe of camel apolactoferrin is similar to the N-lobe of human apolactoferrin, the C-lobe of camel apolactoferrin, however, showed similarity with the C-lobe of hen and duck apoovotransferrins. These findings were validated by performing iron binding and release studies on the proteolyzed, independent lobes, where it was seen that the N-lobe of camel lactoferrin loses iron at a low pH (4.0–2.0), which matched with other lactoferrins. On the other hand, the iron-binding and iron-release behaviors of the C-lobe of camel lactoferrin is similar to that in transferrins, as it loses iron at a much higher pH of 6.0 and above [13].

The first crystal structure of a proteolytically generated functional iron-saturated C-lobe of bovine lactoferrin (Fe-C-Lobe) revealed 2593 protein atoms (residues 342–676 and 681–685), 124 carbohydrate atoms (from ten monosaccharide

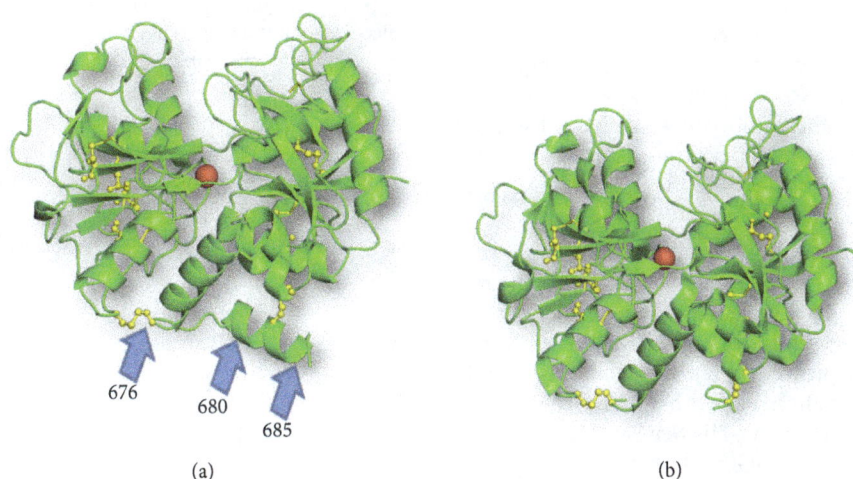

(a)

(b)

FIGURE 5: A comparison of the (a) C-lobe of intact lactoferrin and (b) C-lobe which has been cleaved from the intact lactoferrin using proteinase K. Apart from one single cut at the N-terminus, the enzyme cleaves the C-terminal part of the protein at three points, leading to the generation of a pentapeptide which is anchored to the main protein by a disulfide bond between C405 and C684.

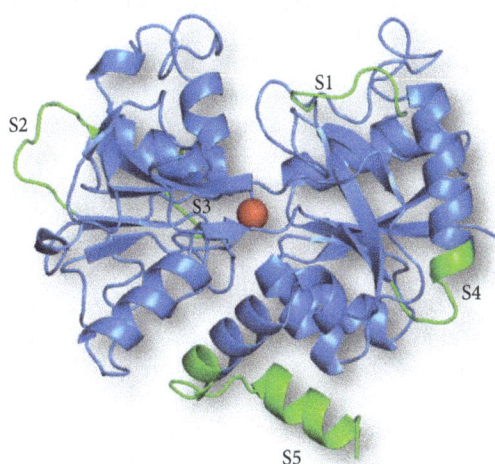

FIGURE 6: Schematic diagrams of C-lobe of bovine lactoferrin (in blue), showing the sequences which are most variable across species in green. The sequences have been labeled S1–S5.

units, in three glycan chains), one Fe^{3+}, one CO_3^{2-}, two Zn^{2+}, and 230 water oxygen molecules [20] (Figure 4).

The structure of bovine C-lobe began from residue 342 and ended with residue 676, with a cleaved pentapeptide fragment 681–685 which remained attached to the C-lobe through a disulphide bridge, Cys405–Cys684. Though the overall folding of the C-lobe was found to be similar to that of C-lobe in the native bovine lactoferrin, there were interesting differences in the conformations of some of the loops. It also showed the presence of a number of novel interactions which seemed to have been generated to stabilize the structure after proteolysis and loss of stabilizing interaction with the N-lobe. The secondary structure topology diagram of Fe-C-Lobe is depicted in Figure S1 in supplementary material available online at http://dx.doi.org/10.1155/2013/271641.

The most interesting observation was the presence of 6 interdomain hydrogen bonds in the structure of Fe-C-Lobe,

which was more than the 4 interdomain hydrogen bonds reported in the structure of Fe-Lf [20]. Two zinc ions, which were found to be situated at sites other than the iron-binding cleft, seemed to be significant in providing stability to the crystal packing.

The action of proteinase K seemed to be on the most exposed two protein segments 334–344 and 674–689, both of which cross over from the N-lobe to the C-lobe and from the C-lobe to the N-lobe, respectively. These two segments are responsible for providing interlobe interactions between the two lobes in lactoferrin. Hence, many interactions were observed to be lost due to cleavage of N- and C-lobes, leading to a rearrangement of side chains of C-lobe in the absence of interacting N-lobe side chains. The C-terminus of Fe-C-Lobe showed that Proteinase K had cleaved the lactoferrin at three points at the C-terminus, Ser676, Leu680, and Ala 685, leading to a loss of helix, designated as $\alpha12$ in native lactoferrin, and generation of pentapeptide loop, that is, still anchored to the protein by a disulfide bond between Cys684 and Cys405 (Figure 5). In conclusion, the structure of C-lobe, starting from residue 342 and ending with residue 676, with a pentapeptide, 681–685 anchored to the main structure, was a fairly stable structure which was able to bind and release iron and also perform other functions.

4. C-Lobe of Lactoferrin: Sequence Comparison

The sequence comparison of bovine C-lobe with the corresponding N-lobe showed 30% sequence identity between the two lobes (Figure S2). Three C-lobe fragments >5 residues were found to be absent in the N-lobe. The three fragments were: Ser-Ser-Leu-Asp-Cys-Val-Leu-Arg, Thr-Asn-Gly-Glu-Ser-Thr-Ala-Asp, and the C-terminal fragment, Ala-Cys-Ala-Phe-Leu-Thr-Arg.

Interestingly, all these three fragments have a functional significance within the C-lobe. The first two fragments have been found to bind to the HA(2) region of viral hemagglutinin, an action which is not performed by N-lobe [21]. The third fragment, the highly hydrophobic C-terminal fragment, is the peptide, that is nicked by proteinase K [20] and is found to be attached to the C-lobe molecule by a disulfide bond even after its cleavage by the nonspecific proteinase K, indicating that it may also have a functional relevance in the antimicrobial role of lactoferrin.

The sequence comparison of bovine C-lobe with the corresponding C-lobe of buffalo, caprine, camel, human, equine, and porcine milk revealed a sequence identity of 96%, 95%, 74%, 73%, 72%, and 72%, respectively. There were at least 5 stretches of sequences >5 residues that were found to be remarkably different from each other (Figure S3). They corresponded to sequence Ser-Lys-His-Ser-Ser-Leu, Lys-Lys-Ala-Asn-GluGly-Leu, Asp-Asp-Gln-Gly-Leu-Asp, Arg-Ser-Asp-Arg-Ala-Ala-His-Val-Glu-Gln, and Ala-Asn-Leu-Lys-Lys-Cys-Ser-Thr-Ser-Pro-Leu-Leu-Glu-Ala-Cys-Ala-Phr-Leu-Thr-Arg, designated as S1, S2, S3, S4, and S5, respectively (Figure 6).

All these sequences were found to be flexible loops exposed to the surface, while the core of the C-lobe remains constant, indicating that there has been an independent evolution of the core and the exposed regions of C-lobe. S1 and S2 were part of the sequences that bind to viral hemagglutinin [22] and are absent in N-lobe. These loops are highly variable in C-lobe across species and are absent in N-lobe, indicating that these sequences could be specific to C-lobe of lactoferrin but seem to have evolved with the C-lobe for a functional need. In case of porcine lactoferrin, S5, the C-terminal sequence starting from residue 673 is entirely missing. Interestingly, this is the longest sequence among the most variable sequences of C-lobe and is also the same that is found anchored to the C-lobe after cleavage by proteinase K, indicating that there has been a structural evolution of C-terminus of this protein, and it may be required for some specialized function, since it remains anchored to the protein despite cleavage by an aggressive protease.

5. Therapeutic Applications of C-Lobe

C-lobe of bovine lactoferrin was found to exert a protective role in the prevention of nonsteroidal anti-inflammatory drug- (NSAID-) induced gastropathy [23]. The structural basis of this action of C-lobe was discovered after observing the bindings of NSAIDs to C-lobe and by determining the structures of complexes of bovine C-lobe with four commonly prescribed NSAIDs, indomethacin, diclofenac, aspirin, and ibuprofen, as well as COX-2 specific NSAIDs (Figure 7(a)). This therapeutic action of C-lobe was also detected in mouse models, where it was found that while the addition of C-lobe led to the effective reversal of NSAID-induced gastropathy, its coadministration with NSAIDs prevented it significantly [23]. This effect was mediated by a novel drug-binding site in lactoferrin which effectively sequesters the unbound NSAIDs in the gut. A similar study of C-lobe with COX-2 specific NSAIDs showed that the C-lobe was able

TABLE 1: The binding constants of various NSAIDs and sugar molecules with the C-lobe of lactoferrin.

S. no.	Compound	Binding constant (10^{-4} M)
NSAIDs		
1	Indomethacin	2.6
2	Aspirin	3.3
3	Ibuprofen	4.8
Sugars		
1	Glucose	1.8
2	Galactose	1.7
3	Mannose	1.7
4	Xylose	1.8
5	Maltose	1.6
6	Cellobiose	1.7
7	Lactose	1.5
8	Sucrose	1.9
9	Dextrin	1.4

to bind and sequester COX-2 specific NSAIDs with the same efficiency (Figure 7(b)) [24].

As observed by the binding studies (Table 1) [24, 25] and the structure determination studies, the same binding site in C-lobe also functions as a binding site for sugars. C-lobe has also been shown to reduce the levels of glucose due to its binding with various edible sugars [25] which bind at the sugar-binding site in the protein. While the administration of C-lobe to human serum showed reduction in the glucose levels, it was shown to be mediated by the binding of C-lobe with various sugar molecules, such as glucose, galactose, mannose, xylose, maltose, cellobiose, lactose, sucrose, and dextrin. All the sugars bound to the sugar-binding site of the C-lobe with varying potencies ranging around 10–4 M but were chelated by the C-lobe effectively. It may be mentioned here that C-lobe sequesters the NSAIDs only temporarily to release them gradually on demand as the binding affinities to these compounds vary between 100 micromolar to 1 millimolar. In fact, here the role of C-lobe is to protect the mucus of the intestinal lining from the interactions of NSAIDs that cause injury to tissues. In order to prevent NSAID-induced gastropathy, it is important that the target molecule should bind to the NSAIDs with low affinity, since if it binds with a high affinity, it would compete with specific enzymes such as cyclooxygenases in the gut for NSAID molecules which would reduce the efficacy for NSAIDs and defeat the purpose of these drugs. Hence, the administration of NSAIDs with the coadministration of C-lobe would be the right strategy for alleviating NSAID-induced gastropathy, as this would ensure that NSAIDs could bind to cyclooxygenases in the gut and reduce inflammation, and yet the excess of NSAIDs would be chelated by the C-lobe, so that damage to the gut could be reduced. For this purpose, the moderate binding affinities as found between C-lobe and NSAIDs are appropriate.

In this context, it may be mentioned here that NSAID and sugar-binding site in lactoferrin is slightly obstructed by the C-terminal helix, $\alpha 12$ which is removed from the C-lobe by Proteinase K on proteolysis, making this functional site

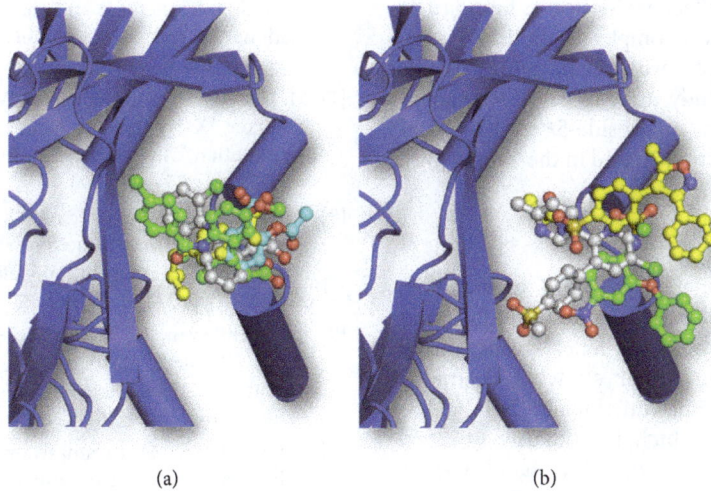

FIGURE 7: Schematic diagram of complex of C-lobe of lactoferrin with (a) nonsteroidal anti-inflammatory drugs (NSAIDS), aspirin (cyan), diclofenac (white), indomethacin (green), and ibuprofen (yellow) and (b) COX-2 specific inhibitors, etoricoxib (white), parecoxib (yellow), and nimesulide (green).

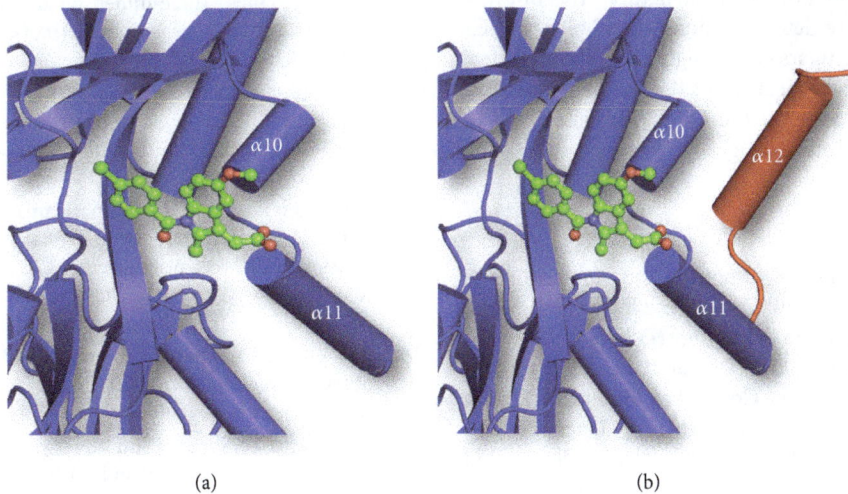

FIGURE 8: Schematic diagram of complex of (a) C-lobe of lactoferrin (blue) with indomethacin (green), showing the accessible drug-binding site in between the two helices, $\alpha10$ and $\alpha11$. The $\alpha12$ helix has been cleaved by proteinase K to make this site more accessible and (b) C-lobe of lactoferrin (blue) with indomethacin (green), with superimposed $\alpha12$ helix (in red) from native bovine lactoferrin, which shows that this helix blocks the passage to the binding site.

more accessible (Figure 8). This makes the C-lobe a superior molecule against gastropathy caused by NSAIDs than the intact lactoferrin.

C-lobe was shown to have a role in anti-influenza therapeutics as it showed inhibition of influenza virus hemagglutination and cell infection, which the N-lobe was not able to perform [22]. This action of C-lobe was attributed to three C-lobe fragments which strongly bind to the HA(2) region of viral hemagglutinin which is the highly conserved region containing the fusion peptide.

Bovine C-lobe was also found to be effective in the corneal epithelial wound closure in vitro at concentrations of $\geq 6.4 \, \mu M$. On comparison of this effect with the native BLF or N-lobe, C-lobe was almost twice as effective which indicated that BLF C-lobe may be a novel treatment for corneal lesions with delayed healing [26].

In another study, the C-lobe of camel lactoferrin was found to have antiviral activity against hepatitis C virus in a study which involved evaluation of inhibitory effects on hepatitis C virus into Huh7.5 cells after incubation of hepatitis C virus with the C-lobe of lactoferrin prior to infection [21].

Bovine C-lobe was demonstrated to promote bifidobacterial growth despite the fact that it did not bind with surface proteins from bifidobacteria, unlike the N-lobe and native lactoferrin which showed binding to those proteins, indicating that C-lobe may be enhancing bifidobacterial growth using mechanisms other than direct binding with the surface proteins from bifidobacteria [27, 28].

Upon investigating the effects of the two lobes of bovine lactoferrin on the contractile activity of collagen gels, it was found that the C-lobe had a greater effect on collagen gel contractile activity as compared to native bovine lactoferrin or its N-lobe [29]. While this effect of C-lobe was found to be markedly dose dependent, it was completely abolished upon further hydrolysis of the C-lobe with serine proteases like pepsin or trypsin. In another study, using in-gel ATPase activity assays, it was shown that the nucleoside-5'-triphosphate-hydrolyzing activity of lactoferrin is found in the C-lobe [30].

6. Conclusions

Lactoferrin is a multifunctional protein that mediates its various functions using its two molecular halves N-lobe and C-lobe. Though structurally, both the lobes are organized in a similar way, they contain certain stretches of unique sequences, which are significant functionally. N-lobe has been established as the lobe which is primarily involved in the antimicrobial function. On the other hand, C-lobe displayed varied therapeutic functions, due to which this lobe has a potential to be used for diseases like gastropathy, diabetes, and wound healing. The molecule of C-lobe produced proteolytically by cleaving the lactoferrin by proteinase K allowed the determination of its structure. The binding studies and the structure determinations of complexes of C-lobe with NSAIDs and sugars gave insights into the chemical properties and structural aspects of its functional properties. More studies need to done on the C-lobe of lactoferrin in order to harness it as a future drug.

Acknowledgments

The authors acknowledge the financial support from the Department of Biotechnology (DBT), New Delhi. T. P. Singh thanks the Department of Biotechnology (DBT) for the award of Distinguished Biotechnology Research Professorship to him. S. Kaushik thanks Indian Council of Medical Research (ICMR), New Delhi, and M. Sinha thanks Department of Science and Technology (DST), Ministry of Science and Technology, New Delhi for the award of fellowships.

References

[1] P. L. Masson and J. F. Heremans, "Metal-combining properties of human lactoferrin (red milk protein). 1. The involvement of bicarbonate in the reaction," *European Journal of Biochemistry*, vol. 6, no. 4, pp. 579–584, 1968.

[2] P. L. Masson and J. F. Heremans, "Lactoferrin in milk from different species," *Comparative Biochemistry and Physiology B*, vol. 39, no. 1, pp. 119–129, 1971.

[3] R. R. Arnold, M. F. Cole, and J. R. McGhee, "A bactericidal effect for human lactoferrin," *Science*, vol. 197, no. 4300, pp. 263–265, 1977.

[4] J. Brock, "Lactoferrin in human milk: its role in iron absorption and protection against enteric infection in the newborn infant," *Archives of Disease in Childhood*, vol. 55, no. 6, pp. 417–421, 1980.

[5] B. F. Anderson, H. M. Baker, G. E. Norris, D. W. Rice, and E. N. Baker, "Structure of human lactoferrin: crystallographic structure analysis and refinement at 2.8 Å resolution," *Journal of Molecular Biology*, vol. 209, no. 4, pp. 711–734, 1989.

[6] G. B. Jameson, B. F. Anderson, G. E. Norris, D. H. Thomas, and E. N. Baker, "Structure of human apolactoferrin at 2.0 Å resolution. Refinement and analysis of ligand-induced conformational change," *Acta Crystallographica Section D: Biological Crystallography*, vol. 54, no. 6, pp. 1319–1335, 1998.

[7] M. Haridas, B. F. Anderson, H. M. Baker, G. E. Norris, and E. N. Baker, "X-ray structural analysis of bovine lactoferrin at 2.5 Å resolution," *Advances in Experimental Medicine and Biology*, vol. 357, pp. 235–238, 1994.

[8] S. A. Moore, B. F. Anderson, C. R. Groom, M. Haridas, and E. N. Baker, "Three-dimensional structure of diferric bovine lactoferrin at 2.8 Å resolution," *Journal of Molecular Biology*, vol. 274, no. 2, pp. 222–236, 1997.

[9] S. Karthikeyan, S. Yadav, M. Paramasivam, A. Srinivasan, and T. P. Singh, "Structure of buffalo lactoferrin at 3.3 Å resolution at 277 K," *Acta Crystallographica Section D: Biological Crystallography*, vol. 56, no. 6, pp. 684–689, 2000.

[10] S. Karthikeyan, M. Paramasivam, S. Yadav, A. Srinivasan, and T. P. Singh, "Structure of buffalo lactoferrin at 2.5 Å resolution using crystals grown at 303 K shows different orientations of the N and C lobes," *Acta Crystallographica Section D: Biological Crystallography*, vol. 55, no. 11, pp. 1805–1813, 1999.

[11] A. K. Sharma, M. Paramasivam, A. Srinivasan, M. P. Yadav, and T. P. Singh, "Three-dimensional structure of mare diferric lactoferrin at 2.6 Å resolution," *Journal of Molecular Biology*, vol. 289, no. 2, pp. 303–317, 1999.

[12] A. K. Sharma, K. R. Rajashankar, M. P. Yadav, and T. P. Singh, "Structure of mare apolactoferrin: the N and C lobes are in the closed form," *Acta Crystallographica Section D: Biological Crystallography*, vol. 55, no. 6, pp. 1152–1157, 1999.

[13] J. A. Khan, P. Kumar, M. Paramasivam et al., "Camel lactoferrin, a transferrin-cum-lactoferrin: crystal structure of camel apolactoferrin at 2.6 Å resolution and structural basis of its dual role," *Journal of Molecular Biology*, vol. 309, no. 3, pp. 751–761, 2001.

[14] J. H. Brock, F. Arzabe, F. Lampreave, and A. Pineiro, "The effect of trypsin on bovine transferrin and lactoferrin," *Biochimica et Biophysica Acta*, vol. 446, no. 1, pp. 214–225, 1976.

[15] D. Legrand, J. Mazurier, and M. H. Metz Boutigue, "Characterization and localization of an iron-binding 18-kDa glycopeptide isolated from the N-terminal half of human lactotransferrin," *Biochimica et Biophysica Acta*, vol. 787, no. 1, pp. 90–96, 1984.

[16] C. L. Day, K. M. Stowell, E. N. Baker, and J. W. Tweedie, "Studies of the N-terminal half of human lactoferrin produced from the cloned cDNA demonstrate that interlobe interactions modulate iron release," *Journal of Biological Chemistry*, vol. 267, no. 20, pp. 13857–13862, 1992.

[17] B. Sheth, K. M. Stowell, C. L. Day, E. N. Baker, and J. W. Tweedie, "Cloning and expression of the C-terminal lobe of human lactoferrin," *Advances in Experimental Medicine and Biology*, vol. 357, pp. 259–263, 1994.

[18] S. Sharma, T. P. Singh, and K. L. Bhatia, "Preparation and characterization of the N and C monoferric lobes of buffalo lactoferrin produced by proteolysis using proteinase K," *Journal of Dairy Research*, vol. 66, no. 1, pp. 81–90, 1999.

[19] W. S. Kim, K. I. Shimazaki, and T. Tamura, "Expression of bovine lactoferrin C-lobe in Rhodococcus erythropolis and its purification and characterization," *Bioscience, Biotechnology and Biochemistry*, vol. 70, no. 11, pp. 2641–2645, 2006.

[20] S. Sharma, J. Jasti, J. Kumar, A. K. Mohanty, and T. P. Singh, "Crystal structure of a proteolytically generated functional monoferric C-lobe of bovine lactoferrin at 1.9 Å resolution," *Journal of Molecular Biology*, vol. 331, no. 2, pp. 485–496, 2003.

[21] Y. Liao, E. El-Fakkarany, B. Lönnerdal, and E. M. Redwan, "Inhibitory effects of native and recombinant full-length camel lactoferrin and its N and C lobes on hepatitis C virus infection of Huh7.5 cells," *Journal of Medical Microbiology*, vol. 61, part 3, pp. 375–383, 2012.

[22] M. G. Ammendolia, M. Agamennone, A. Pietrantoni et al., "Bovine lactoferrin-derived peptides as novel broad-spectrum inhibitors of influenza virus," *Pathogens and Global Health*, vol. 106, no. 1, pp. 12–19, 2012.

[23] R. Mir, N. Singh, G. Vikram et al., "The structural basis for the prevention of nonsteroidal antiinflammatory drug-induced gastrointestinal tract damage by the C-lobe of bovine colostrum lactoferrin," *Biophysical Journal*, vol. 97, no. 12, pp. 3178–3186, 2009.

[24] R. Mir, N. Singh, G. Vikram et al., "Structural and binding studies of C-terminal half (C-lobe) of lactoferrin protein with COX-2-specific non-steroidal anti-inflammatory drugs (NSAIDs)," *Archives of Biochemistry and Biophysics*, vol. 500, no. 2, pp. 196–202, 2010.

[25] R. Mir, R. P. Kumar, N. Singh et al., "Specific interactions of C-terminal half (C-lobe) of lactoferrin protein with edible sugars: binding and structural studies with implications on diabetes," *International Journal of Biological Macromolecules*, vol. 47, no. 1, pp. 50–59, 2010.

[26] B. Ashby, G. Qian, and W. Mark, "Bovine lactoferrin structures promoting corneal epithelial wound healing *in vitro*," *Investigative Ophthalmology and Visual Science*, vol. 52, no. 5, pp. 2719–2726, 2011.

[27] M. M. Rahman, W. S. Kim, T. Ito, H. Kumura, and K. I. Shimazaki, "Growth promotion and cell binding ability of bovine lactoferrin to *Bifidobacterium longum*," *Anaerobe*, vol. 15, no. 4, pp. 133–137, 2009.

[28] M. Rahman, W. S. Kim, H. Kumura, and K. I. Shimazaki, "Bovine lactoferrin region responsible for binding to bifidobacterial cell surface proteins," *Biotechnology Letters*, vol. 31, no. 6, pp. 863–868, 2009.

[29] Y. Takayama, K. Mizumachi, and T. Takezawa, "The bovine lactoferrin region responsible for promoting the collagen gel contractile activity of human fibroblasts," *Biochemical and Biophysical Research Communications*, vol. 299, no. 5, pp. 813–817, 2002.

[30] S. E. Babina, D. V. Semenov, V. N. Buneva, and G. A. Nevinsky, "Human milk lactoferrin hydrolyzes nucleoside-5'-triphosphates," *Molekulyarnaya Biologiya*, vol. 39, no. 3, pp. 513–520, 2005.

14

Association between Human Plasma Chondroitin Sulfate Isomers and Carotid Atherosclerotic Plaques

Elisabetta Zinellu,[1] Antonio Junior Lepedda,[1] Antonio Cigliano,[1] Salvatore Pisanu,[1] Angelo Zinellu,[2] Ciriaco Carru,[2] Pietro Paolo Bacciu,[3] Franco Piredda,[3] Anna Guarino,[4] Rita Spirito,[4] and Marilena Formato[1]

[1] Dipartimento di Scienze Fisiologiche, Biochimiche e Cellulari, Università delgi Studi di Sassari, 07100 Sassari, Italy
[2] Dipartimento di Scienze Biomediche, Università delgi Studi di Sassari, 07100 Sassari, Italy
[3] Servizio di Chirurgia Vascolare, Clinica Chirurgica Generale, Università delgi Studi di Sassari, 07100 Sassari, Italy
[4] Centro Cardiologico "F. Monzino," IRCCS, Università delgi Studi di Milano, 20122 Milano, Italy

Correspondence should be addressed to Marilena Formato, formato@uniss.it

Academic Editor: Timothy Douglas

Several studies have evidenced variations in plasma glycosaminoglycans content in physiological and pathological conditions. In normal human plasma GAGs are present mainly as undersulfated chondroitin sulfate (CS). The aim of the present study was to evaluate possible correlations between plasma CS level/structure and the presence/typology of carotid atherosclerotic lesion. Plasma CS was purified from 46 control subjects and 47 patients undergoing carotid endarterectomy showing either a soft or a hard plaque. The concentration and structural characteristics of plasma CS were assessed by capillary electrophoresis of constituent unsaturated fluorophore-labeled disaccharides. Results showed that the concentration of total CS isomers was increased by 21.4% ($P < 0.01$) in plasma of patients, due to a significant increase of undersulfated CS. Consequently, in patients the plasma CS charge density was significantly reduced with respect to that of controls. After sorting for plaque typology, we found that patients with soft plaques and those with hard ones differently contribute to the observed changes. In plasma from patients with soft plaques, the increase in CS content was not associated with modifications of its sulfation pattern. On the contrary, the presence of hard plaques was associated with CS sulfation pattern modifications in presence of quite normal total CS isomers levels. These results suggest that the plasma CS content and structure could be related to the presence and the typology of atherosclerotic plaque and could provide a useful diagnostic tool, as well as information on the molecular mechanisms responsible for plaque instability.

1. Introduction

Atherosclerosis is a progressive disease characterized by the accumulation of lipids and fibrous elements in medium and large arteries. Plaque rupture and thrombosis are the most important clinical complication in the pathogenesis of acute coronary syndromes and peripheral vascular disease [1, 2]. Although numerous risk factors such as hypertension, diabetes, and hyperlipidemia are thought to play a role in the development and progression of this pathology [3], the mechanisms underlying plaque formation and progression are still largely unknown. Abnormal expression and structural modifications of arterial chondroitin sulfate proteoglycans (CS-PGs) have been implicated in atherosclerosis

progression [4–6]. Arterial CS-PGs are markedly increased in early atherosclerotic lesions, participating in lipid retention, modification, and accumulation. Furthermore, CS-PGs play a key role in inflammation processes associated with atherosclerosis [6]. Numerous *in situ* lines of evidence indicate that plaque instability is caused by a substantial increase of both proteolytic activity and inflammatory state [7]. However, nowadays specific systemic markers for early diagnosis and prognosis of atheromatous lesions have not yet been identified.

Several studies have evidenced variations in plasma glycosaminoglycan (GAG) levels associated with both physiological and pathological conditions, such as strong physical training [8], chronic lymphocytic leukaemia and essential thrombocythaemia [9], lupus erythematosus [10], and mucopolysaccharidosis [11, 12].

Chondroitin sulfate (CS) is the main GAG in normal human plasma. It consists of 12–18 repeating disaccharide units, each containing a hexuronic acid linked by β $(1 \to 3)$ bond to a N-acetyl-D-galactosamine residue, of which about 30% is sulfated at C-4 hydroxyl group of the hexosamine. It circulates covalently linked to the proteoglycan bikunin [13], a light subunit carrying the antiproteinase activity of plasma serine-proteinase inter-α-inhibitor (IαI). Several data suggest that bikunin plays an important role in inflammation. The IαI family molecules are synthesized in hepatocytes where one or two polypeptides, called the heavy chains (HCs), are linked to the CS chain of bikunin. After a stimulus, the IαI molecules leave the circulation and, in extravascular sites, the HCs are transferred from CS chain to the locally synthesized hyaluronic acid (HA) to form the serum-derived hyaluronan-associated-protein- (SHAP-) HA complex, which plays important roles in stabilizing extracellular matrices and it is often associated with inflammatory conditions [14]. Moreover, it has been described that bikunin also acts as a growth factor for endothelial cells, regulates the intracellular calcium levels, and inhibits kidney stone formation and smooth muscle cell contraction [15].

The aim of the present study was to evaluate possible correlations between plasma chondroitin sulfate (CS) level/structure and the presence/typology of carotid atherosclerotic lesion.

2. Materials and Methods

2.1. Sample Collection. CS isomers analyses were conducted on preoperative plasma samples from 47 patients (stenosis > 70%, 70.4 ± 7.9 years old) undergoing carotid endarterectomy, showing either a soft (26 patients) or a hard (21 patients) plaque, and from 46 healthy normolipidemic volunteers (control group), aged from 20 to 75 (47.8 ± 14.4 years old). The main clinical parameters of the patients under study are reported in Table 1. Lipid profiles and homocysteine levels in patients and controls are reported in Table 2. Plaque typology was assessed by ultrasonography using a Mylab 70 Xvision ecocolordoppler equipped with a LA332 AppleProbe 11–3 MHz (Esaote). Plaques were classified according to the Gray-Weale classification [16] in soft, with hypoechoic features (types 1 and 2), and hard, with hyperechoic features (types 3, 4, and 5). Informed consent was obtained before enrolment. The study was approved by the local ethical committees in accordance with institution guidelines.

2.2. Plasma CS Isomers Analysis. GAG purification was performed by a microanalytic preparative method, as previously described [17]. Briefly, 500 μL of plasma samples were subjected to proteolytic treatment with papain. Plasma CS was purified by anion exchange chromatography and precipitated with 5 volumes of ethanol at −20°C for 24 h. Subsequently, purified plasma CS isomers were subjected to depolymerization by using chondroitin ABC lyase (0.1 U per

TABLE 1: Main clinical parameters of the 47 patients undergoing carotid endarterectomy according to plaque typology.

	Hard (21)	Soft (26)	All (47)
Age	70.9 ± 5.1	70.2 ± 9.3	70.4 ± 7.9
Sex ratio (m/f)	2/1	1.3/1	1.6/1
Body mass index (kg/m^2)	26.9 ± 3.4	27.9 ± 3.8	27.5 ± 3.6
Symptomatic (%)	44.4	56.2	51
Transient ischemic attack (%)	33.3	25	28
Ictus (%)	11.1	31.2	22
Diabetes (%)	44.4	25	34
On therapy (%)	75	75	75
HbA$_{1C}$ (%)	4.5 ± 2.6	4.5 ± 3.0	4.5 ± 2.8
Hypertension (%)	100	93.75	97
On therapy (%)	100	100	100
Dyslipidemic (%)	77.8	75	76
On therapy (%)	100	100	100
Smokers (%)	77.8	68.7	73.0
CRP (mg/dL)	1.16 ± 1.05	1.34 ± 0.84	1.26 ± 0.92

100 μg hexuronic acid) and the unsaturated disaccharides were derivatized with 0.1 mol/L 2-aminoacridone (AMAC) [18].

Separation and quantitation of chondroitin sulfate-derived disaccharides was obtained by capillary electrophoresis (CE) analysis by using a P/ACE capillary electrophoresis system (Beckman) equipped with a 75 mm id and 47 cm length uncoated fused-silica capillary using a 60 mmol/L sodium acetate buffer, containing 0.05% methylcellulose [17]. Separations were carried out at 25°C and monitored with a laser-induced fluorescence (LIF) detector at 488 nm excitation and 520 nm emission wavelengths. For quantitative analyses a homemade standard CS was purified from human plasma pools by a preparative approach. Briefly, standard CS isomers were obtained from papain-treated plasma by anion exchange chromatography, assayed for hexuronic acid content [19], lyophilized into aliquots, and stored at −20°C. For Δ-disaccharide analyses, a calibration curve was determined by submitting plasma-purified CS isomers to chondroitin ABC lyase treatment and derivatization procedure.

CS levels were expressed as μg of hexuronic acid per mL of plasma (μg$_{UA}$/mL), and CS charge density was evaluated as ratio of 4-sulfated Δ-disaccharides (ΔDi-4S) and total unsaturated disaccharides (ΔDi-4S + ΔDi-nonS).

2.3. Statistical Analyses. Statistical analyses were conducted by using SigmaStat software (Systat Software, Inc.). Student's t-test was used for evaluating differences between normally distributed data, while the Mann-Whitney Rank Sum test was applied to nonparametric ones. Values of $P < 0.05$ were considered to be significant. Correlations between CS content, CS charge density, and age were determined by the Pearson Product Moment Correlation test with 95% confidence intervals.

TABLE 2: Lipid profiles and homocysteine levels in patients and controls.

	Hard (21)	Soft (26)	All patients (47)	Controls (46)
Age	70.9 ± 5.1	70.2 ± 9.3	70.4 ± 7.9	47.8 ± 14.4
Sex ratio (m/f)	2/1	1.3/1	1.6/1	1/1.5
Triglycerides (mg/dL)	119.9 ± 55.1	112.7 ± 67.1	115.4 ± 61.7	86.6 ± 28.7
Total cholesterol (mg/dL)	187.9 ± 43.3	164.5 ± 38.3	173.2 ± 41.0	175.7 ± 17.5
LDL cholesterol (mg/dL)	102.4 ± 35.9	90.9 ± 35.6	95.2 ± 35.4	108.3 ± 12.6
HDL cholesterol (mg/dL)	61.2 ± 17.0	51.5 ± 10.9	55.1 ± 14.0	50.1 ± 10.9
Homocysteine (μmol/L)	12.5 ± 4.7	11.5 ± 4.1	11.8 ± 4.3	10.32 ± 3.27

TABLE 3: Total plasma CS, ΔDi-4S, and ΔDi-nonS levels and CS charge density in control subjects, in the totality of patients and according to plaque typology.

	Total CS (μg_{UA}/mL)	ΔDi-4S (μg_{UA}/mL)	ΔDi-nonS (μg_{UA}/mL)	#CS charge density (%)
Controls ($n = 46$)	5.17 ± 1.48	1.61 ± 0.56	3.56 ± 0.99	30.8 ± 4.5
Patients ($n = 47$)	6.28 ± 2.28	1.82 ± 0.77	4.46 ± 1.59	28.6 ± 4.6
Soft ($n = 26$)	6.76 ± 2.16	2.03 ± 0.75	4.73 ± 1.48	29.7 ± 4.0
Hard ($n = 21$)	5.69 ± 2.34	1.57 ± 0.74	4.12 ± 1.69	27.1 ± 4.8
Patient versus control	**0.009***	0.132	**0.002**	**0.022**
Soft versus control	**<0.001**	**0.017**	**0.001**	0.337
Hard versus control	0.321*	0.81	0.169	**0.004**
Hard versus soft	0.114	**0.04**	0.202	0.055

P values obtained by Student's t-tests for normally distributed parameters or by *the Mann-Whitney Rank Sum tests for nonparametric ones are reported. #CS charge density was evaluated as ratio between ΔDi-4S and the sum of ΔDi-nonS and ΔDi-4S. Significant differences are reported in bold ($P < 0.05$).

3. Results

Plasma samples from atherosclerotic patients and from healthy volunteers were analysed. Patients were sorted in two groups according to the typology of their carotid lesion (either soft or hard) as evaluated by ecocolordoppler ultrasonography. The two groups of patients did not differ for the main clinical (Table 1) and biochemical parameters (Table 2), neither as a whole nor subsorting for gender. Lipid profiles and homocysteine levels (Table 2) were similar in patient and control groups ($P > 0.05$).

CE was used to analyze the quantity and fine structure of plasma CS isomers. The adopted method allows the simultaneous determination of hyaluronan- and CS-derived disaccharides with a good reproducibility of both the migration times (CV%, 0.25) and the peak areas (CV%, 1.4). Intra- and interassay CVs were 5.37 and 7.23%, respectively, and analytical recovery was about 86% [17]. This method has been validated by a comparison with a reference assay (fluorophore-assisted carbohydrate gel electrophoresis) using the Passing and Bablock regression analysis and the Bland-Altman test as specific statistical methods for measurement comparison [17].

CE analyses of plasma CS isomers showed the presence of two main unsaturated disaccharides, the nonsulfated (ΔDi-nonS) and the 4-monosulfated (ΔDi-4S).

To rule out any influence of age on our analytical and structural results, we evaluated its association with both content and charge density of plasma CS in control group. In this respect, no correlation was found even after sorting

controls for gender. The age did not correlate neither with ΔDi-nonS ($r = -0.219$; $P = 0.143$) nor with ΔDi-4S ($r = -0.002$; $P = 0.143$) levels.

Statistical analyses allowed us to evidence significant differences between the whole group of patients and the group of controls (Table 3) consisting in an increase of undersulfated CS levels (4.46 ± 1.59 versus $3.56 \pm 0.99 \mu g_{UA}$/mL, $P = 0.002$), with consequent increase of total CS content (6.28 ± 2.28 versus $5.17 \pm 1.48 \mu g_{UA}$/mL, $P = 0.009$) and reduction in its charge density (28.6 ± 4.6 versus $30.8 \pm 4.5\%$, $P = 0.022$). Interestingly, after sorting for plaque typology, we evidenced significant differences in total CS concentration only in patients with a soft plaque with respect to healthy subjects (6.76 ± 2.16 versus $5.17 \pm 1.48 \mu g_{UA}$/mL, $P < 0.001$), due to significantly higher levels of both ΔDi-nonS (4.73 ± 1.48 versus $3.56 \pm 0.99 \mu g_{UA}$/mL, $P = 0.001$) and ΔDi-4S (2.03 ± 0.75 versus 1.61 ± 0.56, $P = 0.017$), whereas patients with hard plaques showed quite normal levels. In plasma from patients with a soft plaque both normosulfated and undersulfated CS isomers were significantly increased and their relative proportions were unchanged. So, no significant changes in CS charge density were detected. On the contrary, in plasma from patients with hard plaques, although the levels of both CS isomers were quite normal, a significant difference in their relative proportions was found with respect to controls, producing a significantly lower CS charge density (-12%).

No differences by Student's t-test in both total CS content and CS charge density emerged after subsorting both patients and controls for gender.

4. Discussion

Human normal plasma contains principally an undersulfated form of chondroitin at a concentration of about $4\,\mu g_{UA}/mL$ [20], whose origin and physiological roles have not yet been fully elucidated. It circulates covalently linked to bikunin [13, 14], or as a main product of tissue catabolism or produced by blood cells, such as lymphocytes, associated with a variety of plasma proteins [21, 22]. It is known that the plasma GAG association with low-density lipoproteins could affect some of their physicochemical properties [23] and that some physiological and pathological conditions could lead to an increase in plasma GAG levels [8–12]. Moreover, mediators of inflammation such as some cytokines and growth factors are able to modulate the size, the degree, and the pattern of sulfation, as well as the degree of epimerization of the GAG chains [24–26].

In this work we studied possible correlations between plasma CS level/structure and presence/typology of carotid atherosclerotic lesion. In a previous paper [17], we developed a sensitive and reproducible analytical method for the quantitative and structural evaluation of human plasma CS isomers using capillary electrophoresis (CE). Herein, this analytical approach was adopted to evaluate CS concentrations and structural characteristics in 46 healthy human subjects and in 47 atherosclerotic patients having either a soft or a hard plaque.

Plaque typology was assessed by duplex ultrasonography which represents a noninvasive method for the carotid plaque characterization. This technique allows to detect areas with different shades of grey that provide information on plaque consistency. In particular, hypoechoic features are associated with lipid-rich carotid plaques, while hyperechoic features are with fibrous or fibrocalcific ones [16]. In this respect, several recent ultrasound studies have demonstrated that hypoechoic plaques, with a low GSM (gray-scale median) value, were associated with an increased risk of cerebrovascular ischemic events [27–29]. So, the soft plaque shows characteristics of instability and propension to rupture. It is generally held that plaque instability is caused by a substantial increase in proteolytic activity and inflammatory status.

To rule out any influence of age on our results, we evaluated its association with plasma CS content and sulfation pattern by means of Pearson's correlation. In this respect, no correlation was found. Few conflicting data regarding the relationship between plasma GAG content and ageing are present in the literature. Some investigations found that total plasma GAG content does not vary with age [30, 31]. Conversely, a positive correlation between total plasma GAG levels and age has been reported in males [32]. Qualitative analysis of intact plasma GAGs by using cellulose acetate electrophoresis has shown a decline of CS with proceeding age [33], while the structural analysis of plasma CS after depolymerization with specific chondro-/dermatolyases has revealed a significant increase of CS amount and its charge density depending on age [34]. On the basis of our results, both plasma CS levels and its charge density seem to be

unaffected by age. Moreover, no correlation was found with gender.

With regard to the influence of atherosclerotic lesion presence, we found significant differences in total plasma CS content (+21.4%), between the group of patients and the controls. These differences were ascribable to significantly higher levels of undersulfated CS in these patients (+25.3%).

The statistical analyses of the influence of atherosclerotic lesion typology on plasma CS content and sulfation pattern showed that in presence of soft plaques CS content was significantly increased, without significant changes in its charge density, whereas in presence of hard plaques its charge density was reduced, without changes in its content.

The significance of the observed modifications in plasma CS isomers of patients undergoing endarterectomy could be related to different PG/GAG metabolism in vascular tissue in presence of soft/hard atherosclerotic plaques. Atherosclerosis has been associated with a biosynthetic imbalance of chondroitin sulfate proteoglycans [4–6]. It has been described that the ratio of 6-sulfated to 4-sulfated disaccharides is increased in atherosclerotic type II aortas and significantly decreased in atherosclerotic type V, indicating that vascular concentration of GAGs is differently affected during the progression of the disease [4]. Increased levels of both undersulfated and normosulfated CS have been described in postoperative serum samples of patients submitted to coronary artery bypass surgery and proposed as indicative of an inflammatory state of the patient [35].

In this regard, it has been reported that in inflammatory diseases the CS chains carried by bikunin increase in size proportionally to the severity of the inflammatory response [36], while their sulfation degree decreases [37].

Moreover, several studies reported that sulfation pattern of CS isomers may be important for their protective role from oxidative damage [38–43]. The antioxidant properties of GAGs could be explained by both GAGs chelating properties on divalent cations (such as Cu^{2+} and Fe^{2+}, responsible for the initiation of hydroxyl radical reactions) and their improving effects on endogenous antioxidant defences. In particular, the anti-oxidant properties of CS isomers seem to be related to the sulfation at position 4 of galactosamine residue of disaccharide units.

On a whole, our data show that plasma CS level and structure are significantly different in atherosclerotic patients with respect to controls and that these differences do not depend on age. Moreover, the obtained results suggest that both content and distribution of the circulating CS isomers are associated with the presence as well as with the typology of carotid plaque. If plasma CS levels and structure reflect, almost partly, the CS-PGs metabolism in affected vascular tissues, the determination of plasma CS isomers may provide information on molecular mechanisms of atherosclerosis progression.

Therefore, evaluating content and distribution of plasma CS isomers could be useful in diagnosis of carotid atherosclerosis. Further studies on subjects with preclinical atherosclerosis would be advisable. Indeed, finding an association between these molecules and early atherosclerosis stages could provide an important tool for the follow-up of patients in attempting to prevent inauspicious cerebrovascular events.

Acknowledgment

This study was supported by grants from "Fondazione Banco di Sardegna," Sassari, Italy.

References

[1] P. Libby, "Inflammation in atherosclerosis," *Nature*, vol. 420, no. 6917, pp. 868–874, 2002.

[2] E. Lutgens, R. J. Van Suylen, B. C. Faber et al., "Atherosclerotic plaque rupture: local or systemic process?" *Arteriosclerosis, Thrombosis, and Vascular Biology*, vol. 23, no. 12, pp. 2123–2130, 2003.

[3] A. J. Lusis, "Atherosclerosis," *Nature*, vol. 407, no. 6801, pp. 233–241, 2000.

[4] A. D. Theocharis, D. A. Theocharis, G. De Luca, A. Hjerpe, and N. K. Karamanos, "Compositional and structural alterations of chondroitin and dermatan sulfates during the progression of atherosclerosis and aneurysmal dilatation of the human abdominal aorta," *Biochimie*, vol. 84, no. 7, pp. 667–674, 2002.

[5] A. D. Theocharis, I. Tsolakis, G. N. Tzanakakis, and N. K. Karamanos, "Chondroitin sulfate as a key molecule in the development of atherosclerosis and cancer progression," *Advances in Pharmacology*, vol. 53, pp. 281–295, 2006.

[6] D. E. Karangelis, I. Kanakis, A. P. Asimakopoulou et al., "Glycosaminoglycans as key molecules in atherosclerosis: the role of versican and hyaluronan," *Current Medicinal Chemistry*, vol. 17, no. 33, pp. 4018–4026, 2010.

[7] M. Formato, M. Farina, R. Spirito et al., "Evidence for a proinflammatory and proteolytic environment in plaques from endarterectomy segments of human carotid arteries," *Arteriosclerosis, Thrombosis, and Vascular Biology*, vol. 24, no. 1, pp. 129–135, 2004.

[8] M. Contini, S. Pacini, L. Ibba-Manneschi et al., "Modification of plasma glycosaminoglycans in long distance runners," *British Journal of Sports Medicine*, vol. 38, no. 2, pp. 134–137, 2004.

[9] L. Calabrò, C. Musolino, G. Spatari, R. Vinci, and A. Calatroni, "Increased concentration of circulating acid glycosaminoglycans in chronic lymphocytic leukaemia and essential thrombocythaemia," *Clinica Chimica Acta*, vol. 269, no. 2, pp. 185–199, 1998.

[10] C. Friman, D. Nordstrom, and I. Eronen, "Plasma glycosaminoglycans in systemic lupus erythematosus," *Journal of Rheumatology*, vol. 14, no. 6, pp. 1132–1134, 1988.

[11] S. L. Ramsay, P. J. Meikle, and J. J. Hopwood, "Determination of monosaccharides and disaccharides in mucopolysaccharidoses patients by electrospray ionisation mass spectrometry," *Molecular Genetics and Metabolism*, vol. 78, no. 3, pp. 193–204, 2003.

[12] S. Tomatsu, M. A. Gutierrez, T. Ishimaru et al., "Heparan sulfate levels in mucopolysaccharidoses and mucolipidoses," *Journal of Inherited Metabolic Disease*, vol. 28, no. 5, pp. 743–757, 2005.

[13] H. Toyoda, S. Kobayashi, S. Sakamoto, T. Toida, and T. Imanari, "Structural analysis of a low-sulfated chondroitin sulfate chain in human urinary trypsin inhibitor," *Biological and Pharmaceutical Bulletin*, vol. 16, no. 9, pp. 945–947, 1993.

[14] L. Zhuo, V. C. Hascall, and K. Kimata, "Inter-α-trypsin inhibitor, a covalent protein-glycosaminoglycan-protein complex," *Journal of Biological Chemistry*, vol. 279, no. 37, pp. 38079–38082, 2004.

[15] E. Fries and A. M. Blom, "Bikunin—not just a plasma proteinase inhibitor," *International Journal of Biochemistry and Cell Biology*, vol. 32, no. 2, pp. 125–137, 2000.

[16] A. C. Gray-Weale, J. C. Graham, J. R. Burnett, K. Byrne, and R. J. Lusby, "Carotid artery atheroma: comparison of preoperative B-mode ultrasound appearance with carotid endarterectomy specimen pathology," *Journal of Cardiovascular Surgery*, vol. 29, no. 6, pp. 676–681, 1988.

[17] A. Zinellu, S. Pisanu, E. Zinellu et al., "A novel LIF-CE method for the separation of hyalurnan- and chondroitin sulfate-derived disaccharides: application to structural and quantitative analyses of human plasma low- and high-charged chondroitin sulfate isomers," *Electrophoresis*, vol. 28, no. 14, pp. 2439–2447, 2007.

[18] F. Lamari, A. Theocharis, A. Hjerpe, and N. K. Karamanos, "Ultrasensitive capillary electrophoresis of sulfated disaccharides in chondroitin/dermatan sulfates by laser-induced fluorescence after derivatization with 2-aminoacridone," *Journal of Chromatography B*, vol. 730, no. 1, pp. 129–133, 1999.

[19] T. Bitter and H. M. Muir, "A modified uronic acid carbazole reaction," *Analytical Biochemistry*, vol. 4, no. 4, pp. 330–334, 1962.

[20] N. Volpi and F. Maccari, "Microdetermination of chondroitin sulfate in normal human plasma by fluorophore-assisted carbohydrate electrophoresis (FACE)," *Clinica Chimica Acta*, vol. 356, no. 1-2, pp. 125–133, 2005.

[21] F. Pasquali, C. Oldani, M. Ruggiero, L. Magnelli, V. Chiarugi, and S. Vannucchi, "Interaction between endogenous circulating sulfated-glycosaminoglycans and plasma proteins," *Clinica Chimica Acta*, vol. 192, no. 1, pp. 19–27, 1990.

[22] A. Calatroni, R. Vinci, and A. M. Ferlazzo, "Characteristics of the interactions between acid glycosaminoglycans and protein in normal human plasma as revealed by the behaviour of the protein-polysaccharide complexes in ultrafiltration and chromatographic procedures," *Clinica Chimica Acta*, vol. 206, no. 3, pp. 167–180, 1992.

[23] G. M. Cherchi, M. Formato, P. Demuro, M. Masserini, I. Varani, and G. DeLuca, "Modifications of low density lipoprotein induced by the interaction with human plasma glycosaminoglycan-protein complexes," *Biochimica et Biophysica Acta*, vol. 1212, no. 3, pp. 345–352, 1994.

[24] A. Bassols and J. Massague, "Transforming growth factor β regulates the expression and structure of extracellular matrix chondroitin/dermatan sulfate proteoglycans," *Journal of Biological Chemistry*, vol. 263, no. 6, pp. 3039–3045, 1988.

[25] E. Schonherr, H. T. Jarvelainen, L. J. Sandell, and T. N. Wight, "Effects of platelet-derived growth factor and transforming growth factor-β1 on the synthesis of a large versican-like chondroitin sulfate proteoglycan by arterial smooth muscle cells," *Journal of Biological Chemistry*, vol. 266, no. 26, pp. 17640–17647, 1991.

[26] E. Tufvesson and G. Westergren-Thorsson, "Alteration of proteoglycan synthesis in human lung fibroblasts induced by interleukin-1β and tumor necrosis factor-α," *Journal of Cellular Biochemistry*, vol. 77, no. 2, pp. 298–309, 2000.

[27] M. L. M. Grønholdt, B. G. Nordestgaard, T. V. Schroeder, S. Vorstrup, and H. Sillesen, "Ultrasonic echolucent carotid plaques predict future strokes," *Circulation*, vol. 104, no. 1, pp. 68–73, 2001.

[28] J. F. Polak, L. Shemanski, D. H. O'Leary et al., "Hypoechoic plaque at US of the carotid artery: an independent risk factor for incident stroke in adults aged 65 years or older," *Radiology*, vol. 208, no. 3, pp. 649–654, 1998.

[29] A. D. Giannoukas, G. S. Sfyroeras, M. Griffin, V. Saleptsis, G. A. Antoniou, and A. N. Nicolaides, "Association of plaque echostructure and cardiovascular risk factors with symptomatic carotid artery disease," *Vasa - Journal of Vascular Diseases*, vol. 38, no. 4, pp. 357–364, 2009.

[30] K. Sames, *The Role of Proteoglycans and Glycosaminoglycans in Aging*, Karger, Hamburg, Germany, 1994.

[31] "Glycosaminoglycans of blood," in *Mucopolysaccharides (glycosaminoglycans) of Body Fluids in Health and Disease*, R. Varma and R. S. Varma, Eds., pp. 449–508, Walter de Gruyter, Berlin, Germany, 1983.

[32] P. W. Larking, "Total glycosaminoglycans in the plasma of adults: effects of age and gender, and relationship to plasma lipids: A preliminary study," *Biochemical Medicine and Metabolic Biology*, vol. 42, no. 3, pp. 192–197, 1989.

[33] K. B. Komosińska-Vassev, K. Winsz-Szczotka, K. Kuznik-Trocha, P. Olczyk, and K. Olczyk, "Age-related changes of plasma glycosaminoglycans," *Clinical Chemistry and Laboratory Medicine*, vol. 46, no. 2, pp. 219–224, 2008.

[34] N. Volpi and F. Maccari, "Chondroitin sulfate in normal human plasma is modified depending on the age. Its evaluation in patients with pseudoxanthoma elasticum," *Clinica Chimica Acta*, vol. 370, no. 1-2, pp. 196–200, 2006.

[35] D. Karangelis, A. Asimakopoulou, I. Kanakis et al., "Monitoring serum chondroitin sulfate levels in patients submitted to coronary artery bypass surgery," *Biomedical Chromatography*, vol. 25, no. 7, pp. 748–750, 2011.

[36] C. Mizon, C. Mairie, M. Balduyck, E. Hachulla, and J. Mizon, "The chondroitin sulfate chain of bikunin-containing proteins in the inter-α-inhibitor family increases in size in inflammatory diseases," *European Journal of Biochemistry*, vol. 268, no. 9, pp. 2717–2724, 2001.

[37] C. Capon, C. Mizon, J. Lemoine, P. Rodié-Talbère, and J. Mizon, "In acute inflammation, the chondroitin-4 sulphate carried by bikunin is not only longer; It is also undersulphated," *Biochimie*, vol. 85, no. 1-2, pp. 101–107, 2003.

[38] R. Albertini, P. Ramos, A. Giessauf, A. Passi, G. De Luca, and H. Esterbauer, "Chondroitin 4-sulphate exhibits inhibitory effect during Cu2+-mediated LDL oxidation," *FEBS Letters*, vol. 403, no. 2, pp. 154–158, 1997.

[39] R. Albertini, G. De Luca, A. Passi, R. Moratti, and P. M. Abuja, "Chondroitin-4-sulfate protects high-density lipoprotein against copper- dependent oxidation," *Archives of Biochemistry and Biophysics*, vol. 365, no. 1, pp. 143–149, 1999.

[40] H. Arai, S. Kashiwagi, Y. Nagasaka, K. Uchida, Y. Hoshii, and K. Nakamura, "Oxidative modification of apolipoprotein E in human very-low-density lipoprotein and its inhibition by glycosaminoglycans," *Archives of Biochemistry and Biophysics*, vol. 367, no. 1, pp. 1–8, 1999.

[41] R. Albertini, A. Passi, P. M. Abuja, and G. De Luca, "The effect of glycosaminoglycans and proteoglycans on lipid peroxidation," *International journal of molecular medicine*, vol. 6, no. 2, pp. 129–136, 2000.

[42] G. M. Campo, A. Avenoso, S. Campo et al., "Hyaluronic acid and chondroitin-4-sulphate treatment reduces damage in carbon tetrachloride-induced acute rat liver injury," *Life Sciences*, vol. 74, no. 10, pp. 1289–1305, 2004.

[43] G. M. Campo, A. Avenoso, A. D'Ascola et al., "Purified human plasma glycosaminoglycans limit oxidative injury induced by iron plus ascorbate in skin fibroblast cultures," *Toxicology in Vitro*, vol. 19, no. 5, pp. 561–572, 2005.

15

The Role of the Cullin-5 E3 Ubiquitin Ligase in the Regulation of Insulin Receptor Substrate-1

Christine Zhiwen Hu,[1] **Jaswinder K. Sethi,**[2] **and Thilo Hagen**[1]

[1] Department of Biochemistry, Yong Loo Lin School of Medicine, National University of Singapore, Singapore 117597, Singapore
[2] Institute of Metabolic Science, Metabolic Research Laboratories, and Department of Clinical Biochemistry, University of Cambridge, Addenbrooke's Hospital, Cambridge CB20QQ, UK

Correspondence should be addressed to Thilo Hagen, bchth@nus.edu.sg

Academic Editor: Emil Pai

Background. SOCS proteins are known to negatively regulate insulin signaling by inhibiting insulin receptor substrate-1 (IRS1). IRS1 has been reported to be a substrate for ubiquitin-dependent proteasomal degradation. Given that SOCS proteins can function as substrate receptor subunits of Cullin-5 E3 ubiquitin ligases, we examined whether Cullin-5 dependent ubiquitination is involved in the regulation of basal IRS1 protein stability and signal-induced IRS1 degradation. *Findings.* Our results indicate that basal IRS1 stability varies between cell types. However, the Cullin-5 E3 ligase does not play a major role in mediating IRS1 ubiquitination under basal conditions. Protein kinase C activation triggered pronounced IRS1 destabilization. However, this effect was also independent of the function of Cullin-5 E3 ubiquitin ligases. *Conclusions.* In conclusion, SOCS proteins do not exert a negative regulatory effect on IRS1 by functioning as substrate receptors for Cullin-5-based E3 ubiquitin ligases both under basal conditions and when IRS1 degradation is induced by protein kinase C activation.

1. Introduction

Insulin signaling is an important cellular process which regulates glucose uptake and utilization, lipid and protein synthesis as well as transcriptional responses. Binding of insulin to the insulin receptor (IR) leads to autophosphorylation of the tyrosine kinase domain as well as other proteins which interact with the IR tyrosine kinase. One of the major IR substrates is the insulin receptor substrate (IRS) protein, which is able to dock onto the IR. There are six IRS related proteins, including IRS1-4, Gab1, and p62dok. Upon insulin stimulation, these proteins are phosphorylated on tyrosine residues and subsequently act as a multisite docking protein for src homology 2 (SH2) domain proteins, such as p85 phosphatidylinositol 3-kinase (PI3-kinase). Consequently, the activation of these SH2 domain proteins initiates the insulin dependent signaling cascade. One well-characterized downstream pathway is AKT-dependent translocation and activation of glucose transporters. Another is AKT-dependent activation of the mTOR complex 1

(mTORC1) and the downstream serine/threonine kinase p70 S6 kinase 1 (S6K1). Functionally, this pathway is implicated in regulating transcription, autophagy, ribosome biogenesis, and protein stability. S6K1 has also been shown to directly phosphorylate IRS1 and consequently exerts a negative feedback regulation on IRS1.

Another family of SH2 domain containing proteins is the SOCS (Suppressor Of Cytokine Signaling) protein family. These proteins have originally been implicated in the inhibition of the cellular response to cytokine stimulation. However, SOCS proteins are also known to play a role in regulating the insulin signaling pathway. SOCS protein expression in several tissues and cell lines is increased by insulin [1], and the induced SOCS proteins in turn were shown to negatively regulate insulin signaling [2], suggesting that they also mediate negative feedback regulation of the insulin signaling pathway.

There are eight members in the SOCS protein family, SOCS1 to SOCS7 and (cytokine-inducible SH2 domain-containing protein) CIS [3]. The sequence of all SOCS proteins is similar, with a variable N-terminal domain, a central SH2 domain and a conserved C terminal SOCS box. While the importance of the SH2 domain for the inhibitory action of SOCS proteins is well established, it is currently not clear to what extent and via what mechanisms the SOCS box contributes to the activity. Interestingly, the SOCS box of these proteins is similar to the α domain of the pVHL protein [4]. It is well known that pVHL associates with elongin B/C and Cullin-2, forming an E3 ubiquitin ligase complex which facilitates the ubiquitination and subsequent degradation of hypoxia inducible factor-1α (HIF-1α), an important mediator of cellular oxygen sensing [5]. Indeed, SOCS proteins have been shown to bind to elongin B/C in vitro and in vivo [6–9]. Therefore, it can be hypothesized that SOCS proteins may also be able to regulate important proteins in the insulin signaling pathway at the level of their protein stability. Additional clues for the possible function of SOCS proteins come from the finding that, similar to pVHL, SOCS proteins contain a Cullin box, which mediates the binding to Cullin proteins. The Cullin box confers specificity for Cullin proteins, as shown by pVHL binding to CUL2 and SOCS protein binding to Cullin-5 [10].

Cullins are scaffold proteins for the assembly of Cullin RING domain E3 ubiquitin ligases. There are seven mammalian cullin proteins (Cullin-1 to Cullin-7), which bind to adaptor proteins and substrate receptor subunits via their N-terminus. This substrate receptor module is responsible for recruiting E3 ligase substrates. For instance, Cullin-5 acts as a scaffold protein which recruits the adaptor proteins elongin B/C and different substrate receptors including SOCS proteins. Rbx2 is a RING domain-containing protein which binds to the C-terminus of Cullin-5 and recruits the E2 conjugating enzyme [10], to facilitate the transfer of ubiquitin onto the substrate. Cullin E3 ligase-mediated polyubiquitination subsequently leads to recognition and degradation of the substrate by the 26S proteasome.

Interestingly, it has been reported that SOCS1 and SOCS3 bind to IRS1 and promote the ubiquitination and degradation of the IRS1 protein [11]. Therefore, the aim of this study is to determine whether Cul5 E3 ubiquitin ligases, utilizing SOCS proteins as adaptor proteins, are involved in the basal and signal induced degradation of IRS1.

2. Results

2.1. Measurement of Basal IRS1 Protein Stability in Different Cell Lines. To measure basal rates of IRS1 protein stability, several cell lines were treated with a proteasome inhibitor (MG-132) and an inhibitor of protein synthesis (cycloheximide). Treatment with cycloheximide for 6 hours resulted in a marked reduction in IRS1 protein concentrations in HEK293T, HEK293, and HeLa cells, whereas the effect in MCF7 and 3T3-L1 cells was less pronounced. Similarly, treatment with MG-132 caused a moderate increase in IRS1 protein levels in HEK293T, HEK293, and Hela cell lines but was without effect in MCF7 and 3T3-L1 cell lines. Thus, the IRS1 protein in HEK293T, HEK293 and Hela cells is less stable than in MCF7 and 3T3-L1 cells (Figure 1(a)).

To address the potential involvement of Cullin E3 ligases in regulating basal IRS1 stability, we used the Nedd8 E1 activating enzyme (NAE) inhibitor MLN4924 which inhibits all members of the Cullin E3 ligase family [12, 13]. Cullin E3 ligases require the modification of the cullin protein with the ubiquitin-like protein Nedd8 for their activity. Treatment of cells with MLN4924 is known to result in rapid cullin deneddylation and hence Cullin E3 ligase inhibition. The inhibitory effect of MLN4924 on the Cullin E3 ligase family was confirmed by the marked increase in the protein concentration of p27, HIF-1α, and Nrf2, which are bona fide substrates for Cul1, Cul2, and Cul3 E3 ligases, respectively [14], upon treatment with MLN4924 (Figure 1(b)). As expected, stabilization of these Cullin E3 ligase substrates was not affected by treatment with the transcription inhibitor Actinomycin D (Figure 1(b)).

Upon treatment with MLN4924, the IRS1 protein concentration was observed to increase only in HEK293T cells but not in the other cell lines. We also measured the protein abundance of IRS1 in HEK293 cells in the presence or absence of MLN4924 upon inhibition of protein synthesis with cycloheximide. However, no significant differences in the protein abundance of IRS1 was observed over the time course, supported by densitometry measurements (Figure 1(c)). Even in HEK293T cells, where MLN4924 treatment increased the IRS1 protein to similar levels compared to MG-132 (Figure 1(a)), the contribution of Cullin E3 ligases to IRS1 protein stability is likely to be only partial. Thus, in comparison, MLN4924 increased the protein level of the well characterized Cul1 E3 ligase substrate p27, to even higher levels than MG-132 (Figure 1(a)).

To investigate the involvement of Cul5 in basal IRS1 degradation, we tested whether overexpression or siRNA-mediated silencing of this Cullin homologue in HEK293 cells affects IRS1 protein expression. As shown in Figure 2(a), IRS1 protein levels were not significantly altered upon overexpression or knockdown of Cul5. The above results were confirmed by using cycloheximide treatment in cells with Cul5 knockdown. Upon inhibition of new protein synthesis by cycloheximide, the turnover of IRS1 was very similar in untransfected cells and in cells transfected with control or Cul5 siRNA (Figure 2(b)), strongly suggesting that degradation of basal IRS1 is not mediated by Cul5. Taken together, the results suggest that Cullin E3 ligases do not play a major role in regulating basal IRS1 protein stability in these cells. Given that it has been reported that different physiological signals are able to induce IRS1 degradation, we investigated in further experiments whether Cullin E3 ligase-dependent ubiquitination is involved in these regulatory mechanisms.

2.2. Effect of mTOR/S6K1 and TPA on IRS1 Protein Stability. It is well known that chronic treatment with growth factors induces a negative feedback regulation on IRS1 via the mTOR/S6K1 pathway, leading to S6K1-dependent phosphorylation of IRS1. We therefore tested the effect of inhibiting basal mTOR activity in the presence of serum on IRS1 protein concentrations. As expected, upon addition of the mTOR inhibitor rapamycin, phosphorylation of

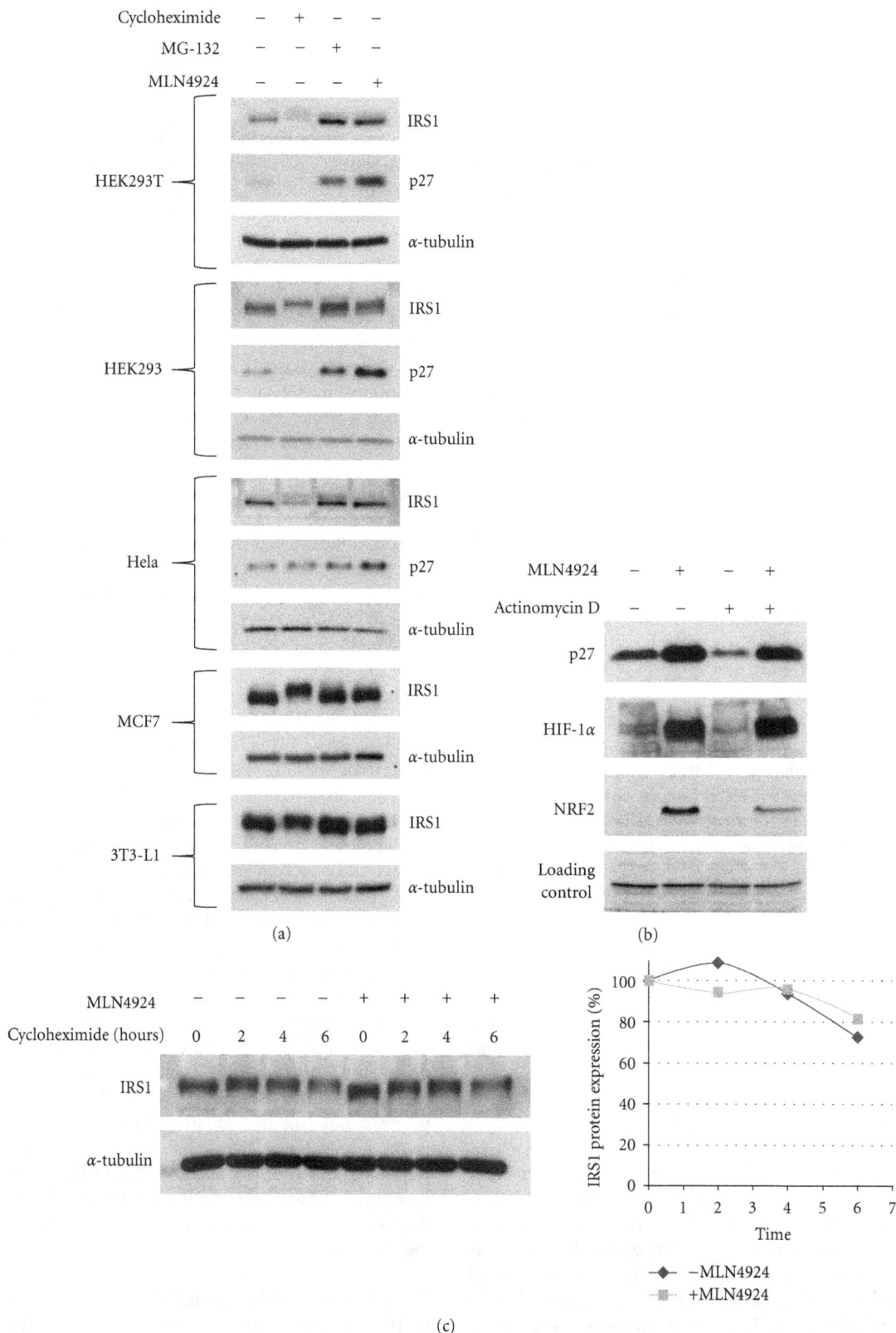

(a)

(b)

(c)

FIGURE 1: Measurement of basal IRS1 protein stability in different cell lines. (a) Cells were treated for 6 hours with the following drugs prior to cell lysis: cycloheximide (40 μM), MG-132 (20 μM), MLN4924 (3 μM). Subsequently, Western blot analysis was performed. (b) HEK293T cells were pretreated with actinomycin D (5 μg/mL) for 25 min before adding MLN4924 (3 μM) for 4 hours, as indicated. (c) HEK293 cells were preincubated with MLN4924 (1 μM) for 2 hours before adding cycloheximide (40 μM) for the respective time points. Cell lysates were analyzed using Western blotting.

Untransfected	+	−	−	−	−	−	−	−
Cul5-V5	−	−	+	+	−	+	+	+
Control siRNA	−	+	−	+	−	−	−	−
Cul5 siRNA-3	−	−	−	−	+	+	−	+
Cul5 siRNA-4	−	−	−	−	+	−	+	+

IRS1

Cul5

V5

GAPDH

(a)

Treatment	DMSO				Cycloheximide			
Untransfected	+	−	−	−	+	−	−	−
Control siRNA	−	+	−	−	−	+	−	−
Cul5 siRNA-3	−	−	+	−	−	−	+	−
Cul5 siRNA-4	−	−	−	+	−	−	−	+

IRS1

Cul5

α-tubulin

(b)

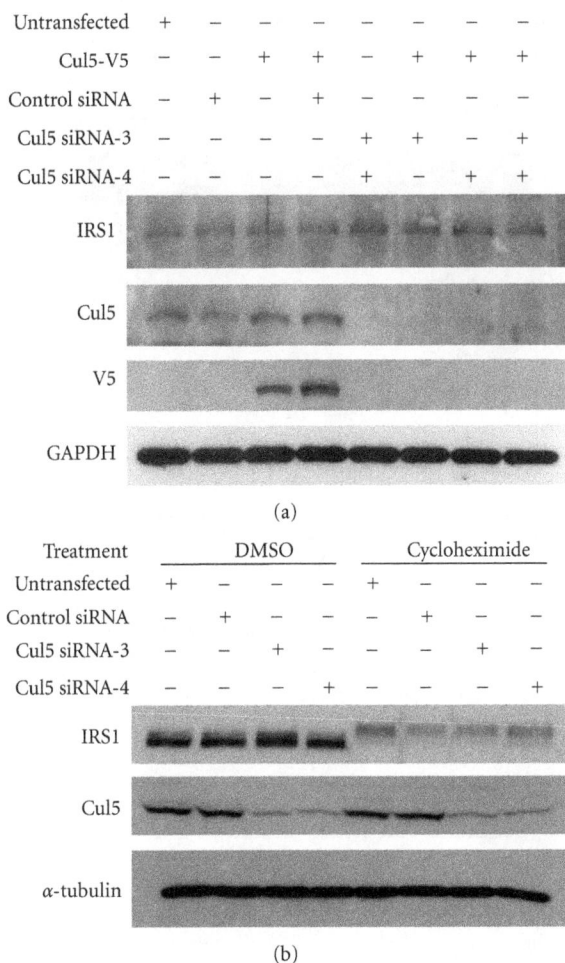

FIGURE 2: Effect of Cullin-5 knockdown on basal IRS1 protein concentrations. (a) Negative control siRNA (NC) or two different Cullin-5 siRNAs were transfected into HEK293 cells, followed by transfection of Cul5-V5 after 24 hours. After further 24 hours, the cells were lysed with detergent lysis buffer and analyzed by Western blotting. (b) HEK293 cells were transfected with negative control or the respective Cul5 siRNA. After three days, the cells were treated with cycloheximide (40 μM) for 6 hours, lysed, and analyzed by Western blotting.

the mTORC1 downstream target p70 S6K was completely prevented (Figure 3(a)). Rapamycin treatment also resulted in an increased mobility of IRS1, indicating that S6K1 contributes to the basal phosphorylation of IRS1. However based on densitometry results from three independent experiments, rapamycin had no effect on IRS1 protein steady-state levels. This result suggests that mTOR/S6K1 does not regulate IRS1 basal protein stability. As previously reported, cycloheximide treatment caused a robust activation of mTORC1 as detected by an increase in S6K activity [15]. Cycloheximide was also found to lower the mobility of IRS1 compared to the untreated control indicating that the phosphorylation of IRS1 was induced. Upon coincubation of cycloheximide with rapamycin, phosphorylation of S6K was fully inhibited and likewise the phosphorylation of IRS1

was inhibited (Figure 3(a), lane 4). The IRS1 protein steady-state level after 6 hours of cycloheximide was decreased to 46% compared to the control. Interestingly, inhibition of mTOR partially prevented the decrease in IRS1 protein (decrease to 74% compared to control). In conclusion, our results suggest that under basal conditions in the presence of growth factors, mTOR/S6K is not involved in regulating IRS1 protein stability. In contrast, upon activation to higher levels, mTOR/S6K may be involved in inducing IRS1 degradation.

Phorbol ester (TPA) has been reported to negatively regulate IRS1 protein concentrations via PKC activation. With this in mind, we used TPA to activate PKC in HEK293T cells. As expected, IRS1 protein concentrations markedly decreased upon addition of TPA, and this effect was reversed in the presence of the PKC inhibitor Gö8963 (Figure 3(b)). Downregulation of IRS1 protein by TPA was prevented in the presence of the proteasome inhibitor MG-132 (see Figure 4(a)), indicating that PKC activation induces IRS1 degradation. To confirm that the TPA-induced IRS1 downregulation is due to an effect on protein stability, a cycloheximide chase experiment was conducted in the presence or absence of TPA. As shown in Figure 3(c), the degradation rate of IRS1 under TPA treatment is faster than in the control.

We also investigated the effect of TPA on transfected IRS1. Unlike the TPA-induced degradation of endogenous IRS1, we observed that transfected IRS1 protein showed a consistent increase in abundance upon treatment with TPA (Figure 3(d)). To determine whether this effect was due to a nonspecific effect of TPA on the CMV promoter of the transfected plasmid, experiments were performed to determine the effect of TPA on endogenous and transfected β-catenin expression. As expected, the endogenous β-catenin protein levels remained unchanged upon treatment with TPA (Figure 3(e)). When β-catenin expression plasmids with either a CMV or a EF2 promoter were transfected into cells, a marked upregulation of transfected β-catenin was observed with both promoters (Figure 3(f)), albeit the CMV promoter showed a greater effect compared to the EF2 promoter. This confirmed that TPA has a nonspecific effect on the promoters of plasmids and therefore, in subsequent experiments we only studied the effects of TPA on endogenous IRS1 protein.

2.3. Cul5 Is Not Involved in TPA-Induced IRS1 Degradation. To determine whether the TPA-induced decrease in endogenous IRS1 is due to Cullin-dependent proteasomal degradation, the NAE inhibitor MLN4924 was added to TPA treated cells. As shown in Figure 4(a), the proteasome inhibitor MG-132 prevented the degradation of IRS1 almost completely, whereas MLN4924 (lane 5) only partially rescued the IRS1 protein expression. Therefore, these results show that TPA-induced IRS1 degradation is only partially dependent on Cullin E3 ligases.

Given the suggested role of SOCS proteins acting as Cul5 substrate receptors in regulating IRS1 stability, we sought to determine the role of the Cul5-based Cullin E3 ligase in TPA-induced IRS1 degradation. To this end, control and Cul5 siRNA duplexes were transfected into cells and 2 days after transfection, TPA was added for

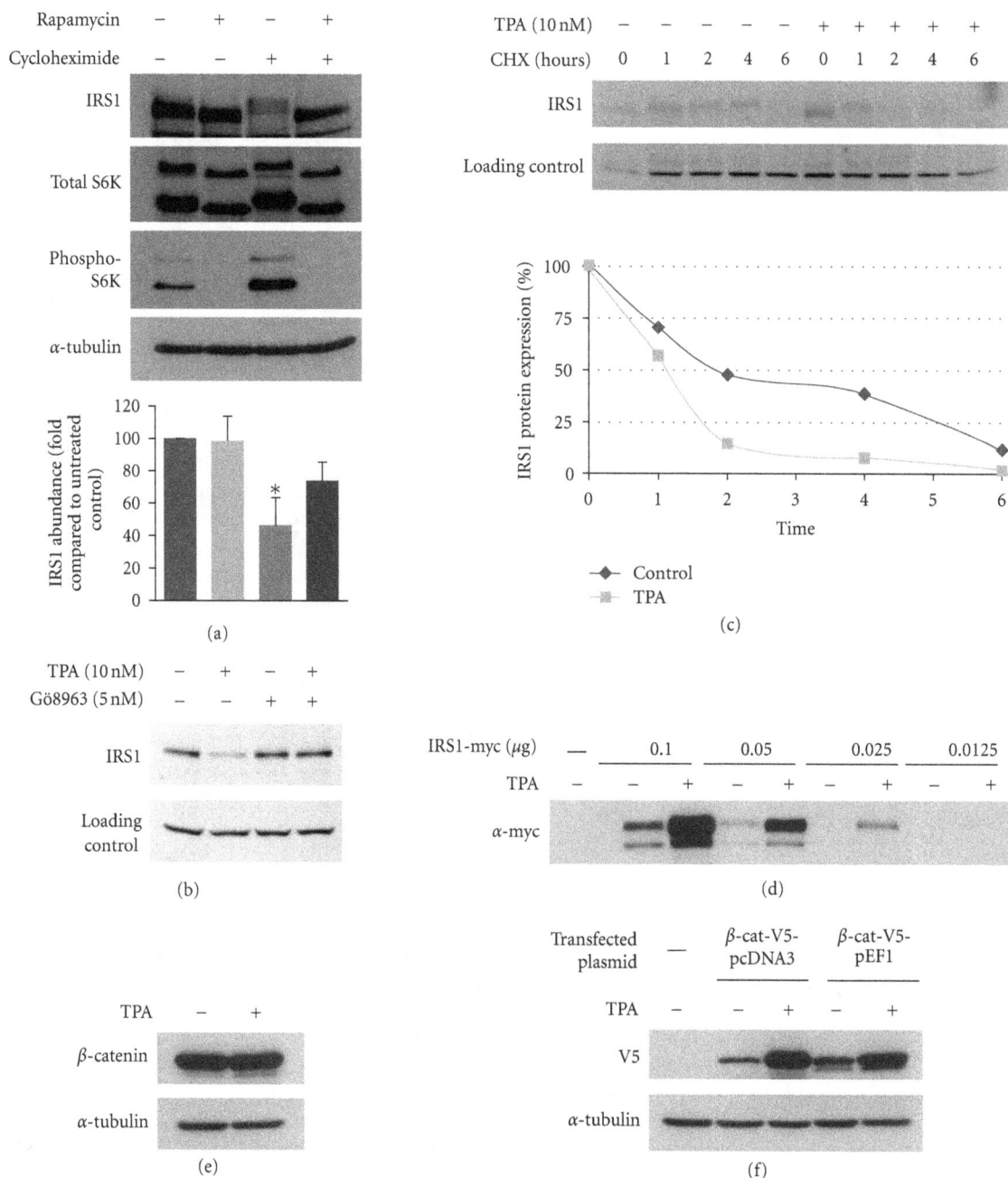

FIGURE 3: Effect of mTOR/S6K1 and TPA on IRS1 protein stability. (a) HEK293T cells were treated with rapamycin (20 nM) and cycloheximide (40 μM), as indicated, for 6 h followed by cell lysis and Western blot analysis. IRS1 protein abundance was determined by densitometry analysis indicated above the band as fold compared to the untreated control. All forms of IRS1 (shifted and not shifted) were included in the densitometry analysis. (b) HEK293T cells were preincubated with Gö8963 (5 nM) for one hour before 24 hour incubation with TPA (10 nM), as indicated. (c) HEK293 cells were pretreated with 10 nM TPA for 1 hour before cycloheximide chase was conducted using the indicated time points. Immunoblot analysis of lysates was then carried out. (d) HEK293T cells were transfected with the specified amounts of IRS1. After the cells reached subconfluence, TPA (10 nM) was added to the medium for 24 hours before cell lysis. (e) HEK293 cells were treated with TPA (10 nM) for 24 hours followed by Western blot analysis of cell lysates for endogenous β-catenin. (f) HEK293 cells were transfected with 0.5 μg of the indicated β-catenin plasmids. After two days, TPA (10 nM) was added for 24 hours followed by Western blot analysis of cell lysates with the indicated antibodies.

24 hours. Consistent with results shown above, Western blot analysis revealed that basal IRS1 protein levels remained unchanged upon Cul5 siRNA-mediated silencing (Figure 4(b)). Importantly, Cul5 knockdown was without effect on the TPA-induced degradation of IRS1, despite a marked reduction in Cul5 protein levels. Furthermore, TPA treatment did not have an effect on IRS2 protein levels.

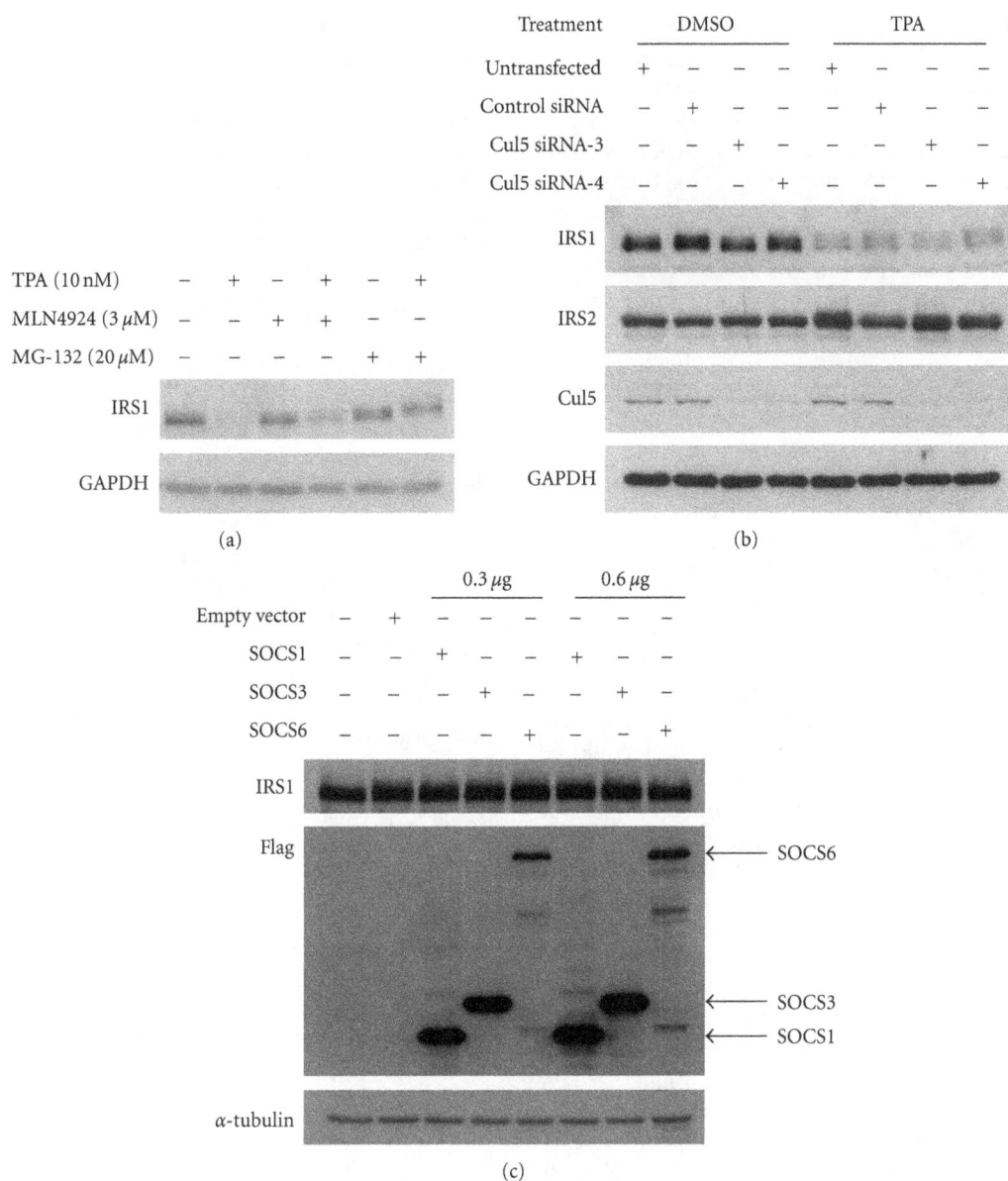

FIGURE 4: Role of the Cullin-5 E3 ligase in TPA-induced IRS1 degradation. (a) HEK293 cells were treated TPA (10 nM) for the last 24 hours, with MLN4924 (3 μM) (24 hours) and MG-132 (20 μM) (6 hours), as indicated. (b) HEK293 cells were transfected with negative control or Cullin-5 siRNA. After two days, TPA (10 nM) was added for 24 hours. Subsequently, the cells were lysed for Western blot analysis. (c) HEK293T cells were transfected with the respective amounts of SOCS plasmid as indicated. Immunoblot analysis of lysates was subsequently carried out.

Finally, to directly test whether SOCS proteins are able to promote the degradation of IRS1, overexpression plasmids for SOCS1, SOCS3, and SOCS6, which have been implicated in inhibiting IRS1 [11], were generated and expressed in HEK293T cells. As shown in Figure 4(c), robust overexpression levels of SOCS1, SOCS3, and SOCS6 did not induce the degradation of IRS1 protein. Coexpression of human insulin receptor and SOCS proteins in both HEK293 and HEK293T cells was also without significant effect on IRS1 protein levels (data not shown). In conclusion, our data suggest that SOCS proteins do not exert a negative regulatory effect on IRS1 by functioning as substrate receptors for the

Cul5-based Cullin E3 ligase, both under basal conditions and when IRS1 degradation is induced by treatment with TPA.

2.4. Discussion. SOCS proteins have been shown to negatively regulate insulin signalling by inhibiting IRS1. The IRS1 protein has been reported to be a substrate for ubiquitin-dependent proteasomal degradation [16, 17]. When measuring basal protein turnover rates of IRS1 in different cell lines, we found that the protein was more unstable in HEK293, HEK293T, and HeLa cells compared to MCF7 and 3T3-L1 cells. The different basal degradation rates may be due to differences in posttranslational modifications in the

IRS1 protein that are important to induce its ubiquitination and degradation or due to differences in the expression levels or activities of E3 ubiquitin ligases involved in IRS1 ubiquitination. SOCS1 and SOCS3 have been reported to induce IRS1 degradation [11]. Given that SOCS proteins can function as substrate receptor subunits for Cullin-5 E3 ubiquitin ligases, it is possible that they function to recruit IRS1 for ubiquitination. However, our results using Cullin-5 knockdown and overexpression of SOCS proteins argue against this possibility. Because the original study by Rui et al. [11] used a cellular system where insulin receptor was overexpressed, we also measured the IRS1 protein stability in cells transfected with a human insulin receptor plasmid. However, overexpression of insulin receptor alone or with various SOCS proteins did not significantly affect IRS1 degradation. Therefore, it is likely that other cullin and non-cullin based E3 ligases exist which can mediate basal IRS1 protein turnover. However, we cannot rule out that there are cell-type specific IRS1 ubiquitination mechanisms that involve Cullin-5 or other Cullin-based E3 ubiquitin ligases.

The phosphorylation of IRS1 is likely to be important for the regulation of its protein stability. For instance, mTOR/S6K exerts a negative feedback on insulin signalling via IRS1 phosphorylation. This negative feedback could be due to inhibition of IRS1 activity and/or induction of IRS1 degradation. Our results suggest that mTOR/S6K might not be involved IRS1 degradation under basal conditions in the presence of serum. However when increased mTOR/S6K signalling occurs, for example, in the presence of cycloheximide, IRS1 protein becomes more unstable.

The phorbol ester TPA is a well known activator of protein kinase C α (PKCα) [18]. PKCα phosphorylates a wide range of substrates, including IRS proteins. It has been reported that PKCα-mediated phosphorylation causes inhibition of IRS1 activity [19]. In our studies we found that TPA decreased the IRS1 protein half-life. Furthermore, the PKC inhibitor Gö8963 was able to restore the IRS1 protein level in the presence of TPA. In order to characterize the molecular basis of TPA-induced IRS1 protein degradation, that is, to identify involved phosphorylation sites and IRS1 ubiquitin modification, we intended to use transfected versions of IRS1. However, our experimental results indicated that the transfected IRS1 protein was consistently upregulated upon treatment with TPA, and this effect is likely due to a nonspecific effect of TPA on the promoter of the transfected plasmid.

Since TPA induced a robust degradation of the endogenous IRS1 protein, the role of Cul5 mediated ubiquitination was determined using NAE inhibitor MLN4924 and siRNA-mediated gene silencing. Cul5 knockdown did not restore IRS1 protein levels in the presence of TPA, suggesting that Cullin-5 might not be the Cullin E3 ligase responsible for the degradation of IRS1. We observed that MLN4924 partially inhibited TPA-induced IRS1 degradation. However, it appears unlikely that TPA induces IRS1 degradation via more than one cellular E3 ubiquitin ligase. Therefore, the observed effect of MLN4924 may be due to inhibition of basal IRS1 protein turnover, while TPA induced degradation is likely mediated via a different, non-cullin based cellular E3 ubiquitin ligase.

In conclusion, our results indicate that the inhibition of IRS1 by SOCS proteins is not primarily mediated via their function as substrate receptor subunits of Cullin-5 based E3 ubiquitin ligases. Thus, other non-Cullin-5 based cellular E3 ligases are likely to be responsible for basal and signal induced IRS1 protein degradation.

3. Materials and Methods

3.1. Cell Culture. HEK293, HEK293T, HeLa, and 3T3-L1 cells were grown in DMEM medium and MCF7 cells in RPMI medium. All media were supplemented with penicillium/streptomycin, 10% fetal bovine serum (FBS), and L-glutamine.

3.2. Immunoblotting. Cells were washed with ice-cold 1X PBS and lysed in lysis buffer with 0.1% β-mercaptoethanol. Lysates were precleared using centrifugation and equal amounts of proteins were loaded using the Bradford protein assay. The following antibodies were used: monoclonal anti-glyceraldehyde-3-phosphate dehydrogenase (G8140-04; U.S. Biological), monoclonal anti-tubulin (236-10501; Molecular Probes), monoclonal anti-myc (2276; Cell Signaling), polyclonal anti-IRS1 (sc-7200; Santa Cruz Biotechnology), polyclonal anti-IRS2 (06-506; Upstate), monoclonal anti-p27 (610241; BD Biosciences), polyclonal anti-Nrf2 (sc722; Santa Cruz Biotechnology), monoclonal anti-V5 (Serotec), polyclonal anti-p70 S6 Kinase (9202; Cell Signaling), monoclonal anti-phospho-p70 S6 Kinase (Thr389) (9234; Cell Signaling), monoclonal anti-HIF-1α (610959; BD Pharmingen), and polyclonal anti-Cul5 (sc-13014; Santa Cruz Biotechnology). Western blots shown are representative of at least two independent experiments.

3.3. siRNA-Mediated Gene Silencing. siRNA transfection was carried out using RNAi Max lipofectamine (Invitrogen) according to the manufacturer's instruction. Cells were lysed three days after siRNA knockdown for Western blot analysis, as described above. Cullin-5 siRNAs were obtained from Integrated DNA Technologies (HSC.RNAI.N3478.10.3 and HSC.RNAI.N3478.10.4).

3.4. Plasmid Constructs and Transfection of Cells. The plasmids pcDNA3.1-Myc-his-mIRS1 and pcDNA3.1-Myc-his-mIRS2 were generated as previously reported [19]. To generate the β-catenin plasmids, β-catenin coding sequence including a C-terminal V5 tag, was inserted into KpnI and XbaI sites of two vectors, pcDNA3 and pEF1. The human SOCS3 and SOCS6 cDNA clone was purchased from Geneservice (I.M.A.G.E ID 30333577 and 3917519). To generate C-terminally FLAG tagged SOCS3 and SOCS6, clones was PCR amplified and inserted into pcDNA3. The human SOCS1 was PCR amplified from cDNA and inserted into modified pcDNA3.1 with N-terminal FLAG tag. Subconfluent cells were transfected using Genejuice (Novagen) according to the manufacturer's instructions.

Conflict of Interests

The authors declare that they have no conflict of interests.

Acknowledgments

MLN4924 was a kind gift from Millennium: The Takeda Oncology Company. This project was supported by a Grant from Singapore National Medical Research Council (Grant no. NMRC/EDG/0069/2009).

References

[1] A. S. Banks, J. Li, L. McKeag et al., "Deletion of SOCS7 leads to enhanced insulin action and enlarged islets of Langerhans," *Journal of Clinical Investigation*, vol. 115, no. 9, pp. 2462–2471, 2005.

[2] B. Emanuelli, P. Peraldi, C. Filloux, D. Sawka-Verhelle, D. Hilton, and E. Van Obberghen, "SOCS-3 is an insulin-induced negative regulator of insulin signaling," *The Journal of Biological Chemistry*, vol. 275, no. 21, pp. 15985–15991, 2000.

[3] D. J. Hilton, R. T. Richardson, W. S. Alexander et al., "Twenty proteins containing a C-terminal SOCS box form five structural classes," *Proceedings of the National Academy of Sciences of the United States of America*, vol. 95, no. 1, pp. 114–119, 1998.

[4] J. J. Babon, J. K. Sabo, J. G. Zhang, N. A. Nicola, and R. S. Norton, "The SOCS box encodes a hierarchy of affinities for Cullin5: implications for ubiquitin ligase formation and cytokine signalling suppression," *Journal of Molecular Biology*, vol. 387, no. 1, pp. 162–174, 2009.

[5] M. Ohh, C. W. Park, M. Ivan et al., "Ubiquitination of hypoxia-inducible factor requires direct binding to the β-domain of the von Hippel-Lindau protein," *Nature Cell Biology*, vol. 2, no. 7, pp. 423–427, 2000.

[6] A. Kibel, O. Iliopoulos, J. A. DeCaprio, and W. G. Kaelin, "Binding of the von Hippel-Lindau tumor suppressor protein to Elongin B and C," *Science*, vol. 269, no. 5229, pp. 1444–1446, 1995.

[7] T. Aso, D. Haque, R. J. Barstead, R. C. Conaway, and J. W. Conaway, "The inducible elongin A elongation activation domain: structure, function and interaction with the elongin BC complex," *EMBO Journal*, vol. 15, no. 20, pp. 5557–5566, 1996.

[8] T. Kamura, S. Sato, D. Haque et al., "The Elongin BC complex interacts with the conserved SOCS-box motif present in members of the SOCS, ras, WD-40 repeat, and ankyrin repeat families," *Genes and Development*, vol. 12, no. 24, pp. 3872–3881, 1998.

[9] J. G. Zhang, A. Farley, S. E. Nicholson et al., "The conserved SOCS box motif in suppressors of cytokine signaling binds to elongins B and C and may couple bound proteins to proteasomal degradation," *Proceedings of the National Academy of Sciences of the United States of America*, vol. 96, no. 5, pp. 2071–2076, 1999.

[10] T. Kamura, K. Maenaka, S. Kotoshiba et al., "VHL-box and SOCS-box domains determine binding specificity for Cul2-Rbx1 and Cul5-Rbx2 modules of ubiquitin ligases," *Genes and Development*, vol. 18, no. 24, pp. 3055–3065, 2004.

[11] L. Rui, M. Yuan, D. Frantz, S. Shoelson, and M. F. White, "SOCS-1 and SOCS-3 block insulin signaling by ubiquitin-mediated degradation of IRS1 and IRS2," *The Journal of Biological Chemistry*, vol. 277, no. 44, pp. 42394–42398, 2002.

[12] T. A. Soucy, P. G. Smith, M. A. Milhollen et al., "An inhibitor of NEDD8-activating enzyme as a new approach to treat cancer," *Nature*, vol. 458, no. 7239, pp. 732–736, 2009.

[13] J. E. Brownell, M. D. Sintchak, J. M. Gavin et al., "Substrate-assisted inhibition of ubiquitin-like protein-activating enzymes: the NEDD8 E1 inhibitor MLN4924 forms a NEDD8-AMP mimetic in situ," *Molecular Cell*, vol. 37, no. 1, pp. 102–111, 2010.

[14] J. Lee and P. Zhou, "Cullins and cancer," *Genes and Cancer*, vol. 1, no. 7, pp. 690–699, 2010.

[15] D. Finlay, A. J. Ruiz-Alcaraz, C. Lipina, S. Perrier, and C. D. Sutherland, "A temporal switch in the insulin-signalling pathway that regulates hepatic IGF-binding protein-1 gene expression," *Journal of Molecular Endocrinology*, vol. 37, no. 2, pp. 227–237, 2006.

[16] R. Zhande, J. J. Mitchell, J. Wu, and X. J. Sun, "Molecular mechanism of insulin-induced degradation of insulin receptor substrate 1," *Molecular and Cellular Biology*, vol. 22, no. 4, pp. 1016–1026, 2002.

[17] I. Briaud, L. M. Dickson, M. K. Lingohr, J. F. McCuaig, J. C. Lawrence, and C. J. Rhodes, "Insulin receptor substrate-2 proteasomal degradation mediated by a mammalian target of rapamycin (mTOR)-induced negative feedback down-regulates protein kinase B-mediated signaling pathway in β-cells," *The Journal of Biological Chemistry*, vol. 280, no. 3, pp. 2282–2293, 2005.

[18] M. Bouché, F. Zappelli, M. Polimeni et al., "Rapid activation and down-regulation of protein kinase C α in 12-O-tetradecanoylphorbol-13-acetate-induced differentiation of human rhabdomyosarcoma cells," *Cell Growth and Differentiation*, vol. 6, no. 7, pp. 845–852, 1995.

[19] R. Nawaratne, A. Gray, C. H. Jørgensen, C. P. Downes, K. Siddle, and J. K. Sethi, "Regulation of insulin receptor substrate 1 pleckstrin homology domain by protein kinase C: role of serine 24 phosphorylation," *Molecular Endocrinology*, vol. 20, no. 8, pp. 1838–1852, 2006.

16

Staurosporine Inhibits Frequency-Dependent Myofilament Desensitization in Intact Rabbit Cardiac Trabeculae

Kenneth D. Varian,[1] Brandon J. Biesiadecki,[1] Mark T. Ziolo,[1]
Jonathan P. Davis,[1] and Paul M. L. Janssen[1,2]

[1] Department of Physiology and Cell Biology, College of Medicine, The Ohio State University, 1645 Neil Avenue, Columbus, OH 43210, USA
[2] Department of Physiology and Cell Biology, College of Medicine, The Ohio State University, 304 Hamilton Hall, 1645 Neil Avenue, Columbus, OH 43210-1218, USA

Correspondence should be addressed to Paul M. L. Janssen, janssen.10@osu.edu

Academic Editor: W. Glen Pyle

Myofilament calcium sensitivity decreases with frequency in intact healthy rabbit trabeculae and associates with Troponin I and Myosin light chain-2 phosphorylation. We here tested whether serine-threonine kinase activity is primarily responsible for this frequency-dependent modulations of myofilament calcium sensitivity. Right ventricular trabeculae were isolated from New Zealand White rabbit hearts and iontophoretically loaded with bis-fura-2. Twitch force-calcium relationships and steady state force-calcium relationships were measured at frequencies of 1 and 4 Hz at 37 °C. Staurosporine (100 nM), a nonspecific serine-threonine kinase inhibitor, or vehicle (DMSO) was included in the superfusion solution before and during the contractures. Staurosporine had no frequency-dependent effect on force development, kinetics, calcium transient amplitude, or rate of calcium transient decline. The shift in the pCa_{50} of the force-calcium relationship was significant from 6.05 ± 0.04 at 1 Hz versus 5.88 ± 0.06 at 4 Hz under control conditions (vehicle, $P < 0.001$) but not in presence of staurosporine (5.89 ± 0.08 at 1 Hz versus 5.94 ± 0.07 at 4 Hz, $P = $ NS). Phosphoprotein analysis (Pro-Q Diamond stain) confirmed that staurosporine significantly blunted the frequency-dependent phosphorylation at Troponin I and Myosin light chain-2. We conclude that frequency-dependent modulation of calcium sensitivity is mediated through a kinase-specific effect involving phosphorylation of myofilament proteins.

1. Introduction

The relationship between heart rate and myocardial contractility has been studied extensively since Bowditch first recognized what we now refer to as the force frequency relationship (FFR) [1]. Modulation of contractility through heart rate is an intrinsic property of the heart that occurs independent of neurohumoral activity and principally through augmentation of calcium handling and the altering of myofilament properties. In patients suffering from congestive heart failure (CHF), a blunted or negative FFR is observed regardless of the underlying etiology [2–4]. This alteration of normal physiology likely contributes to exercise intolerance and general lack of cardiac reserve seen in patients suffering from CHF. Although a robust increase in contractility with an increase in heart rate is a crucial regulatory property of nonfailing myocardium in all mammals [5], its governing underlying mechanisms are still incompletely understood.

Augmentation of the calcium transient amplitude and rate of decline with increased frequency has been well documented [6, 7]. The mechanism underlying altered calcium handling has been the most extensively investigated aspect of the FFR, and several mechanisms have been suggested. It is likely that the enhanced calcium handling is due in part, if not exclusively, to intrinsic properties of the calcium signaling system. An increase in heart rate increases the amount of calcium entering the L-type calcium channels per unit time and increases intracellular sodium both of which can result in an increase in sarcoplasmic reticulum (SR) load [8, 9]. The increase in SR load results in the rise in peak systolic calcium, resulting in enhanced myocardial force production.

SR calcium reuptake rate increases due the sarcoplasmic reticulum calcium ATPase (SERCA2a) pump working higher on its $[Ca^{2+}]_i$-velocity curve. However, it is still possible (calcium-dependent) kinase(s) are activated at higher heart rates which could potentially augment calcium handling through phosphorylation of the L-type calcium channel, phospholamban, SERCA2a itself, or the ryanodine receptor. So far the most likely candidate for a frequency dependent phosphorylation is calcium calmodulin-dependent kinase II (CaMKII) which has been examined in several studies [10–12]. However, a conclusive target has yet to be found. The roles of PKC [13], PKA [14], and PKG [15] in the FFR have been investigated to some extent, but a conclusive mechanism is still lacking. Modulation of myofilament properties with changes in heart rate has been much less investigated, and the few studies that have focused on this potentially contributing mechanism have, until recently, been inconclusive. Previous studies have found myofilament calcium sensitivity to be increased [16], decreased [17], and unchanged [15] with an increase in frequency. To some extent, these differences may reside in the animal model used; *in vivo*, the rat and mouse depend much less on changes in frequency (~20% in mouse, 40–50% in rat) than larger mammals, including humans (~250–350%) [5, 18]. As such, results obtained in larger mammals may better reflect human frequency-dependent behavior. We demonstrated in rabbit right ventricular trabeculae, under near physiological conditions, that myofilament calcium sensitivity indeed decreases with frequency [19]. We then proceeded to show that, in a model of right ventricular hypertrophy [20], this frequency-dependent myofilament desensitization was impaired [21], potentially explaining diastolic dysfunction. Moreover, we showed that frequency-dependent myofilament desensitization was associated with Troponin I and Myosin light chain-2 phosphorylation [19, 21]. Lamberts and coworkers recently showed that, in failing rat myocardium, a frequency-dependent myofilament desensitization also exists [13], although they did not observe significant changes in healthy rat right ventricular trabeculae at low frequencies. Therefore, to what extent, and in particular how, myofilament calcium sensitivity changes with frequency still remains incompletely resolved.

Based on the previous studies, we hypothesize that a phosphorylation event is primarily at the basis of frequency-dependent modulation of myofilament calcium sensitivity. Using the broad-spectrum serine-threonine kinase inhibitor staurosporine, we set out to determine a potential role of kinases involved with the FFR. We show that the frequency-dependent myofilament desensitization is inhibited by staurosporine. This finding associated with an inhibition of frequency-dependent phosphorylation of myofilament proteins.

2. Materials and Methods

All protocols were approved by the institutional laboratory animal care and use committee of The Ohio State University. Male, New Zealand White (NZW) rabbits (1.5–2 kg) were given 5,000 U/kg heparin iv and were anesthetized with 50 mg/kg pentobarbital. The chest was opened via bilateral thoracotomy, and the heart rapidly excised and flushed retrogradely with a cardioplegic Krebs Henseleit solution containing 20 mM 2, 3 butanedione monoxime (BDM) to prevent cutting injury [22]. The right ventricle was opened and thin, uniform trabeculae were carefully dissected and mounted in a setup constructed on the stage of a Nikon-inverted fluorescent microscope as previously described [17, 19, 23, 24].

2.1. Measurement of Intracellular Calcium and Force Frequency. The cardioplegic K-H was replaced with a normal K-H solution, and the muscle was stimulated at 1 Hz, at 37°C at an extracellular calcium concentration of 1.5 mM, till contractile parameters had stabilized. The muscle was then stretched to an optimal length where further stretching raised diastolic and systolic force proportionally. It has been shown that this muscle length corresponds with a sarcomere length of around 2.2-2.3 μm [25]. For loading of the indicator, temperature was temporarily dropped to 23°C to facilitate the loading process. Trabeculae were iontophoretically loaded with the calcium indicator bis-fura 2 as described previously for the indicator fura-2 [24]. The loading procedure typically lasted between 15 and 25 minutes. We have shown that the kinetics of this indicator, even at body temperature, are sufficiently fast to accurately track the calcium transient even at the highest frequency [26]. After returning to 37°C, a control force-frequency relationship was determined on each muscle by stimulating the muscle to twitch for 2 minutes (more than sufficient to reach steady state), at 1, 2, 3, and 4 Hz. Force tracings and fluorescent emissions at 340 nm and 380 nm were collected. The stimulation was returned to 1 Hz and either DMSO (0.01% v/v) or staurosporine (0.1 μM) was added to the superfusion solution and allowed to circulate for 5 minutes before a second force-frequency was measured (same protocol as control). The 0.1 μM concentration of staurosporine was chosen because it is at least 5 times the *in vitro* K_i for the most obvious candidate kinases (PKA 15 nM, PKC 5 nM, PKG 18 nM, CaMKII 20 nM and MLCK 21 nM) [27] while still below the concentration where some of the nonspecific effects of staurosporine have been found to occur [27].

2.2. Measurement of Steady-State Myofilament Activation. To obtain a steady-state myofilament calcium sensitivity relationship at 37°C, we employed potassium-induced contractures as described previously [19, 28, 29]. Immediately after the second force-frequency measurement, trabeculae under the influence of staurosporine or vehicle control were stimulated to contract at 1 or 4 Hz. The superfusion solution was switched from regular Krebs Henseleit solution to one with a modified Na/K balance (6 mM Ca^{2+} 110 mM K^+ and 40 mM Na^+). Bis-fura 2 fluorescent emission ratios were collected along with force till the peak of the contracture. The fluorescence signal ratio of 340/380 was converted into $[Ca^{2+}]_i$ by obtaining the minimum and maximum ratios (R_{min} and R_{max}) as described previously [24]. The $[Ca^{2+}]_i$ values were plotted against force, and the data was fitted using the Hill equation. In each muscle, a potassium

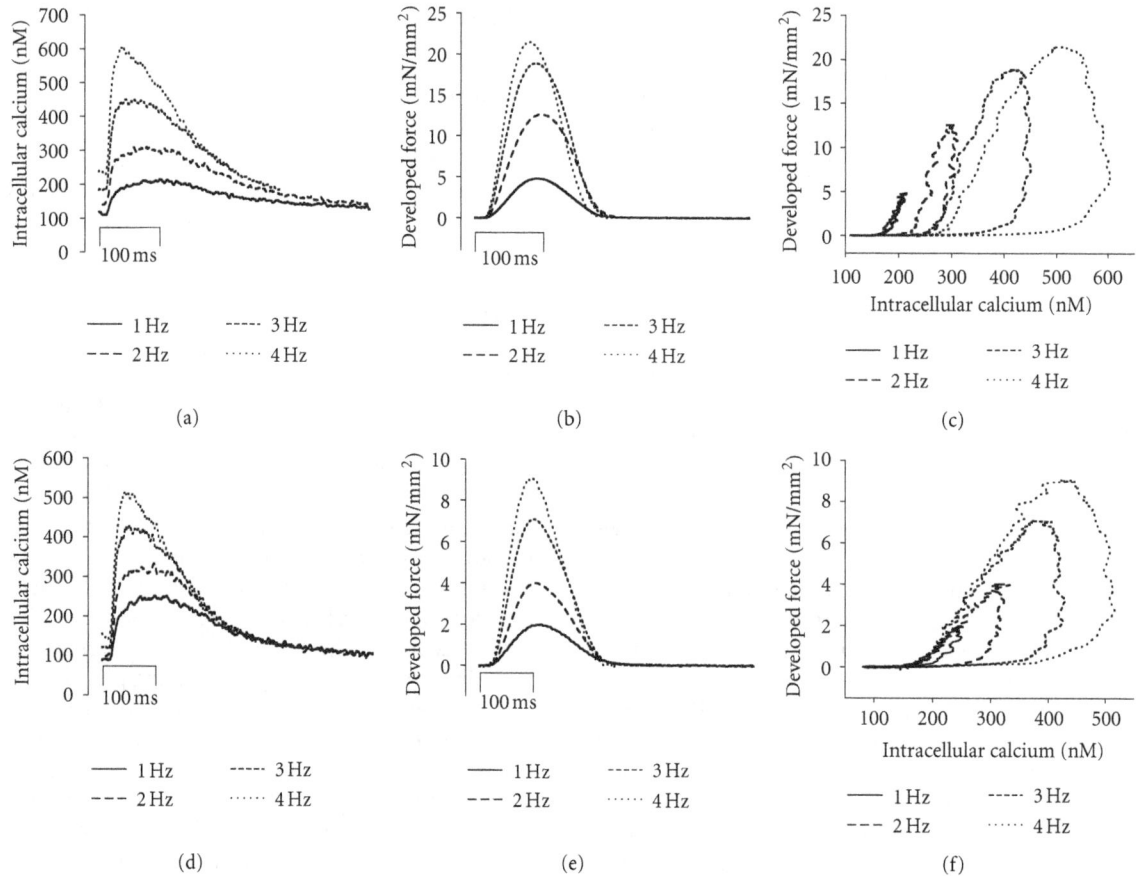

FIGURE 1: Panels (a), (b), and (c) show representative calcium transients (a), force tracings (b), and corresponding phase plane plots (c) from rabbit right ventricular trabeculae stimulated to contract at 1, 2, 3, and 4 Hz (37°C). Note the shift to the right of the relaxation trajectory of the phase plane plots. The same preparation was then treated with 100 nM staurosporine and the protocol repeated. Calcium transients (d), force tracings (e), and the resulting phase planes (f) are shown. Note the elimination of the relaxation trajectory shift after the addition of staurosporine.

contracture was performed at 1 and 4 Hz under the influence of either staurosporine or vehicle. Muscle length was held constant between the two measurements to ensure sarcomere length was excluded as a cause in myofilament calcium sensitivity shift between the two frequencies. Therefore, paired comparisons were made between the two frequencies with or without staurosporine ($n = 10$ DMSO, $n = 9$ staurosporine).

2.3. Measuring Protein Phosphorylation (Pro-Q Diamond Stain). Phosphoprotein analysis via the Pro-Q Diamond Stain was performed as described previously [19, 29]. Briefly, trabeculae twitching at either 1 or 4 Hz with either DMSO or 0.1μM staurosporine in the superfusion bath were doused with liquid nitrogen until frozen and rapidly removed from the experimental setup. The tissue was then homogenized in an SDS protein lysis buffer and loaded on a 15% 8 × 10 cm SDS-PAGE gel (1.5 mm thickness, 4% stacking gel, 15 wells). The gel was then run for 45 minutes at 175 V. The gel was fixed overnight and stained using the Pro-Q Diamond phosphoprotein stain (Invitrogen). Following staining and destaining, the gel was imaged in a Typhoon variable mode scanner (GE Healthcare) using an excitation

wavelength of 532 nm and a 610 nm (BP30) emission filter at a photomultiplier setting of 450 nm. After imaging, the gel was washed in water for 1 hour and stained for total protein with Sypro Ruby protein stain (Invitrogen). Following staining and destaining, the gel was imaged in a Typhoon scanner using an excitation wavelength of 488 nm and a 610 nm (BP30) emission filter at a PMT setting of 425 nm. Densitometric analysis was performed on each band, and the ratio of Pro-Q stain intensity to total protein intensity was calculated.

2.4. Statistics. Data was analyzed using paired two tailed t-tests for sensitivity curves where $P < 0.05$ was considered significant. Average data is presented with error bars showing standard error of the mean. Two- and one-way ANOVAs with post hoc t-tests (Bonferroni correction) were used for analysis of twitch and calcium transient data.

3. Results

Figure 1 shows an experiment where a force frequency relationship was obtained before and after the addition of staurosporine to the superfusate. As previously shown [19, 21],

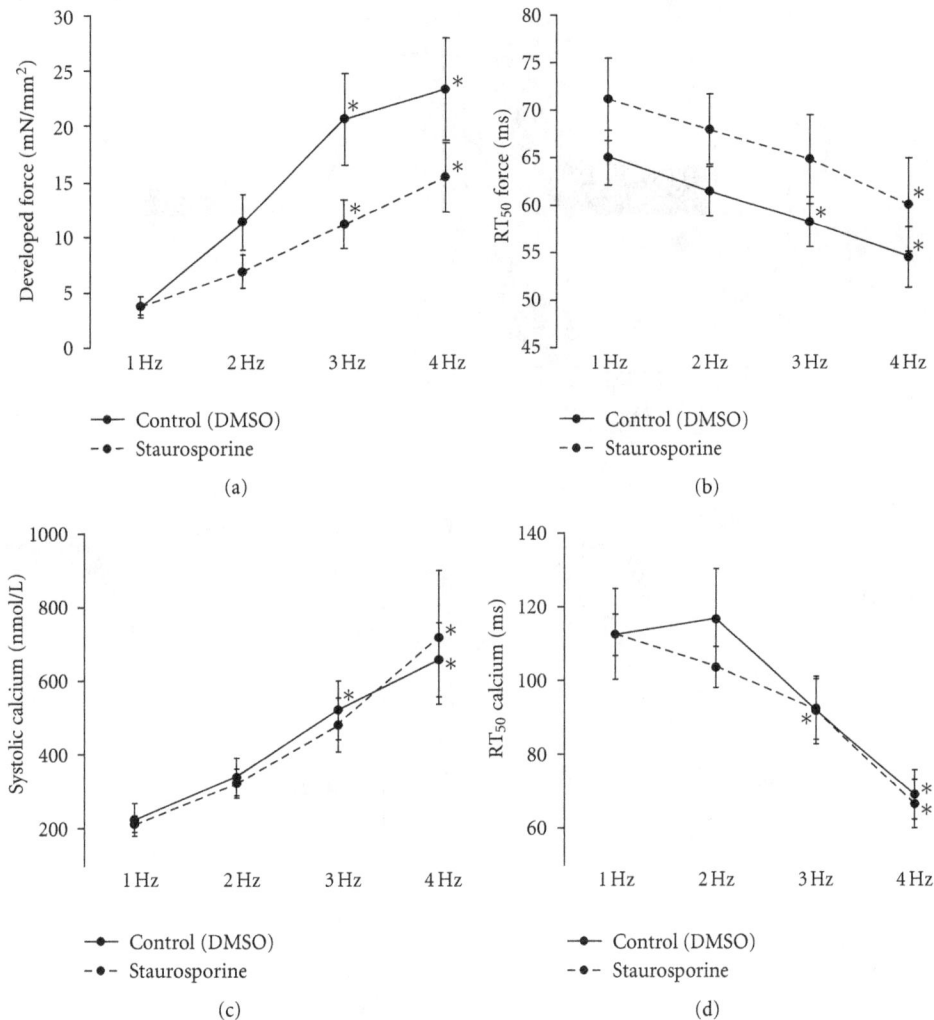

FIGURE 2: Average twitch and calcium transient data from trabeculae treated with staurosporine or vehicle (DMSO) ($n = 8$ for each group). Panels (a) and (b) show the developed force (a) and relaxation time to 90% of peak (RT_{50}) at each frequency 1 through 4 Hz. Analysis by two-way ANOVA (Bonferroni correction, $P < 0.05$) showed that frequency significantly altered force and relaxation, but staurosporine did not significantly alter the FFR or FDAR. Panels (c) and (d) show the average systolic calcium levels (c) and time constant of calcium decline (tau). There was no significant effect of 100 nM staurosporine, nor an interaction between staurosporine and frequency. The effect of frequency within each group was analyzed by one-way AVOVA (*significantly different from 1 Hz).

under control conditions, the relaxation trajectory of the phase plane plots of $[Ca^{2+}]_i$ versus force shifts to the right with an increase in frequency, suggesting myofilament desensitization. The phase-plane plots allow for the determination of calcium versus force relationship trajectories independent of timing [23, 30, 31]. We observed that, with the addition of staurosporine the relaxation trajectories line up, that is no shift is observed. Although this is an indirect assessment, this suggests that staurosporine was able to inhibit frequency-dependent myofilament desensitization.

Figure 2 shows average data for developed force (F_{dev}), time from peak tension to 50% relaxation (RT_{50} force), systolic calcium concentration (nM), and the RT50 of calcium decline (RT_{50} calcium). Developed force increased with frequency in both control and staurosporine groups when stimulation frequency is increased. Statistical analysis (2-way

ANOVA) revealed no significant effect of staurosporine on any force, or calcium-transient parameter.

Next, we set out to determine if staurosporine had an effect on myofilament calcium sensitivity modulations due to frequency. For this, we employed so called "potassium contractures" to introduce a pseudo-steady state between the intracellular calcium concentration and developed force in order to measure the calcium sensitivity of the myofilaments. Figure 3(a) shows the original data of a representative potassium contracture. The superfusate solution is rapidly switched from normal Krebs-Henseleit solution to one containing high potassium and low sodium (indicated by the grey area in the figure). Upon application of the "potassium contraction" solution, calcium enters the cytoplasm slowly, inducing a slowing forming contracture. By rapidly switching between 340 nm and 380 nm excitation wavelengths,

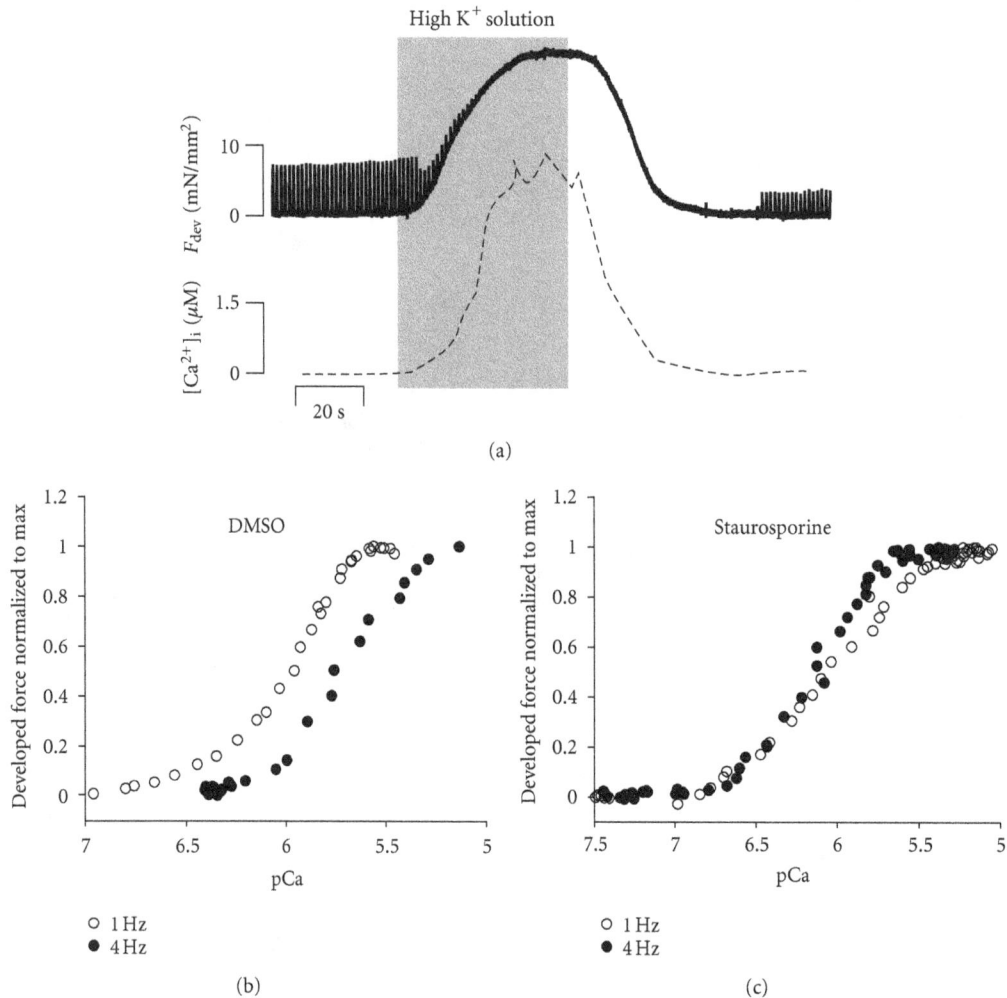

FIGURE 3: Figure 3 shows a representative acquisition and analysis of the force-calcium relationship. Panel (a) shows the chart of a potassium contracture. The trabecula in this example is twitching at 1 Hz (top of panel) when the superfusate solution is rapidly switched from normal Krebs Henseleit solution to that with a higher potassium and lower sodium concentration. Diastolic calcium (lower panel) is measured by rapidly switching the excitation wavelength between 340 and 380 nm. After the potassium solution is washed out, the contracture relaxes and the muscle recovers. Panels (b) and (c) show representative force pCa relationships obtained in trabeculae twitching at 1 or 4 Hz with vehicle (b) or staurosporine (c) in the superfusate solution.

diastolic calcium was measured as this contracture formed, up till peak force had developed. Thereafter, the perfusate was switched back to the normal Krebs-Henseleit solution, and the membrane potential slowly reestablished, while calcium is removed from the cytosol, and subsequently the contracture dissipates. The simultaneously recorded force and bis-fura-2 ratio data allows us to construct a myofilament calcium sensitivity relationship, from which characterizing parameters are determined by fitting this data with the Hill equation. Figures 3(b) and 3(c) show an example of $[Ca^{2+}]_i$ versus. force curves derived from control and staurosporine protocols. A potassium contracture was performed at 1 and 4 Hz for each muscle.

Figure 4 shows the average for parameters pCa_{50} and F_{max} for control and staurosporine. While there was a significant desensitization with respect to the pCa_{50} from 1 to 4 Hz in the control group, in line with our previous observations,

no such frequency-dependent shift was detected with staurosporine. Maximum developed force of each potassium contracture (F_{max}) was not significantly ($P > 0.05$) different between each group although the difference in average F_{max} between 1 and 4 Hz in the control group approached significance ($P < 0.1$). The hill coefficient, an indicator of cooperativity of the myofilament force response, was not significantly different between control and staurosporine sensitivity curved (N_{Hill}; control: 1 Hz 3.99 ± 0.60, 4 Hz 2.82 ± 0.56; staurosporine: 1 Hz 2.28 ± 0.38, 4 Hz 2.0 ± 0.18).

Figure 5 shows a section of an SDS gel stained for total protein (a) (Sypro ruby) and phosphoprotein (b) (Pro-Q Diamond) at the location corresponding with Troponin I (TnI) and Myosin light chain-2 (MLC-2). The average ratio of band density of phosphoprotein to total protein for TnI (c) and MLC-2 (d) is also depicted. A significant increase in the ratio was found for TnI between 1 and 4 Hz in the

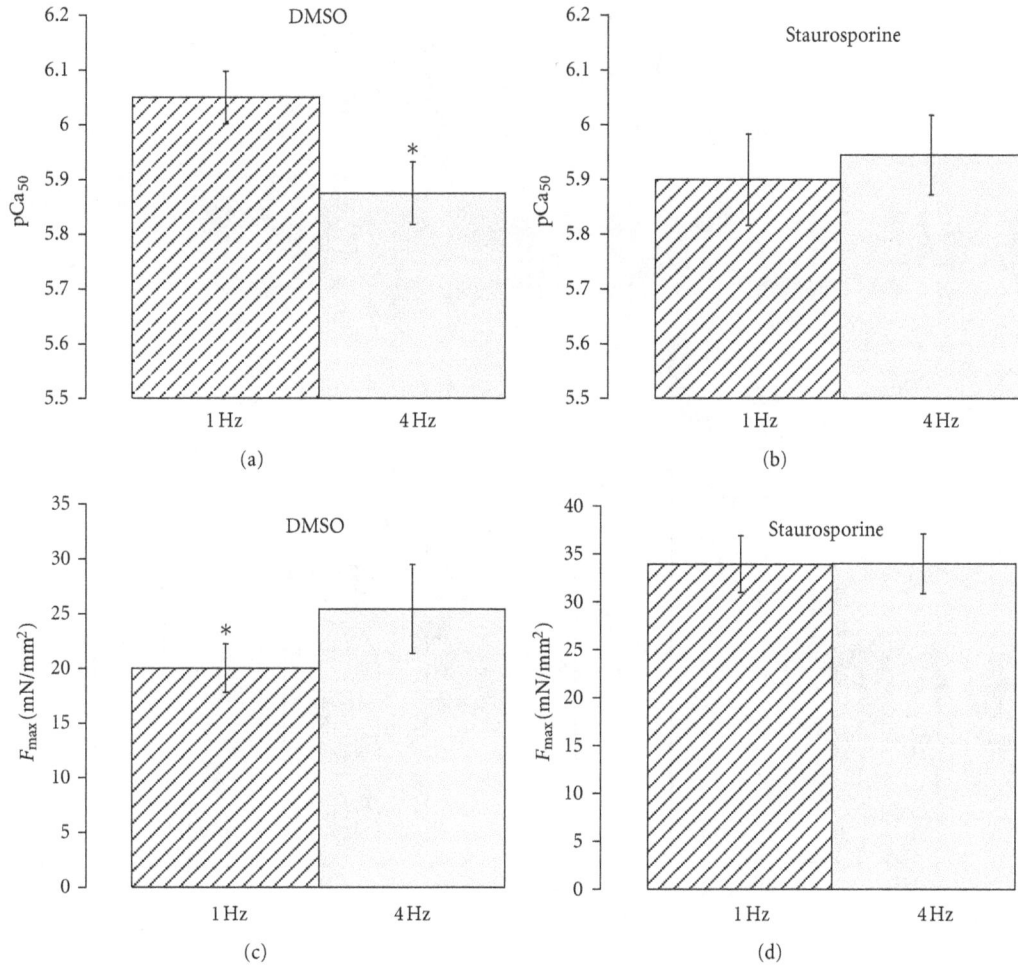

FIGURE 4: Average pCa_{50} and maximum force (F_{max}) data obtained from the force calcium relationships ($n = 10$ for DMS and $n = 9$ for staurosporine). A significant difference ($P < 0.05$ denoted with *) was seen in the pCa_{50} and F_{max} between the 1 and 4 Hz under control conditions (panels (a) and (b)). These shifts in sensitivity and F_{max} were eliminated by staurosporine (panels (c) and (d)).

control group consistent with previous findings [19]. The difference in ratio between 4 Hz with staurosporine and 4 Hz with DMSO was also found to be significantly different. The difference in ratio of phosphoprotein to total protein for MLC-2 approached significance between 1 and 4 Hz for the control group ($P < 0.1$ but >0.05). However, a significant difference in MLC-2 phosphorylation was found between 4 Hz staurosporine and 4 Hz DMSO.

4. Discussion

In this study, we investigated whether frequency-dependent myofilament desensitization is mediated by a kinase dependent pathway and how this phenomenon impacts cardiac trabecular twitch contractions. We now show for the first time that broad-spectrum serine-threonine kinase inhibition inhibits frequency-dependent myofilament desensitization. Staurosporine was able to inhibit frequency-induced phosphorylation of TnI which is likely (at least partially) responsible for frequency-dependent myofilament desensitization. We also show that staurosporine does not, at the

concentration used in our studies, affect the dynamics of the calcium transient or its relationship to stimulation rate. From this, we conclude serine threonine kinase pathways activated with changes in frequency mainly, if not exclusively, modulate myofilament function, but not calcium transient regulation. Although changes in phosphatase activity cannot be excluded at this point, the near-complete abolishment of frequency-dependent changes in myofilament calcium sensitivity indicate that phosphatases likely play only a minor, if any, role.

Despite extensive research, the molecular mechanisms of frequency-dependent augmentation calcium handling and modulations of myofilament calcium sensitivity have eluded full understanding. This is not entirely surprising, as frequency modulation occurs *in vivo* during a highly dynamic response. Governing factors such as calcium concentration, external mechanical load, internal passive-elastic elements, and ionic fluxes never reach a steady state balance. Thus, a steady-state snap shot of these parameters or interactions can never fully describe the prevailing dynamical situation. Using our approach, we found that staurosporine, at a

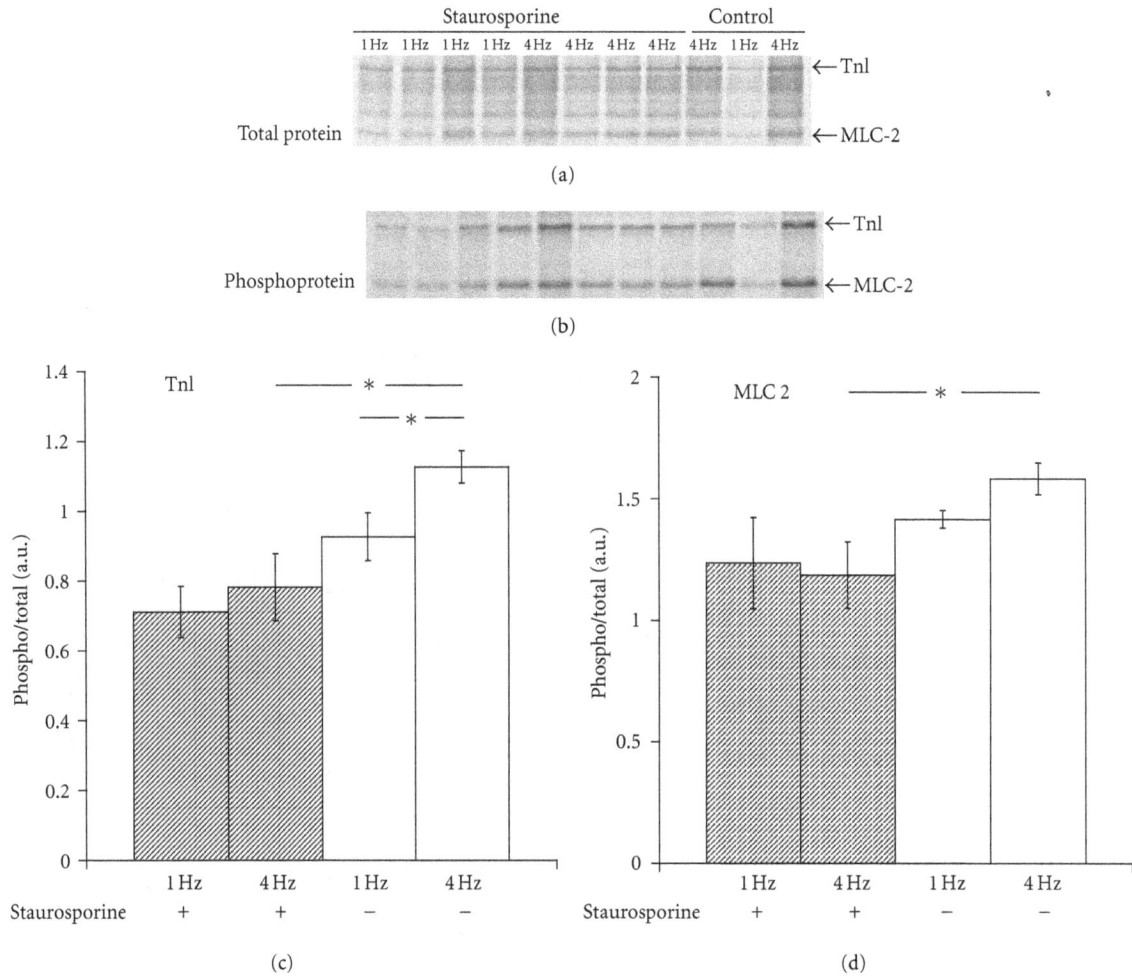

FIGURE 5: Analysis of TnI and MLC-2 phosphorylation status in muscles stimulated to twitch at 1 or 4 Hz, with and without staurosporine. Panel (a) shows a representative SDS gel stained for total protein using the Sypro Ruby protein stain. Panel (b) shows the same gel counterstained for phosphoprotein using the Pro-Q Diamond Stain. Densitometric analysis of bands corresponding to TnI and MLC-2 was obtained from each stain and the ratio of phosphoprotein to total protein calculated for each group. TnI phospho/total protein ratio is shown in panel (c) and MLC-2 shown in panel (d) ($n = 4$ for each group, other gel not shown); error bars are SEM, *$P < 0.05$ considered significant.

concentration where several major kinases are inhibited, had no effect on the frequency dependency of the calcium transient amplitude or decline, nor force amplitude or decline. This data suggests that changes in frequency do not induce phosphorylations that affect the calcium transient itself. This finding is consistent with Kassiri et al. where the use of broad spectrum kinase inhibitor K252-a yielded no change in the frequency-dependent augmentation of calcium handling [15]. Although we cannot exclude phosphorylation events do occur with frequency on key calcium handling proteins, our results indicate that, if they do occur, their functional effect on the cytosolic calcium transient appears to have little or no functional significance. Other studies have focused on the specific effect of CaMKII [11, 32] as a potential regulator of calcium handling dynamics. Most recently, Picht and coworkers [12] found that CaMKII inhibition directed to the SR inhibits frequency-dependent acceleration of calcium decline by inhibiting, an increase in SERCA2a V_{max}. While,

in the present study we did not measure or derive SERCA velocity measurements, we did find that the RT_{50} of calcium decline was not different between control and staurosporine. This thus suggests that intracellular calcium decline, which can be impacted by the calcium transient amplitude itself [33], was not significantly blunted with kinase inhibition under the present experimental conditions, which were very close as they prevail *in vivo* regarding mechanical load, temperature, and stimulation range. Clearly, our data here suggest that the intrinsic properties, rather than frequency-induced phosphorylations, of the calcium handling system play a major (if not the only) role in augmentation of amplitude and rate of decline.

A striking effect of staurosporine was the complete elimination of the frequency-dependent shift in myofilament calcium sensitivity from 1 to 4 Hz. In addition, there appears to (possibly) be an effect of frequency on maximal developed force. Although this shift in F_{max} did not quite reach significance ($P < 0.1$ but >0.05), there may still be a kinase-dependent effect on F_{max}. Supporting this speculation is the fact that in the staurosporine group, there was almost no variability in the myofilament sensitivity curve F_{max}. These data show that changes in stimulation frequency activate signaling pathways that lead to the activation of protein kinases, which in turn affect myofilament calcium sensitivity and possibly F_{max}. Since these effects were completely eliminated by staurosporine, it is likely that phosphorylation events comprise most, if not all, of the mechanism responsible for frequency-dependent changes in myofilament function. This data was verified using a phosphoproteomics protocol. We show that TnI is hyperphosphorylated at higher frequency, in line with our previous observation [19]. Staurosporine inhibited this hyperphosphorylation, and we observed that the relative TnI phosphorylation levels were lower than those in the absence of staurosporine at 1 Hz, indicating that even, at the lowest frequency baseline, phosphorylation of TnI is present. Despite the clear effect of staurosporine on myofilament calcium sensitivity, staurosporine induced no significant interaction between frequency and the twitch parameters measured. Frequency-dependent acceleration of relaxation (FDAR) was still mostly intact. This paradigm may be explained by the fact that staurosporine may have inhibited more than one pathway which can have opposite effects on myofilament calcium sensitivity and twitch parameters. We found that staurosporine significantly reduced TnI and MLC-2 phosphorylation at 4 Hz, compared to control but not at 1 Hz. It has been shown by many investigators that TnI phosphorylation can decrease myofilament calcium sensitivity [34–36]. Also, it has been shown that MLC-2 phosphorylation can increase myofilament calcium sensitivity [37–39]. Therefore, we should not necessarily expect the effect of staurosporine on the twitch parameters to be straightforward. The pCa_{50} values attained by potassium contractures reveal that staurosporine, while inhibiting the shift in pCa_{50}, set the myofilament sensitivity to level between that of 1 and 4 Hz in the control group 4 Hz. With sensitivity set to a slightly lower value, the rate-limiting step of relaxation may begin to favor other factors such as the rate of calcium decline, including Troponin-C calcium off-rate kinetics [40, 41] and calcium-regulated SERCA-activity [33]. In addition, MLC-2 phosphorylation has been shown to slow cross bridge kinetics [42] which would counter the acceleratory effect of TnI phosphorylation. Abolishment of both of these may have a null effect on relaxation. Finally, it has been previously speculated shifts in myofilament calcium sensitivity do not influence the actual rate of relaxation to a large extent [43], and recent views on contraction-relaxation coupling have postulated an overall governing mechanical rate that may impact contractile kinetics as a whole and not be specific to relaxation [44, 45].

Although for certain compounds specific inhibitors exist that work well in a test tube with a limited number of substrates, recent pilot experiments revealed that compounds such as for instance ML7 (an inhibitor of MLC2) have drastic side effects on the calcium transients, and thus likely have multiple targets in the myocyte, to the extent that the muscle barely contracts, and prohibited an unambiguous investigation of contractile properties in present of this compound. As the study of frequency-dependent activation requires an intact preparation, future studies in which the specificity of kinases is extensively documented in such an integrated preparation may aid to reveal the exact identification of the specific kinase responsible. For instance, CaMKII could be a potential mediator, since the increase in calcium during high frequencies may activate this pathway. On the other hand, the time course of CaMKII activation may be too slow [46] to be critically involved in frequency-dependent desensitization, but, at present, the complete elucidation of pathway and specific amino acid targets is deemed beyond the scope of this study. Furthermore, we and others showed the importance of frequency-dependent as well as the role of TnI in hypertrophy [21] and aging [47], and thus future elucidation of specific kinases and specific targets would be of significant importance, as would be to resolve the temporal resolution of the changes in myofilament responsiveness upon a change in frequency.

In summary, we have shown that broad-spectrum kinase inhibition has no significant effect on calcium transient amplitude and rate of decline, or how these parameters change with frequency. Frequency-dependent shifts in myofilament calcium sensitivity are however virtually abolished with staurosporine, suggesting the kinase-dependent augmentation of the force frequency appears to primarily target the myofilaments, whereas the frequency-dependent augmentation of the calcium transient mainly relies on its intrinsic properties.

Acknowledgments

This investigation was supported by National Heart, Lung, and Blood Institute Grants R01 746387 and KO2 83957 (PMLJ), Established Investigator Award 0740040 of the American Heart Association (PMLJ), and American Heart Association Ohio Valley affiliate predoctoral fellowship 0615288B (KDV).

References

[1] H. P. Bowditch, "Ueber die Eigenthuemlichkeiten der Reizbarkeit, welche die Muskelfasern des Herzens zeigen," *Sächsische Akademie der Wissenschaften*, vol. 23, pp. 652–689, 1871.

[2] L. A. Mulieri, G. Hasenfuss, B. Leavitt, P. D. Allen, and N. R. Alpert, "Altered myocardial force-frequency relation in human heart failure," *Circulation*, vol. 85, no. 5, pp. 1743–1750, 1992.

[3] E. I. Rossman, R. E. Petre, K. W. Chaudhary et al., "Abnormal frequency-dependent responses represent the pathophysiologic signature of contractile failure in human myocardium," *Journal of Molecular and Cellular Cardiology*, vol. 36, no. 1, pp. 33–42, 2004.

[4] P. M. L. Janssen, S. E. Lehnart, J. U. Prestle, and G. Hasenfuss, "Preservation of contractile characteristics of human myocardium in multi-day cell culture," *Journal of Molecular and Cellular Cardiology*, vol. 31, no. 8, pp. 1419–1427, 1999.

[5] P. M. L. Janssen and M. Periasamy, "Determinants of frequency-dependent contraction and relaxation of mammalian myocardium," *Journal of Molecular and Cellular Cardiology*, vol. 43, no. 5, pp. 523–531, 2007.

[6] H. C. Lee and W. T. Clusin, "Cytosolic calcium staircase in cultured myocardial cells," *Circulation Research*, vol. 61, no. 6, pp. 934–939, 1987.

[7] J. Layland and J. C. Kentish, "Positive force- and $[Ca^{2+}](i)$-frequency relationships in rat ventricular trabeculae at physiological frequencies," *American Journal of Physiology*, vol. 276, no. 1, pp. H9–H18, 1999.

[8] G. D. Mills, D. M. Harris, X. Chen, and S. R. Houser, "Intracellular sodium determines frequency-dependent alterations in contractility in hypertrophied feline ventricular myocytes," *American Journal of Physiology*, vol. 292, no. 2, pp. H1129–H1138, 2007.

[9] G. Antoons, K. Mubagwa, I. Nevelsteen, and K. R. Sipido, "Mechanisms underlying the frequency dependence of contraction and $[Ca^{2+}]_i$ transients in mouse ventricular myocytes," *Journal of Physiology*, vol. 543, pp. 889–898, 2002.

[10] R. A. Bassani, J. W. M. Bassani, and D. M. Bers, "Relaxation in ferret ventricular myocytes: role of the sarcolemmal Ca ATPase," *Pflugers Archiv European Journal of Physiology*, vol. 430, no. 4, pp. 573–578, 1995.

[11] J. DeSantiago, L. S. Maier, and D. M. Bers, "Frequency-dependent acceleration of relaxation in the heart depends on CamKII, but not phospholamban," *Journal of Molecular and Cellular Cardiology*, vol. 34, no. 8, pp. 975–984, 2002.

[12] E. Picht, J. DeSantiago, S. Huke, M. A. Kaetzel, J. R. Dedman, and D. M. Bers, "CaMKII inhibition targeted to the sarcoplasmic reticulum inhibits frequency-dependent acceleration of relaxation and Ca^{2+} current facilitation," *Journal of Molecular and Cellular Cardiology*, vol. 42, no. 1, pp. 196–205, 2007.

[13] R. R. Lamberts, N. Hamdani, T. W. Soekhoe et al., "Frequency-dependent myofilament Ca^{2+} desensitization in failing rat myocardium," *Journal of Physiology*, vol. 582, pp. 695–709, 2007.

[14] E. Takimoto, D. G. Soergel, P. M. L. Janssen, L. B. Stull, D. A. Kass, and A. M. Murphy, "Murphy: frequency- and afterload-dependent cardiac modulation in vivo by troponin I with constitutively active protein kinase A phosphorylation sites," *Circulation Research*, vol. 94, no. 4, pp. 496–504, 2004.

[15] Z. Kassiri, R. Myers, R. Kaprielian, H. S. Banijamali, and P. H. Backx, "Rate-dependent changes of twitch force duration in rat cardiac trabeculae: a property of the contractile system," *Journal of Physiology*, vol. 524, pp. 221–231, 2000.

[16] W. D. Gao, N. G. Perez, and E. Marban, "Calcium cycling and contractile activation in intact mouse cardiac muscle," *Journal of Physiology*, vol. 507, no. 1, pp. 175–184, 1998.

[17] C. W. Tong, R. D. Gaffin, D. C. Zawieja, and M. Muthuchamy, "Roles of phosphorylation of myosin binding protein-C and troponin I in mouse cardiac muscle twitch dynamics," *Journal of Physiology*, vol. 558, pp. 927–941, 2004.

[18] D. Georgakopoulos and D. A. Kass, "Minimal force-frequency modulation of inotropy and relaxation of in situ murine heart," *Journal of Physiology*, vol. 534, no. 2, pp. 535–545, 2001.

[19] K. D. Varian and P. M. L. Janssen, "Frequency-dependent acceleration of relaxation involves decreased myofilament calcium sensitivity," *American Journal of Physiology*, vol. 292, no. 5, pp. H2212–H2219, 2007.

[20] S. C. Gupta, K. D. Varian, N. C. Bal, J. L. Abraham, M. Periasamy, and P. M. L. Janssen, "Pulmonary artery banding alters the expression of Ca^{2+} transport proteins in the right atrium in rabbits," *American Journal of Physiology*, vol. 296, no. 6, pp. H1933–H1939, 2009.

[21] K. D. Varian, A. Kijtawornrat, S. C. Gupta et al., "Impairment of diastolic function by lack of frequency-dependent myofilament desensitization in rabbit right ventricular hypertrophy," *Circulation*, vol. 2, no. 5, pp. 472–481, 2009.

[22] L. A. Mulieri, G. Hasenfuss, F. Ittleman, E. M. Blanchard, and N. R. Alpert, "Protection of human left ventricular myocardium from cutting injury with 2,3-butanedione monoxime," *Circulation Research*, vol. 65, no. 5, pp. 1441–1444, 1989.

[23] P. M. L. Janssen, L. B. Stull, and E. Marbán, "Myofilament properties comprise the rate-limiting step for cardiac relaxation at body temperature in the rat," *American Journal of Physiology*, vol. 282, no. 2, pp. H499–H507, 2002.

[24] P. H. Backx and H. E. D. J. Ter Keurs, "Fluorescent properties of rat cardiac trabeculae microinjected with fura-2 salt," *American Journal of Physiology*, vol. 264, no. 4, pp. H1098–H1110, 1993.

[25] P. M. L. Janssen and W. C. Hunter, "Force, not sarcomere length, correlates with prolongation of isosarcometric contraction," *American Journal of Physiology*, vol. 269, no. 2, pp. H676–H685, 1995.

[26] M. M. Monasky, K. D. Varian, J. P. Davis, and P. M. L. Janssen, "Dissociation of force decline from calcium decline by preload in isolated rabbit myocardium," *Pflugers Archiv European Journal of Physiology*, vol. 456, no. 2, pp. 267–276, 2008.

[27] F. Meggio, A. Donella Deana, M. Ruzzene et al., "Different susceptibility of protein kinases to staurosporine inhibition. Kinetic studies and molecular bases for the resistance of protein kinase CK2," *European Journal of Biochemistry*, vol. 234, no. 1, pp. 317–322, 1995.

[28] K. D. Varian, S. Raman, and P. M. L. Janssen, "Measurement of myofilament calcium sensitivity at physiological temperature in intact cardiac trabeculae," *American Journal of Physiology*, vol. 290, no. 5, pp. H2092–H2097, 2006.

[29] M. M. Monasky, B. J. Biesiadecki, and P. M. L. Janssen, "Increased phosphorylation of tropomyosin, troponin I, and myosin light chain-2 after stretch in rabbit ventricular myocardium under physiological conditions," *Journal of Molecular and Cellular Cardiology*, vol. 48, no. 5, pp. 1023–1028, 2010.

[30] W. D. Gao, P. H. Backx, M. Azan-Backx, and E. Marban, "Myofilament Ca^{2+} sensitivity in intact versus skinned rat ventricular muscle," *Circulation Research*, vol. 74, no. 3, pp. 408–415, 1994.

[31] L. E. Dobrunz, P. H. Backx, and D. T. Yue, "Steady-state $[Ca^{2+}]$(i)-force relationship in intact twitching cardiac muscle: direct evidence for modulation by isoproterenol and EMD 53998," *Biophysical Journal*, vol. 69, no. 1, pp. 189–201, 1995.

[32] C. A. Valverde, C. Mundiña-Weilenmann, M. Said et al., "Frequency-dependent acceleration of relaxation in mammalian heart: a property not relying on phospholamban and SERCA-2a phosphorylation," *Journal of Physiology*, vol. 562, pp. 801–813, 2005.

[33] S. R. Roof, T. R. Shannon, P. M. Janssen, and M. T. Ziolo, "Effects of increased systolic Ca^{2+} and phospholamban phosphorylation during beta-adrenergic stimulation on Ca^{2+} transient kinetics in cardiac myocytes," *American Journal of Physiology*, vol. 301, no. 4, pp. H1570–H1578, 2011.

[34] E. G. Kranias and R. J. Solaro, "Phosphorylation of troponin I and phospholamban during catecholamine stimulation of rabbit heart," *Nature*, vol. 298, no. 5870, pp. 182–184, 1982.

[35] R. Zhang, J. Zhao, and J. D. Potter, "Phosphorylation of both serine residues in cardiac troponin I is required to decrease the Ca^{2+} affinity of cardiac troponin C," *Journal of Biological Chemistry*, vol. 270, no. 51, pp. 30773–30780, 1995.

[36] R. J. Solaro and J. van der Velden, "Why does troponin I have so many phosphorylation sites? Fact and fancy," *Journal of Molecular and Cellular Cardiology*, vol. 48, no. 5, pp. 810–816, 2010.

[37] I. Morano, F. Hofmann, M. Zimmer, and J. C. Ruegg, "The influence of P-light chain phosphorylation by myosin light chain kinase on the calcium sensitivity of chemically skinned heart fibres," *FEBS Letters*, vol. 189, no. 2, pp. 221–224, 1985.

[38] H. Wang, J. E. Grant, C. M. Doede, S. Sadayappan, J. Robbins, and J. W. Walker, "PKC-betaII sensitizes cardiac myofilaments to Ca^{2+} by phosphorylating troponin I on threonine-144," *Journal of Molecular and Cellular Cardiology*, vol. 41, no. 5, pp. 823–833, 2006.

[39] J. van der Velden, Z. Papp, N. M. Boontje et al., "The effect of myosin light chain 2 dephosphorylation on Ca^{2+}-sensitivity of force is enhanced in failing human hearts," *Cardiovascular Research*, vol. 57, no. 2, pp. 505–514, 2003.

[40] C. Norman, J. A. Rall, S. B. Tikunova, and J. P. Davis, "Modulation of the rate of cardiac muscle contraction by troponin C constructs with various calcium binding affinities," *American Journal of Physiology*, vol. 293, no. 4, pp. H2580–H2587, 2007.

[41] J. P. Davis and S. B. Tikunova, "Ca^{2+} exchange with troponin C and cardiac muscle dynamics," *Cardiovascular Research*, vol. 77, no. 4, pp. 619–626, 2008.

[42] M. C. Olsson, J. R. Patel, D. P. Fitzsimons, J. W. Walker, and R. L. Moss, "Basal myosin light chain phosphorylation is a determinant of Ca^{2+} sensitivity of force and activation dependence of the kinetics of myocardial force development," *American Journal of Physiology*, vol. 287, no. 6, pp. H2712–H2718, 2004.

[43] P. P. de Tombe, A. Belus, N. Piroddi et al., "Myofilament calcium sensitivity does not affect cross-bridge activation-relaxation kinetics," *American Journal of Physiology*, vol. 292, no. 3, pp. R1129–R1136, 2007.

[44] P. M. L. Janssen, "54th Bowditch Lecture: myocardial contraction-relaxation coupling," *American Journal of Physiology*, vol. 299, no. 6, pp. H1741–H1749, 2010.

[45] P. M. L. Janssen, "Kinetics of cardiac muscle contraction and relaxation are linked and determined by properties of the cardiac sarcomere," *American Journal of Physiology*, vol. 299, no. 4, pp. H1092–H1099, 2010.

[46] S. Huke and D. M. Bers, "Temporal dissociation of frequency-dependent acceleration of relaxation and protein phosphorylation by CaMKII," *Journal of Molecular and Cellular Cardiology*, vol. 42, no. 3, pp. 590–599, 2007.

[47] B. J. Biesiadecki, K. Tachampa, C. Yuan, J. P. Jin, P. P. de Tombe, and R. J. Solaro, "Removal of the cardiac troponin I N-terminal extension improves cardiac function in aged mice," *Journal of Biological Chemistry*, vol. 285, no. 25, pp. 19688–19698, 2010.

17

ER Stress and Lipid Metabolism in Adipocytes

Beth S. Zha[1] and Huiping Zhou[1,2]

[1] Department of Microbiology and Immunology, School of Medicine, Virginia Commonwealth University, 1217 East Marshall Street, MSB no. 533, Richmond, VA 23298, USA
[2] Department of Internal Medicine, McGuire Veterans Affairs Medical Center, Richmond, VA 23298, USA

Correspondence should be addressed to Huiping Zhou, hzhou@vcu.edu

Academic Editor: Kezhong Zhang

The role of endoplasmic reticulum (ER) stress is a rapidly emerging field of interest in the pathogenesis of metabolic diseases. Recent studies have shown that chronic activation of ER stress is closely linked to dysregulation of lipid metabolism in several metabolically important cells including hepatocytes, macrophages, β-cells, and adipocytes. Adipocytes are one of the major cell types involved in the pathogenesis of the metabolic syndrome. Recent advances in dissecting the cellular and molecular mechanisms involved in the regulation of adipogenesis and lipid metabolism indicate that activation of ER stress plays a central role in regulating adipocyte function. In this paper, we discuss the current understanding of the potential role of ER stress in lipid metabolism in adipocytes. In addition, we touch upon the interaction of ER stress and autophagy as well as inflammation. Inhibition of ER stress has the potential of decreasing the pathology in adipose tissue that is seen with energy overbalance.

1. Introduction

In the last two decades, the complexity of adipose tissue has finally become apparent. Investigations surrounding the biological impact of obesity, insulin resistance, and the metabolic syndrome have surged, resulting in a more intricate understanding of "fat." Adipose tissue (AT) is not only highly specialized to store long-term energy, but is also a central endocrine organ. Therefore, AT is inherently involved in the interplay of inflammatory cascades and energy metabolism, which are important players in metabolic disorders. Even more, sick fat, or adiposopathy, has now been coined an independent endocrine disease [1].

Adiposopathy can occur environmentally through overnutrition. Adipocytes store extra energy in the form of triglycerides (TG) inside cytosolic organelles (lipid droplets, or LD). When there is a continuous need to store TGs, adipocytes must expand in size while continuously being stressed to synthesize more proteins for LD formation. There is an inherent threshold at which adipocytes become too stressed, secrete multiple cytokines, and can no longer expand. The cytokines released activate resident macrophages and call in circulating macrophages, which

begin to attempt to engulf these cells, forming the signature "crownlike structures" found in obese tissue [2].

During this cascade, increased cytokines can increase adipocyte lipolysis. Increased lipolysis leads to an increase of circulating free fatty acids (FFA) that are deposited in muscle and liver ("lipid dumping") and results in a decreased insulin sensitivity in these tissues (reviewed in [3]). Particularly, FFA from visceral AT is directly deposited into the portal vein, increasing the risk of fatty liver disease. This may be the underlying basis of current clinical understanding that increased visceral fat is a high-risk factor for cardiovascular disease [4, 5].

An increase in FFA release is not only induced by an inflammatory state in AT, but also cellular insulin insensitivity. For this reason, most literature focusing on adipocyte dysregulation in metabolic disease concentrates on the nutrient sensing pathways. However, another important pathway involved in adipocyte pathology is the induction of endoplasmic reticulum (ER) stress. In the past, overstimulation of ER stress has been linked to diseases of genetics and aging (reviewed in [6]), but may in fact be involved in more environmentally induced diseases as well. This paper discusses the recent understanding regarding the role of ER stress in regulating lipid metabolism in adipocytes and the clinical consequence therein.

2. ER Stress in the Adipocyte

Numerous cellular pathways can be altered in times of stress, leading to cellular aberrations and dysfunction. However, in the realm of overnutrition, ER stress is arguably the most common and important [7–10]. The ER is central for protein folding, secretions (e.g., cytokines), calcium homeostasis, and lipid synthesis. In the adipocyte, the ER is directly involved with LD formations and maintenance of lipid homeostasis.

Inducing ER stress is relatively effortless *via* depletion of ER calcium stores, changes in ER lipid membrane composition, reactive oxygen species (ROS), or accumulation of misfolded and/or unfolded proteins. When triggered, the ER signals to the cell through the unfolded protein response (UPR) to aid in increased production of proteins needed for protein folding, while decreasing transcription and increasing degradation of other nonessential proteins. If the UPR is unable to return the ER to homeostatic conditions, it will trigger apoptosis.

A central component of the UPR is an ER chaperone protein, BiP/GRP78. In homeostatic conditions, BiP/GRP78 is bound to three ER membrane resident proteins. An insult that alters ATP in the lumen decreases calcium, or increases a demand for protein folding causes GRP78 to unbind. These three proteins, ER transmembrane kinase/endoribonuclease **IRE1**, double-stranded RNA-activated protein kinase-like ER kinase (**PERK**), and activating transcription factor 6 (**ATF-6**), trigger a cascade upon their release, which ultimately leads to the activation of transcription factors that upregulate protein chaperones, proteasome components, and with continuous activation, turns on GADD-153/**CHOP** (C/EBP homologous protein), a major transcriptional factor responsible for ER-stress-induced apoptosis.

2.1. IRE1. Upon release from GRP78, IRE1 transautophosphorylates, activating its RNase activity. The activated IRE1 specifically acts on its downstream target X-box-binding protein 1 (**XBP1**) and removes a 26 base pair intron sequence of XBP1 resulting in the formation of spliced XBP1 (XBP1^s). There are multiple targets of XBP1^s, such as ER protein chaperones and proteins involved in ER-associated degradation (ERAD) [11–13]. However, beyond the traditional genes it activates, the biological function of XBP1^s has now been shown to be more diverse.

In fact, XBP1's ability to induce many ER proteins, and increase expansion of the rough ER [14] has demonstrated its necessity in ER biogenesis. Specific and elaborate knockout models have demonstrated this further; when the ER was poorly developed, secretory cells subsequently failed to function [15, 16]. Sriburi et al. have found that overexpression of XBP1^s in preadipocytes induces upregulation of the rate-limiting enzyme in phosphatidylcholine synthesis (CTP: phosphocholine cytidylyltransferase or CCT) [14, 17]. As this is the major phospholipid found in the ER

membrane, it follows that XBP1 increases ER biogenesis by both stimulation of ER proteins and membrane components.

This activity of XBP1 is most likely not cell specific, due to the already described centrality of this transcription factor in secretory cell types and hepatocytes. What is of interest in adipocytes, however, is the close interplay of ER biogenesis and LD formations. LDs, as mentioned previously, are a central organelle in adipocytes, though they also are found to a much lesser extent in other cells such as hepatocytes and macrophages. LDs are known to contain a core of triacylglycerols and cholesterol, but the multiple proteins found in their phospholipid monolayer are only beginning to be understood [18]. Although it is already known that the ER assembles and processes the lipids and proteins needed for LD formation, it is not fully known how they are transferred. The formation of a naïve LD is hypothesized to occur when neutral lipids accumulate at the ER membrane and then bud off. However, others propose LDs form as a bicelle or vesicular budding. In addition, the ER may in fact remain linked to LDs, allowing free exchange of proteins [19, 20].

Beyond the debate on whether these two organelles are physically linked, there is no dispute on the centrality of CCT. When CCT is limited, LDs begin to fuse due to less phosphatidylcholine on their surface [21]. Even more, when one gene of CCT was knocked down 60% in drosophila, there was a significant increase of triacylglycerol content [21]. This may be a compensation in which diacylglycerols normally utilized in the CCT pathway are now channeled to neutral lipids in the LDs. Nonetheless, the main end is larger and denser LDs with less active CCT.

The link between CCT, LDs, and the UPR is most likely the foundation of the essential nature of the IRE1-XBP1 pathway in adipogenesis. XBP1-shRNA-treated preadipocytes fail to differentiate, and only transduction of the XBP1^s rescued cells [22]. *In vivo* mouse models are more difficult to handle, as the full XBP1 knockout die *in utero* [15]. To circumvent this, one group has placed a liver-specific XBP1 gene into this model, but even these mice die during the neonatal starvation period [16]. These mice are smaller with a negligible white adipose mass, even compared to their heterozygous counterparts.

The mechanism underlying XBP1's significant role may be due to the upregulation of CCAAT/enhancer-binding protein-α (C/EBPα) [22]. CCAAT/enhancer-binding proteins are essential transcription factors in adipogenesis, with β and δ being major players in early differentiation and α essential in mid- to late differentiation. Sha et al. found that XBP1^s upregulates C/EBPα, and C/EBPβ increases transcription of XBP1 [22]. Therefore, XBP1 is integral in the loop of transcriptional activation of adipocyte differentiation as well as the functional maturation of LD formation.

2.2. PERK. The PERK-eIF2α pathway is another UPR leg involved in adipogenesis. When released, PERK transautophosphorylates leading to activation of its kinase domain. The major result of this is phosphorylation of eukaryotic translation initiation factor 2α (eIF2α). In the phosphorylated state, this essential component of the translational machinery cannot recycle GTP, inhibiting general translation but at the same time increasing the translation of mRNAs which contain internal ribosome entry sites, such as ATF-4, BiP/GRP78, and SREBP-1 [23–25].

Activating transcription factor (**ATF**)-4 is a well-studied protein involved in the UPR (reviewed in [26]). This transcription factor is heavily involved in increasing amino acid metabolism and protein transport [27, 28]. Importantly, ATF-4 also upregulates stress-related transcription factors ATF-3 and CHOP. CHOP is a central transcription factor involved in cellular perturbations, including inhibition of adipocyte differentiation [29–31], and ultimately inducing apoptosis. However, there is still necessity of balance as although high induction of ATF-4 will lead to CHOP activation, complete absence will affect AT lipogenesis [32]. More studies are needed to fully understand the role of ATF-4 in lipogenesis in adipocytes.

In contrast, more is understood about **SREBPs** (sterol regulatory element-binding proteins). SREBPs are additional transcription factors found in the ER membrane. There are three isoforms- SREBP-1a, -1c, and -2. SREBP-1c is involved in fatty acid synthesis and lipogenesis, -2 in cholesterol synthesis, and -1a in both pathways. The SREBPs are retained in the ER *via* insulin-induced gene (Insig) binding to SREBP-cleavage-activating-protein-(SCAP-) bound SREBP. At times of sensed decreases in cholesterol or fatty acids, SCAP-SREBP dissociates from Insig and relocates to the Golgi where SREBP is cleaved by two site proteases (S1P and S2P). The mature form of SREBP further translocates to the nucleus, activating genes involved in cholesterol and lipid metabolism, such as 3-hydroxy-3-methylglutaryl-CoA (HMG-CoA) synthase, HMG-CoA reductase, squalene synthase, acetyl-CoA carboxylase (ACC), and fatty acid synthase (FAS). Therefore, disruption of ER homeostasis not only alters protein production, but also affects cholesterol and fatty acid synthesis.

Normally, SREBPs are released when there is a sense of depletion of cholesterol or lipids in the ER membrane. However, SREBP1 processing is also regulated through PERK-eIF2α. In fact, knockout of PERK substantially decreases active SREBP1 in mammary glands [33]. This is most likely a result from the recent finding that SREBP1 contains an internal ribosome entry site [23]. Therefore, activation of ER stress will redundantly lead to active SREBP1 through both upregulation of translation and release of protein from the membrane.

In adipocytes of the SREBPs isoform, -1c is the most highly expressed. SREBP1c is an essential transcription factor during adipogenesis (and thus has a dual name of adipocyte determination and differentiation 1/ADD1). Likewise, the PERK pathway has also been found to be important during differentiation of adipocytes *in vitro* [33]. Overexpression of ADD1/SREBP1c leads to an increase of LD formation in preadipocytes, while conditional overexpression in mouse AT inhibits normal mass growth [34]. In addition, SREBP1c has been shown to directly activate C/EBPβ [35], further supporting its role in adipogenesis. The contradictory results demonstrated with the above mouse models may demonstrate the balance needed by all transcription factors for functional and normal AT.

2.3. ATF-6. There are two genes encoding ATF-6, α, and β. The α isoform is a strong transcriptional activator [36], and the form classically studied during UPR activation. When ATF-6 is released from GRP78, it is translocated to the Golgi *via* a localization signal that was hidden when in the bound form. In the Golgi, ATF-6 is cleaved by the same proteases that process SREBPs, releasing the active cytoplasmic domain, which is a transcription factor. Here, ATF6α heterodimerizes with XBP1 and upregulates genes with the ER stress response element (ERSE) in their promoters, including GRP78 [37] and other ER chaperone proteins, CHOP, and even XBP1 (reviewed in [38]).

In the realm of UPR activation altering lipid metabolism in adipocytes, not much has been noted in the literature concerning ATF6. Knockout mouse models of either ATF6α or β do not show any striking physiological changes, but have allowed for the clarification that ATF6α is the more essential isoform for the ER stress pathway [39], though β is also involved [36]. Some work has recently demonstrated that ATF6 activation plays a role in the liver to control lipid deposition [40, 41] through inhibition of SREBP-2 [42]. What is of more importance in the adipocyte is the direct function of ATF6 to upregulate XBP1, described above as central in adipogenesis.

ATF6α heterodimerizes with XBP1^s in the nucleus to activate genes downstream of UPR activation. However, it is currently not shown if this relationship is also required for upregulation of C/EBPα, or CCT activity. More investigations are needed to completely elucidate the direct function of ATF6 in adipocyte lipid metabolism.

3. Autophagy, the UPR, and Lipid Metabolism Dysregulation

Autophagy is a self-protective cellular pathway activated by multiple stimuli including viral infection, perceived starvation, organelle dysfunction, and ER stress (discussed below). However, just as in the case of UPR, autophagy has the ability to increase cellular damage or cell death when overstimulated. The multifaceted autophagic pathway is continuously being studied, as is the capacity of this process to help regulate metabolism in mammalian cells. In the past few years, an expanding area of research has unfurled around autophagy and lipid metabolism regulation. In hepatocytes, autophagosomes aid in the control of lipid accumulations by delivering LDs to lysosomes [43]. Similarly, in neurons altered autophagy leads to lipid accumulation [44]. Due to its obvious role in lipid metabolism, Singh and colleagues have now coined this leg of autophagy as lipophagy, in which lipid droplets are degraded through autophagy rather than lipolysis [45].

Further, components of the autophagosome may be necessary for lipid droplet formations [46]. This link was found through the microtubule-associated protein 1A/1B light chain 3 (**LC3**), an essential protein in the autophagy pathway. At induction of autophagy, a double membrane sequesters components of the cytoplasm through the coordination of multiple proteins and membrane expansion. During the initial stages, cytosolic LC3-I is activated through other autophagic-specific proteins by cleavage and lipidation, converting it to membrane-bound LC3-II. Shibata et al. have found that LC3-II does not only colocalize to autophagosomes (the specific autophagy sequestering vacuoles), but also to LDs in hepatocytes and cardiac myocytes [46]. This same group has also demonstrated that LC3 colocalizes to LDs in differentiating adipocytes by using LC3-siRNA [47]. The siRNA of LC3 drastically decreased the ability of adipogenesis [47]. LC3-II has been shown to have tethering capacity to help the fusion of autophagosomes to lysosomes [48]. Therefore, there is a hypothesis that LC3-II is acting to bring LDs into the autophagosome pathway for downstream lipid breakdown [43, 49]. This would provide another pathway of lipid flux beyond lipases acting directly on the LD.

Knockout models have demonstrated how essential autophagy is in adipogenesis. Baerga et al. were able to establish this by first showing the significant increase of autophagosome formations during induction of adipogenesis, followed by the inhibition of differentiation in a knockout *atg5* mouse model [50]. *Atg5* encodes a protein that is required similarly to LC3 for the maturation of the pre-autophagosome. Using this model, Baerga et al. saw both *in vitro* and *in vivo* that inhibition of autophagy restrained maturation of preadipocytes, resulting in a marked reduction of WAT in neonatal mice (this mouse model is not able to survive the neonatal starvation period). Of most interest, in the knockout mouse embryonic fibroblasts induced to differentiate, cells that began to mature died through apoptosis, while those in the same culture that did not begin to differentiate remained alive. This study was followed by another with adipose-specific deletion of *atg7* [51], the gene encoding an essential protein upstream of Atg5. Interestingly, WAT tissue of this knockout model was more characteristic of BAT in both morphology (smaller cells and LDs) and enzyme levels. The importance of Atg7 in adipogenesis was confirmed by Singh et al. who knocked down the same gene, but used slightly different cell lines and mouse model [43]. However, both groups came upon the same finding that the autophagic pathway is essential in adipogenesis.

The trigger of autophagy activation during adipogenesis is currently not known. However, PPARγ, an essential transcription factor of adipogenesis, may be involved. In one cancer cell line, it was found that PPARγ agonists can activate the autophagy pathway [52]. Yet, there is another study that contradicts these findings [53], and such investigations have not yet been repeated in an adipocyte model. Nonetheless, the summation of above experiments does demonstrate that autophagy is essential in adipogenesis, and without, may cause a transdifferentiation of WAT to BAT. On the other hand, a decrease of autophagy in the liver leads to lipid overload in hepatocytes. Intuitively, the difference lies in the biology of the two cell types, where adipocytes are normally storing lipids and hepatocytes are not. In metabolic disease states, such as the metabolic syndrome, it is easy to conceive how dysregulation of autophagy could ultimately lead to fatty liver with increased TG storage in the liver and decreased storage in AT.

4. Autophagy and ER Stress

Autophagy and ER stress pathways are not disconnected from one another as previously assumed. In contrast, activation of both can aid in cell survival at times of stress. For one example, autophagy offers an alternative pathway for degradation of proteins when ER-activated proteasomes can no longer handle the load [54–58]. In addition, activation of cell death of each pathway may be interlinked. While classic knowledge is based on ER stress activating apoptosis through CHOP upregulation and autophagy-mediated cell death via a completely separate process, recent findings demonstrate that these two cell death pathways are interlinked.

In more noxious circumstances, it has been shown that cell death through prolonged UPR activation can occur through autophagy-induced cell death [55]. Likewise, inhibition of autophagy increases cell viability with prolonged ER stress [59–61]. However, in nutrient overload and metabolic disorders, impaired autophagy can increase ER stress [62], perhaps due to decreased aberrant protein degradation and energy turnover needed to maintain ER homeostasis. This complex crosstalk of ER stress with the autophagy pathway is not yet well understood. Recently, it was found that ER stress activation can inhibit Akt phosphorylation, the upstream inducer of autophagy at times of perceived starvation [63]. However, the responsible protein(s) are still not known and may even be cell-type specific [64].

Another link is hypothesized to occur through the PERK pathway of the UPR [65, 66]. Some studies have shown that PERK phosphorylation of eIF2α leads to an upregulation of LC3 [58]. Yet, it has not been shown if this is directly from eIF2α phosphorylation inducing LC3 translation, or through ATF-4 activation increasing *Atg12* transcription [67, 68]. In fact, our current studies suggest that HIV Protease inhibitor (PI)-induced activation of autophagy is closely linked to ER stress *via* the ATF-4 pathway. We have found that those HIV PIs that induce metabolic side effects in the clinic also induce ER stress and autophagy in hepatocytes and adipocytes. The corresponding activation of autophagy seems to be one of the underlying factors by which HIV PIs induce dysregulation of lipid metabolism.

Recent studies have shown a strong link between activation of ER stress, increased autophagy induction, and increased SREBP activity leading to lipid overload in hepatocytes [69], although a mechanism remains to be determined. One group of investigators has demonstrated the capability of SREBP-2 to directly upregulate the expression of autophagy essential proteins [70], giving significance to a previous finding that cholesterol depletion leads to autophagy induction in multiple cell lines [71]. Additionally, knockdown of SREBP-2 decreased LC3 association with LDs in hepatocytes [70]. Although SREBPs are not a current forefront of proposed activators of autophagy, it is probable that in times of cellular lipid depletion, LDs are processed for more essential cellular requirements, and this pathway can be activated through ER stress-induced activation of SREBPs. Although these investigations have not been completed in adipocytes, our laboratory has found that in addition to HIV PIs inducing ER-stress and autophagy in adipocytes,

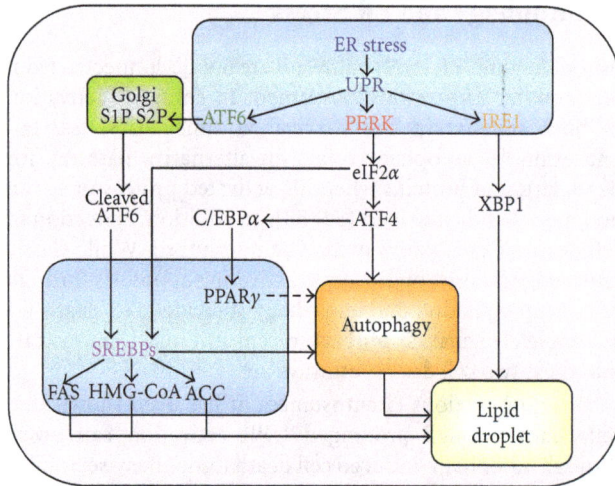

FIGURE 1: Potential link between ER stress signaling pathways and lipid droplet formation in adipocytes.

SREBP-1c activation is also altered. Until more investigations are completed, the exact stream can only be hypothesized (Figure 1).

5. ER Stress and Inflammation

Obesity and resulting metabolic diseases such as insulin resistance are now known to be strongly associated with chronic inflammation, a substantial risk factor for further complications, most notably atherosclerosis. Increased plasma concentrations of IL-6 and TNF-α have been repeatedly noted in obese individuals [72–74]. Investigations into mechanisms underlying obesity and diabetes has demonstrated that inflammation in AT can detrimentally alter human physiology.

With increasing overload, adipocytes begin to hypertrophy. Cells become stressed from the actual expansion and from exceeding an adequate oxygen diffusion distance in tissue [75, 76]. Adipocytes then signal with a release of proinflammatory IL-6 and TNF-α cytokines, which activate resident macrophages as well as induce infiltration of circulating macrophages. Stressed adipocytes are subsequently engulfed, resulting in the formation of characteristic crown-like structures.

During this process, released IL-6 and TNF-α from stressed adipocytes and activated macrophages can inhibit adipogenesis [77]. In fact, TNF-α alone is enough to inhibit induction of PPARγ and C/EBPα [78]. Even more, the induction of inflammation can also lead to insulin resistance in AT, already well known and continuously investigated [79–82]. Taken together, the ability to store excess energy in AT is drastically decreased with the decrease of mature adipocytes and the death of cells.

Even more, ER stress has been shown to be activated at times of overnutrition [8]. In adipocytes, ER stress can be activated due to the need of LD synthesis, enzyme production, and conversion of energy to TG at times of overnutrition. Importantly, ER stress has repeatedly been shown

to induce the cellular inflammatory cascade through the c-Jun N-terminal kinase (JNK) pathway, and JNK has been shown to be upregulated in AT of obese individuals [83, 84]. Additionally, ER stress may trigger the adipocyte inflammatory cascade through PERK activating IKB kinase β (IKKβ) when cells are stimulated with free fatty acids [85]. This pathway is also known to be a heavy regulator of inflammatory cytokine release and, together with JNK activation, would lead to the proinflammatory state seen in AT in metabolic disease states.

Proinflammatory profile at times of overnutrition is not unique to AT, but occurs throughout the body. However, AT is unique in that it is solely responsible for the subsequent decrease of adiponectin secretion. Adiponectin is an adipocyte-specific anti-inflammatory cytokine that negatively correlates with cardiovascular disease and fatty liver disease [86–88], with a decrease of secretion in overexpanded or stressed tissue [89]. It has been found that adiponectin can alleviate ER stress [90]. Zhou et al. have shown that ER stress initiation is sufficient to decrease adiponectin release. In animal models, they demonstrated that stabilization of adiponectin protein can decrease obesity-induced ER stress in AT [90]. *In vitro*, induction of autophagy could alleviate ER stress responses and subsequently stabilize adiponectin secretions [91]. These are promising findings, and more studies are needed to determine if upregulation of autophagy could ultimately lead to therapeutic options for metabolic diseases (Figure 2).

6. Future Directions

We have provided ample references demonstrating that ER stress can induce lipid metabolism dysregulation in adipocytes. Such an assertion is not only important for interested molecular biologists, but for clinicians as well. It has been shown that fat depots of obese patients have increased ER stress [84, 92, 93]. What is more, there may be a link between ER stress upregulation, the inflammatory state of this tissue, and insulin resistance [84, 92, 94, 95].

The cycle of overnutrition, ER stress, and AT pathology is complex. With the information provided here and our own findings, we support the hypothesis that inhibiting ER stress activation may be therapeutically beneficial in the treatment of metabolic diseases. Chaperones, which enhance ER-protein-folding capacity, have shown potential in the laboratory.

Two chaperones already FDA approved have been studied in hepatocytes, adipocytes, and β-cells for their ability to relieve ER-stress-induced dysfunctions, namely, 4-phenylbutyric acid (PBA) and taurine-conjugated ursodeoxycholic acid (TUDCA). Both were shown to relieve insulin resistance in adipocytes at times of ER stress [96, 97]. In addition, they were able to decrease JNK and IKKβ activity when cells were stimulated with ER stress inducers, including free fatty acids [85, 96]. *In vivo*, PBA and TUDCA were able to relieve ER stress activation in obese mice [96]. However, further studies are needed to confirm these beneficial effects and elaborate on the extent that chaperone treatment may aid in nutrition overload-induced ER stress and downstream alterations.

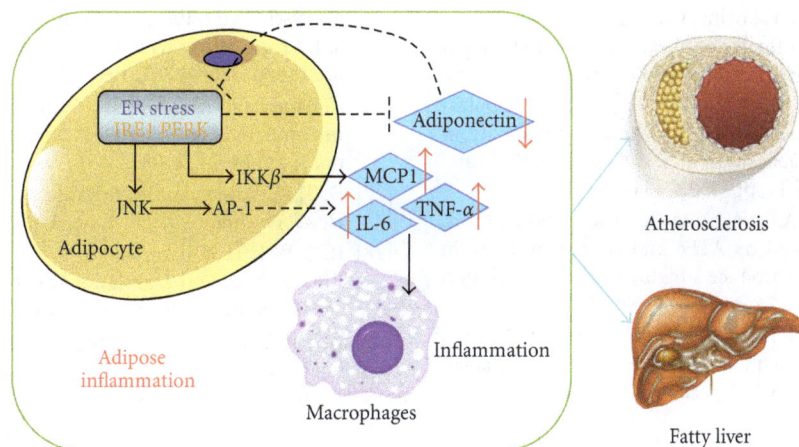

FIGURE 2: ER-stress-induced inflammation in adipocytes and macrophages contributes to atherosclerosis and fatty liver diseases.

Inhibiting ER stress activation may be the key to an approach for metabolic syndrome therapy. However, more questions remain in this field. Namely, the role of all parts of the UPR in adipocyte lipid metabolism needs to be uncovered, and the mechanism intertwining ER stress and autophagy needs to be further elucidated. Understanding these missing components will allow not only further understanding of key lipid pathways in a central metabolic cell type, but also help determine the best approach that can be utilized for clinical metabolic dysfunctions in patients with altered AT physiology.

Abbreviations

ACC: Acetyl-CoA carboxylase
Add1: Adipocyte determination and differentiation 1
AT: Adipose tissue
ATF: Activating transcription factor
CHOP: C/EBP homologous protein
CCT: CTP phosphocholine cytidylyltransferase
eIF2α: Eukaryotic translation initiation factor
ER: Endoplasmic reticulum
FAS: Fatty acid synthase
FFA: Free fatty acids
Insig: Insulin-induced gene
HMG-CoAR: 3-Hydroxy-3-methylgutaryl-CoA reductase
IKKβ: IKB kinase β
IRE1: Inositol requiring enzyme 1
IRS1: Insulin response substrate
JNK: cJun N-terminal kinase
LC3: Microtubule associated light chain protein
LD: Lipid droplet
PBA: Protein-1 namely 4-phenylbutyric acid

PERK: PKR-like eukaryotic initiation factor 2α kinase
SCAP: SREBP cleavage activating protein
SREBP: Sterol regulatory element binding protein
TUDCA: Taurine-conjugated ursodeoxycholic acid
TG: Triglyceride
UPR: Unfolded protein response
XBP1: X-box binding protein.

References

[1] H. E. Bays, J. M. González-Campoy, R. R. Henry et al., "Is adiposopathy (sick fat) an endocrine disease?" *International Journal of Clinical Practice*, vol. 62, no. 10, pp. 1474–1483, 2008.

[2] B. K. Surmi and A. H. Hasty, "Macrophage infiltration into adipose tissue: initiation, propagation and remodeling," *Future Lipidology*, vol. 3, no. 5, pp. 545–556, 2008.

[3] S. Mittra, V. S. Bansal, and P. K. Bhatnagar, "From a glucocentric to a lipocentric approach towards metabolic syndrome," *Drug Discovery Today*, vol. 13, no. 5-6, pp. 211–218, 2008.

[4] M. Kabir, K. J. Catalano, S. Ananthnarayan et al., "Molecular evidence supporting the portal theory: a causative link between visceral adiposity and hepatic insulin resistance," *American Journal of Physiology*, vol. 288, no. 2, pp. E454–E461, 2005.

[5] H. Yoshii, T. K. T. Lam, N. Gupta et al., "Effects of portal free fatty acid elevation on insulin clearance and hepatic glucose flux," *American Journal of Physiology*, vol. 290, no. 6, pp. E1089–E1097, 2006.

[6] H. Yoshida, "ER stress and diseases," *FEBS Journal*, vol. 274, no. 3, pp. 630–658, 2007.

[7] S. Alhusaini, K. McGee, B. Schisano et al., "Lipopolysaccharide, high glucose and saturated fatty acids induce endoplasmic reticulum stress in cultured primary human adipocytes: salicylate alleviates this stress," *Biochemical and Biophysical Research Communications*, vol. 397, no. 3, pp. 472–478, 2010.

[8] G. Boden, X. Duan, C. Homko et al., "Increase in endoplasmic reticulum stress-related proteins and genes in adipose tissue of obese, insulin-resistant individuals," *Diabetes*, vol. 57, no. 9, pp. 2438–2444, 2008.

[9] M. F. Gregor, L. Yang, E. Fabbrini et al., "Endoplasmic reticulum stress is reduced in tissues of obese subjects after weight loss," *Diabetes*, vol. 58, no. 3, pp. 693–700, 2009.

[10] N. K. Sharma, S. K. Das, A. K. Mondal et al., "Endoplasmic reticulum stress markers are associated with obesity in nondiabetic subjects," *Journal of Clinical Endocrinology and Metabolism*, vol. 93, no. 11, pp. 4532–4541, 2008.

[11] H. Yoshida, T. Matsui, A. Yamamoto, T. Okada, and K. Mori, "XBP1 mRNA is induced by ATF6 and spliced by IRE1 in response to ER stress to produce a highly active transcription factor," *Cell*, vol. 107, no. 7, pp. 881–891, 2001.

[12] H. Yoshida, T. Matsui, N. Hosokawa, R. J. Kaufman, K. Nagata, and K. Mori, "A time-dependent phase shift in the mammalian unfolded protein response," *Developmental Cell*, vol. 4, no. 2, pp. 265–271, 2003.

[13] A. H. Lee, N. N. Iwakoshi, and L. H. Glimcher, "XBP-1 regulates a subset of endoplasmic reticulum resident chaperone genes in the unfolded protein response," *Molecular and Cellular Biology*, vol. 23, no. 21, pp. 7448–7459, 2003.

[14] R. Sriburi, S. Jackowski, K. Mori, and J. W. Brewer, "XBP1: a link between the unfolded protein response, lipid biosynthesis, and biogenesis of the endoplasmic reticulum," *Journal of Cell Biology*, vol. 167, no. 1, pp. 35–41, 2004.

[15] A. M. Reimold, N. N. Iwakoshi, J. Manis et al., "Plasma cell differentiation requires the transcription factor XBP-1," *Nature*, vol. 412, no. 6844, pp. 300–307, 2001.

[16] A. H. Lee, G. C. Chu, N. N. Iwakoshi, and L. H. Glimcher, "XBP-1 is required for biogenesis of cellular secretory machinery of exocrine glands," *EMBO Journal*, vol. 24, no. 24, pp. 4368–4380, 2005.

[17] R. Sriburi, H. Bommiasamy, G. L. Buldak et al., "Coordinate regulation of phospholipid biosynthesis and secretory pathway gene expression in XBP-1(S)-induced endoplasmic reticulum biogenesis," *Journal of Biological Chemistry*, vol. 282, no. 10, pp. 7024–7034, 2007.

[18] R. V. Farese Jr. and T. C. Walther, "Lipid droplets finally get a little R-E-S-P-E-C-T," *Cell*, vol. 139, no. 5, pp. 855–860, 2009.

[19] T. C. Walther and R. V. Farese Jr., "The life of lipid droplets," *Biochimica et Biophysica Acta*, vol. 1791, no. 6, pp. 459–466, 2009.

[20] H. Robenek, I. Buers, M. J. Robenek et al., "Topography of lipid droplet-associated proteins: insights from freeze-fracture replica immunogold labeling," *Journal of Lipids*, vol. 2011, Article ID 409371, 10 pages, 2011.

[21] Y. Guo, T. C. Walther, M. Rao et al., "Functional genomic screen reveals genes involved in lipid-droplet formation and utilization," *Nature*, vol. 453, no. 7195, pp. 657–661, 2008.

[22] H. Sha, Y. He, H. Chen et al., "The IRE1α-XBP1 pathway of the unfolded protein response is required for adipogenesis," *Cell Metabolism*, vol. 9, no. 6, pp. 556–564, 2009.

[23] F. Damiano, S. Alemanno, G. V. Gnoni, and L. Siculella, "Translational control of the sterol-regulatory transcription factor SREBP-1 mRNA in response to serum starvation or ER stress is mediated by an internal ribosome entry site," *Biochemical Journal*, vol. 429, no. 3, pp. 603–612, 2010.

[24] D. J. DeGracia, R. Kumar, C. R. Owen, G. S. Krause, and B. C. White, "Molecular pathways of protein synthesis inhibition during brain reperfusion: implications for neuronal survival or death," *Journal of Cerebral Blood Flow and Metabolism*, vol. 22, no. 2, pp. 127–141, 2002.

[25] Q. Yang and P. Sarnow, "Location of the internal ribosome entry site in the 5' non-coding region of the immunoglobulin heavy-chain binding protein (BiP) mRNA: evidence for specific RNA-protein interactions," *Nucleic Acids Research*, vol. 25, no. 14, pp. 2800–2807, 1997.

[26] K. Ameri and A. L. Harris, "Activating transcription factor 4," *International Journal of Biochemistry and Cell Biology*, vol. 40, no. 1, pp. 14–21, 2008.

[27] D. T. Rutkowski and R. J. Kaufman, "All roads lead to ATF4," *Developmental Cell*, vol. 4, no. 4, pp. 442–444, 2003.

[28] R. C. Wek, H. Y. Jiang, and T. G. Anthony, "Coping with stress: EIF2 kinases and translational control," *Biochemical Society Transactions*, vol. 34, no. 1, pp. 7–11, 2006.

[29] N. Batchvarova, X. Z. Wang, and D. Ron, "Inhibition of adipogenesis by the stress-induced protein CHOP (Gadd153)," *EMBO Journal*, vol. 14, no. 19, pp. 4654–4661, 1995.

[30] S. L. Clarke, C. E. Robinson, and J. M. Gimble, "CAAT/Enhancer binding proteins directly modulate transcription from the peroxisome proliferator-activated receptor γ2 promoter," *Biochemical and Biophysical Research Communications*, vol. 240, no. 1, pp. 99–103, 1997.

[31] G. Adelmant, J. D. Gilbert, and S. O. Freytag, "Human translocation liposarcoma-CCAAT/enhancer binding protein (C/EBP) homologous protein (TLS-CHOP) oncoprotein prevents adipocyte differentiation by directly interfering with C/EBPβ function," *Journal of Biological Chemistry*, vol. 273, no. 25, pp. 15574–15581, 1998.

[32] C. Wang, Z. Huang, Y. Du, Y. Cheng, S. Chen, and F. Guo, "ATF4 regulates lipid metabolism and thermogenesis," *Cell Research*, vol. 20, no. 2, pp. 174–184, 2010.

[33] E. Bobrovnikova-Marjon, G. Hatzivassiliou, C. Grigoriadou et al., "PERK-dependent regulation of lipogenesis during mouse mammary gland development and adipocyte differentiation," *Proceedings of the National Academy of Sciences of the United States of America*, vol. 105, no. 42, pp. 16314–16319, 2008.

[34] I. Shimomura, R. E. Hammer, J. A. Richardson et al., "Insulin resistance and diabetes mellitus in transgenic mice expressing nuclear SREBP-1c in adipose tissue: model for congenital generalized lipodystrophy," *Genes and Development*, vol. 12, no. 20, pp. 3182–3194, 1998.

[35] S. L. Lay, I. Lefrère, C. Trautwein, I. Dugail, and S. Krief, "Insulin and sterol-regulatory element-binding protein-1c (SREBP-1c) regulation of gene expression in 3T3-L1 adipocytes: identification of CCAAT/enhancer-binding protein β as an SREBP-1c target," *Journal of Biological Chemistry*, vol. 277, no. 38, pp. 35625–35634, 2002.

[36] D. J. Thuerauf, M. Marcinko, P. J. Belmont, and C. C. Glembotski, "Effects of the isoform-specific characteristics of ATF6α and ATF6β on endoplasmic reticulum stress response gene expression and cell viability," *Journal of Biological Chemistry*, vol. 282, no. 31, pp. 22865–22878, 2007.

[37] K. Haze, H. Yoshida, H. Yanagi, T. Yura, and K. Mori, "Mammalian transcription factor ATF6 is synthesized as a transmembrane protein and activated by proteolysis in response to endoplasmic reticulum stress," *Molecular Biology of the Cell*, vol. 10, no. 11, pp. 3787–3799, 1999.

[38] E. L. Davenport, G. J. Morgan, and F. E. Davies, "Untangling the unfolded protein response," *Cell Cycle*, vol. 7, no. 7, pp. 865–869, 2008.

[39] Y. Adachi, K. Yamamoto, T. Okada, H. Yoshida, A. Harada, and K. Mori, "ATF6 is a transcription factor specializing in the regulation of quality control proteins in the endoplasmic reticulum," *Cell Structure and Function*, vol. 33, no. 1, pp. 75–89, 2008.

[40] D. T. Rutkowski, J. Wu, S. H. Back et al., "UPR pathways combine to prevent hepatic steatosis caused by ER stress-mediated suppression of transcriptional master regulators," *Developmental Cell*, vol. 15, no. 6, pp. 829–840, 2008.

[41] K. Yamamoto, K. Takahara, S. Oyadomari et al., "Induction of liver steatosis and lipid droplet formation in ATF6α-knockout mice burdened with pharmacological endoplasmic reticulum stress," *Molecular Biology of the Cell*, vol. 21, no. 17, pp. 2975–2986, 2010.

[42] M. Schröder and R. J. Kaufman, "ER stress and the unfolded protein response," *Mutation Research*, vol. 569, no. 1-2, pp. 29–63, 2005.

[43] R. Singh, S. Kaushik, Y. Wang et al., "Autophagy regulates lipid metabolism," *Nature*, vol. 458, no. 7242, pp. 1131–1135, 2009.

[44] M. Martinez-Vicente, Z. Talloczy, E. Wong et al., "Cargo recognition failure is responsible for inefficient autophagy in Huntington's disease," *Nature Neuroscience*, vol. 13, no. 5, pp. 567–576, 2010.

[45] R. Singh, "Autophagy and regulation of lipid metabolism," *Results and Problems in Cell Differentiation*, vol. 52, pp. 35–46, 2010.

[46] M. Shibata, K. Yoshimura, N. Furuya et al., "The MAP1-LC3 conjugation system is involved in lipid droplet formation," *Biochemical and Biophysical Research Communications*, vol. 382, no. 2, pp. 419–423, 2009.

[47] M. Shibata, K. Yoshimura, H. Tamura et al., "LC3, a microtubule-associated protein1A/B light chain3, is involved in cytoplasmic lipid droplet formation," *Biochemical and Biophysical Research Communications*, vol. 393, no. 2, pp. 274–279, 2010.

[48] H. Nakatogawa, Y. Ichimura, and Y. Ohsumi, "Atg8, a ubiquitin-like protein required for autophagosome formation, mediates membrane tethering and hemifusion," *Cell*, vol. 130, no. 1, pp. 165–178, 2007.

[49] J. Kovsan, N. Bashan, A. S. Greenberg, and A. Rudich, "Potential role of autophagy in modulation of lipid metabolism," *American Journal of Physiology*, vol. 298, no. 1, pp. E1–E7, 2010.

[50] R. Baerga, Y. Zhang, P. H. Chen, S. Goldman, and S. Jin, "Targeted deletion of autophagy-related 5 (atg5) impairs adipogenesis in a cellular model and in mice," *Autophagy*, vol. 5, no. 8, pp. 1118–1130, 2009.

[51] Y. Zhang, S. Goldman, R. Baerga, Y. Zhao, M. Komatsu, and S. Jin, "Adipose-specific deletion of autophagy-related gene 7 (atg7) in mice reveals a role in adipogenesis," *Proceedings of the National Academy of Sciences of the United States of America*, vol. 106, no. 47, pp. 19860–19865, 2009.

[52] J. Zhou, W. Zhang, B. Liang et al., "PPARγ activation induces autophagy in breast cancer cells," *International Journal of Biochemistry and Cell Biology*, vol. 41, no. 11, pp. 2334–2342, 2009.

[53] J. Yan, H. Yang, G. Wang et al., "Autophagy augmented by troglitazone is independent of EGFR transactivation and correlated with AMP-activated protein kinase signaling," *Autophagy*, vol. 6, no. 1, pp. 67–73, 2010.

[54] T. Kawakami, R. Inagi, H. Takano et al., "Endoplasmic reticulum stress induces autophagy in renal proximal tubular cells," *Nephrology Dialysis Transplantation*, vol. 24, no. 9, pp. 2665–2672, 2009.

[55] T. Yorimitsu, U. Nair, Z. Yang, and D. J. Klionsky, "Endoplasmic reticulum stress triggers autophagy," *Journal of Biological Chemistry*, vol. 281, no. 40, pp. 30299–30304, 2006.

[56] W. X. Ding and X. M. Yin, "Sorting, recognition and activation of the misfolded protein degradation pathways through macroautophagy and the proteasome," *Autophagy*, vol. 4, no. 2, pp. 141–150, 2008.

[57] W. X. Ding, H. M. Ni, W. Gao et al., "Linking of autophagy to ubiquitin-proteasome system is important for the regulation of endoplasmic reticulum stress and cell viability," *American Journal of Pathology*, vol. 171, no. 2, pp. 513–524, 2007.

[58] Y. Kouroku, E. Fujita, I. Tanida et al., "ER stress (PERK/eIF2α phosphorylation) mediates the polyglutamine-induced LC3 conversion, an essential step for autophagy formation," *Cell Death and Differentiation*, vol. 14, no. 2, pp. 230–239, 2007.

[59] J. Price, A. K. Zaidi, J. Bohensky, V. Srinivas, I. M. Shapiro, and H. Ali, "Akt-1 mediates survival of chondrocytes from endoplasmic reticulum-induced stress," *Journal of Cellular Physiology*, vol. 222, no. 3, pp. 502–508, 2010.

[60] L. Qin, Z. Wang, L. Tao, and Y. Wang, "ER stress negatively regulates AKT/TSC/mTOR pathway to enhance autophagy," *Autophagy*, vol. 6, no. 2, pp. 239–247, 2010.

[61] M. Ogata, S. I. Hino, A. Saito et al., "Autophagy is activated for cell survival after endoplasmic reticulum stress," *Molecular and Cellular Biology*, vol. 26, no. 24, pp. 9220–9231, 2006.

[62] L. Yang, P. Li, S. Fu, E. S. Calay, and G. S. Hotamisligil, "Defective hepatic autophagy in obesity promotes ER stress and causes insulin resistance," *Cell Metabolism*, vol. 11, no. 6, pp. 467–478, 2010.

[63] A. K. Gupta, B. Li, G. J. Cerniglia, M. S. Ahmed, S. M. Hahn, and A. Maity, "The HIV protease inhibitor nelfinavir downregulates Akt phosphorylation by inhibiting proteasomal activity and inducing the unfolded protein response," *Neoplasia*, vol. 9, no. 4, pp. 271–278, 2007.

[64] S. M. Schleicher, L. Moretti, V. Varki, and B. Lu, "Progress in the unraveling of the endoplasmic reticulum stress/autophagy pathway and cancer: implications for future therapeutic approaches," *Drug Resistance Updates*, vol. 13, no. 3, pp. 79–86, 2010.

[65] W. Jia, R. M. Loria, M. A. Park, A. Yacoub, P. Dent, and M. R. Graf, "The neuro-steroid, 5-androstene 3β,17α diol; induces endoplasmic reticulum stress and autophagy through PERK/eIF2α signaling in malignant glioma cells and transformed fibroblasts," *International Journal of Biochemistry and Cell Biology*, vol. 42, no. 12, pp. 2019–2029, 2010.

[66] K. W. Kim, L. Moretti, L. R. Mitchell, D. K. Jung, and B. Lu, "Endoplasmic reticulum stress mediates radiation-induced autophagy by perk-eIF2α in caspase-3/7-deficient cells," *Oncogene*, vol. 29, no. 22, pp. 3241–3251, 2010.

[67] T. Rzymski, M. Milani, L. Pike et al., "Regulation of autophagy by ATF4 in response to severe hypoxia," *Oncogene*, vol. 29, no. 31, pp. 4424–4435, 2010.

[68] M. Milani, T. Rzymski, H. R. Mellor et al., "The role of ATF4 stabilization and autophagy in resistance of breast cancer cells treated with Bortezomib," *Cancer Research*, vol. 69, no. 10, pp. 4415–4423, 2009.

[69] S. Nishina, M. Korenaga, I. Hidaka et al., "Hepatitis C virus protein and iron overload induce hepatic steatosis through the unfolded protein response in mice," *Liver International*, vol. 30, no. 5, pp. 683–692, 2010.

[70] Y.-K. Seo, T.-I. Jeon, H. K. Chong, J. Biesinger, X. Xie, and T. F. Osborne, "Genome-wide localization of SREBP-2 in hepatic chromatin predicts a role in autophagy," *Cell Metabolism*, vol. 13, no. 4, pp. 367–375, 2011.

[71] J. Cheng, Y. Ohsaki, K. Tauchi-Sato, A. Fujita, and T. Fujimoto, "Cholesterol depletion induces autophagy," *Biochemical and Biophysical Research Communications*, vol. 351, no. 1, pp. 246–252, 2006.

[72] G. P. Van Guilder, G. L. Hoetzer, J. J. Greiner, B. L. Stauffer, and C. A. DeSouza, "Influence of metabolic syndrome on biomarkers of oxidative stress and inflammation in obese adults," *Obesity*, vol. 14, no. 12, pp. 2127–2131, 2006.

[73] C. Tsigos, I. Kyrou, E. Chala et al., "Circulating tumor necrosis factor alpha concentrations are higher in abdominal versus peripheral obesity," *Metabolism*, vol. 48, no. 10, pp. 1332–1335, 1999.

[74] P. A. Kern, S. Ranganathan, C. Li, L. Wood, and G. Ranganathan, "Adipose tissue tumor necrosis factor and interleukin-6 expression in human obesity and insulin resistance," *American Journal of Physiology*, vol. 280, no. 5, pp. E745–E751, 2001.

[75] R. W. O'Rourke, "Inflammation in obesity-related diseases," *Surgery*, vol. 145, no. 3, pp. 255–259, 2009.

[76] G. H. Goossens, "The role of adipose tissue dysfunction in the pathogenesis of obesity-related insulin resistance," *Physiology and Behavior*, vol. 94, no. 2, pp. 206–218, 2008.

[77] B. Gustafson, A. Hammarstedt, C. X. Andersson, and U. Smith, "Inflamed adipose tissue: a culprit underlying the metabolic syndrome and atherosclerosis," *Arteriosclerosis, Thrombosis, and Vascular Biology*, vol. 27, no. 11, pp. 2276–2283, 2007.

[78] L. Meng, J. Zhou, H. Sasano, T. Suzuki, K. M. Zeitoun, and S. E. Bulun, "Tumor necrosis factor α and interleukin 11 secreted by malignant breast epithelial cells inhibit adipocyte differentiation by selectively down-regulating CCAAT/enhancer binding protein α and peroxisome proliferator-activated receptor γ: mechanism of desmoplastic reaction," *Cancer Research*, vol. 61, no. 5, pp. 2250–2255, 2001.

[79] J. P. Bastard, M. Maachi, C. Lagathu et al., "Recent advances in the relationship between obesity, inflammation, and insulin resistance," *European Cytokine Network*, vol. 17, no. 1, pp. 4–12, 2006.

[80] G. S. Hotamisligil, "Inflammation and metabolic disorders," *Nature*, vol. 444, no. 7121, pp. 860–867, 2006.

[81] R. Monteiro and I. Azevedo, "Chronic inflammation in obesity and the metabolic syndrome," *Mediators of Inflammation*, vol. 2010, Article ID 289645, 10 pages, 2010.

[82] H. Xu, G. T. Barnes, Q. Yang et al., "Chronic inflammation in fat plays a crucial role in the development of obesity-related insulin resistance," *Journal of Clinical Investigation*, vol. 112, no. 12, pp. 1821–1830, 2003.

[83] J. Hirosumi, G. Tuncman, L. Chang et al., "A central, role for JNK in obesity and insulin resistance," *Nature*, vol. 420, no. 6913, pp. 333–336, 2002.

[84] G. Boden and S. Merali, "Measurement of the increase in endoplasmic reticulum stress-related proteins and genes in adipose tissue of obese, insulin-resistant individuals," *Methods in Enzymology*, vol. 489, pp. 67–82, 2011.

[85] P. Jiao, J. Ma, B. Feng et al., "FFA-induced adipocyte inflammation and insulin resistance: involvement of ER stress and IKKβ pathways," *Obesity*, vol. 19, no. 3, pp. 483–491, 2011.

[86] S. A. Polyzos, J. Kountouras, C. Zavos, and E. Tsiaousi, "The role of adiponectin in the pathogenesis and treatment of non-alcoholic fatty liver disease," *Diabetes, Obesity and Metabolism*, vol. 12, no. 5, pp. 365–383, 2010.

[87] S. A. Polyzos, J. Kountouras, and C. Zavos, "Nonalcoholic fatty liver disease: the pathogenetic roles of insulin resistance and adipocytokines," *Current Molecular Medicine*, vol. 9, no. 3, pp. 299–314, 2009.

[88] E. Ros, "Nuts and novel biomarkers of cardiovascular disease," *American Journal of Clinical Nutrition*, vol. 89, no. 5, pp. 1649S–1656S, 2009.

[89] P. A. Kern, G. B. Di Gregorio, T. Lu, N. Rassouli, and G. Ranganathan, "Adiponectin expression from human adipose tissue: relation to obesity, insulin resistance, and tumor necrosis factor-α expression," *Diabetes*, vol. 52, no. 7, pp. 1779–1785, 2003.

[90] L. Zhou, M. Liu, J. Zhang, H. Chen, L. Q. Dong, and F. Liu, "DsbA-L alleviates endoplasmic reticulum stress-induced adiponectin downregulation," *Diabetes*, vol. 59, no. 11, pp. 2809–2816, 2010.

[91] L. Zhou and F. Liu, "Autophagy: roles in obesity-induced ER stress and adiponectin downregulation in adipocytes," *Autophagy*, vol. 6, no. 8, pp. 1196–1197, 2010.

[92] M. Kars, L. Yang, M. F. Gregor et al., "Tauroursodeoxycholic acid may improve liver and muscle but not adipose tissue insulin sensitivity in obese men and women," *Diabetes*, vol. 59, no. 8, pp. 1899–1905, 2010.

[93] M. Miranda, X. Escoté, V. Ceperuelo-Mallafré et al., "Relation between human LPIN1, hypoxia and endoplasmic reticulum stress genes in subcutaneous and visceral adipose tissue," *International Journal of Obesity*, vol. 34, no. 4, pp. 679–686, 2010.

[94] N. Marsollier, P. Ferré, and F. Foufelle, "Novel insights in the interplay between inflammation and metabolic diseases: a role for the pathogen sensing kinase PKR," *Journal of Hepatology*, vol. 54, no. 6, pp. 1307–1309, 2011.

[95] G. S. Hotamisligil, "Inflammation and endoplasmic reticulum stress in obesity and diabetes," *International Journal of Obesity*, vol. 32, no. 7, pp. S52–S54, 2008.

[96] U. Özcan, E. Yilmaz, L. Özcan et al., "Chemical chaperones reduce ER stress and restore glucose homeostasis in a mouse model of type 2 diabetes," *Science*, vol. 313, no. 5790, pp. 1137–1140, 2006.

[97] Y. Nakatani, H. Kaneto, D. Kawamori et al., "Involvement of endoplasmic reticulum stress in insulin resistance and diabetes," *Journal of Biological Chemistry*, vol. 280, no. 1, pp. 847–851, 2005.

18

Hyaluronan Regulates Cell Behavior: A Potential Niche Matrix for Stem Cells

Mairim Alexandra Solis,[1,2] **Ying-Hui Chen,**[1,2] **Tzyy Yue Wong,**[1,2]
Vanessa Zaiatz Bittencourt,[1,2] **Yen-Cheng Lin,**[1,2] **and Lynn L. H. Huang**[1,2,3,4]

[1] *Institute of Biotechnology, College of Bioscience and Biotechnology, National Cheng Kung University, Tainan, Taiwan*
[2] *Research Center of Excellence in Regenerative Medicine, National Cheng Kung University, Tainan, Taiwan*
[3] *Institute of Clinical Medicine, College of Medicine, National Cheng Kung University, Tainan, Taiwan*
[4] *Advanced Optoelectronic Technology Center, National Cheng Kung University, Tainan, Taiwan*

Correspondence should be addressed to Lynn L. H. Huang, lynn@mail.ncku.edu.tw

Academic Editor: Manuela Viola

Hyaluronan is a linear glycosaminoglycan that has received special attention in the last few decades due to its extraordinary physiological functions. This highly viscous polysaccharide is not only a lubricator, but also a significant regulator of cellular behaviors during embryogenesis, morphogenesis, migration, proliferation, and drug resistance in many cell types, including stem cells. Most hyaluronan functions require binding to its cellular receptors CD44, LYVE-1, HARE, layilin, and RHAMM. After binding, proteins are recruited and messages are sent to alter cellular activities. When low concentrations of hyaluronan are applied to stem cells, the proliferative activity is enhanced. However, at high concentrations, stem cells acquire a dormant state and induce a multidrug resistance phenotype. Due to the influence of hyaluronan on cells and tissue morphogenesis, with regards to cardiogenesis, chondrogenesis, osteogenesis, and neurogenesis, it is now been utilized as a biomaterial for tissue regeneration. This paper summarizes the most important and recent findings regarding the regulation of hyaluronan in cells.

1. Hyaluronan Properties

Hyaluronan (also known as hyaluronate or hyaluronic acid, HA) is a nonsulfated linear polysaccharide present in the extracellular matrix of every vertebrate's tissue. It is a member of the glycosaminoglycan (GAG) family and is synthesized at the inner leaflet of the plasma membrane [1] as a large, unbranched polymer of repeating disaccharides of glucuronic acid and N-acetylglucosamine. It has a simple chemical structure but differs from other glycosaminoglycans in its high molecular weight, lack of sulfate groups, and absence of covalent attachment to core proteins [1]. It was assumed that its major functions were in joint lubrication, tissue homeostasis, and holding gel-like tissues together. However, the high molecular weight of hyaluronan enables it to acquire a viscous characteristic that goes beyond these functions. Indeed, hyaluronan coregulates numerous physiological processes including embryonic development, inflammation, tissue regeneration, cell migration, and proliferation. Some reports [2, 3] have also suggested that hyaluronan plays some role in cancer invasion and in the promotion of angiogenesis.

Many functions of hyaluronan depend upon its interaction with cell surface receptors, including cluster determinant 44 (CD44), receptor for hyaluronan-mediated motility (RHAMM), lymphatic vessel endothelial hyaluronan receptor (LYVE-1), hyaluronan receptor for endocytosis (HARE), liver endothelial cell clearance receptor (LEC receptor), and toll-like receptor 4 (TLR4) [1]. Although it is well established that hyaluronan-induced signaling occurs through receptor interactions, its signal transduction mechanism in cells has not been fully characterized.

Stem cells have the unique abilities of self-renewal and pluripotency. These characteristics therefore play essential roles in organogenesis during embryonic development and tissue regeneration. The potential utility of stem cells for tissue regeneration and treatment of many formerly incurable diseases has motivated scientists to discover the key influences of stem cell behavior. The identification of hyaluronan in many locations where stem cells are present raised the possibility that hyaluronan influences these potent cells. Indeed, emerging studies [4–6] have proven the veracity of this hypothesis, and hyaluronan is now being studied as a key component that may be manipulated to induce desired stem cell behaviors.

2. Hyaluronan Receptors and Signaling

Hyaluronan interacts with cell surfaces in at least two ways. It can bind cell surface receptors, hyaladherins, such as CD44, LYVE-1, HARE, layilin, TLR4, and RHAMM, to induce the transduction of a range of intracellular signals, either directly or by activating other proteins. Through signal-transducing receptors, hyaluronan influences cell proliferation, survival, motility, and differentiation and might have roles in cancer pathogenesis. The detail signaling mechanisms for each receptor is reviewed as follows.

2.1. HARE. HARE, also known as stabilin-2, is a transmembrane receptor protein that contains a C-type lectin-like HA-binding module, which enables binding and endocytosis of hyaluronan ligand. The function of this receptor has been regarded to be involved in the mediation of normal turnover process of hyaluronan and other GAGs, such as chondroitin sulfate, from the circulatory system. Hyaluronan is internalized via a clathrin-coated pit pathway, leading to hyaluronan degradation in the lysosome [7].

2.2. Layilin. Similar to HARE, layilin is a transmembrane protein homologous to C-type lectins, which is located at the membrane ruffle. It acts as a membrane docking site for talin and can also specifically bind hyaluronan [8, 9] through its lectin-like domain. Similar to CD44, layilin extracellularly binds HA and induces binding of cytoskeletal proteins such as talin and ERM complex through its cytoplasmic domain. After receiving the HA signal, layilin interacts with merlin, a protein of the ERM superfamily, at the amino terminus and modulates cell cytoskeletal structure [10].

2.3. RHAMM. RHAMM is a ubiquitous protein which is present in the nucleus, cytoplasm, and cell plasma membrane [11–13]. RHAMM can be alternatively spliced to produce molecules of different sizes. When the membrane-bound RHAMM interacts with hyaluronan, kinase activity of c-Src, FAK, or PKC is activated [13, 14]; the intracellular RHAMM domain is capable of binding kinase or cytoskeletal components [11], forming a cytoskeleton regulating complex and triggering cytoskeletal rearrangement. Disruption of hyaluronan-RHAMM interaction using either protein mutation or antisense mRNA treatment significantly reduces cell motility rate and formation of stable focal adhesions [14], supporting the model of RHAMM-mediated regulation of FAK activity. It has been reported that RHAMM binding of epiregulin, a novel ligand of EGFR, activates downstream tyrosine kinase-mediated autophosphorylation and regulates cell survival and proliferation of human cementifying fibroma [15] cells (HCF).

2.4. LYVE-1. LYVE-1 is a type I integral membrane protein containing a link domain, similar to the proteolytic hyaluronan binding domain of the Link protein superfamily [16]. LYVE-1 is identified as an endocytic receptor of hyaluronan, specifically expressed in the lymphatic endothelium. LYVE-1 has been widely used as a marker for investigation of lymph vessel structure in both benign and malignant prostate tissues and hence applicable in cancer-related studies. In HS-578T human breast cancer cells, adhesion to COS-7 cells was enhanced by overexpression of LYVE-1. This phenomenon was reduced by enzyme digestion treatment of HS-578T cell surface with bacterial *Streptomyces hyaluronidase* prior to adhesion experiment, suggesting that LYVE-1 enhances tumor cell adhesion through interaction with hyaluronan. A recent study pointed out that in human primary effusion lymphoma (PEL) tumor cells, when LYVE-1 and EMMPRIN/CD147 coexpressed with drug transporter protein, breast cancer resistance protein/ABCG2 (BCRP), EMMPRIN induced upregulation of BCRP, and LYVE-1 colocalized with BCRP on the cell surface, suggesting that LYVE-1 takes part in facilitating hyaluronan and EMMPRIN-mediated chemoresistance.

2.5. CD44. CD44-mediated cell interaction with hyaluronan has been implicated in a variety of physiological events including cell-cell and cell-substrate adhesion, cell migration, cell proliferation, and hyaluronan uptake and degradation [17]. A proposed signaling pathway activated by the CD44 receptor is depicted in Figure 1. Hyaluronan is often bound to CD44 isoforms, which are ubiquitous, abundant, and functionally important cell surface receptors [18] in coordinating intracellular signaling pathways (e.g., Ca^{2+} mobilization [19], Rho signaling [20], PI_3-kinase/AKT activation [21], NHE1-mediated cellular acidification [22], transcriptional upregulation, and cytoskeletal function [23]) and generating the concomitant onset of tumor cell activities (e.g., tumor cell adhesion, growth, survival, migration, and invasion) and tumor progression. CD44-hyaluronan interaction in MSC migration can be found in the rat MSC line Ap8c3 and mouse $CD44^{-/-}$ or $CD44^{+/+}$ bone marrow stromal cells. Adhesion and migration of MSC Ap8c3 cells [24] to hyaluronan was suppressed by anti-CD44 antibody and by CD44 small interfering RNA (siRNA). In some tumor studies, hyaluronan binding to tumor cells promoted Nanog protein in association with CD44 followed by Nanog activation and the expression of pluripotent stem cell regulators such as Rex1 and Sox2. Nanog also formed a complex with the signal transducer and activator of transcription protein 3 (Stat-3) in the nucleus leading to Stat-3-specific transcriptional activation and MDR1 (P-glycoprotein) multidrug transporter gene expression [25].

Scientists have long noticed superficial similarities between stem cells and cancer cells. We believe that CD-44 mediated cell interaction with hyaluronan will provide valuable new insights into previously understood aspects of solid tumor malignancy. In fact, human breast cancer cells strongly expressing CD44 along with low or no expression of CD24 effectively formed tumors. These cells possess the

Hyaluronan —— Molecular weight

LM

CD44 antibody/soluble HABPs —| |— Hyaluronan oligosaccharide/soluble hyaluronan

Glycosylation* — CD44 ———————————————————— Alternative splicing*

RNA interference/null mice —|— Expression*

SS

— MMP-7/MMP-9/MT1-MMP*   Exon V6   Exon V3

LPA   TPA   ADAM17/ADAM10/MT1-MMP ✂   SF/HGF ⟶ c-Met*   ERBB*   TGFβRI*

TM

CT   Ca²⁺ influx ——   Endocytosis ·····> Metabolism   Ras   SMAD

Rho —| Rac   PKC —   γ-secretase ✂ — Clustering/oligomerization*   Ras   CDC37   Nuclear signaling

CaMKII ⟶ Ser325/Ser291 — Phosphorylation —   Raf   Raf

Merlin-P   ERM-P ← PKCδ   Ankyrin   Vav2/Tiam1*   Src* ←   MEK   MEK

⊥   ← PIP₂   ↓   ↓   FAK —Tyr397 → PI3K   ERK   ERK   Cell cycle control?

Ras   F-actin   RhoA   Rac1   Cortactin-P PKC   ⊦ PTEN

Raf   ROK   F-actin   ⊥   ↓   PIP₃

F-actin   F-actin   Raf   Focal adhesion complex   ⊦ PTEN

MEK   MEK   PDK1 ⟶ Akt-P   Proliferation

ERK   ERK   Migration Adhesion   MDR

Proliferation   ● Membrane ruffling/projection   Drug resistance

● Cell motility/contraction

● Cell transformation: invasion and metastasis Proliferation   ⟶ Cell survival

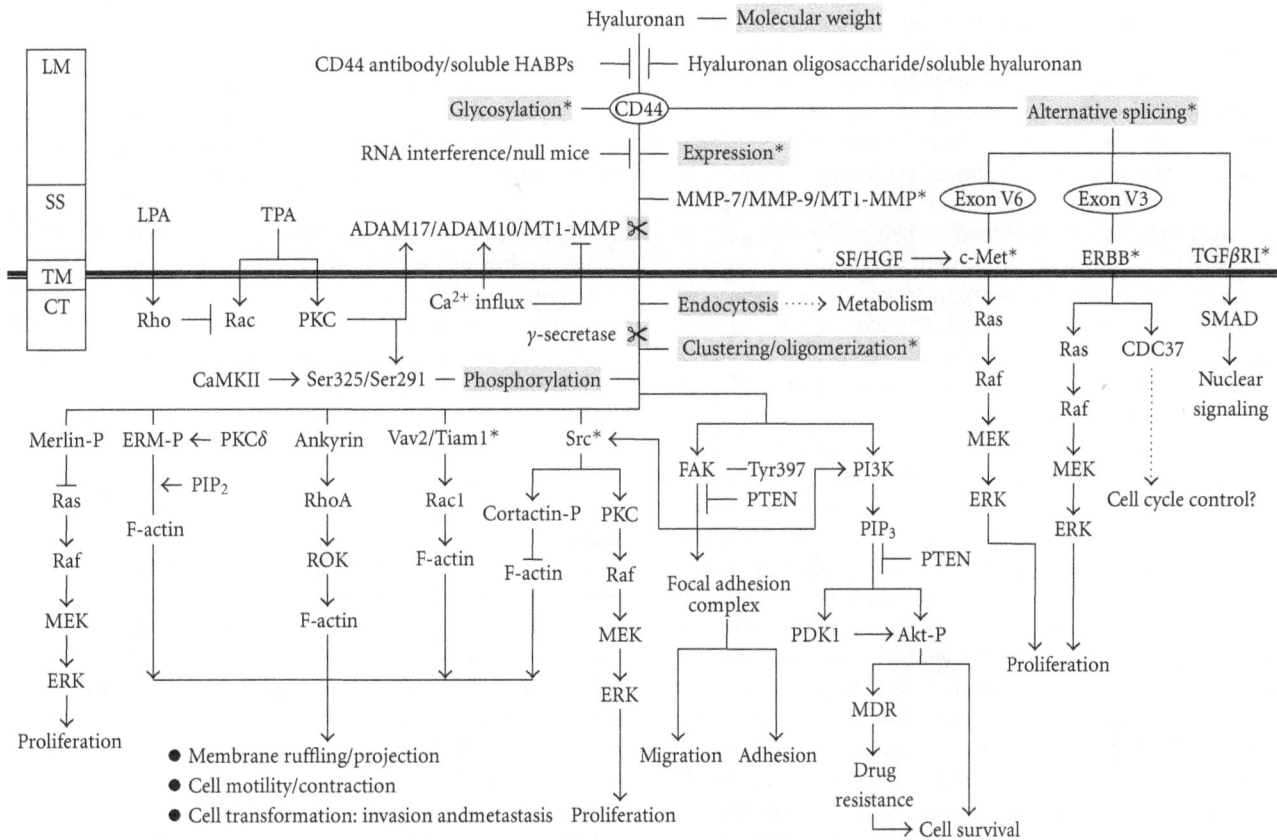

FIGURE 1: Signaling through CD44: hyaluronan binding to a receptor such as CD44 or RHAMM will cause a conformational change in the receptor. In the case of CD44, γ-secretase cleavage of the intracellular fragment can lead to phosphorylation of cellular components such as CaMK (calmodulin kinase). Clustering of the intracellular portion of CD44 with cellular proteins such as merlin, Src, and PKC leads to downstream activation of the Raf/MEK/ERK pathway, enhancing cell proliferation. Alternatively, activation with ankyrin, ERM, or Vav2/Tiam1 leads to F-actin activation and cytoskeletal rearrangement, membrane ruffling, and cell motility. Hyaluronan signaling also regulates cell migration and adhesion through interaction of PI3K and activation of the focal adhesion complex. PI3K activation by hyaluronan signaling, when incorporated with the Akt pathway, may also lead to MDR and cell survival.

ability to differentiate, proliferate, and self-renew, comparable to normal stem/progenitor cells; hence, CD44⁺/CD24^low/− cells were considered to possess the stem/progenitor cell phenotype [26]. Later, this idea was confirmed when epithelial mesenchymal transition (EMT) traits were observed to correlate with the CD44⁺/CD24^low/− stem cell phenotype in human breast cancer [27]. Another marker, ALDH1, further divides the CD44⁺CD24^low/− cell population into 2 fractions, in which the ALDH1⁺CD44⁺CD24^low/− cells are highly tumorigenic [28]. Forced expression of CD24 in CD44⁺CD24^low/− cells resulted in decreased MEK/MAPK signaling, reduced cell proliferation, and enhanced DNA damage-induced apoptosis through attenuated NFκβ signaling [29]. In contrast, siRNA silencing of CD44-encoding genes in CD44⁺CD24^low/− cells enhanced sensitivity to doxorubicin, suggesting CD44 as a suitable target for treating cancer via gene therapy [30].

This new understanding of hyaluronan/CD44-mediated oncogenic signaling events may have important clinical utility and could establish CD44 and its associated signaling components (hyaluronan/CD44-mediated Nanog-Stat-3) as

important tumor markers for early detection and evaluation of oncogenic potentials. This could also serve as groundwork for the future development of new drug targets that may inhibit hyaluronan/CD44-mediated tumor metastasis and cancer progression.

3. Hyaluronan Constitution in the Stem Cell Niche

After the term "niche" was first proposed in 1978 [31], important related studies in cell biology emerged, including a focus on the mysterious microenvironment that supports stem cells. The stem cell niche is not solely the location where these cells are present but involves the surrounding cellular components of the microenvironment and the signals emanating from the support cells [32]. This niche requires an environment that fosters a delicate balance between self-renewal and differentiation. Absence of this balance immediately triggers inappropriate differentiation. Indeed, a decrease in proliferation potential of bone-marrow-derived hematopoietic stem cells (HSC) has been observed in the absence of a "niche" environment [33]. Hematopoiesis, the process by which HSC differentiate into hematopoietic cells in order to generate different blood cell types, is greatly influenced by the surrounding microenvironment. The actual mechanism by which HSCs interact with the niche environment remains largely unknown; however, many studies have shown that the HSC niche is important for attracting and anchoring HSCs. Hyaluronan, being a critical component of the HSC microenvironment and widely distributed in mesenchymal tissue, is thought to take part in the postnatal hematopoietic niche since it is required for in vitro hematopoiesis [34]. Hyaluronan degradation leads to an arrest in HSC proliferation that is necessary for commitment to the maturation of hematopoietic cells [34].

In vertebrates, hyaluronan appears to regulate cell transformation and migration at several embryonic stages, as early as gastrulation. During embryogenesis, the accumulation and organization of hyaluronan plays a role in the epithelial-mesenchymal transition, a critical step in early embryogenesis for differentiation of pluripotent embryonic stem cells (ESC) to mesenchymal stem cells (MSC) for further formation of different tissues [35]. In general, endogenously produced hyaluronan contributes to differentiation of hESC to mesodermal lineage, but more specifically to hematopoietic cells [36]. The likely source of hyaluronan during early embryonic developmental stages is hyaluronan synthase (Has) 2 [37]. Many cardiac or skeletal development anomalies are due to inactivation or upregulation of hyaluronan synthetases. It seems that interaction of hyaluronan with ESC is fundamental to early embryonic development. Failure of this interaction results in abnormal tissue formation. Hence, hyaluronan has a central regulatory role during embryogenesis and morphogenesis [38, 39]. At later stages, Has1, Has2, and Has3 are expressed in both undifferentiated and differentiated hESC [36]. During embryonic development, hyaluronan mostly interacts with cells via the RHAMM and CD44 receptors [15, 34]. High-molecular-weight hyaluronan induces CD44 association with MEKK1 in epicardial cells to promote epithelial mesenchymal transition and differentiation [40]. However, more recent studies have observed that the HA-CD44 signaling pathway is largely associated with activation of ERK, by which the EGF receptor and downstream molecules such as Raf, MEK, and ERK1/2 are activated to promote cell proliferation [15]. ERK and its downstream molecules cyclin D1 and E2F also affect cell cycle progression [41].

In addition to collagen, alginate, and fibrin, hyaluronan is one of the extracellular matrix molecules being used to develop scaffolds for in vitro studies of stem cells, mainly due to its great abundance in the stem cell niche environment. Human ESCs may be maintained in an undifferentiated state by utilizing hyaluronan hydrogels [42], once more indicating the importance of the interactions provided by hyaluronan.

4. Influence of Hyaluronan on Stem Cell Behavior

MSCs are naturally sensitive to their environment, responding to chemical, physical, and mechanical features of their matrices or substrates, as well as the spatial/temporal presentation of biochemical cues [43]. Hyaluronan influence in behaviors such as adhesion, proliferation, differentiation, and migration occurs through several newly discovered genetic signaling mechanisms that involve binding to specific cellular receptors.

4.1. Effect of Hyaluronan on Mitochondrial Function.

The effect of hyaluronan in cell behavior involves a direct influence in cell metabolic activity including bioenergetics. Few studies have underscored this phenomenon in stem cells; however, the first report suggesting an influence of hyaluronan on mitochondrial properties was obtained through studies of chondrocytes from osteoarthritis (OA) patients, who exhibit disrupted cellular behavior [44]. The fact that abundant levels of hyaluronan are present in normal and healthy joint areas and that these levels are dramatically diminished in OA patients [44] suggests that hyaluronan abundance may be a somewhat chondroprotective. Because oxidative stress, disrupted mitochondrial respiration, and mitochondrial damage promote aging, cell death, functional failure, and degeneration in a variety of tissues [45], including joint regions, it is possible that the putative chondroprotective role of hyaluronan may occur through the preservation of mitochondrial function. This was proven true when incubation of primary human chondrocytes from OA patients with hyaluronan significantly increased mtDNA integrity, improved ATP levels, and increased cell viability under normal conditions [44]. Moreover, hyaluronan ameliorated the negative effects of reactive oxygen and nitrogen species on mtDNA integrity, mtDNA repair, ATP production, and cell viability, all of which chondrocytes are exposed to during OA. Anti-CD44 antibody at saturating concentrations abolished the protective effects of hyaluronan, which suggests the mechanism is mediated by this receptor. CD44 promotes apoptotic resistance in colonic epithelium via a mitochondria-controlled pathway [46]. In addition, expression of CD44 in some cell types, such as stem cells, may provide the means to internalize hyaluronan by endocytosis, and one of the functions of internalized hyaluronan is to protect DNA from oxidative metabolism [47]. In this study, high-molecular-weight hyaluronan in culture medium prevented H_2O_2-induced H2AX phosphorylation in 2 cell types. In contrast, the effect of low-molecular-weight hyaluronan was somewhat less pronounced. Indeed, there is evidence that some glycosaminoglycans such as chondroitin-4-sulfate and hyaluronan inhibit lipid peroxidation caused by oxidative stress and thereby decrease inflammatory reactions mediated by oxidants [48]. These conditions mimic the situation in vivo when cells are growing within an intercellular matrix containing hyaluronan and are exposed to an exogenous oxidizing agent. This supports the proposition that one of the biological functions of hyaluronan is to provide protection against cellular damage caused by radicals produced by oxidation or ionizing radiation.

Stem cells thus may have several diverse mechanisms to protect the integrity of their DNA, such as by maintaining low metabolic activity to ensure minimal oxidative damage, having a highly effective efflux pump that rapidly removes genotoxic agents from the cell [49], and possibly allowing internalization of hyaluronan, which protects DNA from oxidants. These mechanisms may be coordinated with each other and with cell cycle status. However, information is sparse concerning how exactly CD44 may affect mitochondrial function. Stem cell mitochondria have recently come under increased scrutiny because new information has revealed their role in numerous cellular processes, beyond ATP production and apoptosis regulation, suggesting mitochondria to serve as a cell fate or lineage determinant [50–52]. Thus, hyaluronan influence on mitochondrial properties may at least to some extent affect stem cell self-renewal and differentiation. Unfortunately, no ongoing studies have been published thus far.

4.2. Cell Migration. Observations made by a very successful research group in Boston in the beginning of the 1970s revealed that hyaluronan accumulation coincides with periods of cellular migration [53]. This event may occur through the physiochemical properties of hyaluronan or via direct interaction with cells. When hyaluronan is synthesized and released to the extracellular environment, its physiochemical characteristics of viscosity and elasticity contribute to local tissue hydration. This results in weakening of cell anchorage to the extracellular matrix, allowing temporal detachment to facilitate cell migration and division [54]. Chemically, this can be explained by the hyaluronan structure. In dilute solutions, where the domain that is occupied by each hyaluronan molecule expands because of mutual repulsion between the carboxyl groups inside its structure, a large volume will eventually be occupied with water trapped inside it. This property provides resilience and malleability to many tissues so that in hyaluronan-rich areas internal pressure may cause the separation of physical structures and create "highways" for cell migration; during fetal development, the migration path through which neural crest cells migrate is rich in hyaluronan [55]. Another example is the migration of mesenchymal cells into the cornea following increased hyaluronan deposition, hydration, and concomitant swelling of the migratory pathway [56]. Beyond the physiochemical interactions with hyaluronan, cells may be able to mediate, direct, and control their migration and locomotor mechanism through specific interactions via cell surface hyaluronan receptors. The principal cell surface receptors include CD44, RHAMM, and ICAM-1 [17]. RHAMM in particular forms links with several protein kinases associated with cell locomotion, for example, extracellular signal-regulated protein kinase (ERK), p125[fak], and pp60[c−src] [17]. Increased cell movement in response to hyaluronan can also be demonstrated experimentally in other cell types, and cell movement can be inhibited, at least partially, by degradation and/or blocking of hyaluronan receptor occupancy.

4.3. Cell Cycle and Proliferation. The formation and repair of mature hard tissue require cell proliferation. Hyaluronan levels have been shown to have a direct influence on this event. Cell proliferation is activated by hyaluronan, which increases volume and surface area for cell migration and cellular activities, and stimulates receptor-mediated events.

Hyaluronan can form a pericellular coat, settle into a cell-poor space, and facilitate cell detachment from the matrix and mitosis in response to mitogenic stimulators such as preinflammatory mediators and growth factors [57]. Human fibroblasts synchronized with colchicine or cytochalasin showed increased hyaluronate synthesis at the time of cell rounding during mitosis but declined sharply as cells entered G_1-phase and resumed the spread morphology [57]. It is reasonable to suggest that basal levels of hyaluronan synthesis during the G_1, S, and G_2 phases could be used for cell migration; during mitosis, synthase activity could be activated at all cellular contact areas to cause detachment and rounding. The high local concentration of hyaluronan causes release of endogenous growth factors and stimulates cell-cell interaction, resulting in faster cell proliferation during early stages of in vitro culture [58]. Although hyaluronan facilitates cell detachment, it has not been shown to possess direct mitogenic activity. However, by facilitating cell mitosis in response to mitogenic factors that are abundant during the early phases of tissue repair, hyaluronan may also have an important, although indirect role in cell proliferation [17].

Many reports have experimentally confirmed this idea, demonstrating the acceleration of stem cell proliferation by hyaluronan. In previous studies using mouse adipose-derived stem cells (mADSCs), we showed that supplementation of the culture medium with minute amounts of high-molecular-weight hyaluronan increases the growth rate of mADSCs at early passages, contributes to the extension of their lifespan with a marked reduction of cellular senescence during subcultivation, and prolongs their differentiation potential [4]. When mADSCs were cultured on a hyaluronan precoated surface, cell aggregates formed with a much more gradual growth profile. Hyaluronan-containing matrix seemed to be a poor attachment substratum for cell growth, as previously observed [59]. At later passages, aggregation limited propagation and contact inhibition, resulting in a slight increase of p16INK4a expression, consistent with previous reports [60]. Our study provided evidence that placenta-derived mesenchymal stem cells (PDMSCs) grown on a hyaluronan-coated surface are maintained in slow-cycling mode and that a prolonged G_1 phase occurs through elevated levels of p27[kip] and p130, which are responsible for suppressing cell entry into S-phase [6]. This in turn may be the main cause of reduced proliferation observed in stem cells cultured on hyaluronan-coated surfaces [4]. This line of evidence was confirmed by another research group who found that a high percentage of primary rat calvarial osteoblasts, in the presence of sulfated hyaluronan (HAS) derivatives, remained in G_1 phase with a concomitant decrease in the number of cells in S and G_2/M phase, eventually leading to a decrease in cell proliferation [61]. Recent studies have shown that prolonged G_1 transit is associated with an increase in p27[Kip1] [62]. Because p27 phosphorylation by cyclin E/Cdk2 is a prerequisite for its ubiquitination and degradation, it is interesting that in CD44-treated cells cyclin E/Cdk2 kinase activity is decreased, suggesting that CD44 might inhibit the ubiquitin-dependent proteolytic pathway of p27, leaving this molecule in an active form [62]. The specific signaling cascade that inhibits p27 degradation via CD44 remains to be characterized. However, it was speculated that blockage of the Ras pathway, which represents a principal force in driving the cell cycle [63], is involved in this process [64–66].

Hyaluronan thus appears to be effective in maintaining mADSCs in a proliferative state, delays senescence, and, more strikingly, directly influences cell proliferation. Different forms of hyaluronan supplementation may cause distinct proliferative behaviors in stem cells. Although the mechanism by which hyaluronan promotes or slows proliferation and preserves the differentiation potential of mADSCs merits further investigation, the addition of hyaluronan to a culture system may be a useful approach for expanding adult stem cells in vitro without losing their replicative and differentiation capabilities.

4.4. Cell Differentiation. In native tissue, MSCs reside in a defined microenvironment that regulates stem cell survival, self-renewal, and differentiation through growth factors, cell-cell contact, and cell-matrix adhesion. This occurs due to the direct influence of cell adhesion to their underlying biomaterials. Although soluble factors are potent regulators of stem cell differentiation, recent discoveries have underscored the importance of the physical and chemical characteristics of the matrices in determining stem cell fate [67]. Long-term culture of undifferentiated ESCs on a hyaluronan-coated surface instead of feeder layers yields pluripotency and differentiation characteristics similar to those of cells cultured on mouse embryonic fibroblasts (mEF) after 1 month of culture [68]. It seems that hyaluronan matrices act as a unique microenvironment for propagation of hESCs, likely due to the regulatory role of hyaluronan in the maintenance of hESCs in their undifferentiated state in vitro and in vivo. Indeed, in humans, the hyaluronan content is greatest in undifferentiated cells during early embryogenesis and decreases at the onset of differentiation [39]. It has been suggested that hESCs are able to take up and degrade hyaluronan through CD44 and thereby remodel hyaluronan matrices, a feature necessary for cell survival and migration [42]. After activation of several environmental cues that dictate the need for fully differentiated cells, hyaluronan supplementation may in turn induce faster cell attachment and enhance cell differentiation, possibly through improved cell-cell communication [33]. How hyaluronan may mediate this event is still unknown but has been confirmed by a new, engineered class of hyaluronan-based hydrogels that provide a natural extracellular matrix environment with a complex mechanical and biochemical interplay. The hydrogel induced osteoblast differentiation of MSCs without the use of osteogenic media. This most likely occurred through the enhancement of cell adhesion [33].

4.5. Multidrug Resistance. "Classic multidrug" resistance (MDR) is caused by increased drug export through ATP-dependent efflux pumps, such as multidrug-resistance proteins (MRPs), and other members of the ATP-binding cassette (ABC) transporter families [69]. The finding that hyaluronan stimulates cell survival signaling and that treatment of tumor cells with hyaluronidase increased the activities of various

chemotherapeutic agents [70] led to the further investigation of the possible role of hyaluronan in drug resistance. Increased hyaluronan production stimulates drug resistance in drug-sensitive cancer cells, whereas disruption of endogenous hyaluronan-induced signaling suppressed resistance to doxorubicin, taxol, vincristine, and methotrexate [71]. Therefore, the effects of hyaluronan on cell-survival signaling might alter drug resistance. Although the antiapoptotic effect of hyaluronan probably contributes to these phenomena, it is also known that lipid products of PI_3K, such as phosphatidylinositol-3,4-diphosphate, and phosphatidylinositol-3,4,5-triphosphate, directly mediate the function of ABC transporters that are involved in bile transport [72]. Because hyaluronan stimulates PI_3K activity, it might also influence drug resistance by stimulating drug transport [2]. Thus, transmembrane pumps may induce MDR by decreasing the intracellular accumulation and retention of drugs. Of particular interest is a recent work [73] indicating that inhibitors of MDR block hyaluronan synthesis and secretion and that hyaluronan might be secreted through multidrug transporters. Another possible regulator of hyaluronan-mediated multidrug resistance is the extracellular-matrix metalloproteinase inducer (EMMPRIN). EMMPRIN also stimulates the production of hyaluronan in mammary carcinoma cells [74]. Consequently, EMMPRIN promotes cell-survival signaling and induces multidrug resistance in a hyaluronan-dependent manner. In previous studies, we provided evidence that PDMSCs grown on a concentrated hyaluronan-coated surface become doxorubicin-resistant and that the interaction between CD44 and hyaluronan is crucial for this drug resistance [5]. Hyaluronan-CD44 interactions upregulate expression of drug transporters, including MDR1 [75], MRP2 [75], and BCRP [76]. In addition, small fragments of hyaluronan may mediate multidrug resistance by binding to CD44 and promoting translocation of a specific transcriptional regulator (YB-1) of the multidrug resistance MDR1 gene. This event simultaneously induces P-glycoprotein, the product of the MDR1 gene, and a broad-spectrum multidrug efflux protein present in both cancerous and healthy cells. Previous studies have reported that MDR is an important regulator of stem cell commitment [77]. Thus, we suggest that P-glycoprotein induced by the hyaluronan-coated surface holds PDMSCs in a primitive state by enabling them to extrude molecules required for differentiation. To the best of our knowledge, no reports have been published on MDR acquisition in mesenchymal stem cells; therefore, our study was the first to show that a hyaluronan substratum induces PDMSCs to acquire MDR as a result of increased P-glycoprotein expression through CD44 signaling. These and previous observations [4, 6] lead us to suggest that hyaluronan may cause PDMSCs to become dormant, the natural state of stem cells, which is consistent with slow cycling and drug resistance [5].

5. Hyaluronan Regulates Tissue Morphogenesis

Hyaluronan plays an important role in many morphogenetic processes during vertebrate development. Interaction of hyaluronan with surrounding stem cells regulates cell differentiation or the fate of cells. After hyaluronan comes in contact with surrounding cells through receptors, hyaluronan degradation by proteolysis releases cell-associated factors which affect cell motility.

5.1. HA in Cardiogenesis. High-molecular-weight hyaluronan (HMW-HA) stimulated epicardial cells, during formation of coronary vasculature in embryonic development, promoted association of hyaluronan receptor CD44 with MEKK1 protein, induced MEKK1 phosphorylation, and activated the ERK-dependent and NFκB-dependent pathways. Two methods of CD44 blockage decreased the HMW-HA-induced invasive response in epicardial cells [40]. Recent development of hyaluronan mixed esters of butyric and retinoic acids to improve the yield of cardiovascular stem cells revealed an enhancement in smad1, 3, and 4 gene expression and further upregulation of the cardiogenic gene Nkx-2.5 expression, which led to high cardiogenesis from stem cells [78].

5.2. HA in Osteogenesis and Chondrogenesis. The application of hyaluronan for bone regeneration is well known. In fact, the importance of hyaluronan in spine development was found in a hyaluronan synthase-2 (Has2) knockout mouse model [79]. Previous reports and our unpublished data have demonstrated the potential of hyaluronan in the enhanced chondrogenic and osteogenic differentiation potentials of mesenchymal stem cells via stimulated expression of specific target genes. These genes include chondrogenic markers such as sulfated glycosaminoglycans, SOX-9, aggrecan, and collagen type II; osteogenic markers alkaline phosphatase (AKP), osterix, runx2, and collagen type I [80–82].

5.3. HA in Neurogenesis. There is substantial evidence of hyaluronan participation in morphogenesis and neural cell differentiation in the central nervous system. Sulfated hyaluronan can increase proliferation of normal human astrocytes and differentiation through enhanced expression of connexin-26, -32, and -43 [83]. Hyaluronan-containing hydrogels provide a suitable environment for neural stem cell growth and differentiation [84, 85].

5.4. HA in Angiogenesis. Previous reports have shown that tumor growth and metastasis are dependent on neovascularization, and oligo-hyaluronan induces angiogenesis in vivo [3]. Oligo hyaluronan-induced proliferation of endothelial cells and tube formation during angiogenesis is mediated by CD44 signaling, followed by phosphorylation of Src, FAK, and ERK1/2 proteins, leading to upregulation of *c-jun* and *c-fos*. This phenomenon is reversed after silencing of CD44 with siRNA [86]. Hyaluronan has been tested for cell therapy in a limb ischemic mouse model and was shown to enhance angiogenesis and improve revascularization [55]. Our unpublished data demonstrated that in vitro culture of PDMSCs with high-molecular-weight hyaluronan slightly increased the expression of angiogenic markers, including KDR, CD31, and vWF.

6. Role of Hyaluronan and Stem Cells on Tissue Regeneration

Hyaluronan plays an important role in preservation of the normal extracellular matrix structure and induces neodermis at the wound bed [87]. It is extensively used as an important component of assorted categories of biomaterials. Hyaluronan, which supports stem cell interaction with extracellular matrix molecules [88], is capable of regulating the inflammatory chemokines, receptors, metalloproteinases, and tissue inhibitors that are necessary to form an efficient scaffold. Modulation of the expression of inflammatory factors (CXCL12, CXCL13, and CXCR5) in mesenchymal stem cells by hyaluronan contributes to regenerative processes [89]. In addition, fragmented hyaluronan stimulates inflammatory gene expression in various immune cells at injury sites via TLR4, TLR2, and CD44 [90]. This occurs through downregulation of the anti-inflammatory A2a receptor [91]. During leukocyte recruitment, hyaluronan interaction with CD44 activates various inflammatory cells, such as macrophages, through CD44-dependent signaling [92]. In contrast, low-molecular-weight hyaluronan induces dendritic cell maturation and promotes dendritic and endothelial cell release of proinflammatory cytokines such as TNFα, IL-1β, and IL-12 through TLR4 [93, 94]

The migratory, proliferative, and differentiation influences of hyaluronan in stem cells provide insights for the development of potent biomaterials for regeneration of full-thickness wounds (e.g., diabetic foot ulcer and burn wounds) through initial restoration of the dermal layer and a later application of differentiated local cells that may lead to faster and better regeneration [95]. For instance, regeneration of bone defects can be achieved by combining cells and osteogenic signals in a suitable scaffold. Combination of human bone-marrow-derived mesenchymal stromal cells and TGFβ growth factor involved in chondrogenesis, on a commercially available hyaluronan biomaterial scaffold showed that cells were capable of proliferating and differentiating, forming a cartilage-like construct in vitro with increased expression of typical chondrogenic markers [88]. Combination of stem cells with bone morphogenetic protein-2 (BMP-2) in hyaluronan-based hydrogel has also been widely used for the same purpose [96].

Neural stem cell (NSC) therapy can be used for nerve and brain regeneration. Hyaluronan is the major extracellular matrix component of the adult central nervous system. Its presence alone is necessary for the differentiation of NSCs to neuronal cells in vitro [84]. Nevertheless, other reports have shown that using a 3D scaffold composed of HA, collagen, and the nerve growth factor neurotrophin-3 provided perfect nerve regeneration in vivo through differentiation of NSCs [97]. The efficacy of hyaluronan in several aspects of tissue regeneration is becoming better understood. This knowledge provides important insights into the therapeutic utility of stem cells in human disease.

7. Conclusion

The physiological role of hyaluronan, especially in regulation of cellular activities, may provide a potent therapeutic alternative with profound advantages. However, this hypothesis cannot be applied to clinical practice without a more complete biological understanding of its mechanisms. Different mechanisms of hyaluronan regulation in stem cells are now emerging but at a very slow pace. The possibility of maintaining stemness or even inducing differentiation by manipulation of hyaluronan concentration/molecular weight, or by targeting genes activated by hyaluronan, may be a start point in the race to find new therapies for fatal diseases. Still, many knowledge gaps need to be filled, opening a window for future research efforts.

Acknowledgments

This work was supported by Research Grants NSC 98~99-3111-B-006-002, NSC 99-2321-B-006-017, and NSC 100-2325-B-006-013 from the National Science Council of Taiwan. The authors also appreciate Dr. C. M. Liu for providing the figure. Mairim Alexandra Solis and Ying-Hui Chen contributed equally to this work.

References

[1] J. B. Park, H. J. Kwak, and S. H. Lee, "Role of hyaluronan in glioma invasion," *Cell Adhesion & Migration*, vol. 2, no. 3, pp. 202–207, 2008.

[2] B. P. Toole, "Hyaluronan: from extracellular glue to pericellular cue," *Nature Reviews Cancer*, vol. 4, no. 7, pp. 528–539, 2004.

[3] P. Rooney, S. Kumar, J. Ponting, and M. Wang, "The role of hyaluronan in tumour neovascularization (review)," *International Journal of Cancer*, vol. 60, no. 5, pp. 632–636, 1995.

[4] P. Y. Chen, L. L. H. Huang, and H. J. Hsieh, "Hyaluronan preserves the proliferation and differentiation potentials of long-term cultured murine adipose-derived stromal cells," *Biochemical and Biophysical Research Communications*, vol. 360, no. 1, pp. 1–6, 2007.

[5] C. M. Liu, C. H. Chang, C. H. Yu, C. C. Hsu, and L. L. H. Huang, "Hyaluronan substratum induces multidrug resistance in human mesenchymal stem cells via CD44 signaling," *Cell and Tissue Research*, vol. 336, no. 3, pp. 465–475, 2009.

[6] C. M. Liu, C. H. Yu, C. H. Chang, C. C. Hsu, and L. L. H. Huang, "Hyaluronan substratum holds mesenchymal stem cells in slow-cycling mode by prolonging G1 phase," *Cell and Tissue Research*, vol. 334, no. 3, pp. 435–443, 2008.

[7] E. N. Harris, J. A. Weigel, and P. H. Weigel, "The human hyaluronan receptor for endocytosis (HARE/stabilin-2) is a systemic clearance receptor for heparin," *Journal of Biological Chemistry*, vol. 283, no. 25, pp. 17341–17350, 2008.

[8] M. L. Borowsky and R. O. Hynes, "Layilin, a novel talin-binding transmembrane protein homologous with C-type lectins, is localized in membrane ruffles," *Journal of Cell Biology*, vol. 143, no. 2, pp. 429–442, 1998.

[9] P. Bono, K. Rubin, J. M. G. Higgins, and R. O. Hynes, "Layilin, a novel integral membrane protein, is a hyaluronan receptor," *Molecular Biology of the Cell*, vol. 12, no. 4, pp. 891–900, 2001.

[10] P. Bono, E. Cordero, K. Johnson et al., "Layilin, a cell surface hyaluronan receptor, interacts with merlin and radixin," *Experimental Cell Research*, vol. 308, no. 1, pp. 177–187, 2005.

[11] V. Assmann, D. Jenkinson, J. F. Marshall, and I. R. Hart, "The intracellular hyaluronan receptor RHAMM/IHABP interacts with microtubules and actin filaments," *Journal of Cell Science*, vol. 112, part 22, pp. 3943–3954, 1999.

[12] C. L. Hall and E. A. Turley, "Hyaluronan: RHAMM mediated cell locomotion and signaling in tumorigenesis," *Journal of Neuro-Oncology*, vol. 26, no. 3, pp. 221–229, 1995.

[13] C. L. Hall, C. Wang, L. A. Lange, and E. A. Turley, "Hyaluronan and the hyaluronan receptor RHAMM promote focal adhesion turnover and transient tyrosine kinase activity," *Journal of Cell Biology*, vol. 126, no. 2, pp. 575–588, 1994.

[14] C. L. Hall, B. Yang, X. Yang et al., "Overexpression of the hyaluronan receptor RHAMM is transforming and is also required for H-ras transformation," *Cell*, vol. 82, no. 1, pp. 19–26, 1995.

[15] H. Hatano, H. Shigeishi, Y. Kudo et al., "RHAMM/ERK interaction induces proliferative activities of cementifying fibroma cells through a mechanism based on the CD44-EGFR," *Laboratory Investigation*, vol. 91, no. 3, pp. 379–391, 2011.

[16] S. Banerji, J. Ni, S. X. Wang et al., "LYVE-1, a new homologue of the CD44 glycoprotein, is a lymph-specific receptor for hyaluronan," *Journal of Cell Biology*, vol. 144, no. 4, pp. 789–801, 1999.

[17] W. Y. J. Chen and G. Abatangelo, "Functions of hyaluronan in wound repair," *Wound Repair and Regeneration*, vol. 7, no. 2, pp. 79–89, 1999.

[18] A. Aruffo, I. Stamenkovic, M. Melnick, C. B. Underhill, and B. Seed, "CD44 is the principal cell surface receptor for hyaluronate," *Cell*, vol. 61, no. 7, pp. 1303–1313, 1990.

[19] O. Nagano, D. Murakami, D. Hartmann et al., "Cell-matrix interaction via CD44 is independently regulated by different metalloproteinases activated in response to extracellular Ca^{2+} influx and PKC activation," *Journal of Cell Biology*, vol. 165, no. 6, pp. 893–902, 2004.

[20] L. Y. W. Bourguignon, G. Wong, C. Earle, K. Krueger, and C. C. Spevak, "Hyaluronan-CD44 interaction promotes c-Src-mediated twist signaling, microRNA-10b expression, and RhoA/RhoC up-regulation, leading to Rho-kinase-associated cytoskeleton activation and breast tumor cell invasion," *Journal of Biological Chemistry*, vol. 285, no. 47, pp. 36721–36735, 2010.

[21] P. A. Singleton and L. Y. W. Bourguignon, "CD44v10 interaction with Rho-Kinase (ROK) activates inositol 1,4,5-triphosphate (IP3) receptor-mediated Ca^{2+} signaling during hyaluronan (HA)-induced endothelial cell migration," *Cell Motility and the Cytoskeleton*, vol. 53, no. 4, pp. 293–316, 2002.

[22] R. Tammi, K. Rilla, J. P. Pienimäki et al., "Hyaluronan enters keratinocytes by a novel endocytic route for catabolism," *Journal of Biological Chemistry*, vol. 276, no. 37, pp. 35111–35122, 2001.

[23] L. Y. W. Bourguignon, V. B. Lokeshwar, X. Chen, and W. G. L. Kerrick, "Hyaluronic acid-induced lymphocyte signal transduction and HA receptor (GP85/CD44)-cytoskeleton interaction," *Journal of Immunology*, vol. 151, no. 12, pp. 6634–6644, 1993.

[24] H. Zhu, N. Mitsuhashi, A. Klein et al., "The role of the hyaluronan receptor CD44 in mesenchymal stem cell migration in the extracellular matrix," *Stem Cells*, vol. 24, no. 4, pp. 928–935, 2006.

[25] L. Y. W. Bourguignon, K. Peyrollier, W. Xia, and E. Gilad, "Hyaluronan-CD44 interaction activates stem cell marker Nanog, Stat-3-mediated MDR1 gene expression, and ankyrin-regulated multidrug efflux in breast and ovarian tumor cells," *Journal of Biological Chemistry*, vol. 283, no. 25, pp. 17635–17651, 2008.

[26] M. Al-Hajj, M. S. Wicha, A. Benito-Hernandez, S. J. Morrison, and M. F. Clarke, "Prospective identification of tumorigenic breast cancer cells," *Proceedings of the National Academy of Sciences of the United States of America*, vol. 100, no. 7, pp. 3983–3988, 2003.

[27] T. Blick, H. Hugo, E. Widodo et al., "Epithelial mesenchymal transition traits in human breast cancer cell lines parallel the CD44^{hi}/CD24$^{lo/-}$ stem cell phenotype in human breast cancer," *Journal of Mammary Gland Biology and Neoplasia*, vol. 15, no. 2, pp. 235–252, 2010.

[28] S. Ricardo, A. F. Vieira, R. Gerhard et al., "Breast cancer stem cell markers CD44, CD24 and ALDH1: expression distribution within intrinsic molecular subtype," *Journal of Clinical Pathology*, vol. 64, no. 11, pp. 937–944, 2011.

[29] J. H. Ju, K. Jang, K. M. Lee et al., "CD24 enhances DNA damage-induced apoptosis by modulating NF-κB signaling in CD44-expressing breast cancer cells," *Carcinogenesis*, vol. 32, no. 10, pp. 1474–1483, 2011.

[30] P. van Phuc, P. L. C. Nhan, T. H. Nhung et al., "Downregulation of CD44 reduces doxorubicin resistance of CD44CD24 breast cancer cells," *Journal of OncoTargets and Therapy*, vol. 4, pp. 71–78, 2011.

[31] R. Schofield, "The relationship between the spleen colony-forming cell and the haemopoietic stem cell. A hypothesis," *Blood Cells*, vol. 4, no. 1-2, pp. 7–25, 1978.

[32] L. Li and T. Xie, "Stem cell niche: structure and function," *Annual Review of Cell and Developmental Biology*, vol. 21, pp. 605–631, 2005.

[33] A. K. Jha, X. Xu, R. L. Duncan, and X. Jia, "Controlling the adhesion and differentiation of mesenchymal stem cells using hyaluronic acid-based, doubly crosslinked networks," *Biomaterials*, vol. 32, no. 10, pp. 2466–2478, 2011.

[34] V. Y. Matrosova, I. A. Orlovskaya, N. Serobyan, and S. K. Khaldoyanidi, "Hyaluronic acid facilitates the recovery of hematopoiesis following 5-fluorouracil administration," *Stem Cells*, vol. 22, no. 4, pp. 544–555, 2004.

[35] S. Shukla, R. Nair, M. W. Rolle et al., "Synthesis and organization of hyaluronan and versican by embryonic stem cells undergoing embryoid body differentiation," *Journal of Histochemistry and Cytochemistry*, vol. 58, no. 4, pp. 345–358, 2010.

[36] I. U. Schraufstatter, N. Serobyan, J. Loring, and S. K. Khaldoyanidi, "Hyaluronan is required for generation of hematopoietic cells during differentiation of human embryonic stem cells," *Journal of Stem Cells*, vol. 5, no. 1, pp. 9–21, 2010.

[37] T. D. Camenisch, A. P. Spicer, T. Brehm-Gibson et al., "Disruption of hyaluronan synthase-2 abrogates normal cardiac morphogenesis and hyaluronan-mediated transformation of epithelium to mesenchyme," *Journal of Clinical Investigation*, vol. 106, no. 3, pp. 349–360, 2000.

[38] P. Vabres, "Hyaluronan, embryogenesis and morphogenesis," *Annales de Dermatologie et de Venereologie*, vol. 137, supplement 1, pp. S9–S14, 2010.

[39] B. P. Toole, "Hyaluronan in morphogenesis," *Journal of Internal Medicine*, vol. 242, no. 1, pp. 35–40, 1997.

[40] E. A. Craig, P. Parker, A. F. Austin, J. V. Barnett, and T. D. Camenisch, "Involvement of the MEKK1 signaling pathway in the regulation of epicardial cell behavior by hyaluronan," *Cellular Signalling*, vol. 22, no. 6, pp. 968–976, 2010.

[41] D. Kothapalli, J. Flowers, T. Xu, E. Puré, and R. K. Assoian, "Differential activation of ERK and Rac mediates the proliferative and anti-proliferative effects of hyaluronan and CD44," *Journal of Biological Chemistry*, vol. 283, no. 46, pp. 31823–31829, 2008.

[42] S. Gerecht, J. A. Burdick, L. S. Ferreira, S. A. Townsend, R. Langer, and G. Vunjak-Novakovic, "Hyaluronic acid hydrogel for controlled self-renewal and differentiation of human embryonic stem cells," *Proceedings of the National Academy of Sciences of the United States of America*, vol. 104, no. 27, pp. 11298–11303, 2007.

[43] D. E. Discher, D. J. Mooney, and P. W. Zandstra, "Growth factors, matrices, and forces combine and control stem cells," *Science*, vol. 324, no. 5935, pp. 1673–1677, 2009.

[44] V. Grishko, M. Xu, R. Ho et al., "Effects of hyaluronic acid on mitochondrial function and mitochondria-driven apoptosis following oxidative stress in human chondrocytes," *Journal of Biological Chemistry*, vol. 284, no. 14, pp. 9132–9139, 2009.

[45] Y. H. Wei, S. B. Wu, Y. S. Ma, and H. C. Lee, "Respiratory function decline and DNA mutation in mitochondria, oxidative stress and altered gene expression during aging," *Chang Gung Medical Journal*, vol. 32, no. 2, pp. 113–132, 2009.

[46] M. Lakshman, V. Subramaniam, U. Rubenthiran, and S. Jothy, "CD44 promotes resistance to apoptosis in human colon cancer cells," *Experimental and Molecular Pathology*, vol. 77, no. 1, pp. 18–25, 2004.

[47] H. Zhao, T. Tanaka, V. Mitlitski, J. Heeter, E. A. Balazs, and Z. Darzynkiewicz, "Protective effect of hyaluronate on oxidative DNA damage in WI-38 and A549 cells," *International Journal of Oncology*, vol. 32, no. 6, pp. 1159–1167, 2008.

[48] G. M. Campo, A. Avenoso, S. Campo, A. M. Ferlazzo, D. Altavilla, and A. Calatroni, "Efficacy of treatment with glycosaminoglycans on experimental collagen-induced arthritis in rats," *Arthritis Research & Therapy*, vol. 5, no. 3, pp. R122–R131, 2003.

[49] M. H. G. P. Raaijmakers, "ATP-binding-cassette transporters in hematopoietic stem cells and their utility as therapeutical targets in acute and chronic myeloid leukemia," *Leukemia*, vol. 21, no. 10, pp. 2094–2102, 2007.

[50] S. Mandal, A. G. Lindgren, A. S. Srivastava, A. T. Clark, and U. Banerjee, "Mitochondrial function controls proliferation and early differentiation potential of embryonic stem cells," *Stem Cells*, vol. 29, no. 3, pp. 486–495, 2011.

[51] C. T. Chen, Y. R. V. Shih, T. K. Kuo, O. K. Lee, and Y. H. Wei, "Coordinated changes of mitochondrial biogenesis and antioxidant enzymes during osteogenic differentiation of human mesenchymal stem cells," *Stem Cells*, vol. 26, no. 4, pp. 960–968, 2008.

[52] C. Nesti, L. Pasquali, F. Vaglini, G. Siciliano, and L. Murri, "The role of mitochondria in stem cell biology," *Bioscience Reports*, vol. 27, no. 1–3, pp. 165–171, 2007.

[53] B. P. Toole and J. Gross, "The extracellular matrix of the regenerating newt limb: synthesis and removal of hyaluronate prior to differentiation," *Developmental Biology*, vol. 25, no. 1, pp. 57–77, 1971.

[54] L. A. Culp, B. A. Murray, and B. J. Rollins, "Fibronectin and proteoglycans as determinants of cell-substratum adhesion," *Journal of Supramolecular and Cellular Biochemistry*, vol. 11, no. 3, pp. 401–427, 1979.

[55] R. M. Pratt, M. A. Larsen, and M. C. Johnston, "Migration of cranial neural crest cells in a cell free hyaluronate rich matrix," *Developmental Biology*, vol. 44, no. 2, pp. 298–305, 1975.

[56] B. P. Toole and R. L. Trelstad, "Hyaluronate production and removal during corneal development in the chick," *Developmental Biology*, vol. 26, no. 1, pp. 28–35, 1971.

[57] M. Brecht, U. Mayer, E. Schlosser, and P. Prehm, "Increased hyaluronate synthesis is required for fibroblast detachment and mitosis," *Biochemical Journal*, vol. 239, no. 2, pp. 445–450, 1986.

[58] X. Zou, H. Li, L. Chen, A. Baatrup, C. Bünger, and M. Lind, "Stimulation of porcine bone marrow stromal cells by hyaluronan, dexamethasone and rhBMP-2," *Biomaterials*, vol. 25, no. 23, pp. 5375–5385, 2004.

[59] M. Fisher and M. Solursh, "The influence of the substratum on mesenchyme spreading in vitro," *Experimental Cell Research*, vol. 123, no. 1, pp. 1–13, 1979.

[60] R. J. Wieser, D. Faust, C. Dietrich, and F. Oesch, "p16[INK4] mediates contact-inhibition of growth," *Oncogene*, vol. 18, no. 1, pp. 277–281, 1999.

[61] R. Kunze, M. Rösler, S. Möller et al., "Sulfated hyaluronan derivatives reduce the proliferation rate of primary rat calvarial osteoblasts," *Glycoconjugate Journal*, vol. 27, no. 1, pp. 151–158, 2010.

[62] Z. Gadhoum, M. P. Leibovitch, J. Oi et al., "CD44: a new means to inhibit acute myeloid leukemia cell proliferation via p27Kip1," *Blood*, vol. 103, no. 3, pp. 1059–1068, 2004.

[63] F. Chang, L. S. Steelman, J. G. Shelton et al., "Regulation of cell cycle progression and apoptosis by the Ras/Raf/MEK/ERK pathway (Review)," *International Journal of Oncology*, vol. 22, no. 3, pp. 469–480, 2003.

[64] A. Vidal, S. S. Millard, J. P. Miller, and A. Koff, "Rho activity can alter the translation of p27 mRNA and is important for RasV12-induced transformation in a manner dependent on p27 status," *Journal of Biological Chemistry*, vol. 277, no. 19, pp. 16433–16440, 2002.

[65] W. Hu, C. J. Bellone, and J. J. Baldassare, "RhoA stimulates p27(Kip) degradation through its regulation of cyclin E/CDK2 activity," *Journal of Biological Chemistry*, vol. 274, no. 6, pp. 3396–3401, 1999.

[66] H. Ponta, L. Sherman, and P. A. Herrlich, "CD44: from adhesion molecules to signalling regulators," *Nature Reviews Molecular Cell Biology*, vol. 4, no. 1, pp. 33–45, 2003.

[67] M. P. Lutolf, P. M. Gilbert, and H. M. Blau, "Designing materials to direct stem-cell fate," *Nature*, vol. 462, no. 7272, pp. 433–441, 2009.

[68] M. Á. Ramírez, E. Pericuesta, M. Yáñez-Mó, A. Palasz, and A. Gutiérrez-Adán, "Effect of long-term culture of mouse embryonic stem cells under low oxygen concentration as well as on glycosaminoglycan hyaluronan on cell proliferation and differentiation," *Cell Proliferation*, vol. 44, no. 1, pp. 75–85, 2011.

[69] M. M. Gottesman, T. Fojo, and S. E. Bates, "Multidrug resistance in cancer: role of ATP-dependent transporters," *Nature Reviews Cancer*, vol. 2, no. 1, pp. 48–58, 2002.

[70] G. Baumgartner, C. Gomar-Höss, L. Sakr, E. Ulsperger, and C. Wogritsch, "The impact of extracellular matrix on the chemoresistance of solid tumors—experimental and clinical results of hyaluronidase as additive to cytostatic chemotherapy," *Cancer Letters*, vol. 131, no. 1, pp. 85–99, 1998.

[71] S. Misra, S. Ghatak, A. Zoltan-Jones, and B. P. Toole, "Regulation of multidrug resistance in cancer cells by hyaluronan," *Journal of Biological Chemistry*, vol. 278, no. 28, pp. 25285–25288, 2003.

[72] S. Misra, P. Ujházy, Z. Gatmaitan, L. Varticovski, and I. M. Arias, "The role of phosphoinositide 3-kinase in taurocholate-induced trafficking of ATP-dependent canalicular transporters in rat liver," *Journal of Biological Chemistry*, vol. 273, no. 41, pp. 26638–26644, 1998.

[73] P. Prehm and U. Schumacher, "Inhibition of hyaluronan export from human fibroblasts by inhibitors of multidrug resistance transporters," *Biochemical Pharmacology*, vol. 68, no. 7, pp. 1401–1410, 2004.

[74] E. A. Marieb, A. Zoltan-Jones, R. Li et al., "Emmprin promotes anchorage-independent growth in human mammary carcinoma cells by stimulating hyaluronan production," *Cancer Research*, vol. 64, no. 4, pp. 1229–1232, 2004.

[75] S. Misra, S. Ghatak, and B. P. Toole, "Regulation of MDR1 expression and drug resistance by a positive feedback loop involving hyaluronan, phosphoinositide 3-kinase, and ErbB2," *Journal of Biological Chemistry*, vol. 280, no. 21, pp. 20310–20315, 2005.

[76] A. G. Gilg, S. L. Tye, L. B. Tolliver et al., "Targeting hyaluronan interactions in malignant gliomas and their drug-resistant multipotent progenitors," *Clinical Cancer Research*, vol. 14, no. 6, pp. 1804–1813, 2008.

[77] K. D. Bunting, "ABC transporters as phenotypic markers and functional regulators of stem cells," *Stem Cells*, vol. 20, no. 3, p. 274, 2002.

[78] M. Maioli, S. Santaniello, A. Montella et al., "Hyaluronan esters drive smad gene expression and signaling enhancing cardiogenesis in mouse embryonic and human mesenchymal stem cells," *PLoS ONE*, vol. 5, no. 11, Article ID e15151, 2010.

[79] P. J. Roughley, L. Lamplugh, E. R. Lee, K. Matsumoto, and Y. Yamaguchi, "The role of hyaluronan produced by Has2 gene expression in development of the spine," *Spine*, vol. 36, no. 14, pp. E914–E920, 2011.

[80] L. Astachov, R. Vago, M. Aviv, and Z. Nevo, "Hyaluronan and mesenchymal stem cells: from germ layer to cartilage and bone," *Frontiers in Bioscience*, vol. 16, no. 1, pp. 261–276, 2011.

[81] Z. Schwartz, D. J. Griffon, L. P. Fredericks, H. B. Lee, and H. Y. Weng, "Hyaluronic acid and chondrogenesis of murine bone marrow mesenchymal stem cells in chitosan sponges," *American Journal of Veterinary Research*, vol. 72, no. 1, pp. 42–50, 2011.

[82] S. C. Wu, J. K. Chang, C. K. Wang, G. J. Wang, and M. L. Ho, "Enhancement of chondrogenesis of human adipose derived stem cells in a hyaluronan-enriched microenvironment," *Biomaterials*, vol. 31, no. 4, pp. 631–640, 2010.

[83] S. Ahmed, T. Tsuchiya, M. Nagahata-Ishiguro, R. Sawada, N. Banu, and T. Nagira, "Enhancing action by sulfated hyaluronan on connexin-26, -32, and -43 gene expressions during the culture of normal human astrocytes," *Journal of Biomedical Materials Research Part A*, vol. 90, no. 3, pp. 713–719, 2009.

[84] T. W. Wang and M. Spector, "Development of hyaluronic acid-based scaffolds for brain tissue engineering," *Acta Biomaterialia*, vol. 5, no. 7, pp. 2371–2384, 2009.

[85] M. Preston and L. S. Sherman, "Neural stem cell niches: roles for the hyaluronan-based extracellular matrix," *Frontiers in Bioscience*, vol. 3, pp. 1165–1179, 2011.

[86] Y. Z. Wang, M. L. Cao, Y. W. Liu, Y. Q. He, C. X. Yang, and F. Gao, "CD44 mediates oligosaccharides of hyaluronan-induced proliferation, tube formation and signal transduction in endothelial cells," *Experimental Biology and Medicine*, vol. 236, no. 1, pp. 84–90, 2011.

[87] R. D. Price, V. Das-Gupta, I. M. Leigh, and H. A. Navsaria, "A comparison of tissue-engineered hyaluronic acid dermal matrices in a human wound model," *Tissue Engineering*, vol. 12, no. 10, pp. 2985–2995, 2006.

[88] G. Lisignoli, S. Cristino, A. Piacentini et al., "Cellular and molecular events during chondrogenesis of human mesenchymal stromal cells grown in a three-dimensional hyaluronan based scaffold," *Biomaterials*, vol. 26, no. 28, pp. 5677–5686, 2005.

[89] G. Lisignoli, S. Cristino, A. Piacentini, C. Cavallo, A. I. Caplan, and A. Facchini, "Hyaluronan-based polymer scaffold

modulates the expression of inflammatory and degradative factors in mesenchymal stem cells: involvement of Cd44 and Cd54," *Journal of Cellular Physiology*, vol. 207, no. 2, pp. 364–373, 2006.

[90] D. Jiang, J. Liang, and P. W. Noble, "Hyaluronan as an immune regulator in human diseases," *Physiological Reviews*, vol. 91, no. 1, pp. 221–264, 2011.

[91] S. L. Collins, K. E. Black, Y. Chan-Li et al., "Hyaluronan fragments promote inflammation by down-regulating the anti-inflammatory A2a receptor," *American Journal of Respiratory Cell and Molecular Biology*, vol. 45, no. 4, pp. 675–683, 2011.

[92] K. R. Taylor and R. L. Gallo, "Glycosaminoglycans and their proteoglycans: host-associated molecular patterns for initiation and modulation of inflammation," *FASEB Journal*, vol. 20, no. 1, pp. 9–22, 2006.

[93] C. C. Termeer, J. Hennies, U. Voith et al., "Oligosaccharides of hyaluronan are potent activators of dendritic cells," *Journal of Immunology*, vol. 165, no. 4, pp. 1863–1870, 2000.

[94] C. Termeer, F. Benedix, J. Sleeman et al., "Oligosaccharides of hyaluronan activate dendritic cells via toll-like receptor 4," *Journal of Experimental Medicine*, vol. 195, no. 1, pp. 99–111, 2002.

[95] S. R. Myers, V. N. Partha, C. Soranzo, R. D. Price, and H. A. Navsaria, "Hyalomatrix: a temporary epidermal barrier, hyaluronan delivery, and neodermis induction system for keratinocyte stem cell therapy," *Tissue Engineering*, vol. 13, no. 11, pp. 2733–2741, 2007.

[96] J. Kim, I. S. Kim, T. H. Cho et al., "Bone regeneration using hyaluronic acid-based hydrogel with bone morphogenic protein-2 and human mesenchymal stem cells," *Biomaterials*, vol. 28, no. 10, pp. 1830–1837, 2007.

[97] H. Zhang, Y. T. Wei, K. S. Tsang et al., "Implantation of neural stem cells embedded in hyaluronic acid and collagen composite conduit promotes regeneration in a rabbit facial nerve injury model," *Journal of Translational Medicine*, vol. 6, p. 67, 2008.

19

Role of Translationally Controlled Tumor Protein in Cancer Progression

Tim Hon Man Chan,[1,2] **Leilei Chen,**[1,2] **and Xin-Yuan Guan**[1,2,3]

[1] *Department of Clinical Oncology, Li Ka Shing Faculty of Medicine, The University of Hong Kong, Hong Kong*
[2] *State Key Laboratory of Liver Research, The University of Hong Kong, Hong Kong*
[3] *State Key Laboratory of Oncology in Southern China, Sun Yat-sen University Cancer Center, Guangzhou 510060, China*

Correspondence should be addressed to Xin-Yuan Guan, xyguan@hku.hk

Academic Editor: Malgorzata Kloc

Translationally controlled tumor protein (TCTP) is a highly conserved and ubiquitously expressed protein in all eukaryotes—highlighting its important functions in the cell. Previous studies revealed that TCTP is implicated in many biological processes, including cell growth, tumor reversion, and induction of pluripotent stem cell. A recent study on the solution structure from fission yeast orthologue classifies TCTP under a family of small chaperone proteins. There is growing evidence in the literature that TCTP is a multifunctional protein and exerts its biological activity at the extracellular and intracellular levels. Although TCTP is not a tumor-specific protein, our research group, among several others, focused on the role(s) of TCTP in cancer progression. In this paper, we will summarize the current scientific knowledge of TCTP in different aspects, and the precise oncogenic mechanisms of TCTP will be discussed in detail.

1. Introduction

Translationally controlled tumor protein (TCTP) is a highly conserved multifunctional protein. Since the discovery of TCTP over two decades ago, the expression level of TCTP has been investigated in more than 500 tissues and cell types. Expression levels have been found to vary by nearly two orders of magnitude between different types of tissues [1] with preferential expression in mitotically active tissues [2]. Two mRNA transcripts have been reported to carry the same 5′UTR but different 3′UTR using alternative polyadenylation signals. The solution structure of TCTP from fission yeast also revealed structural similarity with the Mss4/Dss4 protein family. The expression of TCTP is highly regulated both at transcriptional and translational levels in addition to a wide range of extracellular signals. It has been implicated in many cellular processes, such as cell growth, cell cycle progression, apoptosis, malignant transformation, and the regulation of pluripotency. Although TCTP is not a tumor-specific protein, the downregulation of TCTP was found in tumor reversion [3]. Our research group and others also substantiated the link between TCTP deregulation and cancer progression [4–6].

This paper will mainly focus on the biological functions of TCTP and malignant transformation induced by TCTP. Furthermore, the clinical implications of TCTP in human cancers will also be discussed.

2. Features of TCTP mRNA and Protein

2.1. TCTP mRNA. The human TPT1 gene coding for TCTP spanning about 4.2 kb, consists of six exons and five introns [7]. Two alternative polyadenylation signals at 3′UTR generated two mRNA transcripts that differ in the length of 3′UTR. In all mammalian tissues tested, both types of mRNAs are expressed at different ratios, and the shorter transcript is usually expressed more abundantly [2]. Sequence analyses of TCTP transcripts indicated that the 5′UTR is CG-rich with a high degree of secondary structure. TCTP mRNA has a 5′ terminal oligopyrimidine tract (5′TOP), which is a signature of translationally controlled mRNAs [8]. Although AU-rich regions and AUUA elements have been identified at the 3′UTR, they do not match classical mRNA instability elements [1].

2.2. TCTP Protein. As revealed by sequence alignment of TCTP in more than 30 different species, TCTP is highly conserved over a long-term evolution. A cluster of invariant residues were located on one side of the β-stranded "core" domain that is important for molecular interactions [1]. Thaw et al. found that the "core" domain displays significant similarity to that of the Mss4/Dss4 protein family upon analyzing the solution structure of TCTP from *Schizosaccharomyces pombe* [1, 9]. TCTP is also a novel small molecular weight (23 kDa) heat shock protein that protects cells from thermal shock by functioning as a molecular chaperone [10].

2.3. Transcriptional and Translational Regulation. Previous reports demonstrated that expression of TCTP is regulated at the level of the transcription as well as the translation [2, 11]. TCTP is translationally regulated: abundant TCTP mRNA was found as untranslated mRNP particles [11], an increase in TCTP synthesis under the treatment of transcription inhibitor, actinomycin D [12], a 5′-TOP of TCTP mRNA and its extended secondary structure [8].

Comparing the promoter region of TCTP in human, mouse, rat, rabbit, and dog predicted the transcription factor binding sites of TCTP. Not surprisingly, the promoter regions of TCTP are also highly conserved among these species. Andree et al. further demonstrated that the transcription of TCTP is controlled by cAMP signaling via phosphorylation-dependent activation of CRE/CREB interaction [7]. In our previous study, we utilized chromatin immunoprecipitation-based (ChIP-based) cloning strategy to identify genes potentially regulated by CHD1L (chromodomain helicase/ATPase DNA binding protein 1-like gene) [13]. From this strategy, we isolated 35 CHD1L-binidng loci and characterized a specific CHD1L-binding motif (C/A-C-A/T-T-T-T). Two CHD1L-binding motifs have been identified at −748 bp and −851 bp in the 5′-flanking region of TCTP [13]. Importantly, we have demonstrated that the binding of CHD1L to the promoter region of TCTP dramatically activates the transcription of TCTP.

3. Biological Functions

3.1. Growth and Development. A knockout mouse approach had been used to investigate the role of TCTP in development. As demonstrated by Chen et al., heterozygous mice (TCTP$^{+/-}$) were viable, fertile, and morphologically similar to their wild-type littermates, while the homozygous (TCTP$^{-/-}$) were embryonic lethal. Moreover, TCTP$^{-/-}$ embryo at embryonic stage day 5.5 (E5.5) suffered from reduced cell numbers and increased apoptosis and subsequently died around E9.5–10.5 [14]. This suggests that TCTP is essential for normal development. The human TCTP (hTCTP) protein sequence is 50% identical with *Drosophila* TCTP (dTCTP) [15]. They also indicated that silencing of

dTCTP by RNA interference resulted in the reduced cell size, cell number, and organ size. Also, as Rheb is an important regulator in TSC-mTOR pathway, TCTP might function as a growth-regulating protein by the stimulation of GDP/GTP exchange of human Rheb (hRheb) via binding to hRheb [15]. On the contrary, Rehmann et al. argues that TCTP does not act as a guanine nucleotide exchange factor (GEF) of Rheb [16]. Therefore, additional experiments are essential to further substantiate the role of TCTP in cell growth regulation through the TSC-mTOR pathway.

3.2. Regulation of Cell Cycle and Apoptosis. Gachet et al. demonstrated that TCTP interacts with microtubules during G1, S, G2, and early M phase of the cell cycle [17]. During mitosis, it binds to the mitotic spindle and detaches from the spindle during the metaphase-anaphase transition. When TCTP is overexpressed in bovine mammary epithelial cells, rearrangement of microtubule and growth inhibition can be observed [17]. Two-hybrid screening methods identified TCTP as a substrate of polo-like kinase (Plk), which is involved in the formation and function of bipolar spindles and Plk phosphorylates TCTP on two serine residues [18]. Abolishing Plk phosphorylation on TCTP-induced mitotic defects and high incidences of apoptosis, indicating that phosphorylation of TCTP on two serine residues by Plk plays an important role in cell mitosis [18].

3.3. Regulation of Self-Renewal and Pluripotency. Homeodomain transcription factors Oct4 and Nanog have been identified as master regulators of stem cell self-renewal and pluripotency [19]. Oct4 appears to regulate cell fates in a quantitative fashion and maintain a critical concentration to sustain embryonic stem (ES) cell self-renewal [20]. Proteins associated with the regulatory region of the mouse *oct4* gene can be isolated and identified by mass spectrometry [21]. Using this strategy, TCTP was found to bind to the promoter region of *oct4*. Although *tpt1* transcript depletion can inactivate the transcription of both *oct4* and *nanog*, TCTP binds only to the *oct4* promoter as determined by ChIP analysis in mouse ES cells [21]. It has also been suggested that TCTP could regulate *oct4* by directly binding to its proximal promoter. However, TCTP might also regulate *nanog* indirectly or binds to its distal promoter. These data suggest that TCTP is a potential regulator of self-renewal and pluripotency.

4. Malignant Transformation by TCTP during Cancer Progression

Although there are many distinct types of human cancers, six essential alterations to normal cells are believed to define the progression of most human malignancies: they are evasion of apoptosis, sustained angiogenesis, accelerated cell cycle progression, tissue invasion and metastasis, self-sufficiency in growth signals, and insensitivity to antigrowth signals.

Therefore, the information described below provides us with some insights into the oncogenic role of TCTP.

4.1. Differential Expression of TCTP in Cancer. Independent studies indicated that TCTP is preferentially expressed in cancer. In human colon cancer, the level of TCTP mRNA was detected in three human colon carcinoma cell lines (SNU-C2A, SNU-C4, and SNU-C5). The expression levels were not equal among these cell lines. SNU-C5 with the highest expression grew at the fastest rate; however, SNU-C2A with the lowest expression grew at the slowest rate [4]. Higher expression level of TCTP was also observed in prostate cancer specimens compared to normal prostate tissues [5]. In hepatocellular carcinoma (HCC), the expression level of TCTP was detected in a retrospective cohort of 118 HCC patients. As a result, TCTP was found to be significantly upregulated in tumor tissues when compared to their adjacent noncancerous tissues. Overexpression of TCTP (defined as 2 fold increase) was detected in 40.7% of HCC specimens [6].

4.2. Antiapoptosis. Overexpression of TCTP was detected in many types of tumors and its downregulation decreases the viability of those cells [3]. These suggest that TCTP is a prosurvival factor in normal and cancer cells.

Myeloid cell leukemia 1 (Mcl-1) is an antiapoptotic protein identified as an early gene induced during differentiation of ML-1 myeloid leukemia cells. It is a member of Bcl-2 family that plays a pivotal role in animal development. Zhang et al. found that TCTP interacts with Mcl-1, but not any other Bcl-2 family member. Further, the depletion of Mcl-1 rapidly destabilized TCTP in an osteosarcoma cell line U2OS, supporting the conclusion that Mcl-1 serves as a chaperone of TCTP, binding and stabilizing TCTP *in vivo* [22]. On the contrary, Liu et al. suggest that TCTP may also serve as a molecular chaperone and cofactor of Mcl-1, in which the association between TCTP and Mcl-1 is essential for both to function [23].

It has been reported that TCTP interacts with Bcl-xL, an antiapoptotic protein that maintains the integrity of the mitochondrial membrane [24]. They found that the N-terminal region of TCTP is responsible for its interaction with the Bcl-xL BH3 domain, which is critical for eliciting antiapoptotic properties. They also proposed that TCTP might inhibit T-cell apoptosis by preventing the phosphorylation/inactivation of Bcl-xL. More recently, the crystal structure of TCTP provides new insights into its antiapoptotic activity. The H2-H3 helices of TCTP share a structural similarity to the H5-H6 helices of Bax [25]. Mutation of residues (E109 and K102) close to the turn between the two helices H2 and H3 of TCTP reduces the antiapoptotic effect of TCTP on Bax-induced apoptosis, indicating that H2-H3 helices of TCTP play an important role in the inhibition of apoptosis. Despite the lack of evidence to support the binding between TCTP and Bax [22, 23, 25], Susini et al. suggested that the anchorage of TCTP into the mitochondrial membrane could inhibit the dimerization of Bax and subsequent Bax-induced apoptosis.

4.3. Mitotic Defects and Chromosome Missegregation. In mitosis, APC is activated by binding to Cdc20, and this is dependent on high Cdk1 activity [26]. Subsequently, the active APC recognizes securin and cyclin B, thereby provoking their degradation. Degradation of cyclin B inactivates Cdk1, which subsequently permits mitotic exit [27]. The microtubule binding activity and Plk phosphorylation sites indicate that TCTP is an important gene in the regulation of mitotic progression [17, 18]. As reported in our previous study, the role of TCTP in cell cycle progression has been fully investigated by overexpressing TCTP in HCC cell lines. As a result, TCTP has no obvious effect on G1/S transition; however, when cells were released after synchronization at the prometaphase, an accelerated mitotic exit was observed in TCTP-overexpressing cells [6]. Mechanistic study demonstrated that TCTP promoted the ubiquitin-proteasome degradation of Cdc25C during mitotic progression, which caused the failure in the dephosphorylation of Cdk1-Tyr15 and decreased Cdk1 activity. As a consequence, the sudden drop of Cdk1 activity in mitosis induced a faster mitotic exit and chromosome missegregation, which led to chromosomal instability (Figure 1). We did not observe any obvious difference in cyclin B1 expression level between control and TCTP overexpressing cells, suggesting that the TCTP-mediated faster mitotic exit might not be related to APC-mediated degradation of cyclin B1. Xenograft experiments further supported our notion that the overexpression of TCTP could induce mitotic defects and chromosome missegregation [6].

4.4. Migration and Metastasis. Metastasis is the final step in solid tumor progression and is the most common cause of death in cancer patients [28]. Metastasis is a multistep process, all of which must be successfully completed before giving rise to a metastatic tumor. It has been reported that TCTP is preferentially expressed in colon cancer cell lines (LoVo, SW620) with highly metastatic potentials. Depletion of TCTP by shncRNA-TCTP in LoVo cells significantly reduced the number of the hepatic surface metastases in nude mice [29]. Recently, our group has studied the motile and invasive capabilities of TCTP *in vitro* and *in vivo*. As a result, the number of invaded cells was significantly increased in TCTP-overexpressing cells. An experimental metastasis assay was used to examine the metastatic nodules formed in the livers of SCID mice after inoculation with TCTP-overexpressing cells. Cells were injected through the tail-vein of SCID mice, metastatic nodules were counted at 8 weeks after injection. The number of metastatic nodules on the surface of the liver was significantly higher in mice injected with TCTP-transfected cells.

5. Clinical Implication of TCTP in Human Cancers

Overexpression of TCTP was found in different types of cancers, including colon cancer [4], prostate cancer [5], and liver cancer [6]. In our previous study, overexpression of TCPT was detected in 40.7% (48 of 118) of HCC cases. Clinically, overexpression of TCTP was significantly

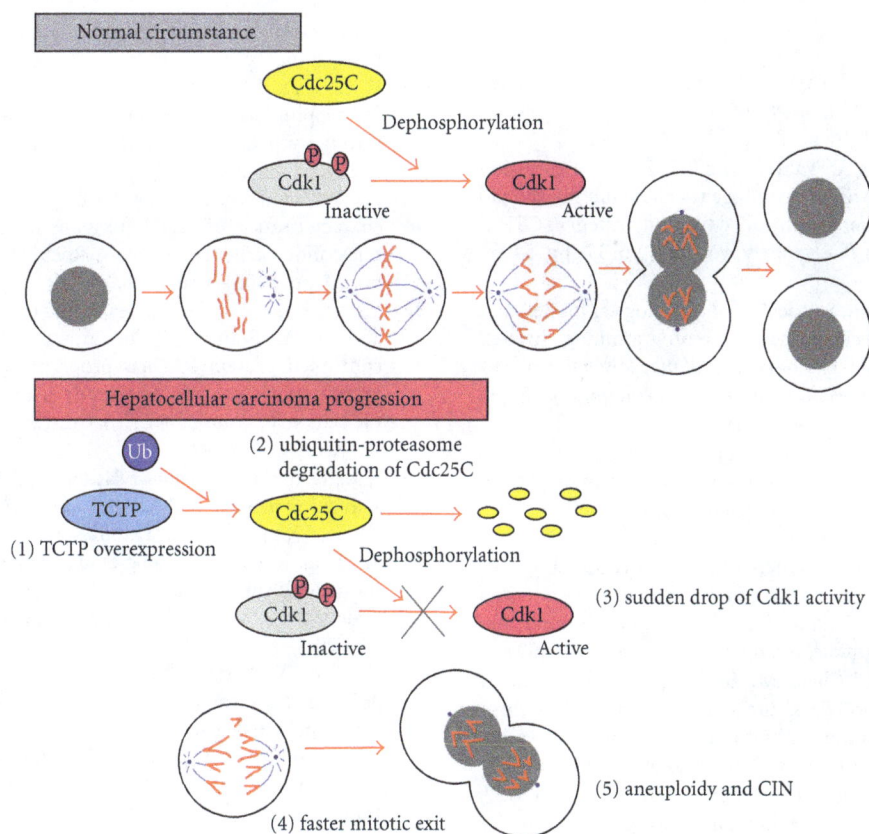

FIGURE 1: Mechanistic diagram showing the effect of abnormal regulation of TCTP/Cdc25C/Cdk1 pathway in HCC development. (Upper panel) Under the normal mitotic progression, Cdc25C activates Cdk1 by the dephosphorylation of Thr14 and Tyr15 in Cdk1. The level of active Cdk1 is a key factor for maintaining the mitotic state and functions as a key switch for cell division. (Lower panel) During the HCC development, TCTP is overexpressed in over 40% of HCC cases. Overexpression of TCTP promotes the ubiquitin-proteasome degradation of Cdc25C, which leads to the failure in the dephosphorylation of Cdk1 on Tyr15 and decreases Cdk1 activity. As a consequence, the sudden drop of Cdk1 activity in mitosis induces a faster mitosis exit and chromosome missegregation, which leads to aneuploidy and CIN, finally causing cancer development.

associated with the advanced tumor stage and overall survival time of HCC patients. TCTP was also determined as an independent marker associated with poor prognostic outcomes [6]. Moreover, our recent study also indicated that the overexpression of TCTP was significantly associated withextrahepatic metastases (e.g., bone, lymph node, and kidney) among HCC patients.

By comparing the proteome of a melanoma cell line (MeWo) and their chemoresistant counterpart, TCTP was also found to be one of the proteins preferentially expressed in chemoresistant melanoma cell lines [30].

6. Conclusions and Future Directions

Due to the ubiquitous expression and high-degree conservation, TCTP protein underlines its important functions in the cell. An increasing number of research investigations are being conducted in this area, particularly into the effect of TCTP during cancer progression. It is implicated in cell growth, cell cycle progression, apoptosis, and regulation in pluripotency. Although TCTP is not tumor-specific, preferential expression of TCTP in different types of cancer

underlines the importance of TCTP in cancer progression. As summarized in this paper, TCTP mainly exerts its tumorigenic function via inhibiting apoptosis, accelerating mitotic exit, inducing invasion and metastasis, and so on. By using molecular biological techniques, we demonstrated a molecular pathway, TCTP/Cdc25c/Cdk1, which plays an important role in hepatocarcinogenesis by accelerating mitotic progression and inducing CIN (Figure 1). CIN is a hallmark of many types of human cancers and is significantly associated with poor prognosis. Thus, characterization of this novel pathway will greatly facilitate our insights into the link between aneuploidy cancer development. To better understand the oncogenic mechanism of TCTP, our current work is focusing on the DNA-binding activity of TCTP and identifying its specific binding motifs. Future research on the regulatory network of TCTP will improve our understanding of this oncogene and may ultimately contribute to the development of more accurate treatment modalities.

Conflict of Interests

The authors declare that they have no conflict of interests.

References

[1] U. A. Bommer and B. J. Thiele, "The translationally controlled tumour protein (TCTP)," *International Journal of Biochemistry and Cell Biology*, vol. 36, no. 3, pp. 379–385, 2004.

[2] H. Thiele, M. Berger, A. Skalweit, and B. J. Thiele, "Expression of the gene and processed pseudogenes encoding the human and rabbit translationally controlled turnout protein (TCTP)," *European Journal of Biochemistry*, vol. 267, no. 17, pp. 5473–5481, 2000.

[3] M. Tuynder, L. Susini, S. Prieur et al., "Biological models and genes of tumor reversion: cellular reprogramming through tpt1/TCTP and SIAH-1," *Proceedings of the National Academy of Sciences of the United States of America*, vol. 99, no. 23, pp. 14976–14981, 2002.

[4] S. Chung, M. Kim, W. J. Choi, J. K. Chung, and K. Lee, "Expression of translationally controlled tumor protein mRNA in human colon cancer," *Cancer Letters*, vol. 156, no. 2, pp. 185–190, 2000.

[5] F. Arcuri, S. Papa, A. Carducci et al., "Translationally controlled tumor protein (TCTP) in the human prostate and prostate cancer cells: expression, distribution, and calcium binding activity," *Prostate*, vol. 60, no. 2, pp. 130–140, 2004.

[6] T. H. M. Chan, L. Chen, M. Liu et al., "Translationally controlled tumor protein induces mitotic defects and chromosome missegregation in hepatocellular carcinoma development," *Hepatology*, vol. 55, no. 2, pp. 491–505, 2012.

[7] H. Andree, H. Thiele, M. Fähling, I. Schmidt, and B. J. Thiele, "Expression of the human TPT1 gene coding for translationally controlled tumor protein (TCTP) is regulated by CREB transcription factors," *Gene*, vol. 380, no. 2, pp. 95–103, 2006.

[8] U. A. Bommer, A. V. Borovjagin, M. A. Greagg et al., "The mRNA of the translationally controlled tumor protein P23/TCTP is a highly structured RNA, which activates the dsRNA-dependent protein kinase PKR," *RNA*, vol. 8, no. 4, pp. 478–496, 2002.

[9] P. Thaw, N. J. Baxter, A. M. Hounslow, C. Price, J. P. Waltho, and C. J. Craven, "Structure of TCTP reveals unexpected relationship with guanine nucleotide-free chaperones," *Nature Structural Biology*, vol. 8, no. 8, pp. 701–704, 2001.

[10] M. Gnanasekar, G. Dakshinamoorthy, and K. Ramaswamy, "Translationally controlled tumor protein is a novel heat shock protein with chaperone-like activity," *Biochemical and Biophysical Research Communications*, vol. 386, no. 2, pp. 333–337, 2009.

[11] S. T. Chitpatima, S. Makrides, R. Bandyopadhyay, and G. Brawerman, "Nucleotide sequence of a major messenger RNA for a 21 kilodalton polypeptide that is under translational control in mouse tumor cells," *Nucleic Acids Research*, vol. 16, no. 5, p. 2350, 1988.

[12] H. Bohm, R. Benndorf, M. Gaestel et al., "The growth-related protein p23 of the Ehrlich ascites tumor: translational control, cloning and primary structure," *Biochemistry International*, vol. 19, no. 2, pp. 277–286, 1989.

[13] L. Chen, T. H. M. Chan, Y. F. Yuan et al., "CHD1L promotes hepatocellular carcinoma progression and metastasis in mice and is associated with these processes in human patients," *Journal of Clinical Investigation*, vol. 120, no. 4, pp. 1178–1191, 2010.

[14] H. C. Sung, P. S. Wu, C. H. Chou et al., "A knockout mouse approach reveals that TCTP functions as an essential factor for cell proliferation and survival in a tissue- or cell type-specific manner," *Molecular Biology of the Cell*, vol. 18, no. 7, pp. 2525–2532, 2007.

[15] Y. C. Hsu, J. J. Chern, Y. Cai, M. Liu, and K. W. Choi, "Drosophila TCTP is essential for growth and proliferation through regulation of dRheb GTPase," *Nature*, vol. 445, no. 7129, pp. 785–788, 2007.

[16] H. Rehmann, M. Brüning, C. Berghaus et al., "Biochemical characterisation of TCTP questions its function as a guanine nucleotide exchange factor for Rheb," *FEBS Letters*, vol. 582, no. 20, pp. 3005–3010, 2008.

[17] Y. Gachet, S. Tournier, M. Lee, A. Lazaris-Karatzas, T. Poulton, and U. A. Bommer, "The growth-related, translationally controlled protein P23 has properties of a tubulin binding protein and associates transiently with microtubules during the cell cycle," *Journal of Cell Science*, vol. 112, no. 8, pp. 1257–1271, 1999.

[18] F. R. Yarm, "Plk phosphorylation regulates the microtubule-stabilizing protein TCTP," *Molecular and Cellular Biology*, vol. 22, no. 17, pp. 6209–6221, 2002.

[19] T. Wang, K. Chen, X. Zeng et al., "The histone demethylases Jhdm1a/1b enhance somatic cell reprogramming in a vitamin-C-dependent manner," *Cell Stem Cell*, vol. 9, no. 6, pp. 575–587, 2011.

[20] B. Liao, X. Bao, L. Liu et al., "MicroRNA cluster 302–367 enhances somatic cell reprogramming by accelerating a mesenchymal-to-epithelial transition," *Journal of Biological Chemistry*, vol. 286, no. 19, pp. 17359–17364, 2011.

[21] M. J. Koziol, N. Garrett, and J. B. Gurdon, "Tpt1 activates transcription of oct4 and nanog in transplanted somatic nuclei," *Current Biology*, vol. 17, no. 9, pp. 801–807, 2007.

[22] D. Zhang, F. Li, D. Weidner, Z. H. Mnjoyan, and K. Fujise, "Physical and functional interaction between myeloid cell leukemia 1 protein (MCL1) and fortilin. The potential role of MCL1 as a fortilin chaperone," *Journal of Biological Chemistry*, vol. 277, no. 40, pp. 37430–37438, 2002.

[23] H. Liu, H. W. Peng, Y. S. Cheng, H. S. Yuan, and H. F. Yang-Yen, "Stabilization and enhancement of the antiapoptotic activity of Mcl-1 by TCTP," *Molecular and Cellular Biology*, vol. 25, no. 8, pp. 3117–3126, 2005.

[24] Y. Yang, F. Yang, Z. Xiong et al., "An N-terminal region of translationally controlled tumor protein is required for its antiapoptotic activity," *Oncogene*, vol. 24, no. 30, pp. 4778–4788, 2005.

[25] L. Susini, S. Besse, D. Duflaut et al., "TCTP protects from apoptotic cell death by antagonizing bax function," *Cell Death and Differentiation*, vol. 15, no. 8, pp. 1211–1220, 2008.

[26] J. M. Peters, "The anaphase-promoting complex: proteolysis in mitosis and beyond," *Molecular Cell*, vol. 9, no. 5, pp. 931–943, 2002.

[27] R. Wäsch and D. Engelbert, "Anaphase-promoting complex-dependent proteolysis of cell cycle regulators and genomic instability of cancer cells," *Oncogene*, vol. 24, no. 1, pp. 1–10, 2005.

[28] B. Parker and S. Sukumar, "Distant metastasis in breast cancer: molecular mechanisms and therapeutic targets," *Cancer Biology & Therapy*, vol. 2, no. 1, pp. 14–21, 2003.

[29] M. Qiang, G. Yan, X. Weiwen et al., "The role of translationally controlled tumor protein in tumor growth and metastasis of colon adenocarcinoma cells," *Journal of Proteome Research*, vol. 9, no. 1, pp. 40–49, 2010.

[30] P. Sinha, S. Kohl, J. Fischer et al., "Identification of novel proteins associated with the development of chemoresistance in malignant melanoma using two-dimensional electrophoresis," *Electrophoresis*, vol. 21, no. 14, pp. 3048–3057, 2000.

20

Length and PKA Dependence of Force Generation and Loaded Shortening in Porcine Cardiac Myocytes

Kerry S. McDonald,[1] Laurin M. Hanft,[1] Timothy L. Domeier,[1] and Craig A. Emter[2]

[1] *Department of Medical Pharmacology & Physiology, School of Medicine, University of Missouri, Columbia, MO 65212, USA*
[2] *Department of Biomedical Sciences, College of Veterinary Medicine, University of Missouri, Columbia, MO 65212, USA*

Correspondence should be addressed to Kerry S. McDonald, mcdonaldks@missouri.edu

Academic Editor: John Konhilas

In healthy hearts, ventricular ejection is determined by three myofibrillar properties; force, force development rate, and rate of loaded shortening (i.e., power). The sarcomere length and PKA dependence of these mechanical properties were measured in porcine cardiac myocytes. Permeabilized myocytes were prepared from left ventricular free walls and myocyte preparations were calcium activated to yield ~50% maximal force after which isometric force was measured at varied sarcomere lengths. Porcine myocyte preparations exhibited two populations of length-tension relationships, one being shallower than the other. Moreover, myocytes with shallow length-tension relationships displayed steeper relationships following PKA. Sarcomere length-K_{tr} relationships also were measured and K_{tr} remained nearly constant over ~2.30 μm to ~1.90 μm and then increased at lengths below 1.90 μm. Loaded-shortening and peak-normalized power output was similar at ~2.30 μm and ~1.90 μm even during activations with the same [Ca^{2+}], implicating a myofibrillar mechanism that sustains myocyte power at lower preloads. PKA increased myocyte power and yielded greater shortening-induced cooperative deactivation in myocytes, which likely provides a myofibrillar mechanism to assist ventricular relaxation. Overall, the bimodal distribution of myocyte length-tension relationships and the PKA-mediated changes in myocyte length-tension and power are likely important modulators of Frank-Starling relationships in mammalian hearts.

1. Introduction

The primary role of cardiac myocytes is to develop force and power. In the isovolumic phase of the cardiac cycle, left ventricular myocytes develop (near)-isometric force against the enclosed ventricular chamber and in doing so increases ventricular pressure. When ventricular pressure exceeds aortic pressure the aortic valve opens, and myocyte force production is accompanied by loaded shortening (i.e., power) as blood is ejected from the ventricle into the systemic circulation. The rate of myocyte force generation determines the duration of isovolumic ventricular contraction and, consequently, the amount of the cardiac cycle devoted to ejection. The blood volume ejected per beat is determined by chamber compression, which is governed by (i) systolic ejection time, (ii) the number of force generating cross-bridges (which controls where on the force-velocity relation the

ensemble of cross-bridges will work), and (iii) the inherent rate of loaded cross-bridge cycling. Hence, ventricular ejection is highly dependent on three myofibrillar mechanical properties: (1) force, (2) rate of force development, and (3) rate of loaded shortening (i.e., myocyte power output). There is considerable information related to these cardiac myofibrillar contractile properties in rodents [1–11], and these biophysical properties, in part, underlie the unique ventricular function of these species when compared to larger animals and humans [12–14]. However, there are fewer studies that have focused on cardiac myofibrillar mechanics in pig, a species of high translational relevance given its anatomic similarities to humans. Most investigations of porcine myofibrillar preparations have focused on steady-state properties at a single sarcomere length [15, 16] or examined stretch activation [17]. In this study we investigated three key myofibrillar mechanical properties (i.e., force, rate of force development, and loaded shortening) and their dependence on sarcomere length and PKA in porcine left ventricular cardiac myocytes.

2. Methods

2.1. Animal Model. Adult male yucatan miniature swine (14 months old) weighing 30–40 kg were obtained from the breeder (Sinclair Research Center; Columbia, MO). Animal care and use procedures complied with the *Guide for the Care and Use of Laboratory Animals* issued by the National Research Council and were approved by the University of Missouri Animal Care and Use Committee.

2.2. In Vivo Cardiovascular Function. Animals ($n = 4$) were initially anesthetized with a telazol (5 mg/kg)/xylazine (2.25 mg/kg) mix and maintained using inhaled isoflurane ($\approx 1.75\%$). Heparin was given with an initial loading dose of 300 U/kg IV, followed by maintenance of 100 U/kg each hour. A median sternotomy was performed and the pericardium was opened near the apex for insertion of the pressure-volume (*P-V*) loop catheter. *P-V* loops were measured utilizing a calibrated 7F admittance-based ADVantage catheter (SciSense; London, Ontario, Canada) positioned in the LV. A 14F balloon occlusion catheter was advanced to the inferior vena cava at the level of the apex of the heart via the deep femoral vein. Peripheral systemic MAP was measured via a fluid filled 6F LCB SH guide catheter (Boston Scientific) introduced through a 7F sheath placed in the right femoral artery and positioned in the descending aorta distal to the aortic band. Catheter placement was visualized and confirmed using angiography (InfiMed software) and Visipaque contrast medium. Following placement of the catheter, animals were brought to a peripheral MAP of 80 mmHg using phenylephrine (I.V. 1–3 μg/kg/min) and allowed to stabilize for 5 minutes. *P-V* loops were collected before and after a single dose of dobutamine (5 μg/kg/min I.V.) administered for 5 minutes under conditions of reducing preload achieved through transient occlusion of the inferior vena cava via inflation of the balloon catheter. Our admittance based *P-V* loop system requires input of baseline stroke volume (SV), which was determined one week prior to the terminal studies using ultrasound and calculated as previously reported [18] using the equation SV = $\pi(r)^2 *$ VTI where r is the radius and VTI is the velocity time interval (measured from apical four-chamber view). Aortic radius was calculated from the aortic left ventricular outflow track (measured in parasternal 2D view).

2.3. Isolation of Cardiac Myocytes. The heart was excised from the experimental animal following administration of a preanesthetic mixture of telazol (5 mg/kg)/xylazine (2.25 mg/kg) and permeabilized myocytes were isolated as previously described [19]. Briefly, a section of left ventricular free wall ($\sim$10 cm^3) near the left anterior descending (LAD) coronary artery was removed and half was rapidly frozen in liquid nitrogen for biochemical analyses, and the other half was placed in ice cold relaxing solution for myocyte experiments.

The piece in relaxing solution was cut into smaller pieces (2-3 mm) and homogenized with a Waring blender. The resultant slurry was centrifuged 75 sec at 165× g and the pellet was suspended for 3 min in 0.5% ultrapure Triton X-100 (Pierce Chemical Co.) in relaxing solution. The permeabilized myocytes were washed and centrifuged twice with cold relaxing solution with the final suspension kept on ice during the day of the experiment.

For intact myocyte isolation, a section of the left-ventricular free wall was perfused via cannulation of the LAD. The tissue was perfused with a nominally calcium-free saline solution containing heparin for 10 minutes, followed by a minimal essential medium (MEM) solution containing 45 μg/mL Liberase Blendzyme TH (Roche Applied Science, Indianapolis, IN, USA) for 30 minutes at 37°C. Digested tissue was minced and filtered, and the dissociated myocytes were washed and maintained in an MEM solution with 50 μM calcium at room temperature until experimental procedures.

2.4. Solutions. Relaxing solution in which the ventricles were disrupted, skinned, and suspended contained (in mmol/L): EGTA 2, MgCl$_2$ 5, ATP 4, imidazole 10, and KCl 100 at pH 7.0. Compositions of relaxing and activating solutions used in mechanical measurements were as follows (mmol/L): EGTA 7, MgCl$_2$ 5, imidazole 20, ATP 4, creatine phosphate 14.5, pH 7.0, Ca^{2+} concentrations of 10^{-9} M (relaxing solution) and $10^{-4.5}$ M (maximal activating solution), and sufficient KCl to adjust ionic strength to 180 mM. The final concentrations of each metal, ligand, and metal-ligand complex were determined with the computer program of Fabiato [20]. Immediately preceding activations, muscle preparations were immersed for 60 s in a solution of reduced Ca^{2+}-EGTA buffering capacity, identical to normal relaxing solution except that EGTA is reduced to 0.5 mM. This protocol resulted in more rapid steady-state force development and helped preserve the striation pattern during activation. Intact cardiomyocyte experiments were performed in a physiological saline solution containing (in mM). 135 NaCl, 4 KCl, 2 CaCl$_2$, 1 MgCl$_2$, 10 D-glucose, 10 Hepes, pH 7.4 with NaOH.

2.5. Experimental Apparatus. The experimental apparatus for physiological measurements of myocyte preparations was similar to one previously described in detail [21] and modified for cardiac myocyte preparations [7]. Myocyte preparations were attached between a force transducer and torque motor by placing the ends of the myocyte preparation into stainless steel troughs (25 gauge). The ends of the myocyte preparations were secured by overlaying a 0.5 mm length of 3–0 monofilament nylon suture (Ethicon, Inc.) onto each end of the myocyte, and then tying the suture into the troughs with two loops of 10–0 monofilament (Ethicon, Inc). The attachment procedure was performed under a stereomicroscope ($\sim$100x magnification) using finely shaped forceps.

Prior to mechanical measurements, the experimental apparatus was mounted on the stage of an inverted microscope (model IX-70, Olympus Instrument Co., Japan), which was placed upon a pneumatic vibration isolation table having

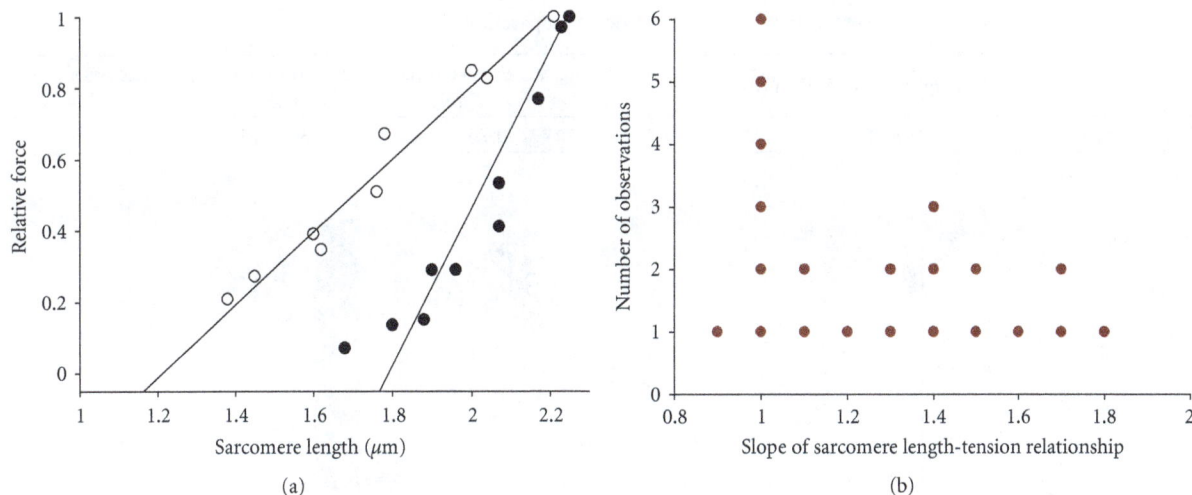

FIGURE 1: Sarcomere length-tension relationships in porcine skinned ventricular cardiac myocyte preparations. (a) Muscle cell preparations were mounted between a force transducer and motor, calcium activated to yield ∼50% maximal force, then isometric force was measured over a range of sarcomere lengths monitored using an IonOptix SarLen system. (b) Histogram showing the slopes of length-tension relationships obtained in porcine cardiac myocytes.

a cut-off frequency of ∼1 Hz. Mechanical measurements were performed using a capacitance-gauge transducer (Model 403-sensitivity of 20 mV/mg (plus a 10x amplifier) and resonant frequency of 600 Hz; Aurora Scientific, Inc., Aurora, ON, Canada). Length changes were introduced using a DC torque motor (model 308, Aurora Scientific, Inc.) driven by voltage commands from a personal computer via a 12-bit D/A converter (AT-MIO-16E-1, National Instruments Corp., Austin, TX, USA). Force and length signals were digitized at 1 kHz and stored on a personal computer using LabView for Windows (National Instruments Corp.). Sarcomere length was monitored simultaneous with force and length measurements using IonOptix SarcLen system (IonOptix, Milton, MA), which used a fast Fourier transform algorithm of the video image of the myocyte. Microscopy was done using a 40x objective (Olympus UWD 40) and a 2.5x intermediate lens.

2.6. Sarcomere-Length Tension Measurements.

All mechanical measurements on cardiac myocytes were performed at 13 ± 1°C. For sarcomere length-tension measurements, an experimental protocol was performed similar to previously described [22]. Following attachment of myocyte preparation to the apparatus, the relaxed preparation was adjusted to a sarcomere length of ∼2.35 μm and then the preparation was maximally Ca^{2+} activated in pCa 4.5 solution. For sarcomere length-tension measurements, the cell preparation was transferred to a pCa solution that yielded ∼50% maximal (i.e., pCa 4.5 or $P_{4.5}$) force and then isometric force was measured over a range of sarcomere lengths monitored by the IonOptix SarcLen system (IonOptix, Milton, MA). Isometric force and sarcomere length were measured simultaneously. Sarcomere length was adjusted in a range between ∼2.35 μm and to ∼1.4 μm by manual manipulation of the length micrometer while the preparation was Ca^{2+} activated. After

each sarcomere length change, ∼10–15 seconds were provided to allow for development of steady-state force. Force at each sarcomere length was obtained via a slack-restretch maneuver (see below for description). For analysis, force at each sarcomere length was normalized to the force obtained at sarcomere length ∼2.35 μm (during the submaximal Ca^{2+} activation). Since force during submaximal Ca^{2+} activations invariably rose slightly during the sustained activation, normalized forces were calculated by interpolating force measurements at sarcomere length 2.35 μm, which were performed at the beginning and end of the series of force measurements. At the end of each experiment, preparations were activated a second time in pCa 4.5 solutions and only experiments in which maximal tension remained >80% of initial were used for analysis. To assess the effects of PKA, length-tension relationships were performed before and after 45 min incubation with PKA (Sigma, 0.125 U/μL). The pCa solution for length tension curves was adjusted to yield the same forces before and after PKA due to decreased Ca^{2+} sensitivity of force following PKA.

2.7. Measurement of the Rate of Force Redevelopment, Loaded Shortening, and Power.

Force redevelopment rates were obtained using a procedure previously described for skinned cardiac myocyte preparations [23–25]. While in Ca^{2+} activating solution, the myocyte preparation was rapidly shortened by 15–20% of initial length (L_0) to yield zero force. The myocyte preparation was then allowed to shorten for ∼20 ms; after 20 ms the preparation was rapidly restretched to ∼105% of its initial length (L_0) for 2 ms and then returned to L_0. Tension redevelopment following a slack-restretch maneuver was fit by a single exponential equation:

$$F = F_{max}\left(1 - e^{-k_{tr}t}\right) + F_{res}, \qquad (1)$$

TABLE 1: Porcine cardiac myocyte preparation characteristics.

	n	Length (μm)	Width (μm)	Sarcomere length (μm)	Passive force (μN)	Maximum force ($P_{4.5}$) (μN)	Maximum force ($P_{4.5}$) (kN m^{-2})
Cardiac myocytes	36	129 ± 28	28 ± 9	2.29 ± 0.06	0.70 ± 0.45	30 ± 14	75 ± 32

Values are means ± S.D.

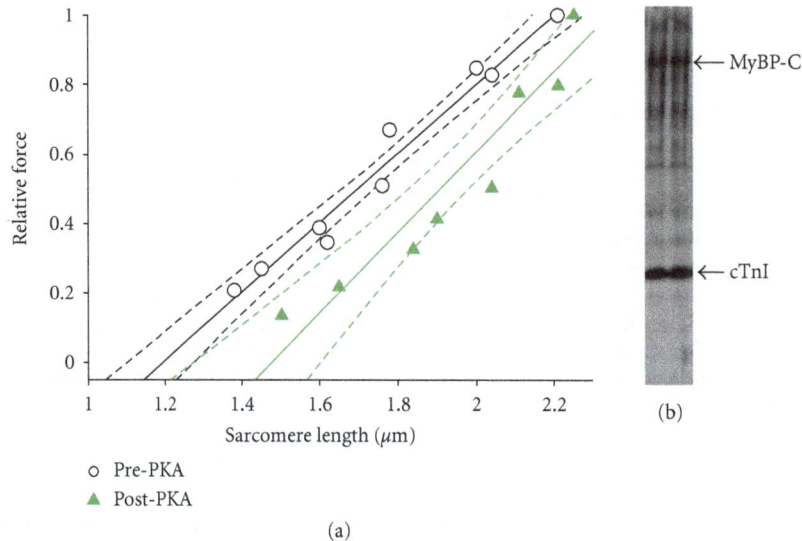

(a)

○ Pre-PKA
▲ Post-PKA

FIGURE 2: (a) Pig cardiac myocyte sarcomere length-tension relationships before and after PKA treatment. PKA-induced phosphorylation markedly steepened the length-tension relationship. (b) An autoradiogram showing radiolabeled phosphate incorporation into pig cardiac myofibrillar proteins (MyBP-C and cTnI) upon PKA treatment. Without PKA treatment, there was no radiolabelled ATP incorporation (data not shown).

where F is force at time t, F_{max} is maximal force, k_{tr} is the rate constant of force development, and F_{res} represents any residual tension immediately after the slack-restretch maneuver.

Power output of single skinned myocyte preparations was determined at varied loads as described earlier [26]. Briefly, myocytes were placed in activating solution and once steady-state force developed, a series of force clamps (less than steady-state force) were performed to determine isotonic shortening velocities. Using a servo-system, force was maintained constant for a designated period of time (200 to 250 msec) while the length change was continuously monitored. Following the force clamp, the myocyte preparation was slackened to reduce force to near zero to allow estimation of the relative load sustained during isotonic shortening; the myocyte was subsequently re-extended to its initial length.

Myocyte preparation length traces during loaded shortening were fit to a single decaying exponential equation:

$$L = Ae^{-kt} + C, \quad (2)$$

where L is cell length at time t, A, and C are constants with dimensions of length, and k is the rate constant of shortening ($k_{shortening}$). Velocity of shortening at any given time, t, was determined as the slope of the tangent to the fitted curve at that time point. In this study, velocities of shortening were calculated by extrapolation of the fitted curve to the onset of the force clamp (i.e., $t = 0$).

Hyperbolic force-velocity curves were fit to the relative force-velocity data using the Hill equation [27]:

$$(P + a)(V + b) = (P_0 + a)b, \quad (3)$$

where P is force during shortening at velocity V, P_0 is the peak isometric force, and a and b are constants with dimensions of force and velocity, respectively. Power-load curves were obtained by multiplying force x velocity at each load on the force-velocity curve. The optimum force for mechanical power output (F_{opt}) was calculated using [28]:

$$F_{opt} = (a^2 + a * P_0)^{1/2} - a. \quad (4)$$

Curve fitting was performed using a customized program written in Qbasic, as well as commercial software (Sigmaplot).

2.8. Intracellular Calcium Measurements. Intact myocytes were plated on laminin coated coverslips and loaded with 5 μM of the calcium indicator dye fluo-4/AM for 10 minutes, followed by a 20-minute wash. 2-dimensional laser-scanning confocal fluorescence microscopy was performed using the resonance scanhead of a Leica SP5 (Leica Microsystems, Buffalo Grove, IL, USA), with excitation at 488 nm and emission collected from 510–550 nm. Field stimulation (0.5 Hz) with a pair of platinum electrodes was used to induce action potentials and intracellular calcium transients. To analyze

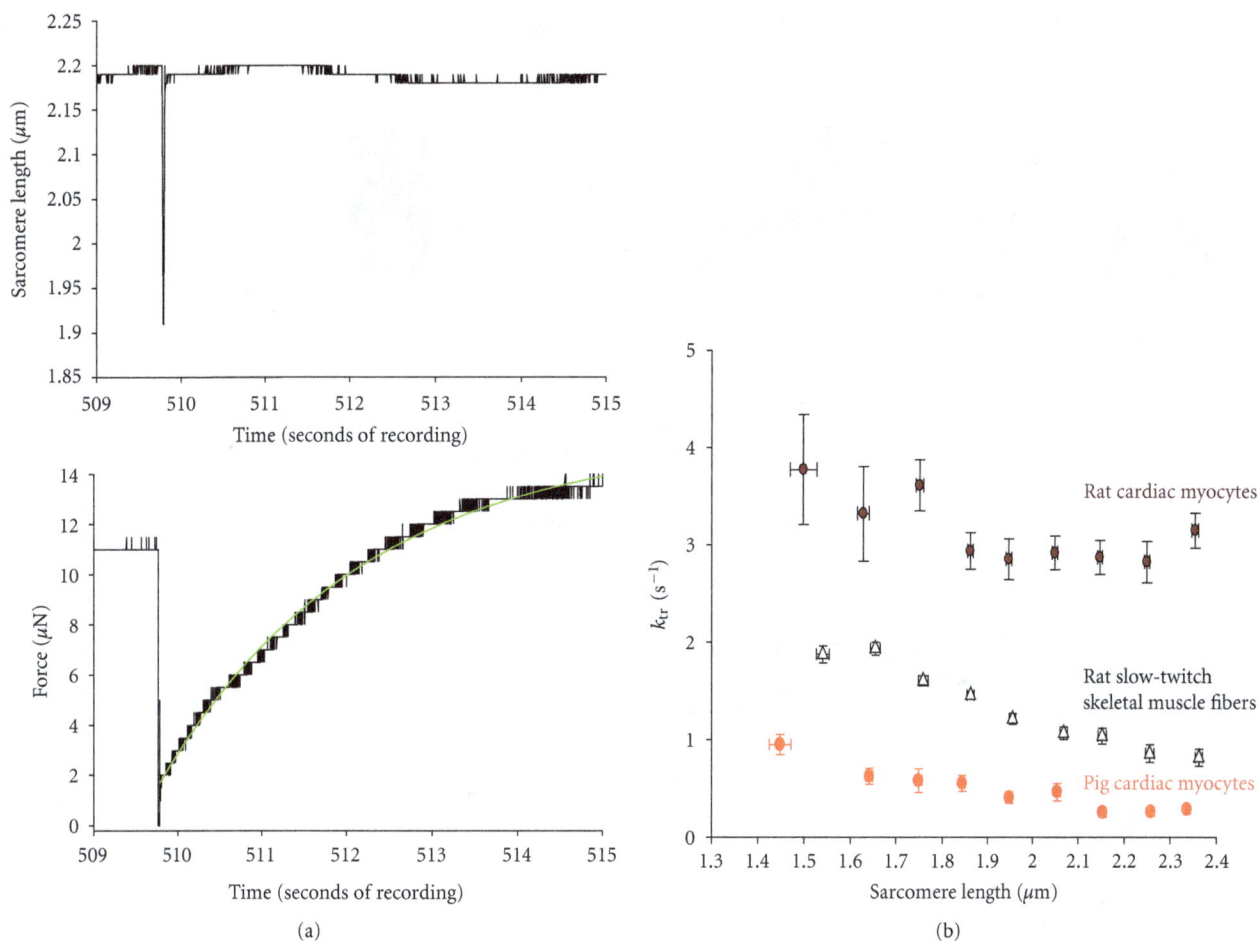

FIGURE 3: Sarcomere length dependence of the rate constant of force redevelopment (k_{tr}). (a) Slow time-based recordings of sarcomere length and force obtained using an IonOptix SarLen system during a slack restretch maneuver during submaximal Ca^{2+} activation. (b) Sarcomere length-dependence of k_{tr} for rat slow-twitch skeletal muscle fibers, rat cardiac myocytes, and pig cardiac myocytes. Although pig cardiac myocyte k_{tr} was much slower at all sarcomere lengths compared to rat cardiac myocytes, both pig and rat cardiac cell preparations showed that k_{tr} increased at short sarcomere lengths despite reductions in force implicating that sarcomere length overrides the Ca^{2+} activation dependence of k_{tr}.

recovery kinetics, calcium transients were normalized using the following formula: $[(F - F_{baseline})/(F_{peak} - F_{baseline})]$.

2.9. SDS-PAGE and Autoradiography. The gel electrophoresis procedure was similar to one previously described [12, 29]. The gels for SDS-PAGE were prepared with a 3.5% acrylamide stacking gel and a 12% acrylamide resolving gel. Samples were separated by SDS-PAGE at constant voltage (250 V) for 8 h. Gels were initially fixed in a 10% acetic acid-50% ethanol solution, followed by 2% glutaraldehyde. MyHC isoforms were visualized by ultrasensitive silver staining, and gels were subsequently dried between mylar sheets.

PKA-induced phosphate incorporation into myofibrillar substrates was determined as described previously [30]. Briefly, skinned cardiac myocytes (10 μg) were incubated with the catalytic subunit of PKA (5 μg/mL) and 50 μCi [γ-^{32}P] ATP at room temperature (21–23°C) for 45 minutes. The reaction was stopped by the addition of electrophoresis sample buffer and heating at 95°C for 3 minutes. The samples were then separated by SDS-PAGE for 2.5 hrs at 12 mA, silver

stained, dried, and subsequently exposed to X-ray film for visualization.

2.10. Statistics. A mixed model incorporating linear regression and analysis of covariance was used to compare response variable (stroke volume) slopes plotted versus end diastolic volume, using treatment (baseline versus dobutamine) as the independent variable. Slopes of length-tension relationships were determined by linear regression. Paired t tests were used to determine whether there were significant differences in length-tension slopes and force-velocity parameters at two different sarcomere lengths or before and after PKA treatments. $P < 0.05$ was chosen as indicating significance. All values are expressed as means ± SD unless, otherwise, noted.

3. Results

3.1. Sarcomere Length Dependence of Force. The characteristics of porcine left ventricular cardiac myocyte preparations

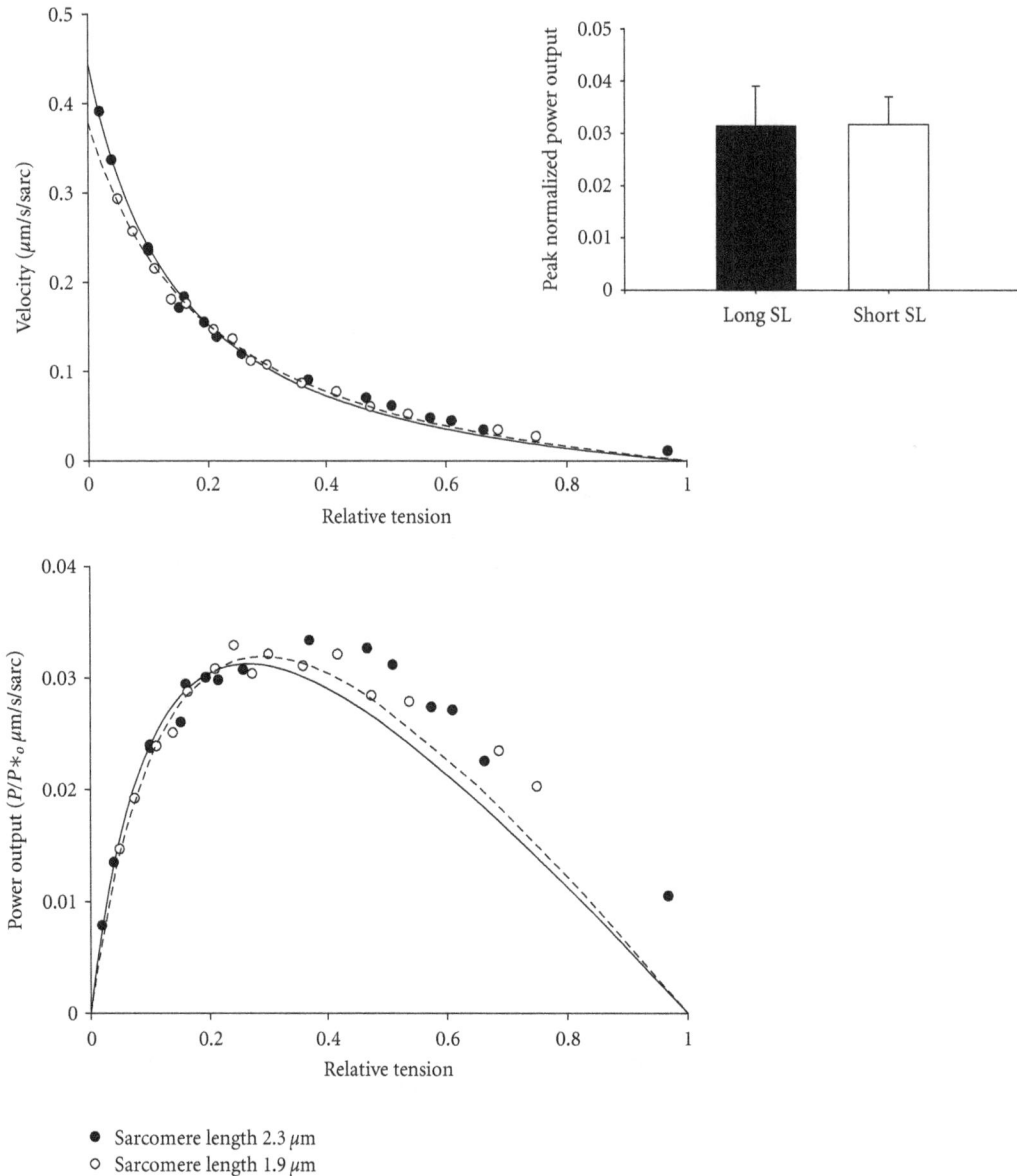

Figure 4: Normalized force-velocity and power-load curves from a pig left ventricular myocyte preparation at long and short sarcomere length obtained during half-maximal Ca^{2+} activations. Pig cardiac myocyte preparations exhibited little sarcomere length dependence of loaded shortening and power output. Inset shows bar plot (mean $\pm$ SD) of peak normalized power output at long ($\sim$2.30 μm) and short ($\sim$1.90 μm) sarcomere length (n = 8 myocyte preparations).

are provided in Table 1. Steady-state sarcomere length-tension relationships were examined in myocyte preparations during near-half-maximal Ca^{2+} activations. Interestingly, porcine cardiac myocyte preparations exhibited a dichotomy of sarcomere length-tension relationships, some had shallow sarcomere length-tension relations while others displayed steep relationships (Figure 1). Histogram analysis of the length-tension relationship slopes indicates near bimodal distribution with one population of cells having a slope near 1.0 and another population with a slope near 1.5 (Figure 1), which has been similarly reported in rat and ferret myocyte preparations [22, 31]. We next examined whether PKA-induced phosphorylation of myofilament proteins may mediate the distribution of length-tension populations. PKA

shifted a shallow length tension relationship to a steep length tension relationship implicating phosphorylation of myosin binding protein-C (MyBP-C) and/or cardiac troponin I (cTnI) as molecular modulators of sarcomere length-tension curves in porcine cardiac myocytes (Figure 2), as was previously observed in rat cardiac myocyte preparations [22].

3.2. Sarcomere Length Dependence of Rates of Force Development (k_{tr}). The rate of force development is thought to mediate pressure development rates in mammalian ventricles. We examined the sarcomere length-dependence of force development rates in porcine cardiac myocytes. Force redevelopment was measured after a slack re-stretch maneuver, and the rate constant of force development (k_{tr}) was

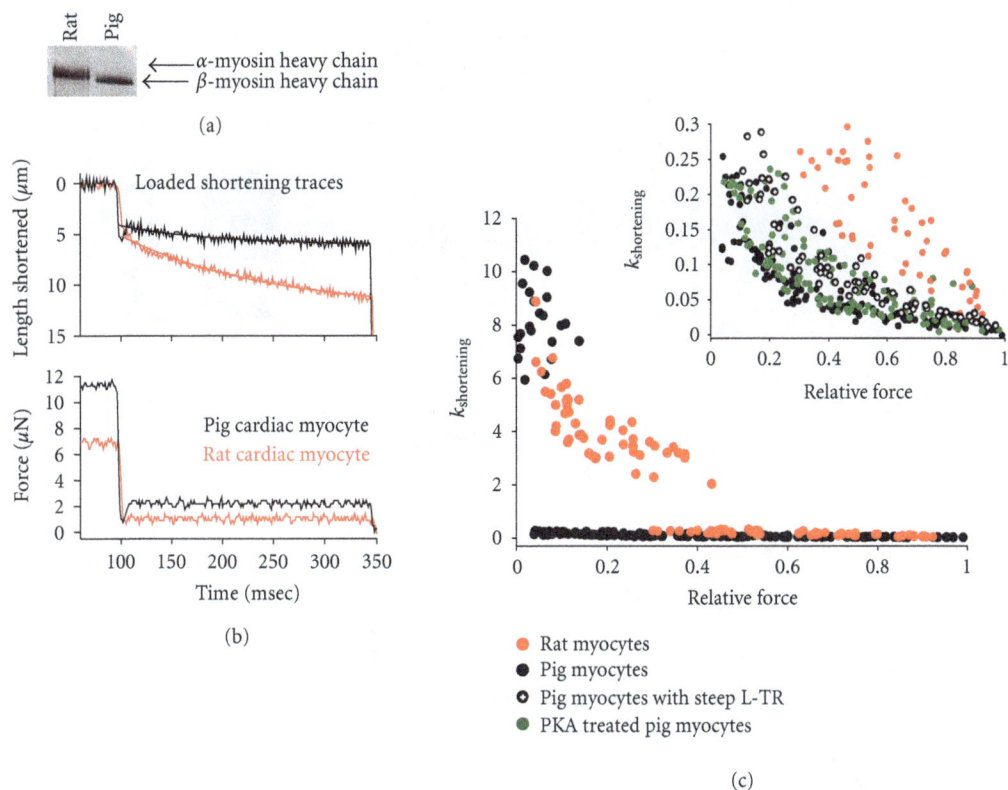

FIGURE 5: (a) Silver-stained gel showing the myosin heavy chain isoforms contained in a rat cardiac myocyte preparation compared to a pig cardiac myocyte preparation. (b) Representative length and force traces during a lightly loaded force clamp in a rat cardiac myocyte preparation (red) and a pig cardiac myocyte preparation (black) during a submaximal Ca^{2+} activation. (c) Length traces exhibited considerably greater curvature (greater $k_{shortening}$) in rat myocytes compared to pig myocyte preparations at all relative loads. Inset in C shows an expanded plot of $k_{shortening}$ versus relative load, which clarifies slope constants below 0.30. Interestingly, PKA-mediated phosphorylation increased the curvature ($k_{shortening}$) of length traces towards those of rat myocyte preparations. In addition, pig cardiac myocyte preparations that exhibited steep length-tension relationships (L-T R) also had more curved length traces. This is consistent with PKA-mediated phosphorylation of myofilaments yielding greater responsiveness to changes in sarcomere length, in this case exhibited by greater shortening-induced cooperative deactivation.

calculated by fitting a single concave exponential equation to the force trace. At sarcomere length $\sim2.30\,\mu m$, k_{tr} was $\sim0.3\,s^{-1}$ during half-maximal activation, which was similar to previously reported for pig myocytes [32], nearly an order of magnitude lower than that measured in rat cardiac myocyte preparations, and only 30% of k_{tr} values in rat slow-twitch skeletal muscle fibers, which like porcine cardiac myocytes contain the β-myosin heavy chain isoform. As sarcomere length was reduced from $\sim2.30\,\mu m$ to $1.90\,\mu m$, k_{tr} remained relatively constant, and then at sarcomere lengths below $1.90\,\mu m$, k_{tr} progressively increased. This k_{tr}-SL relationship was qualitatively similar to that observed in rat slow-twitch skeletal muscle fibers (Figure 3). Since force falls as sarcomere length is decreased but k_{tr} increased with shorter sarcomere lengths, this implicates that sarcomere length *per se* can override the well-described Ca^{2+}-activated force dependence of rates of force redevelopment in cardiac muscle [33–35], that is, sarcomere length plays a dominant role in the kinetics of myofibrillar mechanical properties.

3.3. Sarcomere Length Dependence of Force-Velocity and Power-Load Curves. Previous work has shown a tight regulation between isometric force and normalized force-velocity relationships in rat-skinned cardiac myocyte preparations [26]. However, in porcine cardiac myocyte preparations there was no force dependence of normalized force-velocity and power-load curves when force was altered by changing sarcomere length (i.e., force fell $\sim50\%$ when sarcomere length was shortened from $\sim2.30\,\mu m$ to $1.90\,\mu m$ at the same submaximal activator $[Ca^{2+}]$, Figure 4). The finding that normalized myocyte power did not change over this sarcomere length range in pig myocytes differs from rat cardiac myocyte preparations where normalized force-velocity relationships were shifted downward at short sarcomere length (i.e., $\sim1.90\,\mu m$ versus $\sim2.30\,\mu m$) at the same activator $[Ca^{2+}]$ [36]. The reason for this species difference is not known. One possibility is differences in cardiac myosin heavy chain; rat myocytes contain predominantly α-MyHC while pig myocytes contain mostly β-MyHC (Figure 5(a)). Interestingly, porcine β-MyHC has been shown to have a very slow actin-activated ATPase activity [37], which would prolong the duty cycle (i.e., cross-bridge cycle time spent strongly attached to actin). These strongly attached cross-bridges would tend to keep the thin filament activated [38–40]

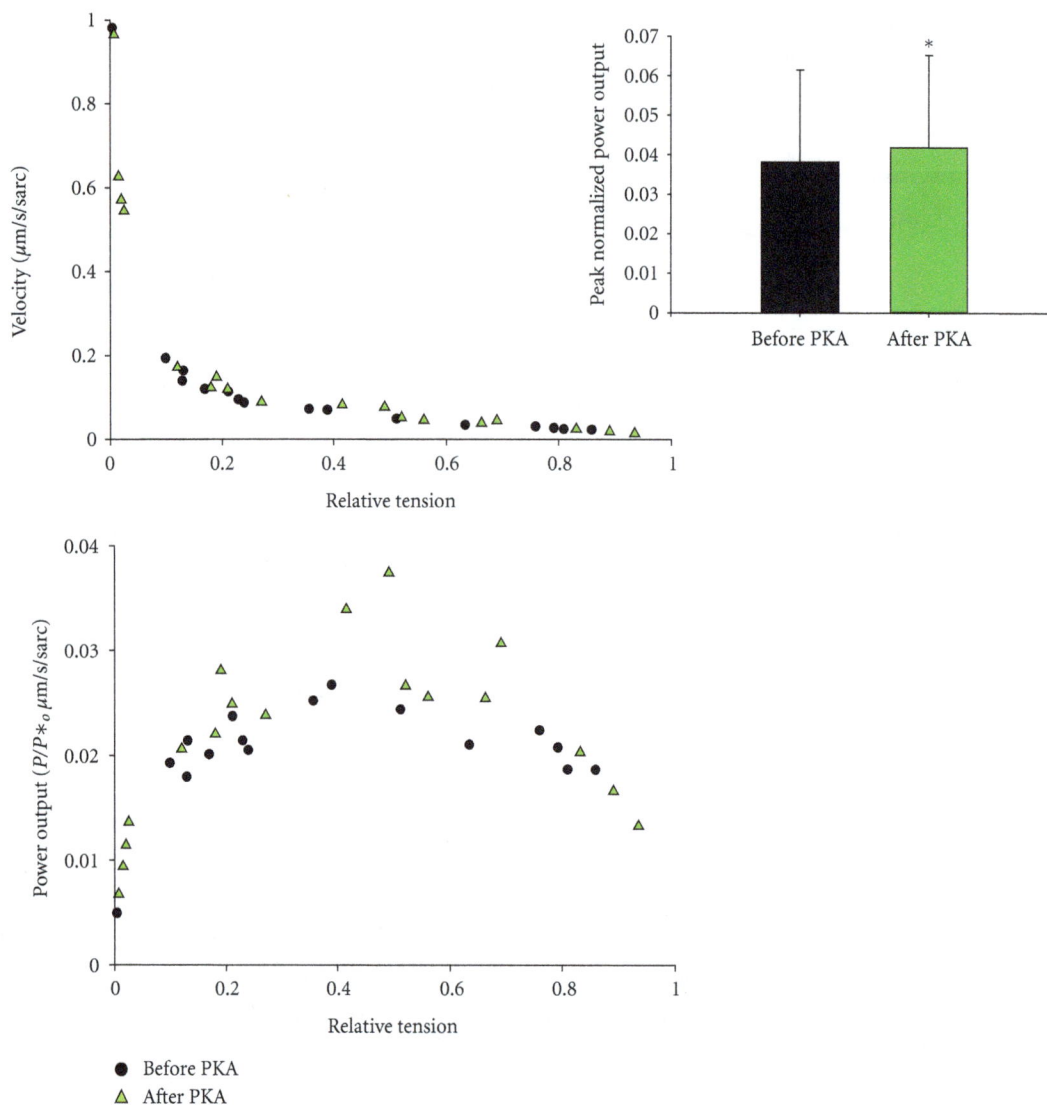

FIGURE 6: Normalized force-velocity and power-load curves from a pig left ventricular myocyte preparation before and after PKA treatment during half-maximal Ca^{2+} activations. Pig cardiac myocyte preparations exhibited more power after PKA treatment. Inset shows bar plot (mean $\pm$ SD) of peak normalized power output before and after PKA ($n = 10$ myocyte preparations).

throughout the duration of the force clamp. This would sustain a relatively high number of cross-bridges to work against the load(s). This is consistent with linear length traces during load clamps in pig cardiac myocyte preparations (Figure 5(b)). The extent of curvature of length traces during load clamps is quantified by $k_{shortening}$ values, which were much lower in pig myocytes than rat myocytes (Figure 5(c)). In pig myocytes, length traces were nearly linear (as indexed by $k_{shortening}$ values) for most load clamps greater than 10% isometric force. This differs markedly from rat cardiac myocyte preparations, whereby length traces during force clamps deviate from linear at load clamps near 40% isometric force during submaximal Ca^{2+} activations (Figure 5(c)).

3.4. PKA Effects on Loaded Shortening, Power Output, and $k_{shortening}$.

We have previously shown that peak power generating capacity increases after PKA-mediated phosphorylation in rat cardiac myocyte preparations [25, 30] and such a change may contribute to augmented contractility in working hearts (i.e., more stroke volume at a given enddiastolic volume) [12]. We examined if a similar biophysical response would occur in pig cardiac myocyte preparations as a potential means for physiological changes in ventricular contractility in response to β-adrenergic stimulation and its downstream signaling molecule, protein kinase A (PKA). We observed a statistically significant increase in peak normalized power output after PKA treatment in pig myocyte preparations (Figure 6), however, the increase was considerably smaller than observed in rat myocyte preparations (a 10% increase in pig myocytes versus a 33% increase in rat myocytes [25, 30]). This small increase in myocyte power after PKA is consistent with a relatively small leftward shift in ventricular function illustrated by an ∼15% increase in stroke volume for a given end diastolic volume that we observed in anesthetized pigs in response to a 5 μg/kg/min dose of dobutamine at a mean arterial pressure of 80 mmHg (Figure 7).

Interestingly, PKA-mediated phosphorylation increased the curvature ($k_{shortening}$) of length traces towards those of rat myocyte preparations (Figure 5(c)). In addition, pig cardiac myocyte preparations that exhibited steep length-tension relationships also had more curved length traces. This is consistent with the idea that PKA-mediated phosphorylation of myofilaments yields both greater force responsiveness to sarcomere length and greater shortening-induced cooperative deactivation.

In summary, pig cardiac myocyte preparations showed two populations of sarcomere length-tension relationships, which appear to be modulated by PKA. Sarcomere length overrode the Ca^{2+}-activated-force dependence of k_{tr} and loaded shortening. PKA treatment also slightly sped loaded shortening especially at loads optimal for power and yielded more curvilinear length traces during force clamps.

4. Discussion

In order to better understand the intricacies of heart function, it is necessary to determine the intermolecular control of myofibrillar contraction. In this study, we focused on three key myofibrillar functional properties (i) force, (ii) rate of force development, and (iii) power generating capacity, which together dictate ventricular stroke volume. We systematically examined these properties in porcine myofibrillar preparations. The study used pig ventricular myocardium for two main reasons: (1) pig hearts have many similarities to human hearts including heart size, heart rate, coronary circulation, responsiveness to many pharmacologic agents, and expression of mostly -myosin heavy chain (MyHC), and (2) to make comparisons with rat myocardium, which have been more extensively studied [7, 8, 12, 22, 25, 26, 30, 36, 41]. Overall, pigs likely provide an advantageous model to study cellular mechanisms of ventricular function and provide further basic insight into potential defects in cardiomyopathic states more related to the human condition.

We observed that sarcomere-length dependence of force in pig myocyte preparations was very similar to that previously observed in rat cardiac myocyte preparations [22]. There was a dichotomy in the steepness of sarcomere length tension relationship whereby one population was shallower than the other. Interestingly, when myocyte preparations with a shallow length-tension relationship were treated with PKA, the relationships became steeper. While the exact molecular (posttranslational) modification by which PKA steepens length tension relationships remains to be determined, the finding is consistent with steeper ventricular function curves in response to β-adrenergic stimulation, assuming that myocyte length-tension contributes, at least in part, to the cellular basis of the Frank-Starling relationship. PKA also increased loaded shortening especially at loads near peak power and increased the curvature of length traces during

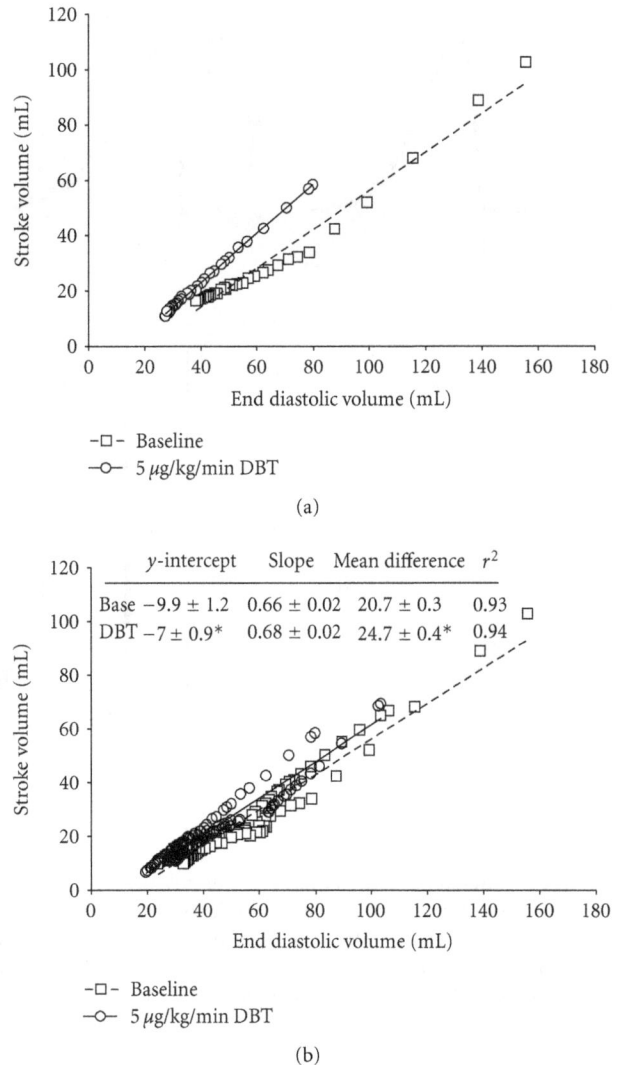

(a)

(b)

FIGURE 7: (a) Representative Frank-Starling relationship from one animal at baseline (Base) and after treatment with dobutamine (DBT). (b) Comprehensive group data from all animals illustrating a significant leftward shift in the Frank-Starling relationship (mixed model, treatment main effect adjusted for EDV covariance, $P < 0.05$). There was no significant interaction or change in slope of the Frank-Starling relationship between treatments (see table inset in (b)), therefore, parallelism was assumed. The y-intercept and marginal mean difference were both significantly increased following the dobutamine treatment (*$P < 0.05$; table inset (b)). The dobutamine challenge resulted in a ∼15% increase in stroke volume (SV) for a given end diastolic volume (EDV) *in vivo*. This increase in ventricular function was similar in magnitude to that observed in our myocyte preparations (∼10%), illustrating the coherence of our whole heart and cardiac myocyte functional data.

force clamps. These PKA-mediated changes in myofibrillar function are consistent with physiological changes induced by β-adrenergic stimulation. β-adrenergic stimulation is known to (i) increase contractility (mediated in part by greater myocyte power at a given sarcomere length), (ii) steepen the Frank-Starling relationship (mediated in part by

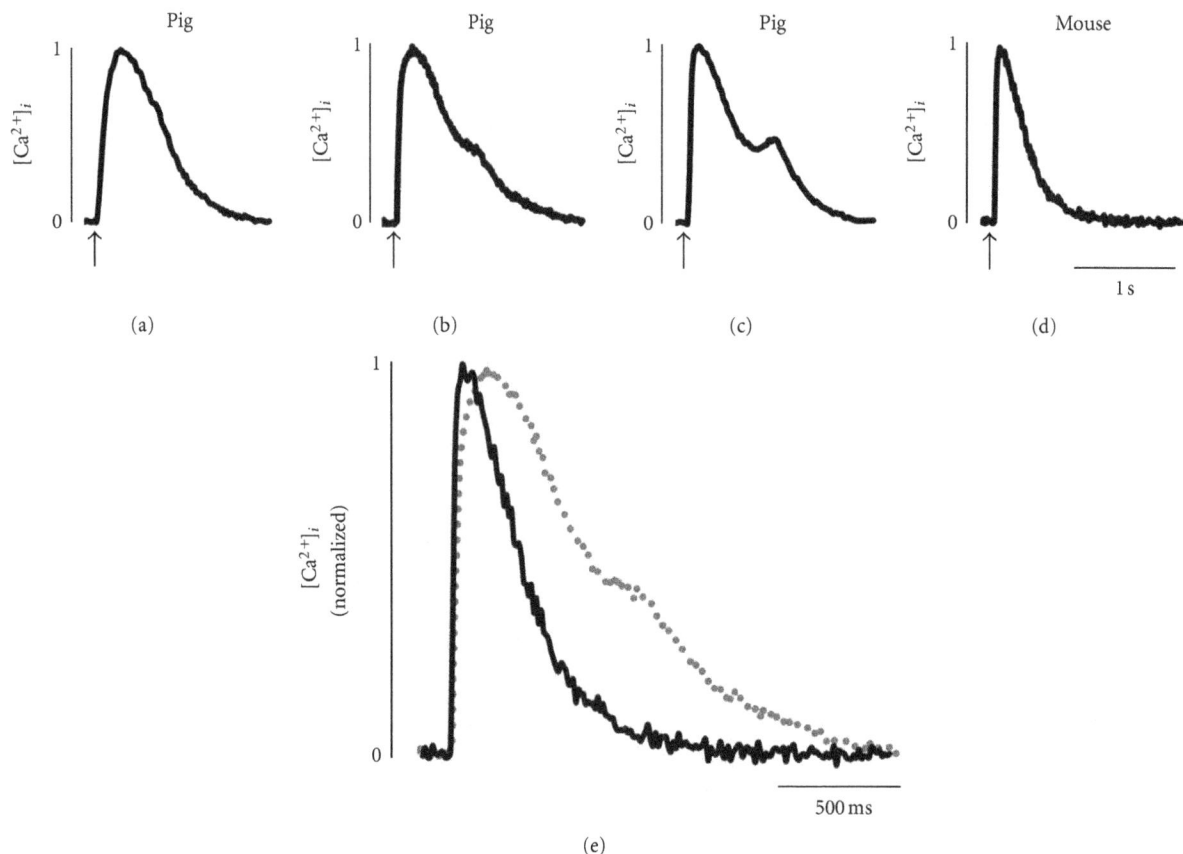

FIGURE 8: Representative amplitude-normalized calcium transients of Pig ((a)–(c)) and Mouse (d) left-ventricular myocytes (0.5 Hz, field stimulus denoted by arrow). Calcium transients from Pig exhibited multiple waveforms, including normal recovery from the transient peak (a, 2 of 14 cells), recovery with a marked shoulder ((b), 8 of 14 cells), and recovery with a secondary increase in calcium (c, 4 of 14 cells). (d) Mouse transients consistently exhibited a rapid transient recovery ($n = 40$). (e) Overlay of transients shown in (b) and (d) illustrating the distinct transient kinetics between pig (gray) and mouse (black) myocytes.

steeper length dependence of force), and (iii) speed relaxation (mediated, in part, by greater extent of shortening-induced cooperative deactivation manifested by more curved length traces). Interestingly, these myofibrillar changes in pig myocytes were all quantitatively less than those observed in rat cardiac myocytes, which is consistent with a slightly lower cardiac reserve that we have observed in pig hearts compared to rat and human hearts [42, 43].

Additional myofibrillar mechanical properties observed in pig myocytes were that at the same activator [Ca^{2+}] there was limited sarcomere length dependence of k_{tr} and force-velocity relationships. The sarcomere length dependence of k_{tr} was similar to that observed in rat slow-twitch skeletal muscle fibers in which k_{tr} was similar over sarcomere length range of ~2.30 to 1.90 um and then increased at shorter sarcomere lengths. Since force falls over this entire sarcomere length range, this indicates that sarcomere length overrides the force dependence of k_{tr} previously reported in cardiac muscle [33–35]. The mechanistic reasons for sarcomere length dominance of k_{tr} is unclear but may indicate that cooperative activation of thin filaments is progressively reduced at shorter sarcomere lengths perhaps by more compliant thin filaments (i.e., shorter persistence length, which

is the length that a mechanical force is transmitted along a functional entity), which would result in less recruitment of cross-bridges from the noncycling pool into the cycling pool, which has been proposed to limit rate of force development [35, 44]. Conversely, the lack of sarcomere length dependence of loaded shortening and power in pig myocytes differs from rat cardiac myocytes, where power decreased at short sarcomere length at the same activator [Ca^{2+}] [36]. This may arise due to the very slow actin-activated ATPase activity reported for porcine β-MyHC [37]. Slow cross-bridge detachment would increase the population of strongly bound cross-bridges, which are thought to shift the thin filament equilibrium towards the open state by direct interaction with the actin-tropomyosin interface [40] and, at least in cardiac muscle, by increased affinity of cTnC for Ca^{2+} [45, 46]. Interestingly, we observed a marked shoulder in Ca^{2+} transients from intact pig myocytes (Figure 8). This shoulder was not observed in mouse myocytes that contain α-MyHC, which has a relatively short duty cycle. Mechanistically, the Ca^{2+} transient shoulder may arise from delayed Ca^{2+} dissociation from cTnC due to prolonged strongly bound attachment state(s) inherent to the long duty cycle of β-MyHC cross-bridges expressed in pig cardiac myocytes.

Additionally, significantly delayed dissociation of Ca^{2+} from cTnC may elevate cytosolic Ca^{2+} at a time when ryanodine receptors have recovered from inactivation, yielding secondary Ca^{2+} induced-Ca^{2+} release from the sarcoplasmic reticulum. It does, however, appear that the long duty cycle of pig β-MyHC can be modulated by PKA mediated phosphorylation of myofibrillar proteins since PKA elevated power and yielded more curved length traces during load clamps (i.e., greater $k_{shortening}$ values). These PKA mediated effects appear qualitatively similar across species and are thus likely to be of important physiological significance. In terms of ventricular contraction, PKA tends to increase the cooperative activation of thin filaments [47–49] providing more force-generating cross-bridges to work against an afterload yielding faster loaded shortening and, thus, more ejection. In terms of relaxation, the greater curvature of length traces implies greater shortening-induced cooperative deactivation, which would assist in ventricular relaxation to allow more time for filling and, thus, preserve stroke volume in the midst of higher heart rates. This mechanical signaling paradigm provides a myofibrillar mechanism to optimize cardiac output in response to high peripheral demands associated with stress (e.g., exercise).

Acknowledgments

This work is supported by a National Heart, Lung, and Blood Institute grant (R01-HL-57852) to K. S. M., and American Heart Association (Heartland Affiliate) Postdoctoral Fellowship (0825725G) to L. M. H., an NIH grant (HL52490) to D. K. Bowles, an NIH grant (HL R01-HL086483) to S. S. Segal, and an NIH grant (HL09398-01) to C. A. E.

References

[1] A. Fabiato and F. Fabiato, "Dependence of the contractile activation of skinned cardiac cells on the sarcomere length," *Nature*, vol. 256, no. 5512, pp. 54–56, 1975.

[2] G. H. Rossmanith, J. F. Y. Hoh, A. Kirman, and L. J. Kwan, "Influence of V1 and V3 isomyosins on the mechanical behaviour of rat papillary muscle as studied by pseudo-random binary noise modulated length perturbations," *Journal of Muscle Research and Cell Motility*, vol. 7, no. 4, pp. 307–319, 1986.

[3] P. P. De Tombe and H. E. D. J. Ter Keurs, "Force and velocity of sarcomere shortening in trabeculae from rat heart. Effects of temperature," *Circulation Research*, vol. 66, no. 5, pp. 1239–1254, 1990.

[4] N. K. Sweitzer and R. L. Moss, "The effect of altered temperature on Ca^{2+}-sensitive force in permeabilized myocardium and skeletal muscle. Evidence for force dependence of thin filament activation," *Journal of General Physiology*, vol. 96, no. 6, pp. 1221–1245, 1990.

[5] M. Puceat, O. Clement, P. Lechene, J. M. Pelosin, R. Ventura-Clapier, and G. Vassort, "Neurohormonal control of calcium sensitivity of myofilaments in rat single heart cells," *Circulation Research*, vol. 67, no. 2, pp. 517–524, 1990.

[6] K. T. Strang, N. K. Sweitzer, M. L. Greaser, and R. L. Moss, "β-Adrenergic receptor stimulation increases unloaded shortening velocity of skinned single ventricular myocytes from rats," *Circulation Research*, vol. 74, no. 3, pp. 542–549, 1994.

[7] K. S. McDonald, M. R. Wolff, and R. L. Moss, "Force-velocity and power-load curves in rat skinned cardiac myocytes," *Journal of Physiology*, vol. 511, no. 2, pp. 519–531, 1998.

[8] D. P. Fitzsimons, J. R. Patel, and R. L. Moss, "Cross-bridge interaction kinetics in rat myocardium are accelerated by strong binding of myosin to the thin filament," *Journal of Physiology*, vol. 530, no. 2, pp. 263–272, 2001.

[9] O. Cazorla, Y. Wu, T. C. Irving, and H. Granzier, "Titin-based modulation of calcium sensitivity of active tension in mouse skinned cardiac myocytes," *Circulation Research*, vol. 88, no. 10, pp. 1028–1035, 2001.

[10] M. Regnier, H. Martin, R. J. Barsotti, A. J. Rivera, D. A. Martyn, and E. Clemmens, "Cross-bridge versus thin filament contributions to the level and rate of force development in cardiac muscle," *Biophysical Journal*, vol. 87, no. 3, pp. 1815–1824, 2004.

[11] P. P. Chen, J. R. Patel, I. N. Rybakova, J. W. Walker, and R. L. Moss, "Protein kinase A-induced myofilament desensitization to Ca^{2+} as a result of phosphorylation of cardiac myosin-binding protein C," *Journal of General Physiology*, vol. 136, no. 6, pp. 615–627, 2010.

[12] F. S. Korte, T. J. Herron, M. J. Rovetto, and K. S. McDonald, "Power output is linearly related to MyHC content in rat skinned myocytes and isolated working hearts," *American Journal of Physiology, Heart and Circulatory Physiology*, vol. 289, no. 2, pp. H801–H812, 2005.

[13] R. J. Belin, M. P. Sumandea, E. J. Allen et al., "Augmented protein kinase C-α-induced myofilament protein phosphorylation contributes to myofilament dysfunction in experimental congestive heart failure," *Circulation Research*, vol. 101, no. 2, pp. 195–204, 2007.

[14] G. Nowak, J. R. Peña, D. Urboniene, D. L. Geenen, R. J. Solaro, and B. M. Wolska, "Correlations between alterations in length-dependent Ca^{2+} activation of cardiac myofilaments and the end-systolic pressure-volume relation," *Journal of Muscle Research and Cell Motility*, vol. 28, no. 7-8, pp. 415–419, 2007.

[15] K. S. McDonald, P. P. A. Mammen, K. T. Strang, R. L. Moss, and W. P. Miller, "Isometric and dynamic contractile properties of porcine skinned cardiac myocytes after stunning," *Circulation Research*, vol. 77, no. 5, pp. 964–972, 1995.

[16] D. J. Duncker, N. M. Boontje, D. Merkus et al., "Prevention of myofilament dysfunction by β-blocker therapy in postinfarct remodeling," *Circulation: Heart Failure*, vol. 2, no. 3, pp. 233–242, 2009.

[17] J. E. Stelzer, H. S. Norman, P. P. Chen, J. R. Patel, and R. L. Moss, "Transmural variation in myosin heavy chain isoform expression modulates the timing of myocardial force generation in porcine left ventricle," *Journal of Physiology*, vol. 586, no. 21, pp. 5203–5214, 2008.

[18] C. A. Emter and C. P. Baines, "Low-intensity aerobic interval training attenuates pathological left ventricular remodeling and mitochondrial dysfunction in aortic-banded miniature swine," *American Journal of Physiology, Heart and Circulatory Physiology*, vol. 299, no. 5, pp. H1348–H1356, 2010.

[19] A. C. Hinken, F. S. Korte, and K. S. McDonald, "Porcine cardiac myocyte power output is increased after chronic exercise training," *Journal of Applied Physiology*, vol. 101, no. 1, pp. 40–46, 2006.

[20] A. Fabiato, "Computer programs for calculating total from specified free or free from specified total ionic concentrations in aqueous solutions containing multiple metals and ligands," *Methods in Enzymology*, vol. 157, no. C, pp. 378–417, 1988.

[21] R. L. Moss, "Sarcomere length-tension relations of frog skinned muscle fibres during calcium activation at short lengths," *Journal of Physiology*, vol. 292, pp. 177–192, 1979.

[22] L. M. Hanft and K. S. McDonald, "Length dependence of force generation exhibit similarities between rat cardiac myocytes and skeletal muscle fibres," *The Journal of Physiology*, vol. 588, no. 15, pp. 2891–2903, 2010.

[23] F. S. Korte, K. S. McDonald, S. P. Harris, and R. L. Moss, "Loaded shortening, power output, and rate of force redevelopment are increased with knockout of cardiac myosin binding protein-C," *Circulation Research*, vol. 93, no. 8, pp. 752–758, 2003.

[24] A. C. Hinken and K. S. McDonald, "Inorganic phosphate speeds loaded shortening in rat skinned cardiac myocytes," *American Journal of Physiology, Cell Physiology*, vol. 287, no. 2, pp. C500–C507, 2004.

[25] L. M. Hanft and K. S. McDonald, "Sarcomere length dependence of power output is increased after PKA treatment in rat cardiac myocytes," *American Journal of Physiology, Heart and Circulatory Physiology*, vol. 296, no. 5, pp. H1524–H1531, 2009.

[26] K. S. McDonald, "Ca^{2+} dependence of loaded shortening in rat skinned cardiac myocytes and skeletal muscle fibres," *Journal of Physiology*, vol. 525, no. 1, pp. 169–181, 2000.

[27] A. V. Hill, "The heat of shortening and the dynamic constants of muscle," *Proceedings of the Royal Society B*, vol. 126, pp. 136–195, 1938.

[28] R. C. Woledge, N. A. Curtin, and E. Homsher, *Energetic Aspects of Muscle Contraction*, Academic Press, London, UK, 1985.

[29] J. M. Metzger, P. A. Wahr, D. E. Michele, F. Albayya, and M. V. Westfall, "Effects of myosin heavy chain isoform switching on Ca^{2+}-activated tension development in single adult cardiac myocytes," *Circulation Research*, vol. 84, no. 11, pp. 1310–1317, 1999.

[30] T. J. Herron, F. S. Korte, and K. S. McDonald, "Power output is increased after phosphorylation of myofibrillar proteins in rat skinned cardiac myocytes," *Circulation Research*, vol. 89, no. 12, pp. 1184–1190, 2001.

[31] O. Cazorla, J. Y. Le Guennec, and E. White, "Length—Tension relationships of sub-epicardial and sub-endocardial single ventricular myocytes from rat and ferret hearts," *Journal of Molecular and Cellular Cardiology*, vol. 32, no. 5, pp. 735–744, 2000.

[32] I. F. Édes, D. Czuriga, G. Csányi et al., "Rate of tension redevelopment is not modulated by sarcomere length in permeabilized human, murine, and porcine cardiomyocytes," *American Journal of Physiology, Regulatory Integrative and Comparative Physiology*, vol. 293, no. 1, pp. R20–R29, 2007.

[33] M. R. Wolff, K. S. McDonald, and R. L. Moss, "Rate of tension development in cardiac muscle varies with level of activator calcium," *Circulation Research*, vol. 76, no. 1, pp. 154–160, 1995.

[34] C. Vannier, H. Chevassus, and G. Vassort, "Ca-dependence of isometric force kinetics in single skinned ventricular cardiomyocytes from rats," *Cardiovascular Research*, vol. 32, no. 3, pp. 580–586, 1996.

[35] D. P. Fitzsimons, J. R. Patel, K. S. Campbell, and R. L. Moss, "Cooperative mechanisms in the activation dependence of the rate of force development in rabbit skinned skeletal muscle fibers," *Journal of General Physiology*, vol. 117, no. 2, pp. 133–148, 2001.

[36] F. S. Korte and K. S. Mcdonald, "Sarcomere length dependence of rat skinned cardiac myocyte mechanical properties:

dependence on myosin heavy chain," *Journal of Physiology*, vol. 581, no. 2, pp. 725–739, 2007.

[37] J. S. Pereira, D. Pavlov, M. Nili, M. Greaser, E. Homsher, and R. L. Moss, "Kinetic differences in cardiac myosins with identical loop 1 sequences," *Journal of Biological Chemistry*, vol. 276, no. 6, pp. 4409–4415, 2001.

[38] D. F. A. McKillop and M. A. Geeves, "Regulation of the interaction between actin and myosin subfragment 1: evidence for three states of the thin filament," *Biophysical Journal*, vol. 65, no. 2, pp. 693–701, 1993.

[39] P. Vibert, R. Craig, and W. Lehman, "Steric-model for activation of muscle thin filaments," *Journal of Molecular Biology*, vol. 266, no. 1, pp. 8–14, 1997.

[40] W. Lehman, A. Galińska-Rakoczy, V. Hatch, L. S. Tobacman, and R. Craig, "Structural basis for the activation of muscle contraction by troponin and tropomyosin," *Journal of Molecular Biology*, vol. 388, no. 4, pp. 673–681, 2009.

[41] G. M. Diffee and E. Chung, "Altered single cell force-velocity and power properties in exercise-trained rat myocardium," *Journal of Applied Physiology*, vol. 94, no. 5, pp. 1941–1948, 2003.

[42] E. Plante, D. Lachance, M. C. Drolet, E. Roussel, J. Couet, and M. Arsenault, "Dobutamine stress echocardiography in healthy adult male rats," *Cardiovascular Ultrasound*, vol. 3, article no. 34, 2005.

[43] H. S. Norman, J. Oujiri, S. J. Larue, C. B. Chapman, K. B. Margulies, and N. K. Sweitzer, "Decreased cardiac functional reserve in heart failure with preserved systolic function," *Journal of Cardiac Failure*, vol. 17, no. 4, pp. 301–308, 2011.

[44] K. Campbell, "Rate constant of muscle force redevelopment reflects cooperative activation as well as cross-bridge kinetics," *Biophysical Journal*, vol. 72, no. 1, pp. 254–262, 1997.

[45] P. A. Hofmann and F. Fuchs, "Evidence for a force-dependent component of calcium binding to cardiac troponin C," *American Journal of Physiology*, vol. 253, pp. C541–C546, 1987.

[46] P. A. Hofmann and F. Fuchs, "Effect of length and cross-bridge attachment on Ca^{2+} binding to cardiac troponin C," *American Journal of Physiology*, vol. 253, no. 1, pp. C90–C96, 1987.

[47] J. Van Der Velden, J. W. De Jong, V. J. Owen, P. B. J. Burton, and G. J. M. Stienen, "Effect of protein kinase A on calcium sensitivity of force and its sarcomere length dependence in human cardiomyocytes," *Cardiovascular Research*, vol. 46, no. 3, pp. 487–495, 2000.

[48] J. P. Konhilas, T. C. Irving, B. M. Wolska et al., "Troponin I in the murine myocardium: influence on length-dependent activation and interfilament spacing," *Journal of Physiology*, vol. 547, no. 3, pp. 951–961, 2003.

[49] L. M. Hanft and K. S. McDonald, "Determinants of loaded shortening in rat cardiac myocytes 2010," *Biophysical Society Meetings*. In press.

21

Mitochondrial Roles and Cytoprotection in Chronic Liver Injury

Davide Degli Esposti,[1,2] **Jocelyne Hamelin,**[1,2] **Nelly Bosselut,**[1,2] **Raphaël Saffroy,**[1,2]
Mylène Sebagh,[3] **Alban Pommier,**[1,2] **Cécile Martel,**[2] **and Antoinette Lemoine**[1,2,4]

[1] AP-HP, Hôpital Paul Brousse, Service de Biochimie et Biologie Moléculaire, 14 Avenue Paul Vaillant Couturier,
  94804 Villejuif Cedex, France
[2] Inserm U1004, Université Paris 11, Institut André Lwoff, PRES Universud-Paris, Institut du Foie/Liver Institute,
  14 Avenue Paul Vaillant Couturier, 94804 Villejuif, France
[3] AP-HP, Inserm U785, Hôpital Paul Brousse, Service d'Anatomie Pathologique, 14 Avenue Paul Vaillant Couturier,
  94804 Villejuif Cedex, France
[4] Université Paris Sud 11, Faculté de Pharmacie, 5 rue Jean-Baptiste Clément, 92296 Châtenay-Malabry Cedex, France

Correspondence should be addressed to Antoinette Lemoine, antoinette.lemoine@pbr.aphp.fr

Academic Editor: Etienne Jacotot

The liver is one of the richest organs in terms of number and density of mitochondria. Most chronic liver diseases are associated with the accumulation of damaged mitochondria. Hepatic mitochondria have unique features compared to other organs' mitochondria, since they are the hub that integrates hepatic metabolism of carbohydrates, lipids and proteins. Mitochondria are also essential in hepatocyte survival as mediator of apoptosis and necrosis. Hepatocytes have developed different mechanisms to keep mitochondrial integrity or to prevent the effects of mitochondrial lesions, in particular regulating organelle biogenesis and degradation. In this paper, we will focus on the role of mitochondria in liver physiology, such as hepatic metabolism, reactive oxygen species homeostasis and cell survival. We will also focus on chronic liver pathologies, especially those linked to alcohol, virus, drugs or metabolic syndrome and we will discuss how mitochondria could provide a promising therapeutic target in these contexts.

1. Introduction

Mitochondria are intracellular double membrane-bound structures that provide energy (ATP) for intracellular metabolism. The intramitochondrial metabolism includes Krebs cycle and beta-oxidation. Mitochondria are also essential for assembly of iron sulfur clusters and regulation of calcium homeostasis. However, mitochondria are not only the cell's powerhouse, organelles whose particular architecture and biochemical composition enable the maximization of energy production by oxidative phosphorylation (OXPHOS), but they also have a second crucial function, namely, the control of cell death following activation of intracellular signaling cascades or death receptor-mediated pathways [1]. Indeed, the mitochondrial membrane permeabilization (MMP) is the decisive event that marks the transition from survival to death. Thus, mitochondrial

membranes integrate proapoptotic and antiapoptotic signals coming from microenvironment or from other intracellular organelles, such as endoplasmic reticulum or lysosomes, defining the ultimate cell fate [1, 2]. The number and functions of mitochondria can vary depending on age, sex, organ, and physiological or pathological conditions that are still unknown [3–5].

Mitochondrial dysfunctions are frequently described as early and initiating events in various chronic pathological conditions in different tissues and organs, such as liver, brain, or heart [6–8]. Most forms of chronic liver diseases are associated with the accumulation of damaged mitochondria responsible for abnormal reactive oxygen species (ROS) formation, glutathione (GSH) depletion, protein alkylation, and respiratory complex alterations. Depending on their nature and severity, the mitochondrial alterations may induce lipid accumulation, apoptosis, and/or necrosis leading to hepatic cytolysis and inflammation. These pathological events can correspond to different clinical features, such as lactacidosis, hypoglycemia, elevated serum transaminases, higher conjugated bilirubinemia, and hyperammonemia. However, a growing body of literature has also shown that demised cells with damaged mitochondria can develop cytoprotective mechanisms to ensure cellular energy homeostasis and limit cell death [9–12]. These mechanisms consist in both activation of intracellular pathways targeting mitochondria function and intercellular and interorgan signaling to coordinate adaptive metabolic responses within the organism as a whole. The regulation of the mitochondrial biogenesis and/or turnover (by general autophagy or specific mitochondria-targeted mitophagy) plays an important role in the balance of cell survival and cell death [13]. This balance is importantly linked to the energy metabolism homeostasis, in particular with ATP synthesis, as it has been reported in some chronic liver pathologies, such as steatosis and nonalcoholic steatohepatitis (NASH) [14].

In this paper, we will focus on the role of mitochondria in liver physiology and pathologies, especially those linked to alcohol, virus, drugs, or metabolic syndrome and we will discuss how mitochondria could provide a promising therapeutic target in these contexts.

2. Mitochondria in Liver Physiology

The liver is one of the richest organs in terms of number and density of mitochondria. The density of mitochondria is different in various tissues depending upon numerous factors, mostly the demands of oxidative phosphorylation. A study showed that in nontumorous liver tissue the copy number of mitochondrial DNA (mtDNA) in male patients affected by hepatocellular carcinoma (HCC) was lower than that of the female patients (5308 ± 484 versus 8027 ± 969, $P < 0.05$) [4]. Since each mitochondrion can host from two to ten copies of mtDNA [5], we can assume that in the liver, the number of mitochondria could range from 500 to 4000 per hepatocyte.

In this chapter, we will review the role of mitochondria in hepatic metabolism, reactive oxygen species (ROS) homeostasis, and cell death regulation.

2.1. Mitochondria Are Essential in the Hepatic Metabolism. The liver is an essential life organ in all mammals and plays a central role in the homeostasis of carbohydrate, lipid, and protein metabolism of the organism. The liver is a main target of insulin and glucagon signaling and contributes to balancing glucose blood levels by regulating glycogen synthesis and gluconeogenesis in hepatocytes [15]. It is also a key organ in maintaining lipid homeostasis: it is the main site of fatty acid oxidation together with the muscle (mainly β-oxidation taking place into the mitochondria) and it is the sole organ able to synthesize fatty acids by *de novo* lipogenesis [16]. Finally, the liver is a key regulator of protein metabolism for the entire organism as hepatocytes synthesize essential proteins such as albumin and lipoproteins and allow ammonia detoxification through the urea cycle [17].

In this context, the mitochondria provide the hub that integrates these pathways, serving as a critical site for the production and exchange of metabolic intermediates (Figure 1). It plays a critical role in orchestrating these complex metabolic networks in order to maintain proper homeostasis.

Mitochondria are largely involved in glucose metabolism, as the pyruvate dehydrogenase (PDH) complex is expressed in the mitochondrial matrix. It is composed by 5 subunits (pyruvate dehydrogenase, E1 alpha and E1 beta; dihydrolipoamide S-acetyl transferase, E2; and lipoamide dehydrogenase, E3 and E3BP). It catalyzes the conversion of pyruvate, the last metabolite of aerobic glycolysis, to Acetyl-CoA and CO_2. In the last ten years evidence has accumulated showing an important involvement of liver mitochondria in insulin resistance. In insulin resistant states, alterations in mitochondrial function, structure, and organization have been described [18]. In particular, a decrease in respiration and ATP production has been frequently described and the decreased efficiency is often attributed to excessive mitochondrial ROS production inducing respiratory chain protein oxidation [14, 18, 19].

Concerning lipid metabolism, few mitochondrial proteins play key roles in catabolism as well as in anabolism. The carnitine palmityl transferases I and II (CPT I-II) are expressed at the mitochondrial outer membrane (MOM) and mitochondrial inner membrane (MIM), respectively, and are essential for acyl-CoA transport and subsequent fatty acid β-oxidation in liver and muscle. A mitochondrial transport protein, the citrate transport protein (CTP), allows acetyl-CoA to be transported from mitochondria to the cytosol in the form of citrate in order to be used as building block in hepatic *de novo* lipogenesis [20]. Hepatic mitochondria are essential also in protein metabolism. Nitrogen enters the liver as free ammonia and amino acids, mostly glutamine and alanine [17]. Enzymes involved in ammonia detoxification and urea synthesis (glutamate dehydrogenase, carbamoyl phosphate synthetase I and ornithine transcarbamylase) are exclusively expressed in the hepatocyte mitochondria. Indeed, the first step in the urea cycle for ammonia detoxification and disposal is located at mitochondria and mediated by the enzyme carbamoyl phosphate synthetase 1 (CPSI). CPSI is allosterically regulated by cytosolic N-acetyl-L-glutamate (NAG) [21]. Ammonia can be also converted to glutamine by the glutamine synthetase (GS) catalyzing the condensation of glutamate and ammonia and, *vice versa*, ammonia can be generated by glutaminase. Therefore, an increase in blood ammonia depends on the activity of the enzyme glutamine synthetase, the glutamine/glutamate cycle, and the tissue capacity to eliminate toxic ammonia. Mitochondria represent a major site of glutamine metabolism, as both glutaminase and GS are mitochondrial processes in the liver. Interestingly, in absence of glucose but with high glutamine concentrations, mitochondrial structure and dynamics change towards a more condensed configuration and extended reticulum [22]. Moreover, urea and glutamine metabolism are differently distributed in the hepatic acinus. Ammonia is taken up by periportal hepatocytes, metabolized to urea via the urea cycle

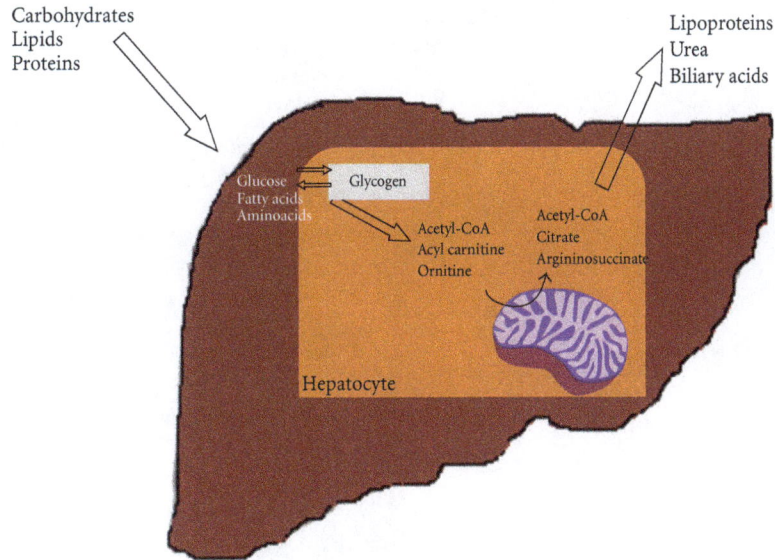

FIGURE 1: The role of hepatocyte mitochondria in liver metabolism. The liver is a central organ for the homeostasis of carbohydrates, lipids and proteins metabolism. In this context, hepatocyte mitochondria are essential in regulating the flux of metabolites in the cell in order to adjust energetic demand, ammonia detoxification, or anabolic pathways. Energy demand is met by complete oxidation of acetyl groups coming from glycolysis through tricarboxyilic acid cycle or of acyl groups coming from lipolysis through β-oxidation. Moreover, mitochondria are a unique site for metabolizing ammonia into the less toxic urea. Then, mitochondria provide shuttle proteins that allow specific addressing to anabolic pathways, as in the case of citrate transport protein (CTP) (see text for details).

and excreted through the kidneys. Any ammonia escaping detoxification is usually trapped by perivenous hepatocytes, where it is converted to glutamine via glutamine synthetase [23]. Indeed, urea synthesis enzymes and glutaminase are expressed in periportal hepatocytes, while glutamine synthetase is expressed in perivenous hepatocytes [17]. Then, the periportal region has a low affinity but a high capacity for ammonia detoxification. Hepatic GS allows ammonia scavenging, and when liver impairment is present, a diminished detoxification capacity is observed. GS has a short half-life and its activity is regulated and modulated by several mediators and hormones. The brain also uses glutamine synthesis for metabolizing ammonia and for deamination in the presynaptic terminals to produce glutamate, an important excitatory neurotransmitter. When it accumulates, it is taken up by the astrocytes and recycled back to glutamine, the "storage centre" for free ammonia [24, 25]. It is interesting to note that the different subcellular localization of GS (mitochondrial in hepatocytes and cytoplasmic in astrocytes) has been considered a partial explanation to the higher toxicity of ammonia in the brain than in the liver [26]. However, the exact role of mitochondrial dysfunctions in hyperammonemia still needs to be addressed, in particular for chronic liver disease. However, liver and mitochondria metabolism are directly involved in the homeostatic balance of brain ammonia, glutamine, and glutamate.

2.2. Mitochondria Are Essential in Reactive Oxygen Species Homeostasis. Mitochondria are the intracellular organelles devolved to energy (ATP) production in all eukaryotic cells through oxidative phosphorylation (OXPHOS). OXPHOS

is allowed by the four multiprotein complexes of the mitochondrial respiratory chain (MRC) and by the ATP synthase. OXPHOS physiologically produces reactive oxygen species (ROS) and *in vitro* estimations lead to considering that up to 2% of oxygen consumption results in superoxide anion generation [27]. Thus, mitochondria are a main source of ROS (Figure 2). ROS are produced during oxidative metabolism mainly by the complexes I, III, or IV of the electron transport chain, where electrons can prematurely reduce oxygen, resulting in the formation of superoxide radical [27–30]. In the normal state, most of the ROS generated by the MRC are detoxified by the mitochondrial antioxidant enzymes, such as SOD2/MnSOD, which convert superoxide to hydrogen peroxide, subsequently detoxified by GSH peroxidase. The remaining nondetoxified ROS diffuse out of mitochondria and serve as signaling molecules vital for normal cellular functions [31]. These physiological ROS are involved in specific cellular pathway aimed to adapt global metabolism to transient or chronic stress conditions. It is interesting to note that ATP synthase may also have a regulating role in ROS production. Actually, in the experimental model of aging provided by the fungus Podospora anserine, characterized by mitochondrial etiology of aging, the alpha subunit of ATP synthase functions as a sensor of oxidative stress and provides an intramolecular quencher (at the residue Trp503) for ROS [32]. Moreover, a recent mechanism that seems to buffer ROS excess has been described in physiological and pathological conditions. The expression of uncoupling proteins (UPCs) promotes a controlled uncoupling of proton flux from the ATP synthase and could lead to decreased ROS production [33].

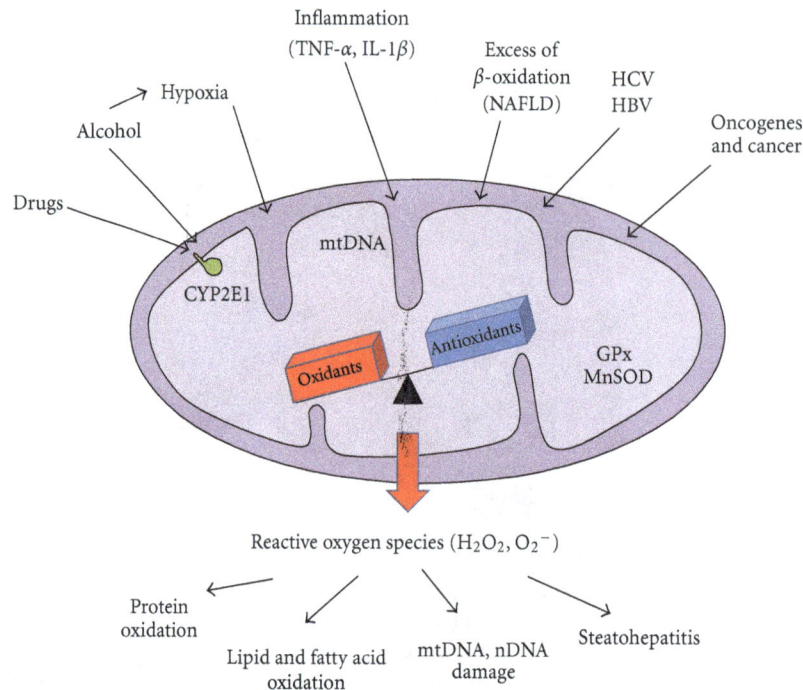

FIGURE 2: The role of hepatocyte mitochondria in reactive oxygen species homeostasis. Mitochondria are a physiological source of reactive oxygen species (ROS). In this context, ROS exert a signaling role in cell proliferation and differentiation. However, different types of stress can target directly or indirectly hepatocyte mitochondria, such as drugs, virus, hypoxia, inflammatory cytokines, excess of β-oxidation, ectopic expression of cytochromes P450. In this case, overproduction of ROS may damage both mitochondrial and other cellular components, such as OXPHOS protein subunits, lipid membranes, mitochondrial, or nuclear DNA. These cellular lesions can favor the development of tissue lesions, such as steatohepatitis or hepatocellular carcinoma.

2.3. Mitochondria Are Essential in Cell Survival. Mitochondria are the essential actor in keeping the balance between cell survival and cell death, in particular in hepatocytes, where they trigger the intrinsic pathway of apoptosis and are also involved in necrotic cell death. The regulation of membrane permeability is the main mechanism that makes the cells shift from survival to cell death. The MOM is permeable to solutes of molecular mass (MM) $\approx$ 6 kDa due to the presence of channels, such as the voltage-dependent anion channel (VDAC), which belongs to the porin subfamily. However, with an estimated pore diameter about 2.6–3 nm, VDAC would not allow the passage of a folded protein like cytochrome c. In contrast, the MIM is almost totally impermeable and transport of ions and solutes is granted by mitochondrial carrier proteins. Most mitochondrial proteins exhibit dual functions, a vital metabolic function, and a lethal pro-apoptotic function. This applies to various channels: voltage-dependent anion channel (VDAC), adenine nucleotide translocase (ANT), Bax, t-Bid, Bak; receptors (e.g., TOM22); chaperones (cyclophilin D, CypD), as well as oxidoreductases (apoptosis-inducing factor, AIF).

During apoptosis, many signals can converge to the mitochondrion to MMP, the rate-limiting step in the execution of the death process [1]. MMP is regulated mainly by the members of Bcl-2 family, members of the PTP complex (VDAC, ANT, CypD) and lipids [1]. Bcl-2 family is composed of pro-apoptotic proteins (e.g., Bax, Bak, Bid, Bik, Bnip3) and anti-apoptotic members (Bcl-2, Bcl-x$_L$,

Bcl-w, Mcl-1, A1). Pro-apoptotic proteins favor MMP by translocating to MOM and forming mega channels, mainly by oligomerization (e.g., Bax-Bak oligomers or Bax-VDAC complexes), while anti-apoptotic members stabilize MOM and tend to prevent MMP [1, 34–36]. Accumulation of modified lipids (e.g., oxidized cardiolipin, ceramide) and ions (e.g., Ca^{2+}) in the mitochondrion can also influence MMP [37]. Moreover, the intracellular milieu, such as pH, ROS, and ATP levels can contribute to define a permissive environment for MMP execution [1]. Multiple mechanisms can mediate MMP, depending on the cell type and the death stimuli. They can affect either the MOM, or both mitochondrial membranes (MOM+MIM). In the MOM model, intermembrane space proteins are released into the cytosol by passage through large proteic/lipidic channels while, in the MOM+MIM model, intermembrane space proteins are freely released into the cytosol through the MOM ruptures. Nevertheless, these two models can coexist in conditions involving on the one hand the translocation of the truncated form of Bid (tBid) to mitochondria, and in the other hand mitochondrial Ca^{2+} accumulation and ROS increase, as observed in conditions of endoplasmic reticulum stress [1]. In the MOM+MIM model, the contribution of the permeability transition pore (PTP) seems to play an important role. The PTP consists of a multiprotein complex (PTPC) and various proteins are involved in its opening. Long lasting opening of PTPC increases MIM permeability and, in the presence of adequate amounts of ATP, would lead to apoptotic cell death [1]. PTPC opening is highly sensitive to Ca^{2+}, prooxidant agents, pro-apoptotic Bcl-2 family members and some chemotherapeutics agents [38]. However, Ca^{2+}-induced PTP opening has been also reported to induce necrotic cell death, in particular when intracellular ATP levels are too low to allow apoptosis execution [39].

Once initiated, MMP leads to the release into the cytosol of caspase-dependent proteins (i.e., cytochrome c or Smac/DIABLO) and caspase-independent proteins (such as apoptosis-inducing factor, AIF, or EndoG) with consequent coordinated cell degradation [40]. Concomitantly, MMP provokes a mitochondrial failure with dissipation of the inner membrane potential ($\Delta\Psi m$), subsequent arrest of OXPHOS and ATP synthesis, and increased ROS level. Therefore, MMP constitutes a point of no return of the activation cascade of cell death [41].

3. Mitochondria in Liver Pathology

Most liver pathologies, including alcoholic liver disease, nonalcoholic fatty liver disease (NAFLD) and nonalcoholic steatohepatitis (NASH), drug-induced hepatotoxicity, viral hepatitis, and HCC, are characterized by mitochondrial dysfunctions. Moreover, during liver surgery, liver cells, in particular hepatocytes and endothelial cells suffer ischemia/reperfusion (I/R) injury. In the liver, as well as in other organs such as brain and heart, I/R injury involved mitochondrial permeability transition [1]. Since these abnormalities affect all the aforementioned physiological functions of mitochondria, we will review their roles in liver pathologies with a particular focus on the aspects of cell death regulation, alteration of hepatocyte metabolism, and disruption of ROS homeostasis.

3.1. Mitochondria in Cell Death Regulation. Mitochondria are key organelles in the development of liver diseases characterized by hepatocyte death and subsequent inflammation (Figure 3). Actually, increased hepatocyte apoptosis has been correlated with inflammation, fibrosis, and cell turnover, conditions that are permissive for the development of HCC [2]. Hepatocyte mitochondria are essential in making effective the extrinsic pathway activated by many ligands, such as Fas, TRAIL or TNF-α [2]. Moreover, constitutive expression of both anti-apoptotic proteins Bcl-x_L and Mcl-1, belonging to the Bcl-2 family, is required to avoid spontaneous caspase 3/7 activation, suggesting essential cytoprotective functions of these proteins in the hepatocyte [42, 43]. Bcl-2 is not constitutively expressed in the liver; however, it can be induced in order to cope with I/R, as shown in ischemic preconditioning during partial hepatectomy [44, 45].

Fas- and TRAIL-mediated apoptosis are involved in viral hepatitis, playing a crucial role in the elimination of infected cells and the hepatitis viral core protein binds Mcl-1 impairing its cytoprotective function [46–48]. TNF-α is secreted by infiltrating cytotoxic T lymphocytes during HBV infection and its apoptotic effect seems to be mediated by HBVx protein [2]. Mitochondrial apoptosis is also involved in the pathogenesis of NAFLD and in NASH [15]. In an

experimental model using mice fed with a methionine and choline deficient diet, apoptosis was induced by an increase hepatic expression of functional p53, with a concomitant increase in the cleavage of Bid to tBid and a decrease expression of Bcl-x_L [49]. Moreover, p53 was also responsible for TRAIL receptor expression, linking intrinsic and extrinsic apoptosis pathway in NASH [49]. Recently, saturated free fatty acids have been shown to activate the proapoptotic proteins Bim and Bax via JNK, thus inducing MMP, and also increase ROS production [50].

Hepatocyte necrosis is usually considered an accidental (nonprogrammed) form of cell death, resulting from metabolic failure and consequent rapid ATP depletion [51]. It has been firstly described during I/R injury following liver transplantation or hepatectomy, but it is also described in NASH. In fact, hepatocytes necrosis is associated with significant inflammatory response, due to the liberation of IL1-β, TNF-α and other newly described proinflammatory proteins, namely, damage-associated molecular-pattern (DAMP) molecules, such as HMGB1, that activate innate immunity response, such liver resident macrophages (Kupffer cells) and polymorphonuclear cells [45, 51–54]. Recently, accumulated evidence indicates that necrosis can also occur in a regulated manner and that the liberation of cytokines from dying cells can function as a sentinel signal alerting to the need for defensive response [51]. This regulated or programmed necrosis (necroptosis) is initiated by death receptors, like apoptosis, but requires activation of specific kinases (receptor interacting proteins 1 and 3) and its execution involves the active disintegration of mitochondrial, lysosomal, and plasma membranes [55]. Interestingly, in the context of I/R injury, the PTPC opening seems to be a common event anticipating both necrotic cell death and apoptosis, reinforcing the idea that programmed necrosis may be involved in clinical and pathological contexts. In an experimental model of orthotopic liver transplantation in rats, inhibition of PTP by cyclosporine A or acidic pH improved mitochondrial and hepatocellular functions, in particular decreasing the percentage of apoptotic cells but not of necrotic cells [56, 57]. These results seem to confirm the concept that apoptosis is typically an early event in hepatocyte injury. Importantly, in steatotic livers submitted to ischemia/reperfusion, necrosis is predominant compared with normal liver in which apoptosis is the main form of cell death [52, 53]. This difference has been partially linked to the metabolic/energetic difference between steatotic livers and normal liver [52] since fatty liver mitochondria have a decreased content of cytochromes c oxidase, produce superoxide anion and H_2O_2 at increased rate and have an increase content in UPC-2 compared with normal livers, resulting in decreased ATP production that affects apoptosis execution, and favors necrosis [58].

3.2. Mitochondria in Alteration of Hepatocyte Metabolism. The aforementioned data suggest that mitochondria may be the convergence point between various metabolic stresses and cell death in hepatocyte. In this context, it merits noting that the cytosolic glucokinase, or hexokinase IV, the

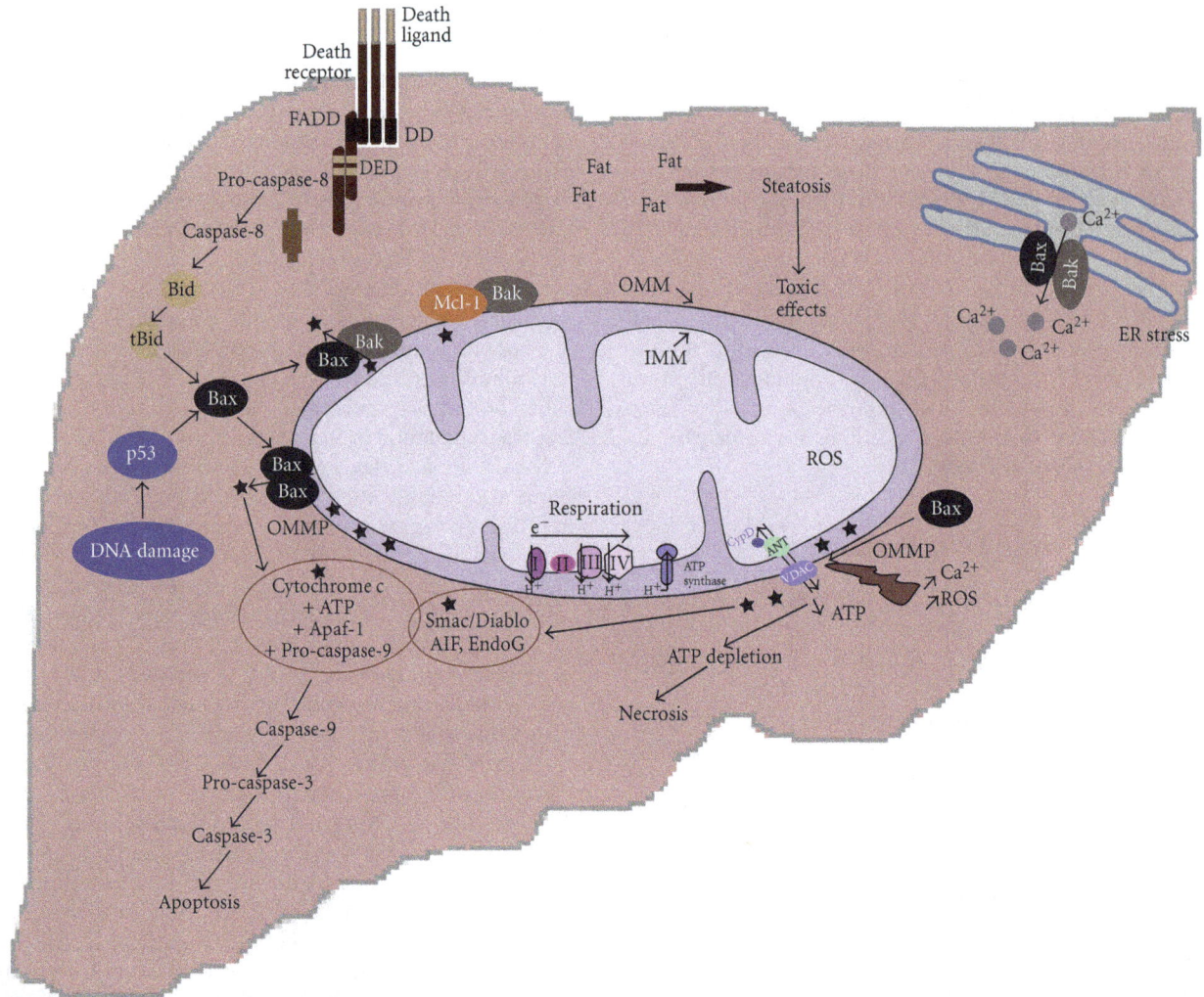

FIGURE 3: The mitochondria are central organelles in determining cell fate in liver diseases. Hepatocyte cell death is common to many liver diseases. Different stress stimuli can induce death signaling, such as toxic free fatty acids, DNA damage, endoplasmic reticulum (ER) stress observed in metabolic disease. In these contexts, mitochondria are essential to determine cell fate, as in hepatocyte the activation of the intrinsic pathway of apoptosis by cell death receptors is not usually sufficient to induce cell death and liberation of proapoptotic factors from mitochondria is a mostly necessary event. Moreover, previous alterations of mitochondrial function causing decreased ATP synthesis can induce a shift from apoptotic to necrotic cell death.

hepatic/pancreatic isoform of hexokinase, has been recently reported to be associated to mitochondrial proteins, such as Bad, at the MOM [59, 60]. The association of the proapoptotic protein Bad with the glucokinase suggests that a close integration exists between the pathways of glucose metabolism and apoptosis [59].

Many studies on obese, diabetic, or NASH patients have shown functional and structural abnormalities in hepatocyte mitochondria, such as OXPHOS impairment or megamitochondria [61]. Interestingly, both increased or decreased β-oxidation in insulin resistant hepatocytes has been reported as characteristic of liver steatosis and insulin resistance [16, 62, 63]. Decrease in β-oxidation activity induces diacylglycerol (DAG) accumulation and steatosis in the hepatocyte with concurrent activation of PKC pathway and inhibition of insulin signaling [62]. In insulin-resistant

patients an increased activity of hepatic β-oxidation was observed and this was correlated to an increase in ROS production [61, 64]. Elevated β-oxidation could be an adaptive mechanism to limit free fatty acid lipotoxicity, thus providing large amounts of reduced equivalents (NADH) regardless of energetic requirements finally promoting ROS production due to impairment of respiratory chain [18]. These results linked mitochondrial metabolic dysfunctions to oxidative stress due to increased ROS production.

3.3. Mitochondria in Disruption of ROS Homeostasis. Increased ROS production has been described in most liver pathologies. Augmented generation of mitochondrial ROS has been reported in various animal models of NASH, namely, genetically obese-diabetic ob/ob mice [58] and rats fed with a choline-deficient diet [65]. Moreover, mitochondria can be an ectopic site of cytochromes P450 2E1 expression [61, 66], where it can produce ROS and induce lipid peroxidation, as shown in the liver of an experimental model of diabetic rat [67].

Mitochondrial dysfunctions and ROS generation have been clearly shown in alcoholic liver disease [68]. Excessive ethanol consumption perturbs sinusoidal blood flow, inducing ischemia regions, and causes increased production of TNF-α, which induces inflammatory cell infiltration and an increase in hepatic O_2 consumption [69, 70]. Chronic ethanol consumption induces profound disruption in mitochondrial metabolism, in particular decreasing the rate of ATP synthesis, thus placing hepatocytes under bioenergetic stress [68]. Under alcohol feeding, mitochondria contribute to the production of ROS in hepatocytes through various mechanisms. Ethanol metabolism increases the availability of NADH, resulting in a more reduced state of complexes I and III of the respiratory chain with a consequent increased probability of superoxide ion production [71]. Moreover, chronic alcohol consumption decreases mitochondrial protein synthesis mainly due to alcohol-mediated damage to mtDNA, contributing to decreased functioning of the oxidative phosphorylation system [72–74].

Mitochondrial ROS also play an important role in viral hepatitis. HCV core protein directly interacts with mitochondria and inhibits complex I activity, inducing an increased production of mitochondrial ROS, reducing threshold for Ca^{2+} and ROS-induced MMP [75]. Moreover, it has been recently shown that during HBV infection, HBx protein interacts with mitochondria, increasing ROS production [76]. The increase in ROS production was necessary, although insufficient, to induce the proinflammatory enzyme cyclooxygenase 2 (COX-2), linking mitochondrial dysfunction with liver inflammation in HBV infection [76]. Numerous investigations have shown that mitochondrial dysfunction is a major mechanism of drug- (or drug-metabolite-) induced liver injury [77]. Different mechanisms of mitochondrial dysfunction have been described in drug-induced hepatotoxicity, including membrane permeabilization, OXPHOS impairment, inhibition of fatty acid oxidation, and mtDNA depletion, and it appears that overproduction of reactive oxygen species by the damaged mitochondria could play a major role [77]. Finally, there is evidence showing a role of ROS in hepatocarcinogenesis [78]. Chemical hepatocarcinogens, such as the mycotoxin aflatoxin B1 and 2-acetylaminofluorene (2-AAF), induced increased ROS production in hepatocytes. In particular, 2-AAF altered mitochondrial redox cycling and it activated NADPH oxidase, an important ROS producing enzyme, through PI3K/Akt pathway [79–81]. Growth factors and activated oncogenes also induce ROS overproduction. Cultured cells treated with epidermal growth factor (EGF) and platelet-derived growth factor (PDGF) showed increased levels of H_2O_2 [82, 83]. Double transgenic mice bearing liver-targeted expression of transforming growth factor and the oncogene *c-myc* develop HCC as early as 4 and 8 months of age and elevated ROS levels associated with lipid peroxidation, mitochondrial damage and

decreased GSH were already observed at 2-3 months of age [84].

Thus, it is clear that even a mild dysfunction of mitochondria in the liver could lead to hepatic and systemic pathological conditions and the identification of type and timing of mitochondrial lesions could allow major advancement in prevention, early diagnosis and treatment of systemic and liver diseases.

4. Mitochondria in the Cytoprotection of Damaged Liver Cells to Ensure Homeostasis in Chronic Liver Diseases

Mitochondrial dysfunction is described in various hepatic diseases or lesions, such as NAFLD, I/R injury, drug toxicity or hepatocellular carcinoma, and it is often detected as an early alteration, suggesting its causative effect [6, 85–87]. Cells have developed different mechanisms to keep mitochondrial integrity or to prevent the effects of mitochondrial lesions, such as disposal of damaged mitochondria by autophagy/mitophagy, increased biogenesis of mitochondria or regulation of signaling pathways to ensure energy metabolism and limit cell death and inflammatory response.

4.1. Increased Biogenesis of Mitochondria. Regulation of mitochondria biogenesis is one of the mechanisms developed by cells to keep mitochondrial integrity or to prevent the effects of mitochondrial lesions. The peroxisome proliferator-activated receptor gamma coactivator-1 alpha (PGC-1 alpha) belongs to the family of PGC-1 transcriptional coactivators (PGC-1 alpha, PGC-1 beta and PRC), which have been shown to be master regulators of mitochondrial biogenesis, and cellular energy metabolism in many organs, including liver [88, 89]. PGC-1 alpha is present at low but inducible levels in the liver where it also regulates most of the metabolic pathways, including gluconeogenesis, fatty acid β-oxidation, ketogenesis and heme biosynthesis (Figure 4) [90–93]. Under stress conditions, such as low temperature, fasting or energy deprivation, PGC-1 alpha is activated both transcriptionally by cAMP response element binding protein (CREB) and post-traductionally by AMP-activated-protein-kinase- (AMPK-) induced phosphorylation and SIRT1-mediated deacetylation [89]. Following PGC-1 alpha activation, different nuclear factors are subsequently activated. In particular, an activation of the nuclear respiratory factors 1 and 2 (NRF-1 and NRF-2) is observed and is followed by increased expression of multiple mitochondrial proteins. Moreover, PGC-1 alpha activates the nuclear receptors peroxisome proliferator-activated receptor alpha (PPAR alpha) and the estrogen-related receptor alpha (ERRalpha) both promoting the transcription of genes involved in β-oxidation, such as medium chain acyl-CoA dehydrogenase and carnitine palmitoyltransferase-1A (CPT-1A) [94, 95]. The absence of adequate levels of PGC-1 alpha is correlated with mice developing fasting hypoglycemia and hepatic steatosis, while mouse models of type 1 and type 2 diabetes showed high hepatic levels of PGC-1 alpha [90]. However, it has been recently shown that the different tissue-specific

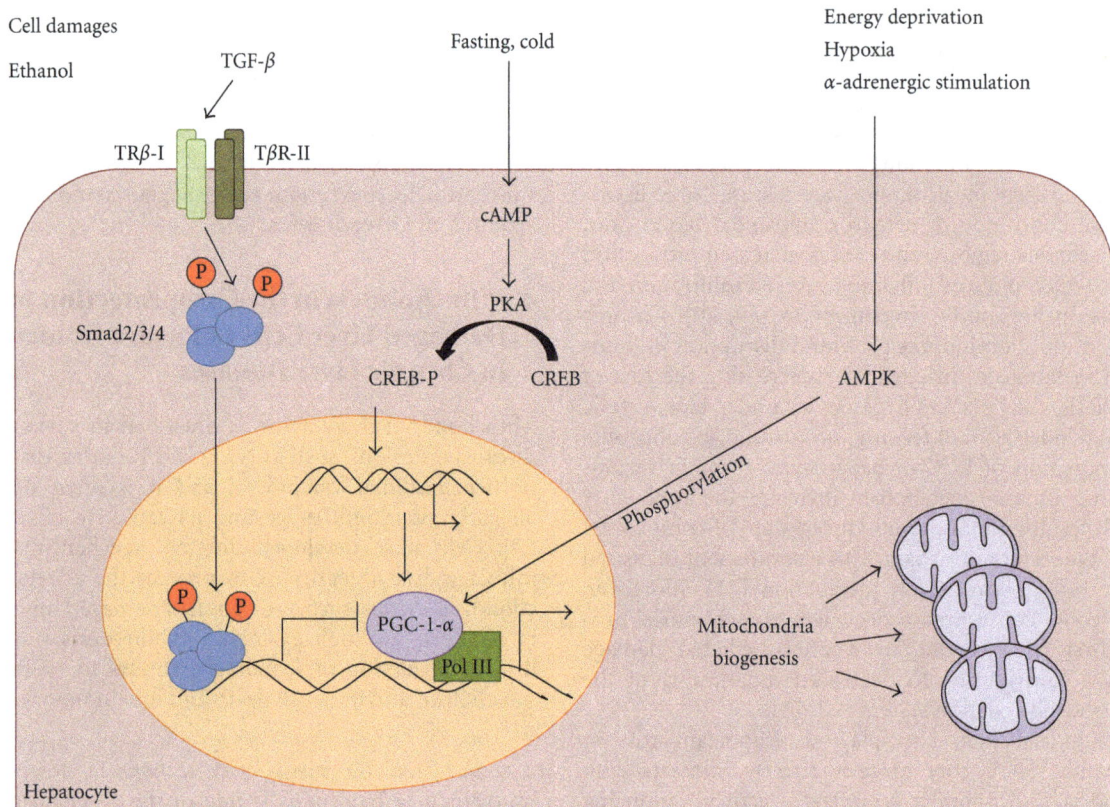

FIGURE 4: Mitochondria biogenesis allows tissue adaption under stress. Mitochondria biogenesis has been recently recognized as a central pathway in the adaptation of stress conditions in the liver, such as fasting, energy deprivation, hypoxia, or alcohol consumption. Different signaling pathways converge on the master regulator of mitochondria biogenesis, PGC1-alpha. In particular, AMPK and PKA signaling may activate gene transcription controlled by PGC1-alpha, while the TGF-β has been shown to inhibit PGC1-alpha-induced gene transcription.

functions of PGC-1 alpha are tightly and independently regulated [96]. In particular, S6 kinase-1 (S6K1), activated in the liver upon feeding, can phosphorylate PGC-1 alpha, decreasing its capacity to turn on genes of gluconeogenesis, while keeping the functions of activator of mitochondrial and fatty acid oxidation genes intact [96]. S6K1, liver kinase B1 and AMPK are key kinases in the regulation of energy metabolism in the liver. Actually, AMPK is emerging as a kinase that links energy metabolism to mitochondrial function and biogenesis since components downstream of AMPK may contribute to stabilize mitochondrial membrane potential for hepatocyte survival, strengthening the relationship between fuel metabolism and cell survival [10]. Actually, in the liver, the activation of AMPK has been shown to decrease gluconeogenesis and fatty acid synthesis, to increase fatty acid oxidation and mitochondrial biogenesis and this could be linked to PGC-1 alpha phosphorylation, as previously observed in skeletal muscle [97, 98]. Interestingly, the hepatitis B virus (HBV) uses the transcriptional machinery involved in the hepatic response to fasting for its own amplification, thus HBV life cycle is under the control of PGC-1 alpha that could be a new target for antiviral therapy [99]. The dynamic changes in mitochondrial morphology, connectivity, and subcellular distribution are also major mechanisms in cellular homeostasis. They are

critically dependent on a highly regulated fusion and fission machinery. Mitochondrial function, dynamics, and quality control are vital for the maintenance of tissue integrity [100]. In the liver, it has been shown that specific protection against hepatocyte mitochondrial dysfunction plays a preventive role in early stages of fibrogenesis, delaying, but not avoiding, its onset [101]. In this context, it is interesting to note that TGF-β/Smad3 signaling pathway, known to be implicated in liver fibrogenesis, has been shown to regulate glucose and energy homeostasis. Smad3-deficient mice are protected from diet-induced obesity and diabetes and Smad3 acts as a repressor of PGC-1α expression, thus suggesting a link between failure in mitochondrial biogenesis, metabolic syndrome, and liver fibrosis [102].

4.2. Autophagy and Mitophagy as Mechanisms to Limit Mitochondrial Lesions. Autophagy is a cellular pathway by which cytoplasmic materials, including organelles, reach lysosomes for degradation. Autophagy may occur either as a general phenomenon, for instance, during nutrient deprivation, or it can specifically target distinct cellular structures, such as damaged mitochondria (mitophagy) [13]. An important interplay exists between induction of autophagy and mitochondria. Actually mitochondria seem to have a key role in general autophagy as they may supply membranes for the biogenesis of autophagosomes during starvation [103]. Moreover, low ATP production or enhanced ROS generation by mitochondria induces general autophagy [104, 105]. The selective removal of mitochondria by mitophagy regulates mitochondrial number to match metabolic demand and is considered a form of quality control to remove damaged mitochondria [106]. Induction of general autophagy by a sublethal stress before a lethal stress can protect cells against cell death [13]. Indeed, we showed that ischemic preconditioning of livers previously treated by chemotherapy or steatotic livers induced autophagy and decreased necrosis without altering apoptosis. [45, 54]. The elimination of damaged mitochondria has been correlated to resistance of residual mitochondria to MMP and opening of PTP, two early events of apoptotic/necrotic cell death. This can be explained either by the removal of mitochondria that have a low threshold for permeabilization or by the fact that MMP or PTP opening occurs in a fraction of mitochondria and may activate autophagic disposal of depolarized mitochondria [106, 107]. Different mechanisms may regulate mitophagy. The dual system PINK1-Parkin is well described especially in neural tissues. The stabilization of the kinase PINK1 occurs at the surface of mitochondria with low $\Delta\Psi_{mito}$ with the subsequent recruitment of the ubiquitin ligase Parkin and ubiquitinylation of outer membrane proteins [108]. Mitophagy can be also stimulated by histone deacetylase 6, which is recruited to mitochondria and catalyzes proautophagic cytoplasmic deacetylation reactions [109].

Interestingly, although hepatic PINK1-expression is described [110], to our knowledge, no reports on PINK1 dependent mitophagy in the liver are published. Another mechanism of mitophagy involves the activation of AMPK [111]. AMPK phosphorylates and activates ULK1, one of the initiators of autophagy and the genetic loss of AMPK or ULK1 results in defective mitophagy in mammalian liver and *C. elegans*. These findings showed a conserved mechanism coupling nutrient status with autophagy and cell survival [111]. Interestingly, mitochondrial degradation by autophagy was also described in the liver of GFP-LC3 transgenic mice following nutrient deprivation, reinforcing the results linking AMPK regulation of mitophagy [112].

4.3. Mitochondria Can Integrate Energy, Nutrient Metabolism, and Oxidative Stress Responses Determining Cell Fate. Insulin, secreted by pancreatic beta cells upon nutrient stimulation, is one of the most important regulators of nutrient utilization and metabolic homeostasis in the liver. Insulin resistance, a hallmark of NASH and more generally of metabolic syndrome and type II diabetes, is accompanied by reduction of mitochondrial OXPHOS activity and increased ROS production [113]. On the other hand, ROS produced during mitochondrial OXPHOS promote insulin signaling through oxidation of insulin receptor and inhibition of phosphatases, such as PTP1B and PTEN [113]. Importantly, recent investigations pointed out a tight molecular crosstalk between cell survival or cell death pathways and energy metabolism. Using *ex vivo* multinuclear NMR-spectroscopy

to study metabolic pathways of [U-(13)C] glucose in mouse liver during Fas-induced apoptosis, Gottschalk et al. found early upregulations in glucose metabolic pathways occurred prior to any visible signs of apoptosis, accompanied by an increased mitochondrial energy production and cellular glutathione synthesis [114]. This metabolic shift seems to potentially contribute to the initiation of apoptosis by mitochondrial energy production and cellular glutathione stores, thus orienting cell fate towards a less pro-inflammatory death. A biochemical analysis using liver mitochondria of two strains of mice (A/J and C57Bl/6, respectively, resistant and susceptible to high-fat diet-induced hepatosteatosis) confirmed a rapid increase by high-fat diet feeding of the respiration rate in A/J but not C57Bl/6 mice. Importantly, ATP production was the same in both types of mitochondria, indicating increased uncoupling of the A/J mitochondria [115]. These results suggest that livers can adapt to high-fat diet feeding by increasing the activity of the oxidative phosphorylation chain and its uncoupling to dissipate the excess of incoming metabolic energy and to reduce the production of ROS [115].

As we mentioned above, liver mitochondria are essential in ammonia detoxification following protein catabolism. In recent years, studies from several laboratories have uncovered a number of factors and pathways that appear to be critically involved in the pathogenesis of hepatic encephalopathy. Foremost is oxidative and nitrosative stress (ONS) and the MMP playing major roles in the mechanism of ammonia-induced astrocyte swelling [116]. The accumulation of intramitochondrial glutamine has been involved. Norenberg et al. [117] were first to describe that the newly synthesized glutamine could be toxic when subsequently metabolized in mitochondria by phosphate-activated glutaminase, yielding glutamate and ammonia. Thus, glutamine can be considered as a carrier of ammonia. The authors propose to consider the intramitochondrial glutamine as a Trojan horse that interferes with mitochondrial function giving rise to excessive production of free radicals and induction of the MPT, two phenomena known to bring about astrocyte dysfunction, including cell swelling.

Moreover, an ammonia-induced increase in intracellular Ca^{2+} has been described which activates a number of enzymes promoting the synthesis of reactive oxygen-nitrogen species, including constitutive nitric oxide synthase, NADPH oxidase and phospholipase A2. ONS subsequently induces the opening of PTP and activates mitogen-activated protein (MAP) kinases and the transcription factor nuclear factor-kappaB (NF-κB). These factors act to generate additional reactive oxygen-nitrogen species, to phosphorylate various proteins and transcription factors, and to cause mitochondrial dysfunction [26]. The pathways and factors described above provide attractive targets for identifying agents potentially useful in the therapy of HE and other hyperammonemic disorders. The most promising of them is the glutamate/glutamine cycle. Indeed, in hyperoxia, glutamine has been described to protect cellular structures, especially mitochondria, from damage. This has been attributed to the activity of the tricarboxylic acid cycle enzyme alpha-ketoglutarate dehydrogenase that was partially protected by its indirect substrate, glutamine, indicating a mechanism of mitochondrial protection [118]. Glutamate dehydrogenase (GDH), a mitochondrial enzyme linking the Krebs cycle to the multifunctional amino acid glutamate could be also an interesting target. Indeed, GDH controls production and consumption of glutamate. GDH activity is under the control of several regulators, conferring to this enzyme energy-sensor property. Indeed, GDH directly depends on the provision of the cofactor NADH/NAD(+), rendering the enzyme sensitive to the redox status of the cell. Moreover, GDH is allosterically regulated by GTP and ADP. GDH is also regulated by ADP-ribosylation, mediated by a member of the energy-sensor family sirtuins, namely, SIRT4. In the brain, GDH ensures the cycling of the neurotransmitter glutamate between neurons and astrocytes. GDH also controls ammonia metabolism and detoxification, mainly in the liver and kidney. Eng and Abraham [119] have described that ammonia, generated from Gln deamination (glutaminolysis) in mitochondria, functions as an autocrine- and/or paracrine-acting stimulator of autophagic flux. Recently, Nissim et al. [120] reported a downregulation of hepatic urea synthesis by oxypurines. Indeed, xanthine and uric acid, both physiologically occurring oxypurines, inhibited the hepatic synthesis of N-acetylglutamate, the key regulator of the first step of mitochondrial urea cycle.

As discussed above, mitochondria are a main source of ROS in hepatocytes and ROS importantly contribute in liver health and disease. While ROS has been commonly associated to lipid, protein, and DNA oxidation and consequent cellular damage, recent studies have shown that mitochondria-generated ROS may be regulated and may regulate many signaling pathways [27, 121]. Oxidative stress may activate prosurvival pathways in hepatocytes, such as NF-κB and NRFs [122–124]. NF-κB regulates a complex network of pathways, as it is known to control the transcription of over 150 genes [122]. Depending on cell type, microenvironmental conditions and eventually costimulated pathways, NF-κB may exert either a pro-survival or a proapoptotic function [122, 125]. In the context of hepatic oxidative stress, it has been shown that NF-κB may induce antiapoptotic factors, such as XIAP, and function like antioxidants in preventing TGF-beta 1-JNK induced-apoptosis [126]. Moreover, NF-κB collaborates with p38 MAP kinase signaling cascade to protect hepatocytes from liver injury induced by TNF-alpha [125]. NRFs also regulates oxidative stress response in the liver. In particular, NRF-1 has been shown to promote cell survival of hepatocytes during development, sustaining the transcription of antioxidant genes and protecting embryonic hepatocytes from TNF-mediated apoptosis [124]. Moreover, NRF-1 has been shown to be induced under prooxidant conditions and to promote the transcription of mitochondrial transcription factor A (Tfam), required for mitochondrial DNA transcription and replication [123]. Hypoxia is another clear example of cell signaling mediated by ROS. Hypoxia is a clinical relevant event both in liver ischemia/reperfusion injury and in hepatocellular carcinoma development. It leads to an increase in production of H_2O_2 from mitochondrial complex III, thus creating a cytosolic signal that stabilizes the hypoxia inducible transcription factors HIF-1 [27, 127]. Moreover, during hypoxia ROS activates AMPK, which in turn phosphorylates Na/K ATPase (in order to promote its endocytosis) and mTOR (in order to decrease protein translation), thus contributing to energy conservation [128, 129]. In addition, hypoxia-induced mitochondrial ROS enhance the DNA binding of NF-κB through a redox-dependent mechanism involving the mitochondrial glutathione (mGSH) pool in cancer cells, including hepatoma cell lines [130, 131]. In this context, mGSH regulates the intensity of ROS diffusion in the cytoplasm, allowing activation of the c-Src kinase, with subsequent phosphorylation of the inhibitory subunit IκB, activation of NF-κB and promotion of cancer cell survival [127, 130]. The liver is one of the organs with the highest content of GSH and mGSH plays a central role in regulating both in antioxidant defense against excessive ROS production and in regulation of ROS signaling in liver physiology and pathology [130]. Alcohol consumption has been shown to sensitize hepatocytes to TNF because of mGSH depletion through impaired transport of GSH to mitochondria [132]. Interestingly, GSH transport impairment and TNF sensitization correlate with free cholesterol accumulation in mitochondrial membranes and seem to be a common pro-inflammatory mechanism in both alcoholic and nonalcoholic steatohepatitis [133]. Similar alterations in mGSH regulation have been reported in liver cirrhosis, in particular in an experimental model of secondary biliary cirrhosis in rats induced by bile-duct ligation [134, 135].

During the past decade, a new family of enzymes, the nicotinamide-adenine-dinucleotide- (NAD-) dependent protein deacetylases named sirtuins, has been described to contribute to extended lifespan many animal models, including mammals [136]. Interestingly of the six mammalian sirtuins, three (SIRT3, 4, and 5) are expressed in the mitochondria where they mediate physiologic adaptation to reduced energy consumption [137]. In the liver, SIRT4 activity was shown to decline during calorie restriction, allowing the consumption of glutamine as a fuel source for glucose synthesis. Moreover, SIRT4 depletion increased fatty acid oxidation [138]. Mitochondrial sirtuins could be also interesting targets in the regulation of ammonia production and disposal. Nakagawa et al. [139] have shown that the sirtuin SIRT5 activates CPS1, which we mentioned before as the first enzyme in the urea cycle. In mice, NAD in liver mitochondria increases during fasting, thereby triggering SIRT5-mediated deacetylation of CPS1 and adaptation to increase in amino acid catabolism. These data indicate SIRT5 also has an emerging role in the metabolic adaptation to fasting, high protein diet and calorie restriction. Finally, recent findings correlate SIRT3 to the production of ROS. In particular, SIRT3$^{-/-}$ cells produce increased levels of ROS and have concomitantly a reduced ATP production [11, 12, 140]. These results suggest that SIRT3-mediated deacetylation of electron transport chain may render OXPHOS more efficient [137]. Moreover, SIRT3 may deacetylate and activate the antioxidant enzyme mitochondrial superoxide dismutase (SOD2) and the isocitrate dehydrogenase 2, which generates NADPH for the glutathione synthesis, in mice [141–143].

The sirtuins' antiaging role and their ability of controlling energy metabolism make them interesting target in cancer and metabolic diseases. Interestingly, in a mouse model of metabolic syndrome-associated liver cancer, overexpression of SIRT1 reduced the susceptibility to liver cancer and improved hepatic protection from both DNA damage and metabolic damage [136]. However, recent studies showed that SIRT1 was upregulated in HCC and it has a role in telomere maintenance [144, 145]. Downregulation of SIRT1 suppressed proliferation of HCC cells and induced cellular senescence or apoptosis [144]. Finally, many recent papers show a possible synergic action of cytosolic and mitochondrial sirtuins in regulating glucose and lipid metabolism in the liver [138, 140, 146–148]. SIRT1 has been shown to regulate hepatic glucose and lipid metabolism by activating AMPK and by inducing gluconeogenic genes via activation of PGC-1 alpha in hepatic cell and mouse liver [146, 147]. Interestingly, SIRT1 did not regulate the PGC-1 alpha effects on mitochondrial biogenesis. The mitochondrial SIRT3 was shown to positively modulate fatty acid oxidation and ATP production, in particular deacetylating the long-chain acyl-CoA dehydrogenase and Complex I of the electron transport chain [140, 148]. Finally, in a recent paper, high-fat diet induced a decrease of hepatic SIRT3, hyperacetylation of mitochondrial proteins and fatty liver in mice [149].

Altogether, the studies reviewed show that mitochondria are much more dynamic organelles than considered traditionally. They are key organelles in the integration and adaptation to external stimuli, such as changing composition of diet (i.e., calorie restriction *versus* high fat diet), hypoxia, cold exposure, or physical exercise [150]. Mitochondrial homeostasis is a highly controlled process balancing organelle biogenesis and degradation (essentially by autophagy/mitophagy) and an alteration of this balance may bring to organelle dysfunction, contributing to the development of liver chronic diseases.

5. Conclusions

The liver is one of the organs richest in mitochondria. Hepatic mitochondria have unique features compared to other organs' mitochondria, since they are the hub that integrates hepatic metabolism of carbohydrates, lipids, and proteins. Thus, correct functioning of hepatic mitochondria is essential not only to prevent liver disease, such as NAFLD, but also to avoid systemic diseases, such as ammonia-induced hepatic encephalopathy. Mitochondria are also essential in hepatocyte survival as mediator of apoptosis and necrosis. Hepatocyte cell death is involved in most liver pathologies, such as alcoholic and nonalcoholic steatohepatitis, viral hepatitis, liver fibrosis, and carcinogenesis. Hepatocytes have developed different mechanisms to keep mitochondrial integrity or to prevent the effects of mitochondrial lesions, in particular regulating organelle biogenesis and degradation. A better knowledge of the mechanisms and pathways involved in mitochondria homeostasis should improve preventive and therapeutic strategies for liver diseases.

Abbreviations

2-AAF:	2-Acetylaminofluorene
DAG:	Diacylglycerol
GS:	Glutamine synthetase
GSH:	Glutathione
HCC:	Hepatocellular carcinoma
MIM:	Mitochondrial inner membrane
MMP:	Mitochondrial membrane permeabilization
MOM:	Mitochondrial outer membrane
MRC:	Mitochondrial respiratory chain
NAG:	N-acetyl-L-glutamate
NASH:	Nonalcoholic steatohepatitis
OXPHOS:	Oxidative phosphorylation
PTP:	Permeability transition pore
ROS:	Reactive oxygen species.

References

[1] G. Kroemer, L. Galluzzi, and C. Brenner, "Mitochondrial membrane permeabilization in cell death," *Physiological Reviews*, vol. 87, no. 1, pp. 99–163, 2007.

[2] H. Malhi, M. E. Guicciardi, and G. J. Gores, "Hepatocyte death: a clear and present danger," *Physiological Reviews*, vol. 90, no. 3, pp. 1165–1194, 2010.

[3] K. L. Veltri, M. Espiritu, and G. Singh, "Distinct genomic copy number in mitochondria of different mammalian organs," *Journal of Cellular Physiology*, vol. 143, no. 1, pp. 160–164, 1990.

[4] P. H. Yin, H. C. Lee, G. Y. Chau et al., "Alteration of the copy number and deletion of mitochondrial DNA in human hepatocellular carcinoma," *British Journal of Cancer*, vol. 90, no. 12, pp. 2390–2396, 2004.

[5] A. Chatterjee, S. Dasgupta, and D. Sidransky, "Mitochondrial subversion in cancer," *Cancer Prevention Research*, vol. 4, no. 5, pp. 638–654, 2011.

[6] R. S. Rector, J. P. Thyfault, G. M. Uptergrove et al., "Mitochondrial dysfunction precedes insulin resistance and hepatic steatosis and contributes to the natural history of non-alcoholic fatty liver disease in an obese rodent model," *Journal of Hepatology*, vol. 52, no. 5, pp. 727–736, 2010.

[7] M. E. Breuer, P. H. G. M. Willems, F. G. M. Russel, W. J. H. Koopman, and J. A. M. Smeitink, "Modeling mitochondrial dysfunctions in the brain: from mice to men," *Journal of Inherited Metabolic Disease*, vol. 35, no. 2, pp. 193–210, 2011.

[8] S. Ong and A. B. Gustafsson, "New roles for mitochondria in cell death in the reperfused myocardium," *Cardiovascular Research*, vol. 94, no. 2, pp. 190–196, 2012.

[9] D. Degli Esposti, M. C. Domart, M. Sebagh et al., "Autophagy is induced by ischemic preconditioning in human livers formerly treated by chemotherapy to limit necrosis," *Autophagy*, vol. 6, no. 1, pp. 172–174, 2010.

[10] Y. M. Yang, C. Y. Han, Y. J. Kim, and S. G. Kim, "AMPK-associated signaling to bridge the gap between fuel metabolism and hepatocyte viability," *World Journal of Gastroenterology*, vol. 16, no. 30, pp. 3731–3742, 2010.

[11] H. S. Kim, K. Patel, K. Muldoon-Jacobs et al., "SIRT3 is a mitochondria-localized tumor suppressor required for maintenance of mitochondrial integrity and metabolism during stress," *Cancer Cell*, vol. 17, no. 1, pp. 41–52, 2010.

[12] K. H. Kim, Y. S. Kum, Y. Y. Park et al., "The protective effect of bee venom against ethanol-induced hepatic injury via

regulation of the mitochondria-related apoptotic pathway," *Basic and Clinical Pharmacology and Toxicology*, vol. 107, no. 1, pp. 619–624, 2010.

[13] D. R. Green, L. Galluzzi, and G. Kroemer, "Mitochondria and the autophagy-inflammation-cell death axis in organismal aging," *Science*, vol. 333, no. 6046, pp. 1109–1112, 2011.

[14] M. Pérez-Carreras, P. Del Hoyo, M. A. Martín et al., "Defective hepatic mitochondrial respiratory chain in patients with nonalcoholic steatohepatitis," *Hepatology*, vol. 38, no. 4, pp. 999–1007, 2003.

[15] E. Bugianesi, A. J. McCullough, and G. Marchesini, "Insulin resistance: a metabolic pathway to chronic liver disease," *Hepatology*, vol. 42, no. 5, pp. 987–1000, 2005.

[16] E. Fabbrini, S. Sullivan, and S. Klein, "Obesity and nonalcoholic fatty liver disease: biochemical, metabolic, and clinical implications," *Hepatology*, vol. 51, no. 2, pp. 679–689, 2010.

[17] L. Boon, W. J. C. Geerts, A. Jonker, W. H. Lamers, and C. J. F. Van Noorden, "High protein diet induces pericentral glutamate dehydrogenase and ornithine aminotransferase to provide sufficient glutamate for pericentral detoxification of ammonia in rat liver lobules," *Histochemistry and Cell Biology*, vol. 111, no. 6, pp. 445–452, 1999.

[18] G. Vial, H. Dubouchaud, and X. M. Leverve, "Liver mitochondria and insulin resistance," *Acta Biochimica Polonica*, vol. 57, no. 4, pp. 389–392, 2010.

[19] G. Serviddio, F. Bellanti, R. Tamborra et al., "Uncoupling protein-2 (UCP2) induces mitochondrial proton leak and increases susceptibility of non-alcoholic steatohepatitis (NASH) liver to ischaemia-reperfusion injury," *Gut*, vol. 57, no. 7, pp. 957–965, 2008.

[20] R. S. Kaplan and J. A. Mayor, "Structure, function and regulation of the tricarboxylate transport protein from rat liver mitochondria," *Journal of Bioenergetics and Biomembranes*, vol. 25, no. 5, pp. 503–514, 1993.

[21] L. Caldovic and M. Tuchman, "N-acetylglutamate and its changing role through evolution," *Biochemical Journal*, vol. 372, no. 2, pp. 279–290, 2003.

[22] J. M. Matés, J. A. Segura, J. A. Campos-Sandoval et al., "Glutamine homeostasis and mitochondrial dynamics," *International Journal of Biochemistry and Cell Biology*, vol. 41, no. 10, pp. 2051–2061, 2009.

[23] S. W. M. Olde Damink, R. Jalan, and C. H. C. Dejong, "Interorgan ammonia trafficking in liver disease," *Metabolic Brain Disease*, vol. 24, no. 1, pp. 169–181, 2009.

[24] J. Albrecht, "Roles of neuroactive amino acids in ammonia neurotoxicity," *Journal of Neuroscience Research*, vol. 51, no. 2, pp. 133–138, 1998.

[25] V. Felipo, C. Hermenegildo, C. Montoliu, M. Llansola, and M. D. Minana, "Neurotoxicity of ammonia and glutamate: molecular mechanisms and prevention," *NeuroToxicology*, vol. 19, no. 4-5, pp. 675–682, 1998.

[26] R. F. Butterworth, "Pathophysiology of hepatic encephalopathy: the concept of synergism," *Hepatology Research*, vol. 38, no. 1, pp. S116–S121, 2008.

[27] R. B. Hamanaka and N. S. Chandel, "Mitochondrial reactive oxygen species regulate cellular signaling and dictate biological outcomes," *Trends in Biochemical Sciences*, vol. 35, no. 9, pp. 505–513, 2010.

[28] T. A. Young, C. C. Cunningham, and S. M. Bailey, "Reactive oxygen species production by the mitochondrial respiratory chain in isolated rat hepatocytes and liver mitochondria: studies using myxothiazol," *Archives of Biochemistry and Biophysics*, vol. 405, no. 1, pp. 65–72, 2002.

[29] Y. Kushnareva, A. N. Murphy, and A. Andreyev, "Complex I-mediated reactive oxygen species generation: modulation by cytochrome c and NAD(P)$^+$ oxidation-reduction state," *Biochemical Journal*, vol. 368, no. 2, pp. 545–553, 2002.

[30] H. Nohl, L. Gille, and K. Staniek, "Intracellular generation of reactive oxygen species by mitochondria," *Biochemical Pharmacology*, vol. 69, no. 5, pp. 719–723, 2005.

[31] D. C. Wallace, W. Fan, and V. Procaccio, "Mitochondrial energetics and therapeutics," *Annual Review of Pathology*, vol. 5, pp. 297–348, 2010.

[32] S. Rexroth, A. Poetsch, M. Rögner et al., "Reactive oxygen species target specific tryptophan site in the mitochondrial ATP synthase," *Biochimica et Biophysica Acta*, vol. 1817, no. 2, pp. 381–387, 2012.

[33] M. O. Dietrich and T. L. Horvath, "The role of mitochondrial uncoupling proteins in lifespan," *Pflugers Archiv European Journal of Physiology*, vol. 459, no. 2, pp. 269–275, 2010.

[34] A. Letai, M. C. Bassik, L. D. Walensky, M. D. Sorcinelli, S. Weiler, and S. J. Korsmeyer, "Distinct BH3 domains either sensitize or activate mitochondrial apoptosis, serving as prototype cancer therapeutics," *Cancer Cell*, vol. 2, no. 3, pp. 183–192, 2002.

[35] S. Shimizu, M. Narita, and Y. Tsujimoto, "Bcl-2 family proteins regulate the release of apoptogenic cytochrome c by the mitochondrial channel VDAC," *Nature*, vol. 399, no. 6735, pp. 483–487, 1999.

[36] M. Zoratti and I. Szabo, "The mitochondrial permeability transition," *Biochimica et Biophysica Acta*, vol. 1241, no. 2, pp. 139–176, 1995.

[37] J. Montero, M. Mari, A. Colell et al., "Cholesterol and peroxidized cardiolipin in mitochondrial membrane properties, permeabilization and cell death," *Biochimica et Biophysica Acta*, vol. 1797, no. 6-7, pp. 1217–1224, 2010.

[38] C. Brenner and S. Grimm, "The permeability transition pore complex in cancer cell death," *Oncogene*, vol. 25, no. 34, pp. 4744–4756, 2006.

[39] A. Rasola and P. Bernardi, "Mitochondrial permeability transition in Ca^{2+}-dependent apoptosis and necrosis," *Cell Calcium*, vol. 50, no. 3, pp. 222–233, 2011.

[40] J. E. Chipuk and D. R. Green, "Do inducers of apoptosis trigger caspase-indipendent cell death?" *Nature Reviews Molecular Cell Biology*, vol. 6, no. 3, pp. 268–275, 2005.

[41] D. R. Green and G. Kroemer, "The pathophysiology of mitochondrial cell death," *Science*, vol. 305, no. 5684, pp. 626–629, 2004.

[42] T. Takehara, T. Tatsumi, T. Suzuki et al., "Hepatocyte-specific disruption of Bcl-xL leads to continuous hepatocyte apoptosis and liver fibrotic responses," *Gastroenterology*, vol. 127, no. 4, pp. 1189–1197, 2004.

[43] B. Vick, A. Weber, T. Urbanik et al., "Knockout of myeloid cell leukemia-1 induces liver damage and increases apoptosis susceptibility of murine hepatocytes," *Hepatology*, vol. 49, no. 2, pp. 627–636, 2009.

[44] F. Charlotte, A. L'Hermine, N. Martin et al., "Immunohistochemical detection of bcl-2 protein in normal and pathological human liver," *American Journal of Pathology*, vol. 144, no. 3, pp. 460–465, 1994.

[45] M. C. Domart, D. Degli Esposti, M. Sebagh et al., "Concurrent induction of necrosis, apoptosis, and autophagy in ischemic preconditioned human livers formerly treated by chemotherapy," *Journal of Hepatology*, vol. 51, no. 5, pp. 881–889, 2009.

[46] P. R. Galle, W. J. Hofmann, H. Walczak et al., "Involvement of the CD95 (APO-1/Fas) receptor and ligand in liver damage,"

Journal of Experimental Medicine, vol. 182, no. 5, pp. 1223–1230, 1995.

[47] B. Mundt, F. Kühnel, L. Zender et al., "Involvement of TRAIL and its receptors in viral hepatitis," *The FASEB Journal*, vol. 17, no. 1, pp. 94–96, 2003.

[48] N. K. Mohd-Ismail, L. Deng, S. K. Sukumaran, V. C. Yu, H. Hotta, and Y. J. Tan, "The hepatitis C virus core protein contains a BH3 domain that regulates apoptosis through specific interaction with human Mcl-1," *Journal of Virology*, vol. 83, no. 19, pp. 9993–10006, 2009.

[49] G. C. Farrell, C. Z. Larter, J. Y. Hou et al., "Apoptosis in experimental NASH is associated with p53 activation and TRAIL receptor expression," *Journal of Gastroenterology and Hepatology*, vol. 24, no. 3, pp. 443–452, 2009.

[50] N. Alkhouri, C. Carter-Kent, and A. E. Feldstein, "Apoptosis in nonalcoholic fatty liver disease: diagnostic and therapeutic implications," *Expert Review of Gastroenterology and Hepatology*, vol. 5, no. 2, pp. 201–212, 2011.

[51] R. S. Hotchkiss, A. Strasser, J. E. McDunn, and P. E. Swanson, "Mechanisms of disease: cell death," *New England Journal of Medicine*, vol. 361, no. 16, pp. 1570–1583, 2009.

[52] M. Selzner, H. A. RüDiger, D. Sindram, J. Madden, and P. A. Clavien, "Mechanisms of ischemic injury are different in the steatotic and normal rat liver," *Hepatology*, vol. 32, no. 6, pp. 1280–1288, 2000.

[53] C. Peralta, R. Bartrons, A. Serafin et al., "Adenosine monophosphate-activated protein kinase mediates the protective effects of ischemic preconditioning on hepatic ischemia-reperfusion injury in the rat," *Hepatology*, vol. 34, no. 6, pp. 1164–1173, 2001.

[54] D. Degli Esposti, M. Sebagh, P. Pham et al., "Ischemic preconditioning induces autophagy and limits necrosis in human recipients of fatty liver grafts, decreasing the incidence of rejection episodes," *Cell Death and Disease*, vol. 2, no. 1, article e111, 2011.

[55] P. Vandenabeele, L. Galluzzi, T. Vanden Berghe, and G. Kroemer, "Molecular mechanisms of necroptosis: an ordered cellular explosion," *Nature Reviews Molecular Cell Biology*, vol. 11, no. 10, pp. 700–714, 2010.

[56] C. Plin, P. S. Haddad, J. P. Tillement, A. Elimadi, and D. Morin, "Protection by cyclosporin A of mitochondrial and cellular functions during a cold preservation-warm reperfusion of rat liver," *European Journal of Pharmacology*, vol. 495, no. 2-3, pp. 111–118, 2004.

[57] J. S. Kim, L. He, T. Qian, and J. J. Lemasters, "Role of the mitochondrial permeability transition in apoptotic and necrotic death after ischemia/reperfusion injury to hepatocytes," *Current Molecular Medicine*, vol. 3, no. 6, pp. 527–535, 2003.

[58] S. Yang, H. Zhu, Y. Li et al., "Mitochondrial adaptations to obesity-related oxidant stress," *Archives of Biochemistry and Biophysics*, vol. 378, no. 2, pp. 259–268, 2000.

[59] N. N. Danial, C. F. Gramm, L. Scorrano et al., "BAD and glucokinase reside in a mitochondrial complex that integrates glycolysis and apoptosis," *Nature*, vol. 424, no. 6951, pp. 952–956, 2003.

[60] C. Arden, S. Baltrusch, and L. Agius, "Glucokinase regulatory protein is associated with mitochondria in hepatocytes," *FEBS Letters*, vol. 580, no. 8, pp. 2065–2070, 2006.

[61] K. Begriche, A. Igoudjil, D. Pessayre, and B. Fromenty, "Mitochondrial dysfunction in NASH: causes, consequences and possible means to prevent it," *Mitochondrion*, vol. 6, no. 1, pp. 1–38, 2006.

[62] D. Zhang, Z. X. Liu, S. C. Cheol et al., "Mitochondrial dysfunction due to long-chain Acyl-CoA dehydrogenase deficiency causes hepatic steatosis and hepatic insulin resistance," *Proceedings of the National Academy of Sciences of the United States of America*, vol. 104, no. 43, pp. 17075–17080, 2007.

[63] D. Pessayre, "Role of mitochondria in non-alcoholic fatty liver disease," *Journal of Gastroenterology and Hepatology*, vol. 22, no. 1, pp. S20–S27, 2007.

[64] D. Pessayre and B. Fromenty, "NASH: a mitochondrial disease," *Journal of Hepatology*, vol. 42, no. 6, pp. 928–940, 2005.

[65] K. Hensley, Y. Kotake, H. Sang et al., "Dietary choline restriction causes complex I dysfunction and increased H_2O_2 generation in liver mitochondria," *Carcinogenesis*, vol. 21, no. 5, pp. 983–989, 2000.

[66] M. A. Robin, H. K. Anandatheerthavarada, J. K. Fang, M. Cudic, L. Otros, and N. G. Avadhani, "Mitochondrial targeted cytochrome P450 2E1 (P450 MT5) contains an intact N terminus and requires mitochondrial specific electron transfer proteins for activity," *Journal of Biological Chemistry*, vol. 276, no. 27, pp. 24680–24689, 2001.

[67] H. Raza, S. K. Prabu, M. A. Robin, and N. G. Avadhani, "Elevated mitochondrial cytochrome P450 2E1 and glutathione S-transferase A4-4 in streptozotocin-induced diabetic rats: tissue-specific variations and roles in Oxidative stress," *Diabetes*, vol. 53, no. 1, pp. 185–194, 2004.

[68] S. K. Mantena, A. L. King, K. K. Andringa, A. Landar, V. Darley-Usmar, and S. M. Bailey, "Novel interactions of mitochondria and reactive oxygen/nitrogen species in alcohol mediated liver disease," *World Journal of Gastroenterology*, vol. 13, no. 37, pp. 4967–4973, 2007.

[69] N. Sato, T. Kamada, and S. Kawano, "Effect of acute and chronic ethanol consumption on hepatic tissue oxygen tension in rats," *Pharmacology Biochemistry and Behavior*, vol. 18, no. 1, pp. 443–447, 1983.

[70] N. Sato, "Central role of mitochondria in metabolic regulation of liver pathophysiology," *Journal of Gastroenterology and Hepatology*, vol. 22, no. 1, pp. S1–S6, 2007.

[71] S. M. Bailey and C. C. Cunningham, "Contribution of mitochondria to oxidative stress associated with alcoholic liver disease," *Free Radical Biology and Medicine*, vol. 32, no. 1, pp. 11–16, 2002.

[72] A. Venkatraman, A. Landar, A. J. Davis et al., "Modification of the mitochondrial proteome in response to the stress of ethanol-dependent hepatotoxicity," *Journal of Biological Chemistry*, vol. 279, no. 21, pp. 22092–22101, 2004.

[73] A. Cahill, X. Wang, and J. B. Hoek, "Increased oxidative damage to mitochondrial DNA following chronic ethanol consumption," *Biochemical and Biophysical Research Communications*, vol. 235, no. 2, pp. 286–290, 1997.

[74] W. B. Coleman and C. C. Cunningham, "Effects of chronic ethanol consumption on the synthesis of polypeptides encoded by the hepatic mitochondrial genome," *Biochimica et Biophysica Acta*, vol. 1019, no. 2, pp. 142–150, 1990.

[75] T. Wang, R. V. Campbell, M. K. Yi, S. M. Lemon, and S. A. Weinman, "Role of Hepatitis C virus core protein in viral-induced mitochondrial dysfunction," *Journal of Viral Hepatitis*, vol. 17, no. 11, pp. 784–793, 2010.

[76] W. Lim, S. H. Kwon, H. Cho et al., "HBx targeting to mitochondria and ROS generation are necessary but insufficient for HBV-induced cyclooxygenase-2 expression," *Journal of Molecular Medicine*, vol. 88, no. 4, pp. 359–369, 2010.

[77] K. Begriche, J. Massart, M. A. Robin, A. Borgne-Sanchez, and B. Fromenty, "Drug-induced toxicity on mitochondria

and lipid metabolism: mechanistic diversity and deleterious consequences for the liver," *Journal of Hepatology*, vol. 54, no. 4, pp. 773–794, 2011.

[78] M. T. Kuo and N. Savaraj, "Roles of reactive oxygen species in hepatocarcinogenesis and drug resistance gene expression in liver cancers," *Molecular Carcinogenesis*, vol. 45, no. 9, pp. 701–709, 2006.

[79] H. M. Shen, C. Y. Shi, Y. Shen, and C. N. Ong, "Detection of elevated reactive oxygen species level in cultured rat hepatocytes treated with aflatoxin B1," *Free Radical Biology and Medicine*, vol. 21, no. 2, pp. 139–146, 1996.

[80] P. C. Klohn, H. Massalha, and H. G. Neumann, "A metabolite of carcinogenic 2-acetylaminofluorene, 2-nitrosofluorene, induces redox cycling in mitochondria," *Biochimica et Biophysica Acta*, vol. 1229, no. 3, pp. 363–372, 1995.

[81] T. K. Macus, Z. Liu, Y. Wei et al., "Induction of human MDR1 gene expression by 2-acetylaminofluorene is mediated by effectors of the phosphoinositide 3-kinase pathway that activate NF-κB signaling," *Oncogene*, vol. 21, no. 13, pp. 1945–1954, 2002.

[82] Y. S. Bae, S. W. Kang, M. S. Seo et al., "Epidermal growth factor (EGF)-induced generation of hydrogen peroxide. Role in EGF receptor-mediated tyrosine phosphorylation," *Journal of Biological Chemistry*, vol. 272, no. 1, pp. 217–221, 1997.

[83] Y. S. Bae, J. Y. Sung, O. S. Kim et al., "Platelet-derived growth factor-induced H_2O_2 production requires the activation of phosphatidylinositol 3-kinase," *Journal of Biological Chemistry*, vol. 275, no. 14, pp. 10527–10531, 2000.

[84] V. M. Factor, A. Kiss, J. T. Woitach, P. J. Wirth, and S. S. Thorgeirsson, "Disruption of redox homeostasis in the transforming growth factor-α/c- myc transgenic mouse model of accelerated hepatocarcinogenesis," *Journal of Biological Chemistry*, vol. 273, no. 25, pp. 15846–15853, 1998.

[85] A. P. Rolo, J. S. Teodoro, C. Peralta, J. Rosello-Catafau, and C. M. Palmeira, "Prevention of I/R injury in fatty livers by ischemic preconditioning is associated with increased mitochondrial tolerance: the key role of ATPsynthase and mitochondrial permeability transition," *Transplant International*, vol. 22, no. 11, pp. 1081–1090, 2009.

[86] D. Pessayre, B. Fromenty, A. Berson et al., "Central role of mitochondria in drug-induced liver injury," *Drug Metabolism Reviews*, vol. 44, no. 1, pp. 34–87, 2012.

[87] G. Serviddio, F. Bellanti, J. Sastre, G. Vendemiale, and E. Altomare, "Targeting mitochondria: a new promising approach for the treatment of liver diseases," *Current Medicinal Chemistry*, vol. 17, no. 22, pp. 2325–2337, 2010.

[88] C. Handschin, "The biology of PGC-1α and its therapeutic potential," *Trends in Pharmacological Sciences*, vol. 30, no. 6, pp. 322–329, 2009.

[89] R. C. Scarpulla, "Metabolic control of mitochondrial biogenesis through the PGC-1 family regulatory network," *Biochimica et Biophysica Acta*, vol. 1813, no. 7, pp. 1269–1278, 2011.

[90] J. C. Yoon, P. Puigserver, G. Chen et al., "Control of hepatic gluconeogenesis through the transcriptional coaotivator PGC-1," *Nature*, vol. 413, no. 6852, pp. 131–138, 2001.

[91] P. Puigserver, J. Rhee, J. Donovan et al., "Insulin-regulated hepatic gluconeogenesis through FOXO1-PGC-1α interaction," *Nature*, vol. 423, no. 6939, pp. 550–555, 2003.

[92] J. Rhee, Y. Inoue, J. C. Yoon et al., "Regulation of hepatic fasting response by PPARγ coactivator-1α (PGC-1): requirement for hepatocyte nuclear factor 4α in gluconeogenesis,"

Proceedings of the National Academy of Sciences of the United States of America, vol. 100, no. 7, pp. 4012–4017, 2003.

[93] C. Handschin, J. Lin, J. Rhee et al., "Nutritional regulation of hepatic heme biosynthesis and porphyria through PGC-1α," *Cell*, vol. 122, no. 4, pp. 505–515, 2005.

[94] J. M. Huss and D. P. Kelly, "Nuclear receptor signaling and cardiac energetics," *Circulation Research*, vol. 95, no. 6, pp. 568–578, 2004.

[95] S. Song, R. R. Attia, S. Connaughton et al., "Peroxisome proliferator activated receptor α (PPARα) and PPAR gamma coactivator (PGC-1α) induce carnitine palmitoyltransferase IA (CPT-1A) via independent gene elements," *Molecular and Cellular Endocrinology*, vol. 325, no. 1-2, pp. 54–63, 2010.

[96] Y. Lustig, J. L. Ruas, J. L. Estall et al., "Separation of the gluconeogenic and mitochondrial functions of pgc-1α through s6 kinase," *Genes and Development*, vol. 25, no. 12, pp. 1232–1244, 2011.

[97] L. F. Yu, B. Y. Qiu, F. J. Nan, and J. Li, "AMPK activators as novel therapeutics for type 2 diabetes," *Current Topics in Medicinal Chemistry*, vol. 10, no. 4, pp. 397–410, 2010.

[98] S. Jäger, C. Handschin, J. St-Pierre, and B. M. Spiegelman, "AMP-activated protein kinase (AMPK) action in skeletal muscle via direct phosphorylation of PGC-1α," *Proceedings of the National Academy of Sciences of the United States of America*, vol. 104, no. 29, pp. 12017–12022, 2007.

[99] A. Shlomai, N. Paran, and Y. Shaul, "PGC-1α controls hepatitis B virus through nutritional signals," *Proceedings of the National Academy of Sciences of the United States of America*, vol. 103, no. 43, pp. 16003–16008, 2006.

[100] L. C. Gomes, G. D. Benedetto, and L. Scorrano, "During autophagy mitochondria elongate, are spared from degradation and sustain cell viability," *Nature Cell Biology*, vol. 13, no. 5, pp. 589–598, 2011.

[101] C. Mitchell, M. A. Robin, A. Mayeuf et al., "Protection against hepatocyte mitochondrial dysfunction delays fibrosis progression in mice," *American Journal of Pathology*, vol. 175, no. 5, pp. 1929–1937, 2009.

[102] H. Yadav, C. Quijano, A. K. Kamaraju et al., "Protection from obesity and diabetes by blockade of TGF-β/Smad3 signaling," *Cell Metabolism*, vol. 14, no. 1, pp. 67–79, 2011.

[103] D. W. Hailey, A. S. Rambold, P. Satpute-Krishnan et al., "Mitochondria supply membranes for autophagosome biogenesis during starvation," *Cell*, vol. 141, no. 4, pp. 656–667, 2010.

[104] C. He and D. J. Klionsky, "Regulation mechanisms and signaling pathways of autophagy," *Annual Review of Genetics*, vol. 43, pp. 67–93, 2009.

[105] G. Kroemer, G. Mariño, and B. Levine, "Autophagy and the integrated stress response," *Molecular Cell*, vol. 40, no. 2, pp. 280–293, 2010.

[106] R. J. Youle and D. P. Narendra, "Mechanisms of mitophagy," *Nature Reviews Molecular Cell Biology*, vol. 12, no. 1, pp. 9–14, 2011.

[107] R. A. Gottlieb and R. M. Mentzer, "Autophagy during cardiac stress: joys and frustrations of autophagy," *Annual Review of Physiology*, vol. 72, pp. 45–59, 2009.

[108] D. Chen, F. Gao, B. Li et al., "Parkin mono-ubiquitinates Bcl-2 and regulates autophagy," *Journal of Biological Chemistry*, vol. 285, no. 49, pp. 38214–38223, 2010.

[109] J. Y. Lee, Y. Nagano, J. P. Taylor, K. L. Lim, and T. P. Yao, "Disease-causing mutations in Parkin impair mitochondrial ubiquitination, aggregation, and HDAC6-dependent mitophagy," *Journal of Cell Biology*, vol. 189, no. 4, pp. 671–679, 2010.

[110] M. d'Amora, C. Angelini, M. Marcoli, C. Cervetto, T. Kitada, and M. Vallarino, "Expression of PINK1 in the brain, eye and ear of mouse during embryonic development," *Journal of Chemical Neuroanatomy*, vol. 41, no. 2, pp. 73–85, 2011.

[111] D. F. Egan, D. B. Shackelford, M. M. Mihaylova et al., "Phosphorylation of ULK1 (hATG1) by AMP-activated protein kinase connects energy sensing to mitophagy," *Science*, vol. 331, no. 6016, pp. 456–461, 2011.

[112] I. Kim and J. J. Lemasters, "Mitochondrial degradation by autophagy (mitophagy) in GFP-LC3 transgenic hepatocytes during nutrient deprivation," *American Journal of Physiology*, vol. 300, no. 2, pp. C308–C317, 2011.

[113] Y. Cheng, M. Zhou, C. H. Tung, M. Ji, and F. Zhang, "Studies on two types of PTP1B inhibitors for the treatment of type 2 diabetes: hologram QSAR for OBA and BBB analogues," *Bioorganic and Medicinal Chemistry Letters*, vol. 20, no. 11, pp. 3329–3337, 2010.

[114] S. Gottschalk, C. Zwingmann, V.-A. Raymond, M. C. Hohnholt, T. S. Chan, and M. Bilodeau, "Hepatocellular apoptosis in mice is associated with early upregulation of mitochondrial glucose metabolism," *Apoptosis*, vol. 17, no. 2, pp. 143–153, 2012.

[115] C. Poussin, M. Ibberson, D. Hall et al., "Oxidative phosphorylation flexibility in the liver of mice resistant to high-fat diet-induced hepatic steatosis," *Diabetes*, vol. 60, no. 9, pp. 2216–2224, 2011.

[116] J. Albrecht and M. D. Norenberg, "Glutamine: a Trojan horse in ammonia neurotoxicity," *Hepatology*, vol. 44, no. 4, pp. 788–794, 2006.

[117] M. D. Norenberg, K. V. Rama Rao, and A. R. Jayakumar, "Mechanisms of ammonia-induced astrocyte swelling," *Metabolic Brain Disease*, vol. 20, no. 4, pp. 303–318, 2005.

[118] S. Ahmad, C. W. White, L. Y. Chang, B. K. Schneider, and C. B. Allen, "Glutamine protects mitochondrial structure and function in oxygen toxicity," *American Journal of Physiology*, vol. 280, no. 4, pp. L779–L791, 2001.

[119] C. H. Eng and R. T. Abraham, "Glutaminolysis yields a metabolic by-product that stimulates autophagy," *Autophagy*, vol. 6, no. 7, pp. 968–970, 2010.

[120] I. Nissim, O. Horyn, I. Nissim et al., "Down-regulation of Hepatic Urea Synthesis by Oxypurines: xanthine and uric acid inhibit N-acetylglutamate synthase," *Journal of Biological Chemistry*, vol. 286, no. 25, pp. 22055–22068, 2011.

[121] D. Han, F. Antunes, R. Canali, D. Rettori, and E. Cadenas, "Voltage-dependent anion channels control the release of the superoxide anion from mitochondria to cytosol," *Journal of Biological Chemistry*, vol. 278, no. 8, pp. 5557–5563, 2003.

[122] H. L. Pahl, "Activators and target genes of Rel/NF-κB transcription factors," *Oncogene*, vol. 18, no. 49, pp. 6853–6866, 1999.

[123] C. A. Piantadosi and H. B. Suliman, "Mitochondrial transcription factor A induction by redox activation of nuclear respiratory factor 1," *Journal of Biological Chemistry*, vol. 281, no. 1, pp. 324–333, 2006.

[124] L. Chen, M. Kwong, R. Lu et al., "Nrf1 is critical for redox balance and survival of liver cells during development," *Molecular and Cellular Biology*, vol. 23, no. 13, pp. 4673–4686, 2003.

[125] T. Luedde and R. F. Schwabe, "NF-κB in the liver-linking injury, fibrosis and hepatocellular carcinoma," *Nature Reviews Gastroenterology and Hepatology*, vol. 8, no. 2, pp. 108–118, 2011.

[126] F. Wang, S. Kaur, L. G. Cavin, and M. Arsura, "Nuclear-factor-κB (NF-κB) and radical oxygen species play contrary roles in transforming growth factor-β1 (TGF-β1)-induced apoptosis in hepatocellular carcinoma (HCC) cells," *Biochemical and Biophysical Research Communications*, vol. 377, no. 4, pp. 1107–1112, 2008.

[127] N. S. Chandel, D. S. McClintock, C. E. Feliciano et al., "Reactive oxygen species generated at mitochondrial Complex III stabilize hypoxia-inducible factor-1α during hypoxia: a mechanism of O_2 sensing," *Journal of Biological Chemistry*, vol. 275, no. 33, pp. 25130–25138, 2000.

[128] L. A. Dada, N. S. Chandel, K. M. Ridge, C. Pedemonte, A. M. Bertorello, and J. I. Sznajder, "Hypoxia-induced endocytosis, of Na,K-ATPase in alveolar epithelial cells is mediated by mitochondrial reactive oxygen species and PKC-ζ," *Journal of Clinical Investigation*, vol. 111, no. 7, pp. 1057–1064, 2003.

[129] L. Liu, T. P. Cash, R. G. Jones, B. Keith, C. B. Thompson, and M. C. Simon, "Hypoxia-induced energy stress regulates mRNA translation and cell growth," *Molecular Cell*, vol. 21, no. 4, pp. 521–531, 2006.

[130] M. Marí, A. Morales, A. Colell, C. García-Ruiz, and J. C. Fernández-Checa, "Mitochondrial glutathione, a key survival antioxidant," *Antioxidants and Redox Signaling*, vol. 11, no. 11, pp. 2685–2700, 2009.

[131] J. M. Lluis, F. Buricchi, P. Chiarugi, A. Morales, and J. C. Fernandez-Checa, "Dual role of mitochondrial reactive oxygen species in hypoxia signaling: activation of nuclear factor-KB via c-SRC- and oxidant-dependent cell death," *Cancer Research*, vol. 67, no. 15, pp. 7368–7377, 2007.

[132] O. Coll, A. Colell, C. García-Ruiz, N. Kaplowitz, and J. C. Fernández-Checa, "Sensitivity of the 2-oxoglutarate carrier to alcohol intake contributes to mitochondrial glutathione depletion," *Hepatology*, vol. 38, no. 3, pp. 692–702, 2003.

[133] M. Marí, F. Caballero, A. Colell et al., "Mitochondrial free cholesterol loading sensitizes to TNF- and Fas-mediated steatohepatitis," *Cell Metabolism*, vol. 4, no. 3, pp. 185–198, 2006.

[134] S. Krahenbuhl, J. Stucki, and J. Reichen, "Reduced activity of the electron transport chain in liver mitochondria isolated from rats with secondary biliary cirrhosis," *Hepatology*, vol. 15, no. 6, pp. 1160–1166, 1992.

[135] S. Krahenbuhl, C. Talos, B. H. Lauterburg, and J. Reichen, "Reduced antioxidative capacity in liver mitochondria from bile duct ligated rats," *Hepatology*, vol. 22, no. 2, pp. 607–612, 1995.

[136] D. Herranz, M. Muñoz-Martin, M. Cañamero et al., "Sirt1 improves healthy ageing and protects from metabolic syndrome-associated cancer," *Nature Communications*, vol. 1, no. 1, 2010.

[137] L. Guarente, "Sirtuins, aging, and medicine," *New England Journal of Medicine*, vol. 364, no. 23, pp. 2235–2244, 2011.

[138] N. Nasrin, X. Wu, E. Fortier et al., "SIRT4 regulates fatty acid oxidation and mitochondrial gene expression in liver and muscle cells," *Journal of Biological Chemistry*, vol. 285, no. 42, pp. 31995–32002, 2010.

[139] T. Nakagawa, D. J. Lomb, M. C. Haigis, and L. Guarente, "SIRT5 deacetylates carbamoyl phosphate synthetase 1 and regulates the urea cycle," *Cell*, vol. 137, no. 3, pp. 560–570, 2009.

[140] B. H. Ahn, H. S. Kim, S. Song et al., "A role for the mitochondrial deacetylase Sirt3 in regulating energy homeostasis," *Proceedings of the National Academy of Sciences of the United States of America*, vol. 105, no. 38, pp. 14447–14452, 2008.

[141] X. Qiu, K. Brown, M. D. Hirschey, E. Verdin, and D. Chen, "Calorie restriction reduces oxidative stress by SIRT3-mediated SOD2 activation," *Cell Metabolism*, vol. 12, no. 6, pp. 662–667, 2010.

[142] R. Tao, M. C. Coleman, J. D. Pennington et al., "Sirt3-mediated deacetylation of evolutionarily conserved lysine 122 regulates MnSOD activity in response to stress," *Molecular Cell*, vol. 40, no. 6, pp. 893–904, 2010.

[143] S. Someya, W. Yu, W. C. Hallows et al., "Sirt3 mediates reduction of oxidative damage and prevention of age-related hearing loss under Caloric Restriction," *Cell*, vol. 143, no. 5, pp. 802–812, 2010.

[144] J. Chen, B. Zhang, N. Wong et al., "Sirtuin 1 is upregulated in a subset of hepatocellular carcinomas where it is essential for telomere maintenance and tumor cell growth," *Cancer Research*, vol. 71, no. 12, pp. 4138–4149, 2011.

[145] H. N. Choi, J. S. Bae, U. Jamiyandorj et al., "Expression and role of SIRT1 in hepatocellular carcinoma," *Oncology Reports*, vol. 26, no. 2, pp. 503–510, 2011.

[146] J. T. Rodgers, C. Lerin, W. Haas, S. P. Gygi, B. M. Spiegelman, and P. Puigserver, "Nutrient control of glucose homeostasis through a complex of PGC-1α and SIRT1," *Nature*, vol. 434, no. 7029, pp. 113–118, 2005.

[147] X. Hou, S. Xu, K. A. Maitland-Toolan et al., "SIRT1 regulates hepatocyte lipid metabolism through activating AMP-activated protein kinase," *Journal of Biological Chemistry*, vol. 283, no. 29, pp. 20015–20026, 2008.

[148] M. D. Hirschey, T. Shimazu, E. Goetzman et al., "SIRT3 regulates mitochondrial fatty-acid oxidation by reversible enzyme deacetylation," *Nature*, vol. 464, no. 7285, pp. 121–125, 2010.

[149] A. A. Kendrick, M. Choudhury, S. M. Rahman et al., "Fatty liver is associated with reduced SIRT3 activity and mitochondrial protein hyperacetylation," *Biochemical Journal*, vol. 433, no. 3, pp. 505–514, 2011.

[150] S. Michel, A. Wanet, A. De Pauw, G. Rommelaere, T. Arnould, and P. Renard, "Crosstalk between mitochondrial (dys)function and mitochondrial abundance," *Journal of Cellular Physiology*, vol. 227, no. 6, pp. 2297–2310, 2012.

Permissions

The contributors of this book come from diverse backgrounds, making this book a truly international effort. This book will bring forth new frontiers with its revolutionizing research information and detailed analysis of the nascent developments around the world.

We would like to thank all the contributing authors for lending their expertise to make the book truly unique. They have played a crucial role in the development of this book. Without their invaluable contributions this book wouldn't have been possible. They have made vital efforts to compile up to date information on the varied aspects of this subject to make this book a valuable addition to the collection of many professionals and students.

This book was conceptualized with the vision of imparting up-to-date information and advanced data in this field. To ensure the same, a matchless editorial board was set up. Every individual on the board went through rigorous rounds of assessment to prove their worth. After which they invested a large part of their time researching and compiling the most relevant data for our readers. Conferences and sessions were held from time to time between the editorial board and the contributing authors to present the data in the most comprehensible form. The editorial team has worked tirelessly to provide valuable and valid information to help people across the globe.

Every chapter published in this book has been scrutinized by our experts. Their significance has been extensively debated. The topics covered herein carry significant findings which will fuel the growth of the discipline. They may even be implemented as practical applications or may be referred to as a beginning point for another development. Chapters in this book were first published by Hindawi Publishing Corporation; hereby published with permission under the Creative Commons Attribution License or equivalent.

The editorial board has been involved in producing this book since its inception. They have spent rigorous hours researching and exploring the diverse topics which have resulted in the successful publishing of this book. They have passed on their knowledge of decades through this book. To expedite this challenging task, the publisher supported the team at every step. A small team of assistant editors was also appointed to further simplify the editing procedure and attain best results for the readers.

Our editorial team has been hand-picked from every corner of the world. Their multi-ethnicity adds dynamic inputs to the discussions which result in innovative outcomes. These outcomes are then further discussed with the researchers and contributors who give their valuable feedback and opinion regarding the same. The feedback is then collaborated with the researches and they are edited in a comprehensive manner to aid the understanding of the subject.

Apart from the editorial board, the designing team has also invested a significant amount of their time in understanding the subject and creating the most relevant covers. They scrutinized every image to scout for the most suitable representation of the subject and create an appropriate cover for the book.

The publishing team has been involved in this book since its early stages. They were actively engaged in every process, be it collecting the data, connecting with the contributors or procuring relevant information. The team has been an ardent support to the editorial, designing and production team. Their endless efforts to recruit the best for this project, has resulted in the accomplishment of this book. They are a veteran in the field of academics and their pool of knowledge is as vast as their experience in printing. Their expertise and guidance has proved useful at every step. Their uncompromising quality standards have made this book an exceptional effort. Their encouragement from time to time has been an inspiration for everyone.

The publisher and the editorial board hope that this book will prove to be a valuable piece of knowledge for researchers, students, practitioners and scholars across the globe.

List of Contributors

Magdalena J. Koziol
Department of Genetics, Yale University School of Medicine, 333 Cedar Street, New Haven, CT 06510, USA

John B. Gurdon
Welcome Trust/Cancer Research UK Gurdon Institute, University of Cambridge, Tennis Court Road, Cambridge CB2 1QN, UK

Jennifer Hurst-Kennedy, Lih-Shen Chin and Lian Li
Department of Pharmacology and Center for Neurodegenerative Disease, Emory University School of Medicine, Atlanta, GA 30322, USA

Monika Cahova, Helena Dankova, Eliska Palenickova, Zuzana Papackova and Ludmila Kazdova
Department of Metabolism and Diabetes, Institute for Clinical and Experimental Medicine, Videnska 1958/9, 14021 Prague 4, Czech Republic

Radko Komers
Diabetes Center, Institute for Clinical and Experimental Medicine, 14021 Prague 4, Czech Republic
Division of Nephrology and Hypertension, Oregon Health and Science University, Portland, OR 97239-3098, USA

Jana Zdychova
Diabetes Center, Institute for Clinical and Experimental Medicine, 14021 Prague 4, Czech Republic
Department of Medicinal and Clinical Chemistry, University of Heidelberg, 69117 Heidelberg, Germany

Eva Sticova
Laboratory of Experimental Hepatology, Institute for Clinical and Experimental Medicine, 14021 Prague 4, Czech Republic

Eun-Kyoung Yim Breuer
Department of Radiation Oncology, Stritch School of Medicine, Loyola University Chicago, Maywood, IL 60153, USA
Department of Molecular Pharmacology and Therapeutics, Stritch School of Medicine, Loyola University Chicago, Maywood, IL 60153, USA

Geun-Hyoung Ha
Department of Radiation Oncology, Stritch School of Medicine, Loyola University Chicago, Maywood, IL 60153, USA

Michiyo Nagano-Ito and Shinichi Ichikawa
Laboratory for Animal Cell Engineering, Niigata University of Pharmacy and Applied Life Sciences (NUPALS), 265-1 Higashijima, Akiha-ku, Niigata-shi, Niigata 956-8603, Japan

Cecile Martel, Le Ha Huynh, Anne Garnier, Renee Ventura-Clapier and Catherine Brenner
Lab Ex LERMIT, INSERM U769, Faculte de Pharmacie, Universite Paris-Sud, 5 Rue J.-B. Clement, 92290 Chatenay-Malabry, France

Cheng Ji
Southern California Research Center for ALPD and Cirrhosis, USC Research Center for Liver Disease, Department of Medicine, Keck School of Medicine, University of Southern California, Los Angeles, CA 90089, USA

Michel Becuwe, Rosine Haguenauer- Tsapis and Sebastien Leon
Institut Jacques Monod, Centre National de la Recherche Scientifique, UMR 7592, Universite Paris Diderot, Sorbonne Paris Cite, 75205 Paris, France

Antonio Herrador and Olivier Vincent
Instituto de Investigaciones Biom´edicas, CSIC-UAM, Arturo Duperier, 4, 28029 Madrid, Spain

Elena Garcıa-Gimenez, Antonio Alcaraz and Vicente M. Aguilella
Laboratory of Molecular Biophysics, Department of Physics, Universitat Jaume I, 12071 Castellon, Spain

Kezhong Zhang
Center for Molecular Medicine and Genetics, The Wayne State University School of Medicine, 540 East Canfield Avenue, Detroit, MI 48201, USA
Department of Immunology and Microbiology, The Wayne State University School of Medicine, Detroit, MI 48201, USA
Karmanos Cancer Institute, The Wayne State University School of Medicine, Detroit, MI 48201, USA

Xuebao Zhang
Center for Molecular Medicine and Genetics, The Wayne State University School of Medicine, 540 East Canfield Avenue, Detroit, MI 48201, USA

Simon A. Young and Terry K. Smith
School of Biology and Chemistry, Biomedical Sciences Research Complex, University of St Andrews, North Haugh, KY16 9ST, UK

John G. Mina
Biophysical Sciences Institute, School of Biological and Biomedical Sciences and Department of Chemistry, University of Durham University Science Laboratories, South Road, Durham DH1 3LE, UK

Paul W. Denny
Biophysical Sciences Institute, School of Biological and Biomedical Sciences and Department of Chemistry, University of Durham University Science Laboratories, South Road, Durham DH1 3LE, UK
School of Medicine and Health, Durham University, Queen's Campus, Stockton-on-Tees TS17 6BH, UK

C. Saldanha
Instituto de Medicina Molecular, Unidade de Biologia Microvascular e Inflamac¸ao, Instituto de Bioquímica Faculdade de Medicina da Universidade de Lisboa. Av Prof. Egas Moniz, 1649-028 Lisboa, Portugal

J. Loureiro
Departmento de Cardiologia, Hospital Fernando da Fonseca. IC19, 2720-276 Amadora, Portugal

C.Moreira
Departmento de Medicina I, Faculdade de Medicina da Universidade de Lisboa. Av Prof. Egas Moniz, 1649-028 Lisboa, Portugal

J. Martins e Silva
Instituto de Biopatologia Quımica Faculdade deMedicina da Universidade de Lisboa. Av Prof. Egas Moniz, 1649-028 Lisboa, Portugal

Sujata Sharma, Mau Sinha, Sanket Kaushik, Punit Kaur and Tej P. Singh
Department of Biophysics, All India Institute of Medical Sciences, New Delhi 110029, India

Elisabetta Zinellu, Antonio Junior Lepedda, Antonio Cigliano, Salvatore Pisanu and Marilena Formato
Dipartimento di Scienze Fisiologiche, Biochimiche e Cellulari, Universit`a delgi Studi di Sassari, 07100 Sassari, Italy

Angelo Zinellu and Ciriaco Carru
Dipartimento di Scienze Biomediche, Universit`a delgi Studi di Sassari, 07100 Sassari, Italy

Pietro Paolo Bacciu and Franco Piredda
Servizio di Chirurgia Vascolare, Clinica Chirurgica Generale, Universit`a delgi Studi di Sassari, 07100 Sassari, Italy

Anna Guarino and Rita Spirito
Centro Cardiologico "F. Monzino," IRCCS, Universita delgi Studi di Milano, 20122 Milano, Italy

Christine Zhiwen Hu and Thilo Hagen
Department of Biochemistry, Yong Loo Lin School of Medicine, National University of Singapore, Singapore 117597, Singapore

Jaswinder K. Sethi
Institute of Metabolic Science, Metabolic Research Laboratories, and Department of Clinical Biochemistry, University of Cambridge, Addenbrooke's Hospital, Cambridge CB20QQ, UK

Paul M. L. Janssen
Department of Physiology and Cell Biology, College of Medicine, The Ohio State University, 1645 Neil Avenue, Columbus, OH 43210, USA
Department of Physiology and Cell Biology, College of Medicine, The Ohio State University, 304 Hamilton Hall, 1645 Neil Avenue, Columbus, OH 43210-1218, USA

Kenneth D. Varian, Brandon J. Biesiadecki, Mark T. Ziolo and Jonathan P. Davis
Department of Physiology and Cell Biology, College of Medicine, The Ohio State University, 1645 Neil Avenue, Columbus, OH 43210, USA

Huiping Zhou
Department of Microbiology and Immunology, School of Medicine, Virginia Commonwealth University, 217 East Marshall Street, MSB no. 533, Richmond, VA 23298, USA
Department of Internal Medicine, McGuire Veterans Affairs Medical Center, Richmond, VA 23298, USA

Beth S. Zha
Department of Microbiology and Immunology, School of Medicine, Virginia Commonwealth University, 1217 East Marshall Street, MSB no. 533, Richmond, VA 23298, USA

Lynn L. H. Huang
Institute of Biotechnology, College of Bioscience and Biotechnology, National Cheng Kung University, Tainan, Taiwan
Research Center of Excellence in Regenerative Medicine, National Cheng Kung University, Tainan, Taiwan
Institute of Clinical Medicine, College of Medicine, National Cheng Kung University, Tainan, Taiwan
Advanced Optoelectronic Technology Center, National Cheng Kung University, Tainan, Taiwan

Mairim Alexandra Solis, Ying-Hui Chen, Tzyy Yue Wong, Vanessa Zaiatz Bittencourt and Yen-Cheng Lin
Institute of Biotechnology, College of Bioscience and Biotechnology, National Cheng Kung University, Tainan, Taiwan
Research Center of Excellence in Regenerative Medicine, National Cheng Kung University, Tainan, Taiwan

Xin-Yuan Guan
Department of Clinical Oncology, Li Ka Shing Faculty of Medicine, The University of Hong Kong, Hong Kong
State Key Laboratory of Liver Research, The University of Hong Kong, Hong Kong
State Key Laboratory of Oncology in Southern China, Sun Yat-sen University Cancer Center, Guangzhou 510060, China

Tim Hon Man Chan and Leilei Chen
Department of Clinical Oncology, Li Ka Shing Faculty of Medicine, The University of Hong Kong, Hong Kong
State Key Laboratory of Liver Research, The University of Hong Kong, Hong Kong

Kerry S. McDonald, Laurin M. Hanft and Timothy L. Domeier
Department of Medical Pharmacology & Physiology, School of Medicine, University of Missouri, Columbia, MO 65212, USA

Craig A. Emter
Department of Biomedical Sciences, College of Veterinary Medicine, University of Missouri, Columbia, MO 65212, USA

Davide Degli Esposti, Jocelyne Hamelin, Nelly Bosselut, Rapha¨el Saffroy and Alban Pommier
AP-HP, Hopital Paul Brousse, Service de Biochimie et Biologie Moleculaire, 14 Avenue Paul Vaillant Couturier, 94804 Villejuif Cedex, France
Inserm U1004, Universite Paris 11, Institut Andre Lwoff, PRES Universud-Paris, Institut du Foie/Liver Institute, 14 Avenue Paul Vaillant Couturier, 94804 Villejuif, France

Antoinette Lemoine
AP-HP, Hopital Paul Brousse, Service de Biochimie et Biologie Moleculaire, 14 Avenue Paul Vaillant Couturier, 94804 Villejuif Cedex, France
Inserm U1004, Universite Paris 11, Institut Andre Lwoff, PRES Universud-Paris, Institut du Foie/Liver Institute, 14 Avenue Paul Vaillant Couturier, 94804 Villejuif, France
Universite Paris Sud 11, Faculte de Pharmacie, 5 rue Jean-Baptiste Clement, 92296 Chatenay-Malabry Cedex, France

Mylene Sebagh
AP-HP, Inserm U785, Hopital Paul Brousse, Service d'Anatomie Pathologique, 14 Avenue Paul Vaillant Couturier, 94804 Villejuif Cedex, France

www.ingramcontent.com/pod-product-compliance
Lightning Source LLC
Chambersburg PA
CBHW080640200326
41458CB00013B/4684